AF749247

Mediterranean Wild Edible Plants

María de Cortes Sánchez-Mata · Javier Tardío
Editors

Mediterranean Wild Edible Plants

Ethnobotany and Food Composition Tables

Springer

Editors
María de Cortes Sánchez-Mata
Departamento de Nutrición y
Bromatología II
Bromatología
Facultad de Farmacia
Universidad Complutense de Madrid
Madrid
Spain

Javier Tardío
Departamento de Investigación
Agroalimentaria
Instituto Madrileño de Investigación y
Desarrollo Rural,
Agrario y Alimentario (IMIDRA)
Alcalá de Henares, Madrid
Spain

ISBN 978-1-4939-3327-3 ISBN 978-1-4939-3329-7 (eBook)
DOI 10.1007/978-1-4939-3329-7
Springer New York Heidelberg Dordrecht London

Library of Congress Control Number: 2015960248

Printed on acid-free paper

Springer Science+Business Media LLC New York is part of Springer Science+Business Media (www.springer.com)

"Schleiden, in his principles of botany states: 'Botany is an indispensable branch of knowledge for the chemist and physiologist'. I think he might have said, with equal truth, chemistry and physiology are indispensable branches of knowledge to the botanist".
Helen C. De Silver Abbott
(Beginning of a lecture delivered before the Franklin Institute of the State of Pennsylvania, USA, 1887).

Preface

This book was born from the close cooperation between two groups led by us: a group of ethnobotanists, working at three different centres (*IMIDRA*, *Universidad Autónoma de Madrid* and *Real Jardín Botánico de Madrid*) and a group of food chemists working at *Universidad Complutense de Madrid,* with the purpose of improving the knowledge of the wild edible plants traditionally used in the Mediterranean area, especially about their nutritional aspects. This collaboration started at the end of 2006, with a research project granted by the Spanish Ministry of Education and Science (CGL2006-0946/BOS), studying 24 of the most important wild plants (20 vegetables and 4 fruits) habitually consumed in central Spain, at least in the last 50 years.

Hence, the idea of writing a book about this topic with the central aim of compiling and resuming the most relevant published nutritional data about the main wild edible plants traditionally consumed in the Mediterranean region. This information, not usually included in nutrient databases of foods, provides an interesting tool to be used with the purpose of the revalorization of wild food species, preservation of their traditional uses and also as alternatives to improve the diversity of modern Mediterranean diets, as they may represent valuable sources of nutrients often lacking in modern societies, such as fibre or vitamin B_9.

With this purpose, the work has expanded from the two initial groups to involve different contributors, including experts from different countries, most of them bordering the Mediterranean Sea. As a result, a multidisciplinary approach has been achieved.

The book is organized into four parts that deal with different aspects of wild edible plants, the last one and largest offering a detailed compilation of ethnobotanical and nutritional information about some of the most important Mediterranean wild edible plants.

The first part, with five chapters, treats about different aspects of the ethnobotany of wild edible plants. Chapter 1 presents an historical perspective of the use of wild food plants in the Mediterranean region. Chapter 2 explains the characteristic of the Mediterranean climate and its influence on the landscape and the abundance of wild edible plants. Chapter 3 analyzes the trends of the wild food plant consumption in Europe, highlighting the herbophilia of the Mediterranean countries.

Chapter 4 presents an ethnobotanical analysis of the wild fruits and vegetables traditionally consumed in Spain, an example that can be extensible to other countries of the Mediterranean region. Lastly, Chapter 5 discusses the availability of these wild resources as well as the possibilities of some of these species for cultivation.

The second part, with four chapters, is devoted to the importance of nutrients and bioactive compounds of the Mediterranean wild edible plants. Chapters 6 and 7 deal with their contribution to dietary intakes of micronutrients (vitamins and minerals, respectively), taking into account the current recommendations. Chapter 8 presents the fatty acid profiles of these plants, whereas Chapter 9 covers their role as sources of carotenoids, fibre, phenolics and other non-nutrient bioactive compounds.

The third part, with three chapters is about the biological activities of wild edible plants, as sources of antioxidants (Chapter 10) and components with antimicrobial actions (Chapter 11), as well as their potential biological–pharmacological activities (Chapter 12).

The fourth part is Chapter 13, a large descriptive dossier of 38 monographs about 41 selected wild edible plants traditionally and widely consumed in different countries of the Mediterranean basin. These monographs have two sections. The first one consists of a botanical and graphical description of the species and a resume of the ethnobotanical data registered in the Mediterranean countries for this edible plant. The second section supplies composition data of their edible parts, in the format of conventional food composition tables, covering the main constituents of proximal composition, minerals, vitamins and other bioactive compounds as well as fatty acid profiles. These analytical data, based on the scientific literature, try to remark the nutritional relevancy of each one of the selected species, and may be a valuable tool to preserve or revalorize their food use.

Finally, we thank all the authors and collaborators who have contributed to the different phases of the elaboration of the present work, making possible the integration of many different branches of science in this book, which we hope could be useful for the valorization of wild edible plants and could help expand the knowledge of this ancient human resource.

May 2015 María de Cortes Sánchez-Mata
Madrid Javier Tardío

Contents

Contributors

Lillian Barros Mountain Research Centre, Polytechnical Institute of Bragança, Bragança, Portugal

Kawaldeep Singh Bharij Centre for Pharmacognosy and Phytotherapy/Research Cluster 'Biodiversity and Medicines', UCL School of Pharmacy, University of London, London, UK

Vito V. Bianco CNR—Institute of Sciences of Food Production (ISPA), Bari, Italy

Noureddine Boussalah Laboratoire de Biomathematique, Biophysique, Biochimie et Scientométrie (L3BS), Université de Bejaia, Bejaia, Algeria

Montaña Cámara Departamento de Nutrición y Bromatología II, Bromatología, Facultad de Farmacia, Universidad Complutense de Madrid, Madrid, Spain

Ana Maria Carvalho Mountain Research Centre, Polytechnical Institute of Bragança, Bragança, Portugal

Donato Di Venere CNR—Institute of Sciences of Food Production (ISPA), Bari, Italy

Carmen Díez-Marqués Departamento de Nutrición y Bromatología II, Bromatología, Facultad de Farmacia, Universidad Complutense de Madrid, Madrid, Spain

Virginia Fernández-Ruiz Departamento de Nutrición y Bromatología II, Bromatología, Facultad de Farmacia, Universidad Complutense de Madrid, Madrid, Spain

Isabel C. F. R. Ferreira Mountain Research Centre, Polytechnical Institute of Bragança, Bragança, Portugal

Patricia García-Herrera Departamento de Nutrición y Bromatología II, Bromatología, Facultad de Farmacia, Universidad Complutense de Madrid, Madrid, Spain

Maria Antonia Gatto CNR—Institute of Sciences of Food Production, Bari, Italy

Jose Luis Guil-Guerrero Food Technology Division, Agronomy Department, Universidad de Almería, Almería, Spain

Michael Heinrich Centre for Pharmacognosy and Phytotherapy/Research Cluster 'Biodiversity and Medicines', UCL School of Pharmacy, University of London, London, UK

Antonio Ippolito Department of Soil, Plant, and Food Sciences, University of Bari Aldo Moro, Bari, Italy

Sarah Kerrouche Centre for Pharmacognosy and Phytotherapy/Research Cluster 'Biodiversity and Medicines', UCL School of Pharmacy, University of London, London, UK

Łukasz Łuczaj Department of Botany, Institute of Applied Biotechnology and Basic Sciences, University of Rzeszów, Rzeszów, Poland

María Cruz Matallana-González Departamento de Nutrición y Bromatología II, Bromatología, Facultad de Farmacia, Universidad Complutense de Madrid, Madrid, Spain

María Molina Departamento de Biología (Botánica), Facultad de Ciencias, Universidad Autónoma de Madrid (UAM), Madrid, Spain

Patricia Morales Departamento de Nutrición y Bromatología II, Bromatología, Facultad de Farmacia, Universidad Complutense de Madrid, Madrid, Spain

Ramón Morales Departamento de Biodiversidad y Conservación, Real Jardín Botánico, CSIC, Madrid, Spain

Manuel Pardo-de-Santayana Departamento de Biología (Botánica), Facultad de Ciencias, Universidad Autónoma de Madrid (UAM), Madrid, Spain

Andrea Pieroni University of Gastronomic Sciences, Pollenzo, Cuneo, Italy

Brígida María Ruiz-Rodríguez Departamento de Nutrición y Bromatología II, Bromatología, Facultad de Farmacia, Universidad Complutense de Madrid, Madrid, Spain

Daniel Sánchez-Mata Departamento de Biología Vegetal II, Facultad de Farmacia, Universidad Complutense de Madrid, Madrid, Spain

María de Cortes Sánchez-Mata Departamento de Nutrición y Bromatología II, Bromatología, Facultad de Farmacia, Universidad Complutense de Madrid, Madrid, Spain

Javier Tardío Departamento de Investigación Agroalimentaria, Instituto Madrileño de Investigación y Desarrollo Rural, Agrario y Alimentario (IMIDRA), Alcalá de Henares, Madrid, Spain

María Esperanza Torija-Isasa Departamento de Nutrición y Bromatología II, Bromatología, Facultad de Farmacia, Universidad Complutense de Madrid, Madrid, Spain

Part I
Ethnobotany of Wild Edible Plants

Chapter 1
A Historical Perspective of Wild Plant Foods in the Mediterranean Area

María Esperanza Torija-Isasa and María Cruz Matallana-González

1.1 Introduction

The human feeding process has experienced constant evolution through history. The first human beings were able to intuitively select from the environment products that could be used for food, preferring those which provided energy and nutrients and rejecting what they thought could be harmful. With the aim of satisfying the most basic need (hunger), they tried to use different natural products for food; later, in times of abundance, they could make a selection, and finally they learnt how to cook, preserve and produce food products (Toussaint-Samat 2009).

The human being is a selective omnivore, which means that his nutrition is not linked to one or several foods. His food habits allow him to choose between food from different origins and even between different varieties of a given food. The primitive food behaviour was influenced exclusively by natural appetite. Whether a certain kind of food is accepted or not depends first on whether a society considers it edible. Then, its acceptance depends on the sensorial exigencies, such as external aspect, odour, taste, flavour, texture and even the noise produced when it is consumed. The consequences derived from that selection are mostly biological since foods provide energy, nutrients and other compounds for human organism. A right selection may have good consequences on health status, while a wrong choice may give rise to diseases; this has been sometimes empirically taken into account during the evolution of human beings to influence the acquisition of food habits. However, sociocultural factors have also produced changes in food behaviour since feeding is influenced by different individual or collective facts such as culture, economy, geography and environment, as well as physiology and personal psychology (Harris 1991).

M. E. Torija-Isasa (✉) · M. C. Matallana-González
Departamento de Nutrición y Bromatología II, Bromatología, Facultad de Farmacia, Universidad Complutense de Madrid, Madrid, Spain
e-mails: metorija@ucm.es; mcmatall@ucm.es

M. de C. Sánchez-Mata, J. Tardío (eds.), *Mediterranean Wild Edible Plants*,
DOI 10.1007/978-1-4939-3329-7_1

From ancient times, humans took from the surrounding environment those products that could be used as food. Plant products known by primitive people were used with different objectives: food, medicine, production of materials (for example clothes) or magic rituals. Wild plant gathering has been a habitual practice since ancient times all over the world, especially in times of famine. Many species now considered as weeds were food for numerous cultures. Today, lots of them have been forgotten in our society even though they have an important nutritive value; in other countries, however, many plant species are still in use.

Phytogenetic resources are essential to improve food security, including seeds and propagation materials from traditional and modern varieties, and also wild species, which are often used as food. The loss of these resources and the absence of suitable links between preservation and utilization are an important danger for food security in the future world. Despite their importance for human survival, plant biodiversity is being lost at an alarming rate. Hundreds of thousands of traditional heterogeneous varieties existing through generations in the fields until the beginning of the twentieth century have been replaced by a small number of commercial modern varieties (Esquinas-Alcázar 2013).

For this reason, the Food and Agriculture Organization of the United Nations and the World Health Organization (FAO/WHO) are devoted to improve the knowledge and preservation of phytogenetic resources, with the aim to assure sustainable food provision in the future and contribute to the optimal utilization of the available genetic resources, including wild plants which may be relatives of crop products, but may contain different genetic material allowing them to live in stress conditions or providing a higher nutritive value. In the past years, the importance of biodiversity has stood out in many campaigns, such as the FAO/WHO World Food Day, with the topic “Harvesting Nature’s Diversity” in 1993 or “Biodiversity for Food Security” in 2004 (FAO 2011a, b).

The wild flora has played an essential role in human feeding, and even nowadays, there are groups depending at a high rate on wild edible plants. In our environment, wild plants have been and are a seasonal complement for our diet, although in times of famine, they have been employed as emergency foods. The most valued species are even subjects of seasonal commerce, usually as consumer products and often in ambulant markets and are also present in some compilations of traditional food recipes.

1.2 Prehistory

The interpretation of the archaeological remains is not an easy task; however, the role of wild plants in the diet of communities during prehistory is supposed to be highly important. Wild plants are abundant resources, easily gathered and collected, which was the reason that, for example, young parts of *Malva parviflora* L., *Rumex* sp., *Silene* sp., *Sisymbrium* sp., *Plantago* sp. or *Calendula* sp. could be used as habitual foods by primitive civilizations (Peña Chocarro 2000). The hunter-gatherer societies had enough resources to survive, although in many cases they were seasonal resources. However, the study of wild vegetables used in the prehistoric period is difficult since most palaeobotanical remains include seeds and even fruits, but not frequently leaves. For example, Olària i Puyoles (2004–2005) mentioned the proofs for the collection of some wild fruits or seeds, such as *Arbutus unedo, Quercus* sp., *Vitis sylvestris, Lathyrus* sp., *Pisum* sp. and *Olea* sp. in Grotta dell'Uzzo (Italy) and also seeds of *Papaver somniferum* L., *Myrtus communis, Rubus idaeus* and acorns of *Quercus* sp. in Cueva del Toro (Antequera, Spain). Lopez et al. (2013) also mentioned the presence of wild *Lathyrus cicera* among others in "Cova des Riuets" (Balearic Islands, Spain). Verde et al. (2004) supposed that many wild vegetables were widely used in the Palaeolithic in spring, especially plants living in wetlands (watercress) or sprouts from different species (asparagus). In the Neolithic, many weeds were still gathered, such as bugloss or bladder campion. Many of these species were weeds growing in cultivated fields, especially grain crops, or in ditches (Alarcón 2013). With the development of agriculture, human beings started to grow a great number of species, beginning with those preferred for food or those more easily adaptable to cultivation, such as cereals and legumes, and also vegetables. However, many vegetables were never domesticated, including leaves, stems, flowers as well as subterranean parts. An increase in the consumption of vegetables, and particularly greens, in this period of the history, might have had a role in the reduction of vitamin deficiencies (Latham 2002).

The adventitious species linked to fields, as well as ruderal species developed in some spaces modified by human beings, were used in the Neolithic period, such as *Plantago* sp., *Chenopodium* sp., *Amaranthus* sp., *Polygonum* sp. as well as some grains of wild species. Precioso-Arévalo (2003) mentioned plants which have been found in palaeobotanical remains belonging to the genus *Silene, Chenopodium* or *Brassica,* as well as *Calendula arvensis* M. Bieb. and *Chenopodium album* in several archaeological sites in the East of Spain.

1.3 Ancient Civilizations

There are not only many evidences about cereal and legume crops in ancient Near Orient cultures but also of the gathering of nuts such as acorns, pistachios and almonds (Redman 1990). García Lenberg (1998) reported some difficulties to identify plants from palaeobotanical remains and cuneiform texts from ancient Mesopotamia. According to this author, references to *Allium* genus, mainly garlic and onion, have been found, as well as different legume seeds, such as vetches. Cucumber seeds may also have been consumed in Nimrud in the seventh century BC, although it is not clear if they correspond to the same product that we know nowadays (Toussaint-Samat 2009).

In Egypt, many vegetables were highly valuable, such as garlic, onion, mint, caraway, fennel or coriander, firstly as wild herbs, but later as cultivated plants. Other vegetables with a double ornamental and food use were lotus and water lily, which became important for Egyptian food habits (Martínez Llopis 1989; Aboelsoud 2010).

Many of the most used plants are mentioned in Ebers Papyrus (sixteenth century BC), one of the most ancient maintained medical documents, compiling about 700 remedies, not only explaining their medicinal applications but also some of their food uses. In this document, reference to lotus appears together with other vegetables such as onions, dates, pomegranates, poppy, elderberry or mint. Lotus, an aquatic plant of the Nymphaeaceae family known as "Egyptian white nenuphar" or "Egyptian white lotus", was a symbol of water and origin of life, as its seeds and tubers were consumed as food in this culture (Aboelsoud 2010; Clifton 1986).

Vegetables were basic components of the Greek diet from ancient times, as shown in different references from ancient Greek literature. For example, Hesiod (eighth century BC) in his manuscript "Works and Days" referred again to lotus, adding also "what a rich treasure is given by asphodel and mallow…" (Vianello de Córdova 1979). Also, some relevant references to lotophagi (lotus eaters) can be found in Homer's manuscripts (eighth century BC), such as *Odyssey* (XI, 76–104; V, 594–608) and *Iliad* (XIV, 346–351). Herodotus (484–425 BC) also mentioned two varieties of lotus: one growing close to the Nile river, and the other one in the so-called lotophagus country. This place is located on the coast of Libya (García González 2008).

Antiphanes (408–334 BC) wrote about people eating leaves (*phyllotrôges*), either wild or cultivated; some comments illustrate how they used to be the main dish of the meal, being consumed either raw or cooked, and also accompanying other foods such as meat or fish, according to García Soler 2001. This author also indicated that the roots and stems of mallows (*maláche*) were eaten either cooked alone or as an ingredient of other dishes.

The Greek historian Polybius (200–118 BC) made the first description of Lusitania, indicating that wild fruits were better preserved there due to the characteristics of the air. He also said that there the asparagus did not stop blooming for more than 3 months (Cubero Corpas 1994; García Quintela 2001).

Many wild plants gathered in ancient Greece also became cultivated, such as cabbage, called *krámbe,* being the flat variety the most widespread. Nicandrus (second century BC) indicated that it sometimes grew wild and presented different colours: brown, purple or "frog colour". There were also different types of lettuces (*thrídax*). Another important wild edible species in classical culture was *skólymos,* which according to Dioscorides' (first century AD) description may be identified as thistle or artichoke (Font Quer 1962). Other Asteraceae belonging to the genus *Sonchus* (*sónkos*) were less appreciated. The "patience herb" (*lápathon*), identified as *Rumex* spp., was usually eaten cooked, being highly appreciated for its pleasant taste and penetrant smell (García Soler 2001; Martínez–Llopis 1989).

Nettle was an appreciated plant in ancient Greece. Aristophanes (444–385 BC) wrote about nettle gathering before the swallows' arrival. Later, Athenaeus of Naucratis (249–150 AD) devoted a short part of the second book of *Deipnosophistae* (meaning "Dinner-table philosophers") to nettle use (*akaléphe*). Teophrastus (372–288 BC) also wrote about cooking of nettles, particularly young stems, and how they became cultivated vegetables. He also indicated that thistle leaves were also eaten in Sicilia (being bitter and suitable for preservation in brine); thistle stems and inflorescences were also used.

Like the Greeks, it is known that Romans were very interested in eating wild nettles or mallow leaves. Many Roman authors have written about these uses: For example, Plautus (251–184 BC) wrote how Roman cooks prepared dishes using a lot of leafy vegetables and herbs dressed with other herbs. Martínez–Llopis (1989) commented about the high appreciation given by Romans to nettle, as well as mallows, vine leaves or polypodium.

Marcus Terentius Varro (116–27 BC) also wrote about chard roots and leaves (white or black chards), cooked together with lentils; this author also mentioned the use of vine or mustard leaves, as well as parsley (called *holisera*). In Rome, thistle stems were also preserved in vinegar, honey or cumin, as indicated by Pliny the Elder (first century AD) in his *Natural History;* also in the widely known ancient roman cooking book *De re coquinaria* written by Apicius (first century AD), several recipes are compiled using thistle midribs cooked, fried or dressed with sauces (Ibáñez Ártica 1995; Martínez–Llopis 1989).

Mention should be made about *De Materia medica,* written by Dioscorides. As the doctor of the army of the Roman emperor Nero, he had the opportunity to travel all around the Mediterranean, compiling the knowledge about the use of natural remedies, including about 600 plant species (López Eire, 2006). But he also described other plant uses, such as food. For example, Dioscorides mentioned the consumption of mallows roots and stems raw in salads, cooked and also as ingredients of several dishes (Font Quer 1962). The same author commented about eating the new growth of an indeterminate thistle, probably *Cynara cardunculus* L., cooked as asparagus.

Other vegetables eaten by Romans: wild *Apium graveolens* L. was appreciated for its leaves, seeds and stems, which were candied the same way as those of *Angelica sylvestris* L. or *A. archangelica* L.; *Umbilicus rupestris* (Salisb.) Dandy (Venus navel) leaves were eaten raw in salads and highly lauded by Hippocrates; and *Inula helenium* L. roots, with a bitter taste, were used either as medicine or food, cooked or preserved in oxicrate (a mixture of water, honey and vinegar), considered useful for gastric diseases (Martínez–Llopis 1989; Grande de Ulierte 2014).

At this moment, the Mediterranean was characterized by its important commercial activity, promoting the exchange of knowledge between different cultures surrounding this area: Phoenician, Greek, Carthaginian and Romans. The Roman Empire expanded to Asia Minor and Africa. Israel was also characterized by the use of herbs for food or seasoning, such as mint, dill or cumin mentioned in the New Testament (Mt 23, 23). After the fall of the Roman Empire, the Byzantine Empire influenced Mediterranean culture. Stephanus of Byzantium (sixth century AD) wrote about the islands Melussa and Kromyusa (thought to be Majorca and Minorca), whose names are derived from the translation of apple and onions, respectively; they may give an idea that those were important plants growing and used in those regions (Cubero Corpas 1994). During all these centuries, the transfer of crops, medicinal and aromatic plants from East to West became very important (Hernández Bermejo and García Sánchez 1998).

1.4 Middle and Modern Age

During the Middle Ages, agriculture was already widely expanded and many wild plants previously eaten had become cultivated vegetables, although many others such as wild nettle leaves, purslane, arugula or borage leaves were still used as foods. The writings of some Hispano-Arabic writers from the Andalusi Agronomic School (tenth to fifteenth centuries AD), such as Arib ben Said, Ibn Wafid, Ibn Hayyay, Abu l-Jayr or Ibn Bassal, reflect the introduction of some Oriental species in the Iberian Peninsula (Hernández Bermejo and García Sánchez 1998).

In medieval literature, such as *The Miracles of Our Lady* (written by Gonzalo of Berceo) or *Codex Aemilianensis* (the first glossary made in the Iberian Peninsula, presumably finished in 964 at the monastery of *San Millán de la Cogolla*), different wild plants growing around the paths and others in gardens are mentioned. The interpretation made by Dutton (1980) of these works indicates the food use of some of the species mentioned, eaten mostly by poor people, who usually had better access to these products; many of them were also included in medieval pharmacopoeias. Some examples of the wild species used according to this author are gooseberries, said to be sweeter than sugar; cooked young nettle leaves; watercress, abundant in boggy places and slow streams and served as fresh salads throughout the year; or wild fennel widely used in the preparation of food and medicines, with a taste similar to anise (Dutton 1980; García-Turza 2004).

The arrival of Europeans to the New World played a crucial role for the introduction and exchange of plant species between Europe and America. Many chronicles reflect plants transported by colonists, farmers, physicians and naturalists on their arrival in America, but also the transport of American plants to Europe, many of which became adapted to cultivation (Hernández-Bermejo and García-Sánchez 1998). Some species, such as wild purslane, are nowadays traditionally consumed in both Europe and America.

During the Renaissance, there was a renewed interest in Europe for classical culture, which was evident in art, literature and also in sciences. In this context, different authors afforded the translation of *De Materia Medica* written by Dioscorides to popular languages: The first one was edited in Venice by Giovanni de Farri in 1542, followed by Pier Andrea Mattioli in 1543, both of them in Italian language. Also important was that of Andrés Laguna into Spanish, commented and published in 1555 (Font Quer 1962).

The habit of gathering plants from the wild has continued during centuries, linked especially to rural societies. They provided a source of food for many people in times of wars and famines, situations that become frequent during the sixteenth to nineteenth centuries. They often represented an important dietary contribution when men were fighting and women were in charge of bringing food to the family. Sometimes, these plants were weeds growing in the crops, and their collection represented both a necessary agricultural practice and, at the same time, a source of food.

But still the habit of gathering plants were often due to the fact that some species were considered as delicacies, as described by Nicolás Monardes, a Spanish physician and botanist from the sixteenth century who described how scorzonera roots preserved in sugar were almost as delicate to taste as those of coriander, which was used to make sweets. Also, the English gardener John Evelyn wrote in seventeenth century about eating some young shoots of scorzonera the same way as asparagus or cooked in pots, indicating the existence of a white and a red variety widely employed in Spain and Italy. In France, Jean-Baptiste de La Quintinie, the gardener of Louis XIV, mentioned the use of purslane for the king's salads, preserved in salt and vinegar (Genders 1998).

Some species, such as *Smyrnium olusatrum* L., consumed as a vegetable and condiment in the times of Dioscorides or Columella (first century AD), became underutilized from the seventeenth century, and no reference to it is found in the works of Spanish agronomists from the nineteenth century, presumably being displaced by other vegetables such as *Apium graveolens*. Some of these species are now plants growing in nitrified environments altered by human activities or are weeds in crops (Morales et al. 2011).

1.5 Contemporary Age

Nowadays, agriculture crops have displaced many of the previously known and appreciated wild species. Some agricultural techniques, such as the use of chemical pesticides, often displace wild greens to places far from cultivated fields (Tardío et al. 2005). However, Mediterranean traditions have made possible that a considerable number of them continue being present in the diet of many people for different reasons: the pleasure of gathering wild plants from the fields, the knowledge about their good properties from either a nutritional or functional point of view, the interest of keeping traditional food habits, or contrarily but not incompatible, the search for "new" (or forgotten) ingredients with unusual flavours, taste or textures in the so-called nouvelle cuisine.

Many ethnobotanical works have been recently published, registering the traditional knowledge about the plants used during the past decades, as can be seen in some of the following chapters, especially in Chap. 13. In Spain, for example, many studies have been carried out in various regions (e.g. Fajardo et al. 2000; Menéndez-Baceta et al. 2012; Pardo-de-Santayana et al. 2005; Tardío et al. 2005; Verde et al. 1998). A compilation of the wild edible plants traditionally consumed in Spain was published by Tardío et al. (2006). This study reflects the consumption of more than 400 species for different food uses, including vegetables (e.g. *Rorippa nasturtium-aquaticum* (L.) Hayek, *Silene vulgaris* (Moench.) Garke, *Asparagus acutifolius* L., *Scolymus hispanicus* (L.) and fruits (*Fragaria vesca* L., *Rubus ulmifolius* Schott, *Castanea sativa* Mill., *Fagus sylvatica* L., *Pinus radiata* D. Don). Also in Italy, during the past 25 years, the consumption of non-cultivated plants has been the focus of a growing number of studies devoted to document traditional knowledge about anthropology in an ethnoecologic/ethnobotanic context. Just to cite a few examples:

Pieroni et al. (2005) informed about the use of unusual species in Central Lucania; Ghirardini et al. (2007) studied 21 communities in Italy where the gathering, processing and consumption of wild edible plants are still important activities despite the socio-economic changes, with species such as *Asparagus acutifolius, Reichardia picroides* (L.) Roth, *Cichorium intybus, Foeniculum vulgare, Sambucus nigra* L., *Silene vulgaris, Taraxacum officinale* (L.) Weber ex F. H. Wigg. and *Urtica dioica* L. as those common in many Italian regions and especially *Borago officinalis* L. as one of the most valued species; Scherrer et al. (2005) and Cornara et al. (2009) studied the use of different species in Campania and Liguria, respectively, to elaborate different typical dishes as *minestra* of *prebuggin.* Other recent studies show a trend of revalorization of wild edible plants in Turkey (Dogan et al. 2004), Jordan (Tukan et al. 1998) and Bulgary (Nedelcheva 2013).

Although some previous studies about wild food plant composition exist (Cowan et al. 1963), from the eighties to the present day, wild edible species in the Mediterranean countries have been more intensively studied, including their biological activity (Heinrich et al. 2006). There have been many surveys about their role in human nutrition, such as those conducted in Spain (Guil-Guerrero and Torija-Isasa 2002; Romojaro et al. 2013; Sánchez-Mata et al. 2012), Portugal (Barros et al. 2010; Martins et al. 2011; Pereira et al. 2011), Greece (Vardavas et al. 2006; Trichopoulou et al. 2000; Zeghichi et al. 2003), Italy (Aliotta and Pollio 1981; Bianco 1998; Gatto et al. 2011) and Turkey (Özcan et al. 2008; Yildrim et al. 2001).

This fact is connected not only with the inscription of the Mediterranean diet in the Representative List of the Intangible Cultural Heritage of UNESCO (UNESCO 2015) but also with all the nutritional studies showing the health benefits induced by Mediterranean diet (Mosconi et al. 2014; Sofi et al. 2010). This Mediterranean diet should be understood not only as food choice recommendations where fruits and vegetables (either wild or cultivated) have a predominant role but also as a lifestyle in which the selection of local traditional products, the traditional culinary activities, the transmission of expertise, the sustainable development of rural communities and the preservation of the biodiversity is an essential part of the human–environment interaction. The Mediterranean diet is characterized by a nutritional model that has remained more or less constant over time and space, from ancient times till nowadays, and its preservation is a challenge of the Mediterranean societies. In this context, all the studies which advance knowledge about traditional foods eaten in this area are quite valuable from both a cultural and a nutritional point of view.

References

Aboelsoud NH (2010) Herbal medicine in ancient Egypt. J Med Plant Res 4(2):82–86

Alarcón Víllora MR (2013) Entre "malas hierbas", criptocultivos y plantas cultivadas: la colleja (*Silene vulgaris*). Ambienta 102:180–188

Aliotta G, Pollio A (1981) Vitamin A and C content in some edible wild plants in Italy. Riv Ital EPPOS 63:47–48

Barros L, Carvalho AM, Morais JS, Ferreira ICFR (2010) Strawberry-tree, blackthorn and rose fruits: detailed characterisation in nutrients and phytochemicals with antioxidant properties. Food Chem 120(1):247–254
Bianco VV, Santamaria P, Elia A (1998) Nutritional value and nitrate content in edible wild species used in Southern Italy. Acta Hort (ISHS) 467:71–87
Clifton C (1986) The sacred Lotus-food and symbol. In: Jaine T (ed) The cooking medium Oxford symposium of food and cookery. Oxford University Press, London
Cornara L, La Rocca A, Marsili S, Mariotti MG (2009) Traditional uses of plants in the Eastern Riviera (Liguria, Italy). J Ethnopharmacol 125(1):16–30
Cowan JW, Sabry ZI, Shadarev SB, Sakr AH (1963) Composition of edible wild plants of Lebanon. J Sci Food Agric 14(7):484–488
Cubero Corpas C (1994) Los recursos vegetales y su aprovechamiento en Hispania según los textos clásicos. Pyrenae 25:117–121
Dogan Y, Baslar S, Ay G, Mert HH (2004) The use of wild edible plants in Western and Central Anatolia (Turkey). Econ Bot 58(4):684–690
Dutton B (1980) Berceo y la Rioja Medieval: Apuntes Botánicos. Berceo 98:3–30
Esquinas Alcázar J (2013) Biodiversidad y Seguridad. Cuadernos de estrategia 161:109–156
Fajardo J, Verde A, Rivera D, Obón C (2000) Las plantas en la cultura popular de la provincia de Albacete. Instituto de Estudios Albacetenses, Albacete
FAO (2011a) El Segundo Informe sobre el Estado de los Recursos Fitogenéticos para la Alimentación y la Agricultura en el Mundo. Comisión de Recursos Genéticos para la Alimentación y la Agricultura, Roma
FAO (2011b) Consulta de expertos sobre indicadores de nutrición para la biodiversidad. 2. Consumo de alimentos. Organización de las Naciones Unidas para la Alimentación y la Agricultura, Roma
Font Quer P (1962) Plantas Medicinales. El Dioscórides renovado. Labor, Barcelona
García González JA (2008) Los lotófagos y el loto homérico. Baetica 30:171–188
García Lenberg J (1998) Vegetales comestibles en Asiria y Babilonia: la evidencia de la documentación escrita y paleobotánica. Rev sobre Oriente Próx y Egypto en la antig 1:193–202
García Quíntela MV (2001) Mitos hispánicos. La Edad Antigua. Akal, Madrid
García Soler MJ (2001) El arte de comer en la Antigua Grecia. Biblioteca Nueva, Madrid
García-Turza C (2004). El Códice Emilianense 31 de la Real Academia de la Historia. Presentación de algunas de las voces de interés para el estudio lingüístico del latín medieval y del iberorromance primitivo. Aemilianense 1:95–170.
Gatto MA, Ippolito A, Linsalata V, Cascarano NA, Nigro F, Vanadia S, Di Venere D (2011) Activity of extracts from wild edible herbs against postharvest fungal diseases of fruit and vegetables. Postharvest Biol Technol 61(1):72–82
Genders R (1998) Plantas Silvestres Comestibles: frutos, bayas, raices, brotes. Blume, Barcelona
Ghirardini M, Carli M, del Vecchio N, Rovati A, Cova O, Valigi F, Agnetti G, Macconi M, Adamo D, Traina M, Laudini F, Marcheselli I, Caruso N, Gedda T, Donati F, Marzadro A, Russi P, Spaggiari C, Bianco M, Binda R, Barattieri E, Tognacci A, Girardo M, Vaschetti L, Caprino P, Sesti E, Andreozzi G, Coletto E, Belzer G, Pieroni A (2007) The importance of a taste. A comparative study on wild food plant consumption in twenty-one local communities in Italy. J Ethnobiol Ethnomed 3(1):22
Grande de Ulierte E (2014) Los alimentos vegetales en la Hispania romana. PhD Thesis. Universidad Complutense de Madrid.
Guarrera PM, Forti G, Marignoli S (2005) Ethnobotanical and ethnomedicinal uses of plants in the district of Acquapendente (Latium, Central Italy). J Ethnopharmacol 96(3):429–444
Guil-Guerrero JL, Torija-Isasa ME (2002) Edible wild plants. In: Majundar DK, Govil JN, Singh VK (eds) Recent progress in medicinal plants, vol VIII. SciTech Publishing LLC, Texas, pp 431–466
Harris M (1991) Introducción a la antropología cultural. Alianza Universidad, Madrid
Heinrich M, Müller WE, Galli C (2006) Local Mediterranean food plants and Nutraceuticals. Karger, Basel

Hernández-Bermejo JE, García-Sánchez E (1998) Economic botany and ethnobotany in al-Andalus (Iberian Peninsula: Tenth-fifteenth centuries), an unknown heritage of mankind. Econ Bot 52(1):15–26

Ibáñez Ártica M (1995) Apicius. De re coquinaria. Gastronomía en la antigua Roma imperial. R & B, Bilbao

Latham MC (2002) Nutrición humana en el mundo en desarrollo. Organización de las Naciones Unidas para la Agricultura y la Alimentación. Colección FAO: Alimentación y nutrición N° 29. FAO, Roma

López Eire A (2006) Dioscorides. De Materia Medica. Manuscrito de Salamanca. Ediciones Universidad de Salamanca, Salamanca

López Gari JM, Pérez Jordá G, Marlascaa Martín R, Farrera Fernández VM Enrich Hoja J (2013) La primera agricultura Pitiusa y Balear: las evidencias de la Cova des Riuets. SAGVNTVM. 45:65–77

Martínez Ezquerro A (1995) Denominaciones populares de plantas silvestres. Revista de Folklore 178:132–142

Martínez–Llopis MM (1989) Historia de la Gastronomía Española. Alianza Editorial, Madrid

Martins D, Barros L, Carvalho AM, Ferreira ICFR (2011) Nutritional and in vitro antioxidant properties of edible wild greens in Iberian Peninsula traditional diet. Food Chem 125(2):488–494

Menendez–Baceta G, Aceituno–Mata L, Tardío J, Reyes–García V, Pardo-de-Santayana M (2012) Wild edible plants traditionally gathered in Gorbeialdea (Biscay, Basque Contry). Genet Resor Crop Evol 59:1329–1347

Morales R, Tardío J, Aceituno L, Molina M, Pardo-de-Santayana M (2011) Biodiversidad y Etnobotánica en España. In: Viejo-Montesinos JL (ed) Biodiversidad. Aproximación a la diversidad botánica y zoológica de España. Real Sociedad Española de Historia Natural, Madrid, pp 157–207

Mosconi L, Murray J, Tsui WH, Li Y, Davies M, Williams S, Pirraglia E, Spector N, Osorio RS, Glodzik L, McHugh P, de Leon MJ (2014) Mediterranean diet and magnetic resonance imaging-assessed brain atrophy in cognitively normal individuals at risk for Alzheimer's disease. J Prev Alzheimers Dis 1(1):23–32

Nedelcheva A (2013) An Ethnobotanical study of wild edible plants in Bulgaria. EurAsian J Biosci 7:77–94

Olària i Puyoles C (2004–2005) El tránsito hacia las economías de producción de las últimas tribus cazadoras-recolectoras del Mediterráneo peninsular. Una reflexión acerca de la validez de ls tesis difusionistas frente a las evolucionistas. Quad Preh Arq Cast 24:43–60

Özcan MM, Unver A, Ucar T, Arslan D (2008) Mineral content of some herbs and herbal teas by infusion and decoction. Food Chem 106(3):1120–1127

Pardo-de-Santayana M, Tardío J, Morales R (2005) The gathering and consumption of wild edible plants in the Campoo (Cantabria, Spain). Int J Food Sci Nutr 56(7):529–542

Peña-Chocarro L (2000) Agricultura y alimentación vegetal en el poblado de la Edad del Bronce de Peñalosa (Baños de la Encina, Jaén). Complutum 11:209–219

Pereira C, Barros L, Carvalho AM, Ferreira IC (2011) Nutritional composition and bioactive properties of commonly consumed wild greens: potential sources for new trends in modern diets. Food Res Int 44(9):2634–2640

Pieroni A, Nebel S, Santoro RF, Heinrich M (2005) Food for two seasons: culinary uses of non-cultivated local vegetables and mushrooms in a south Italian village. Int J Food Sci Nutr 56(4):245–272

Precioso-Arévalo ML (2003) Estudio de los restos paleobotánicos en Murcia. 2. Las plantas recolectadas y sinantrópicas. Revista ArqueoMurcia. No 2. http://www.arqueomurcia.com/revista/n1/htm/semillas.htm. Accesed Feb 2015

Rampelbergh J (2006) Influencia Árabe en algunos aspectos de la vida Occidental. www.caei.com.ar/es/programas/mediooriente/01. Accessed Feb 2015

Redman CL (1990) Los orígenes de la civilización. Desde los primeros agricultores hasta la sociedad urbana en el Próximo Oriente. Ed. Crítica, Barcelona

Rivera D, Obón C, Heinrich M, Inocencio C, Verde A, Fajardo, J (2006) Gathered mediterranean food plants—Etnobotanical investigations and historical development. Forum Nutr 59:18–74

Romojaro A, Botella MA, Obon C, Pretel MT (2013) Nutritional and antioxidant properties of wild edible plants and their use as potential ingredients in the modern diet. Int J Food Sci Nutr 64(8):944–952

Sánchez-Mata MC, Cabrera-Loera RD, Morales P, Fernández-Ruiz V, Cámara M, Díez-Marqués C, Pardo-de-Santayana M, Tardío J (2012) Wild vegetables of the Mediterranean area as valuable sources of bioactive compounds. Genet Resour Crop Evol 59(3):431–443

Scherrer A, Motti R, Weckerle CS (2005) Traditional plant use in the areas of Monte Vesole and Ascea, Cilento National Park (Campania, Southern Italy). J Ethnopharmacol 97:129–143

Sofi F, Abbate R, Franco G, Casini A (2010) Accruing evidence on benefits of adherence to the Mediterranean diet on health: an updated systematic review and meta-analysis. Clin Nutr 92(5):1189–1196

Tardío J, Pascual H, Morales R (2005) Wild food plants traditionally used in the province of Madrid. Econ Bot 59(2):122–136

Tardío J, Pardo-de-Santayana M, Morales R (2006) Ethnobotanical review of wild edible plants in Spain. Bot J Linn Soc 152(1):27–72

Toussaint-Samat M (2009) A history of food. Wiley, West Sussex

Trichopoulou A, Vasilopoulou E, Hollman P, Chamalides Ch, Foufa E, Kaloudis Tr, Kromhout D, Miskaki Ph, Petrochilou I, Poulima E, Stafilakis K, Theophilou D (2000) Nutritional composition and flavonoid content of edible wild greens and green pies: a potential rich source of antioxidant nutrients in the Mediterranean diet. Food Chem 70:319–323

Tukan SK, Takruri HR, Al-Eisawi DM (1998) The use of wild edible plants in the Jordanian diet. Int J Food Sci Nutr 49(3):225–35

UNESCO (2015) http://www.unesco.org/culture/ich/index.php?lg=en&pg=00011. Accessed Feb 2015

Vardavas CI, Majchrzak D, Wagner KH, Elmadfa I, Katafos A (2006) The antioxidant and phylloquinone content of wildly grown greens in Crete. Food Chem 99(4):813–821

Verde A, Rivera D, Obón C (1998) Estudio etnobotánico de las Sierras de Segura y Alcázar: Las plantas y el hombre. I.E.A. Albacete

Verde A, Fajardo J, Rivera D, Obón C, Inocencio C, Heinrich M (2004). La alimentación en Castilla-La Mancha: de la escasez al desperdicio (El valor de los alimentos locales y su utilización sostenible). Azarbe, Murcia

Vianello de Córdova P (1979). Hesiodo. Los trabajos y los dias. UNAM, Mexico

Yildrim E, Dursuna A, Turan M (2001) Determination of the nutrition contents of the wild plants used as vegetables in Upper Çoruh Valley. Turk J Bot 25:367–371

Zeghichi S, Kallithraka S, Simopoulos AP, Kipriotakis Z (2003) Nutritional composition of selected wild plants in the diet of Crete. In: Simopoulos AP, Gopalan C (eds) Plants in human health and nutrition policy. World Review of Nutrition and Dietetics, vol 91. Karger, Basel. pp 22–40

Chapter 2
The Mediterranean Landscape and Wild Edible Plants

Daniel Sánchez-Mata and Ramón Morales

2.1 Introduction

What do we think of when we evoke a Mediterranean environment? How can we define the Mediterranean landscape? From its Latin origin, the word *Mediterranean* means *in the middle of the Earth*, and this geographical name comes from the evidence of an inland sea surrounded by different lands: the Mediterranean Sea.

The main impression of the climate in these lands is their mild and rainy winters contrasting with hot and dry summers. This predominant set of features is a distinctive pattern for the Mediterranean macrobioclimate, typical in the Mediterranean basin between South Europe, North Africa and West Asia, which is also found throughout southern and south-western territories on all the continents on Earth. These areas are often home to some common *Mediterranean plants* and *Mediterranean crops* exported by European settlers, including wheat, asparagus, grapes, olives, citrus fruits, figs, carob, etc.

The lands in these regions around the world are covered by characteristic vegetation types involving dense forests, woodlands and thickets of woody shrubby plants of varying density, generally with evergreen sclerophyllous leaves. There are different vernacular names for these plant formations depending on the territories, languages, physiognomic structures and main species composition of the forests (such as *encinar* and *quejigar* in Spain) to shrublands (*coscojar, matorral, arbusteda, maquia*, etc., in Spain; *maquis* in France and Israel; *macchia* in Italy; *fynbos* or *renosterveld* in South Africa; *chaparral* in California; *matorral* in Chile; and *kwongan* or *mallee* in Australia).

D. Sánchez-Mata (✉)
Departamento de Biología Vegetal II, Facultad de Farmacia,
Universidad Complutense de Madrid, Madrid, Spain
e-mail: dsmata@ucm.es

R. Morales
Departamento de Biodiversidad y Conservación, Real Jardín Botánico, CSIC, Madrid, Spain
e-mail: morales@rjb.csic.es

M. de C. Sánchez-Mata, J. Tardío (eds.), *Mediterranean Wild Edible Plants*,
DOI 10.1007/978-1-4939-3329-7_2

The Mediterranean macrobioclimate and its associated Mediterranean vegetation have a restricted distribution ranging across a wide band of bioclimates, with its geological origins in the Pleistocene era (Axelrod 1973) and a dependence on cold ocean currents (Raven 1973). Thanks to these features, Mediterranean ecosystems show a remarkably high diversity in the heterogeneity of their plant and animal communities, landforms, lithologies and soil types (Di Castri 1981).

A closer look at the Mediterranean Basin regions (all the territories surrounding the Mediterranean Sea) shows that the Iberian Peninsula is the largest geographic area where a true Mediterranean macrobioclimate can be recognised; around 80 % of its area displays this macrobioclimate. Within other geographical territories and areas, such as the Italian, Balkan and Anatolian peninsulas, the Mediterranean macrobioclimate is present to a lesser degree (Fig. 2.1).

A key historical point to remember is that a great part of the European culture was developed in this region of the world. Human activity and management of almost all its ecosystems is therefore an ancient practice in the Mediterranean Basin. Mankind's evolution in this area from hunter-gatherers to farmers occurred around 8000 years ago, and since then the changes in our environment have become ever more pronounced. All the successive cultures who settled in the Mediterranean territories used the natural resources as an important natural legacy. Wild plants were a major source of food and a key nutritional complement to the animal proteins obtained from hunting and fishing. Knowledge of wild edible plants has survived in more advanced farming and livestock societies, and in times of hardship, such as war and famine, it has once again served as a vital source of food and medicinal resources.

2.2 Mediterranean Macrobioclimate and Vegetation Around the World

Bioclimatology is an ecological discipline used widely in phytogeography to determine the relationships between the climate and climatic variables (mainly temperature and rainfall) and the geographic distribution of living organisms and their communities on Earth, focusing mainly on plant taxa and plant communities (Tuhkanen 1980; Rivas-Martínez et al. 2011). Various bioclimatic indexes have been successfully applied to describe, classify and map different vegetation types (Prentice 1990; Blasi et al. 1999; Gavilán 2005; Gavilán et al. 2007; Nakamura et al. 2007).

Bioclimatic models have shown significant reciprocity between climate and vegetation around the world and are compiled and explained in the recently published bioclimatic approach by Rivas-Martínez et al. (2011). This *Worldwide Bioclimatic Classification System* is the most useful tool for understanding the highly diverse climatic reality of our world in relation to the distribution and diversity of natural ecosystems. This classification system also enables the large-scale mapping of a series of bioclimatic indexes and major vegetation types, as adopted and recently

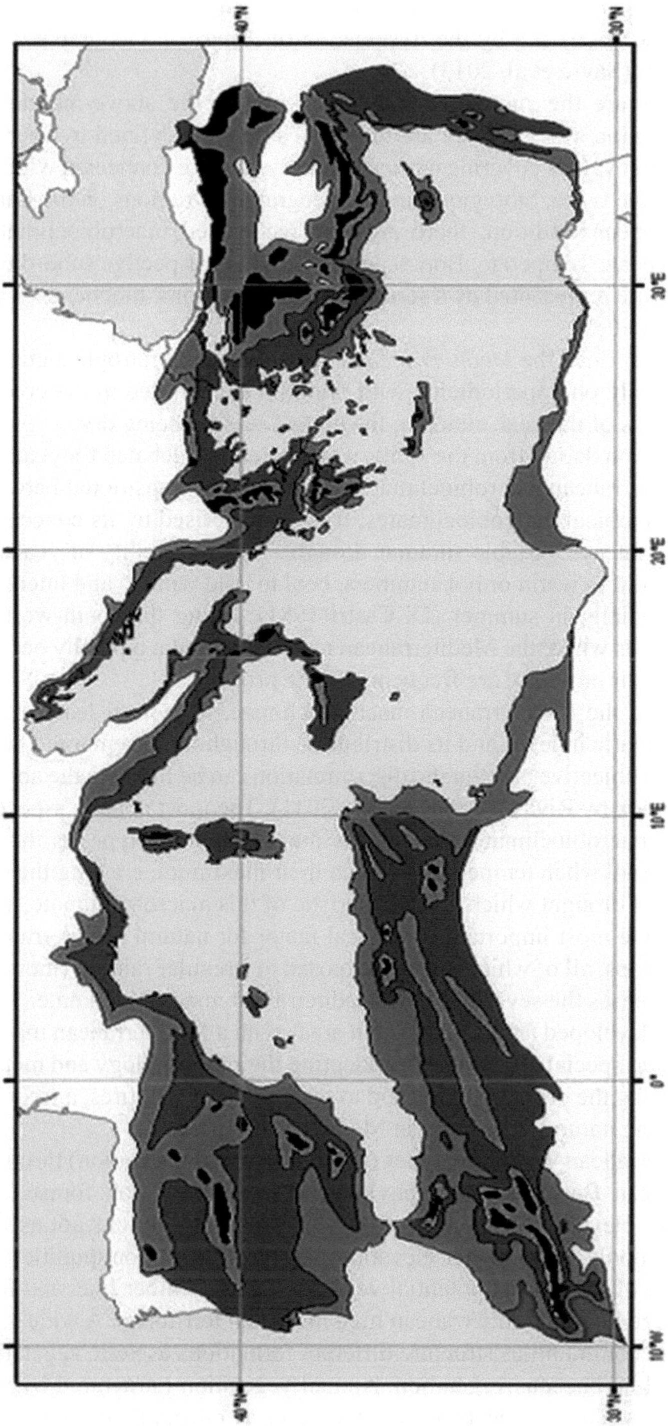

Fig. 2.1 Distribution of the Mediterranean macrobioclimate in the Mediterranean Basin. (Elaborated by Javier Tardío, based on Quézel and Médail 2003)

published by the United States Geological Survey for all the US conterminous territory (Sayre et al. 2009) and by the Association of American Geographers for the African continent (Sayre et al. 2013).

Macrobioclimates are the supreme typological units of the above-mentioned bioclimatic classification system. They are biophysical models defined by specific climate and vegetation values covering a broad territory, and are connected with the major accepted climate types, bioregions and biogeographical regions on the Earth. Following the European tradition, there are five recognised macrobioclimates: Tropical, Mediterranean, Temperate, Boreal and Polar. Their respective subordinate units or bioclimates are represented by a series of plant formations, biocoenoses and typical plant communities.

Emberger (1954) defined the *Mediterranean climate* as 'an extratropical climate with seasonal and daily photoperiodicity, with rainfall concentrated in the cold or relatively cold seasons of the year, summer, the hottest season, being dry'.

This sound definition dating from the 1950s was studied and debated for years. In the 1980s, the Mediterranean macrobioclimate was defined as transitional between the Temperate and Tropical macrobioclimates. It is characterised by its concentration of rainfall in winter, a variable summer drought, high variability of year-on-year precipitation, mild to warm or hot summers, cool to cold winters and intensive solar radiation, especially in summer (Di Castri 1981). Along the south-western coasts of the continents where the Mediterranean macrobioclimate typically occurs, marine fog and high air humidity are frequent climate processes.

A modern view of the Mediterranean macrobioclimate, its typical features, its characteristic bioclimatic indexes and its distribution throughout the emerged lands in the world using an objective bioclimatic discrimination can be found in the above-mentioned compilation by Rivas-Martínez et al. (2011). The most notable aspect of the Mediterranean macrobioclimate affecting its main vegetation types is the restricted summer rainfall when temperatures reach their maximum, creating the real but variable period of drought which is characteristic of this macrobioclimate. This summer drought is the most important ecological factor for natural Mediterranean vegetation and for crops, all of which are also exposed to irregular rainfall (monthly and yearly) that increases the severity of the Mediterranean macrobioclimate.

Plants that have developed and diversified in areas with a Mediterranean macroclimate have acquired special characteristics adapting their morphology and metabolism to survive during the drought period and to resprout after wildfires, a frequent phenomenon affecting natural vegetation in Mediterranean areas.

The main Mediterranean vegetation types (natural potential vegetation) throughout the Mediterranean Basin are sclerophyllous or marcescent oak forests and woodlands at lower elevations, marcescent oak forests in middle elevations and conifer forests or woodlands at higher elevations. Conifer forest communities are always the structure of the natural potential vegetation at the timber line, and they are present as the tree line in Mediterranean high mountain territories. A wide variety of shrubby plant communities structure different formations as xeric vegetation as well as perennial herbaceous vegetation. Annual vegetation (structured by therophytes) develops as pioneer vegetation elsewhere or as nitrophilous formations close to human or animal sites covering a diverse group of habitat types.

Thus, typical Mediterranean shrublands are mainly xeric formations from forest vegetation series, or potential vegetation developed in more xerothermic territories. In the Iberian Peninsula, these potential shrublands cover broad areas of the Ebro River basin and vast areas throughout the south-eastern Iberian territory. In general, we can define Mediterranean shrublands as scrub formations, developed primarily within the more xerothermic range of Mediterranean bioclimates. They are characterised and structured by woody shrubby plants usually with small, stiff, thick, broad evergreen leaves (sclerophyllous leaves), sometimes with an overstory of small scattered trees, with or without an understory of annuals and herbaceous perennials. These usually dense formations may be the natural potential vegetation in extreme xerothermic Mediterranean territories (primary vegetation) or permanent vegetation in sloping or rocky sites, cliffs, etc. They can also represent different successional stages according to the bioclimate and/or human impact (secondary vegetation). These shrublands are always fire prone in territories with a Mediterranean bioclimate.

In the diverse territories with a Mediterranean macrobioclimate around the world, the vegetation shows a similar structure, but with different plant taxa, dynamics and origin. Figure 2.2 shows several selected climatic diagrams from four different places of California, Chile, South Africa and Australia; summer drought is a common feature of these graphs, clearly displaying the Mediterranean character of their bioclimate.

In western North America, from the Pacific Ocean to the Sierra Nevada Mountains, Californian Mediterranean vegetation is very diverse (Barbour et al. 2007). We can distinguish the Temperate macrobioclimate moving north to the Oregon border and the Tropical moving south to the Mexican border or southeast into territories in Nevada and Arizona. West to east, from the Pacific coast to the Sierra Nevada foothills, continentality can be assumed to be the cause of the extreme contrast between the redwood (*Sequoia sempervirens* (D. Don) Endl., Cupressaceae) and coast live oak (*Quercus agrifolia* Née, Fagaceae) forests along the Pacific areas. Continentality increases eastwards towards the Great Central Valley, and these forests are replaced mainly by woodlands of grey pine (*Pinus sabiniana* D. Don, Pinaceae) and blue oak (*Quercus douglasii* Hook. & Arn.) in the foothills and of valley oak *Quercus lobata* Née in the valley bottoms. Ascending through Sierra Nevada, conifer forests appear from the middle elevations to the higher areas, mainly formations structured by *Pinus* sp. pl. (Pinaceae), *Calocedrus decurrens* (Torr.) Florin (Cupressaceae), *Abies lowiana* (Gordon) A. Murray bis (Pinaceae), *Abies magnifica* A. Murray bis and *Tsuga mertensiana* (Bong) Carrière (Pinaceae) in an altitudinal succession. The characteristic *California chaparrals* are typical Mediterranean shrubby vegetation. As in the Mediterranean Basin, these formations may be xeric vegetation from forest communities at disturbed sites, potential vegetation in the more xerothermic areas or permanent plant communities if they grow in special habitats or soils.

The structure of the Mediterranean vegetation in central Chile is comparable to California from the Pacific Ocean to the foothills of the Andean ranges (Dallman 1998). The Mediterranean territory extends north south from La Serena to north of Concepción, including the capital city of Santiago de Chile. *Matorrals* in the

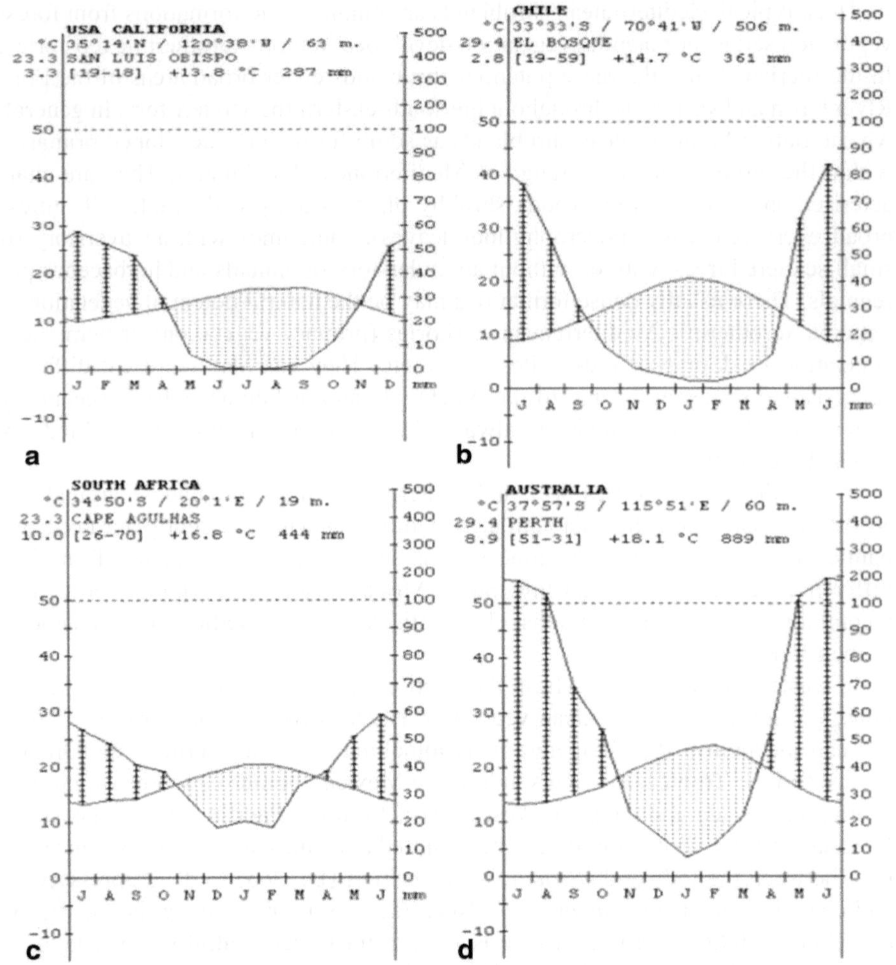

Fig. 2.2 Climatic diagrams of four stations with Mediterranean macrobioclimate around the world, showing their characteristic summer drought period. **a** San Luis Obispo, CA (USA). **b** El Bosque (Chile). **c** Cape Agulhas (South Africa). **d** Perth (Australia)

more xerothermic regions with cacti and bromeliads or along the coastal areas or Andean foothills—*Lithraea caustica* Hook. & Arn. (Anacardiaceae), *Quillaja saponaria* Molina (Rosaceae), *Acacia caven* (Molina) Molina (Fabaceae), *Echinopsis chiloensis* (Colla) H. Friedrich & G.D. Rowley (Cactaceae), etc.—and sclerophyllous woodlands—*Cryptocaria alba* (Molina) Looser (Lauraceae), *Peumus boldus* Molina (Monimiaceae), *Maytenus boaria* Molina (Celastraceae), *Jubaea chilensis* Baill. (Arecaceae), etc.—are the most frequent Mediterranean formations.

Fynbos *(fine bush)* is the most widespread and diverse native formation throughout the Mediterranean macrobioclimate territory in South Africa, located between the Western Cape and Port Elizabeth to the east. Fynbos is dominated mainly by

Fig. 2.3 Several Mediterranean landscapes around the world showing their typical features. **a** A climactic *serpentine chaparral* (xero-edaphic potential vegetation) in California (Napa County). **b** A vegetation mosaic with several shrub formations in an open holm oak woodland (*Quercus ilex* subsp. *ballota* (Desf.) Samp.) in eastern Spain. **c** A *dehesa* formation, with the same species in central Spain. **d** A mosaic of managed Mediterranean vegetation in Crete (Amari Valley)

evergreen shrubs with a high index of biodiversity and endemicity. Proteaceae, Ericaceae and Restionaceae growing alongside succulent plants (*Euphorbia, Aloe*, etc.) structure the woody community characteristic of the Western Cape region of South Africa with a Mediterranean macrobioclimate (Dallman 1998).

The *kwongan* formations in Western and South Australia show a similar structure to the South African fynbos. This is a typical form of sclerophyllous scrub vegetation found mainly between Perth and Adelaide. The *mallee* formations are structured by several species of *Eucalyptus* (Myrtaceae), adopting a multitrunk architecture originating from resprouts after wildfires. These formations can be seen in the drier northern half of the Mediterranean macrobioclimate region of South Australia north of Adelaide. Woodland and forests structured by *Eucalyptus* sp. pl. still remain in the wetter areas of Western and South Australia. There are several plant formations with a well-defined structure characteristic of typical Mediterranean vegetation throughout the Australian landscape (Dallman 1998).

Figure 2.3 shows four different Mediterranean plant formations and communities around the world.

2.3 Mediterranean Basin Territories: Richness and Biodiversity

The Mediterranean Basin territory is the largest area in the world characterised by the Mediterranean macrobioclimate. This territory also has a more complex geography and geomorphology than any of the other Mediterranean macrobioclimate regions. In the north, the land is divided into four large peninsulas: the Iberian Peninsula in the west, the Italian Peninsula in the centre, the Balkan Peninsula in the east and the Anatolian Peninsula in the eastern area, connecting with Asian territories to the east and with Syria, Lebanon and Israel to the south. The southern coastline of the Mediterranean Sea has a less rugged silhouette, with the countries of the Maghreb (Morocco, Algeria and Tunisia) separated from the Middle East by a broad band of Libyan and Egyptian desert. The area of Mediterranean macrobioclimate in the countries of the Mediterranean Basin is primarily coastal, with larger inland extensions in the Iberian and Anatolian peninsulas (Rivas-Martínez et al. 2011).

One of the main features of the climate in the Mediterranean Basin territories is its extremely varied rainfall pattern, a reflection of its vast area and complex geography. Generally, rainfall is greater in the north than that in the south, and its value increases with elevation in the numerous mountain ranges throughout the region. In the northern area, the high-mountain territories sometimes show a true Temperate macrobioclimate, with the summer drought disappearing due to the high altitude. The rainfall pattern and its seasonal timing also vary from one region to another, although summer is always the driest season. The wettest season varies in each country and the values depend on the direction of the prevailing winds every season. As a result, the average summer drought also varies considerably in the territories of the Mediterranean Basin. The average temperature during the year, combined with the rainfall values, defines the high variability of bioclimatic features within the Mediterranean macrobioclimate throughout the Mediterranean Basin countries, north to south and west to east (Rivas-Martínez et al. 2011). Figure 2.4 shows six typical Mediterranean climatic diagrams.

The explanation for the Mediterranean Basin's ancient history is that its lands represent the adjoining portions of three much larger continents (Europe, Africa and Asia). The history and origin of the flora in the Mediterranean Basin has been discussed by many authors, and it is particularly worth highlighting the classic work of Pons and Quézel (1985) and later contributions. A wide diversity of plant species and communities is found throughout the Mediterranean Basin regions, the richest areas being the Iberian, Balkan and Anatolian peninsulas.

The vegetation and plant communities of these territories feature several woody trees usually considered as *true Mediterranean tree species*, and their range of distribution is widely used to define or consider the extent of the Mediterranean macrobioclimate: the cultivated olive tree (*Olea europaea* L., Oleaceae) and several related species of Mediterranean oaks, chiefly the holm oak (*Quercus ilex* subsp. *ballota* (Desf.) Samp.). Other woody species widely distributed in the Mediterranean Basin areas are the Aleppo pine (*Pinus halepensis* Mill.) and the kermes oak (*Quercus coccifera* L.).

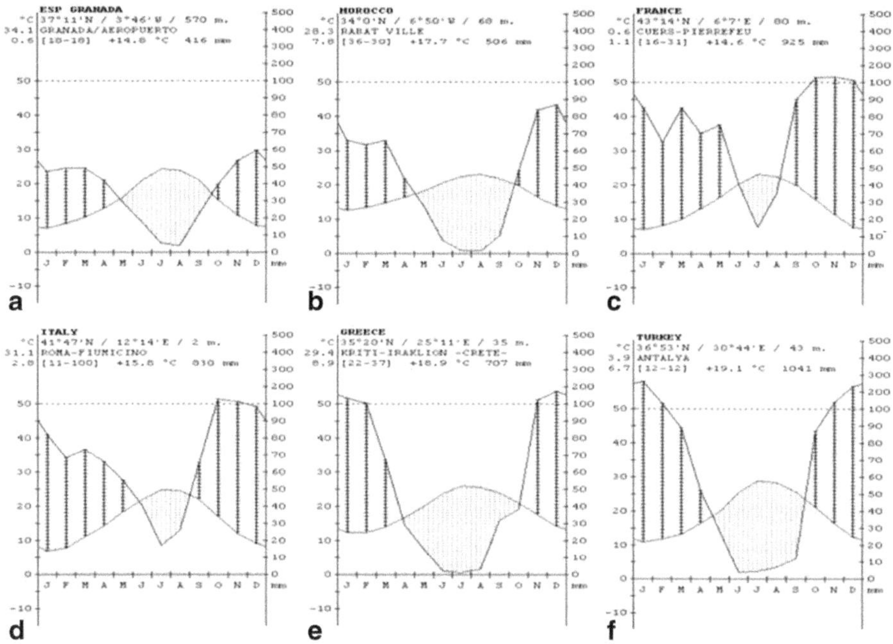

Fig. 2.4 Climatic diagrams of six stations of the Mediterranean Basin. Again, the summer drought period characterises the Mediterranean macrobioclimate in all these places. **a** Granada (Spain). **b** Rabat (Morocco). **c** Cuers-Pierrefeu (France). **d** Rome (Italy). **e** Crete (Greece). **f** Antalya (Turkey)

The natural potential forest vegetation and main vegetation types throughout the Mediterranean Basin are structured by oaks or conifers. The main vegetation types covering the mountain areas are forests and woodlands; above the timber line and tree line, frequent low-growing and cushion-shaped shrubs, spiny cushion plants or perennial grasslands are sometimes the dominant plant formations in high-mountain landscapes. In more xerothermic areas, the so-called *maquis, garriga* or simply *matorral* is the most extensive formation when trees cannot grow; *matorrals* can be also considered as different vegetation types developed on disturbed or altered sites where forest species form the natural potential vegetation and its vegetation series. Other distinctive xeric vegetation types are perennial and annual grasslands with a high biodiversity and grazed by cattle; bulbs and other geophytes play a major role in these plant formations in maintaining the community structure and species diversity. Natural dynamic processes—including deep fire adaptations—and the various systems of managing the natural vegetation used by man since ancient times have shaped the landscape in the Mediterranean Basin for centuries.

Furthermore, most of the food crops cultivated in the area were originated in the Mediterranean Basin and in Middle Eastern regions. Wheat, barley, legumes (such as lentils and chickpeas), vegetables (such as artichokes, asparagus, cabbage, leeks, onions and garlic), and fruits and seeds (such as grapes, olives, almonds and acorns) are only a few good examples.

2.4 History and Management: Changes in Mediterranean Vegetation

The Mediterranean Basin is a vast and ancient territory where different cultures have been active since time immemorial. All the peoples and cultures who settled in these areas were also influenced by cultures from Asia, Sumer, Persia, Babylon, Egypt, etc. This mixture of peoples has often produced a multicultural reality both in ancient times and in the modern day.

The Mediterranean Sea was the most convenient channel for communication and trade throughout this vast territory. Although many ecosystems were common to all the countries of the Mediterranean Basin, others were specific to certain regions. Moreover, the constant exchange of plants and plant materials (including cultivated and horticultural plants) was a frequent practice between the different cultures in the Mediterranean.

Cultural activities always take place in relation to nature, and nature has been changed and managed in order to adapt it to the different cultural conditions and human needs. The various systems of managing nature are also cultural features of each territory and have led to the different characteristics of each geographical area. The present Mediterranean landscape is the complex result of human activity in the management of nature for its own benefit (Morales et al. 2011).

The impact of these activities on nature brought major changes to Mediterranean ecosystems, and the pristine Mediterranean vegetation was transformed through human intervention. This was not only due to agriculture, livestock and the building of cities but also due to mining for metals, a key activity throughout its history. A complex mosaic began to develop as distinctive landscapes evolved as a result of cultivation activities and animal farming around rural settlements (Fig. 2.5). Natural

Fig. 2.5 The transformation of the pristine vegetation for cultivation fields or livestock was the most important human activity in the Mediterranean landscapes. **a** Woodland of holm oak (*Quercus ilex* subsp. *ballota* (Desf.) Samp.) at the *top* and in the *foreground* a barley crop in Guadalajara province (CE Spain). **b** The cork oak woodlands (*Quercus suber* L.) maintain a large black pig population managed in Huelva province (SW Spain)

woods and bushes were transformed into a new cultural space where crops and grasslands could be identified with boundaries defining their ownership. Natural ecosystems survived in singular localities.

All these changes promoted nitrogen-loving (nitrophilous) plants. Many are associated with farmlands and cultivated plants (weeds), and others grow near places of human habitation (ruderals) or live along the borders of paths, roads or cattle tracks, carried by man and domestic animals in their seasonal migration. Annuals like *Matricaria aurea* (Loefl.) Sch. Bip. and *Matricaria discoidea* DC. (Asteraceae) or perennials like *Chamaemelum nobile* (L.) All. (Asteraceae) are widely used chamomiles growing around cattle trails in mountain areas (Pardo-de-Santayana and Morales 2010). Numerous weeds and ruderal plants are used and collected as wild vegetables in the Iberian Peninsula, such as *Scolymus hispanicus* L. (Asteraceae), *Silene vulgaris* (Moench.) Garcke (Caryophyllaceae) and *Silybum marianum* (L.) Gaertn. (Asteraceae) (Tardío 2010).

These species grow in synanthropic plant communities included in several phytosociological vegetation classes. The complete syntaxonomy of this kind of vegetation can be found in several major general compilations, such as Rivas-Martínez et al. (2001, 2002) for Spain and Portugal, Pott (1995) for Germany and Chytrý (2009) for the Czech Republic, among others.

Although the landscape is constantly changing due to new agricultural techniques and the testing of new crops, it is worth noting that the Mediterranean landscape has its own attributes. Although it is changeable, we can usually delimit smallholdings on the plains, sometimes with trees dotted amid the crops, terraces on mountainsides, with verges and sites with natural vegetation where oaks, pines and shrubs grow. The landscape is a part of each geographic territory and comprises a geological scenario covered by a carpet of plants in which the geological formations and lithology determine the appearance of the landscape. Managed forest sites have evolved into traditional Mediterranean formations known as *dehesas* (Spain and influenced countries) or *montados* (Portugal and influenced countries). These formations are structured by scattered trees (usually oaks) with perennial and annual grasslands growing in the understory, and are always used for agriculture and livestock (Fig. 2.5b).

2.5 Seasonality and Wild Plant Food Gathering in the Mediterranean Region

Seasonality is very strong under Mediterranean climate conditions (Fig. 2.6). For this reason, certain wild plants are collected and consumed at particular periods of the year. Late winter, and especially spring, is in general the best time to collect vegetables that reach their higher development while being still tender, though depending on the local climatic conditions, they could be collected in other parts of the year. One example is *Rumex pulcher* L. (Polygonaceae), which is at its best in early spring when the plant has developed its basal leaves and before the growth of the flowering stems, or the young shoots of *Tamus communis* L. (Dioscoreaceae),

Fig. 2.6 Seasonality is a very determinant condition for the life in the Mediterranean areas. The same humanised landscape from Guadalajara province (CE Spain) was pictured **a** in springtime and **b** in fall period of the same year

which are collected in mid-spring. However, nowadays freezing methods make possible the consumption of some wild vegetables all year round.

A frequent process in the Mediterranean region is the alternation of good and bad years as a result of the interrelationships of climate conditions, plant production and the harvest process. Very dry autumns in Spanish continental sites signal the beginning of a bad year for collecting wild vegetables which develop during this season and produce their stems and leaves in late winter and spring. These years frequently led to famines in the past, as rain-fed crops needed more or less the same conditions as these wild edible plants. Other wild edible plants with other conditions were especially sought after in these years. Many wild plants grew around the outer edges of crops and were also associated with some of them. For example, *Asparagus acutifolius* L. (Asparagaceae)—which grows in oak forest communities, in olive groves and, in the past, inside rain-fed farm crops—has very tough underground organs like tuberous roots, and in dry years also produces new shoots or *asparagus*.

However, there are also species that are not so dependent on the yearly climatic variation, such as *Montia fontana* L. (Portulacaceae) and watercress (*Rorippa nasturtium-aquaticum* (L.) Hayek, Brassicaceae) that always grow where there are permanent waters.

However, usually the years of famine are recalled by the population and associated with wild edible plants. Some of the older population near the city of Madrid (Spain) are familiar with a large number of wild edible plants, since their collection was important for survival during a succession of years of shortage after the Spanish Civil War. Various species of *Rumex* (Polygonaceae), *Scolymus hispanicus* L., *Silene vulgaris*, *Silybum marianum, Onopordum* sp. pl. and *Sonchus oleraceus* L. (Asteraceae) and *Anchusa azurea* Mill. (Boraginaceae) have been used as stewed vegetables (Tardío et al. 2005). The young shoots or *asparagus* of *Asparagus acutifolius, Bryonia dioica* Jacq. (Cucurbitaceae), *Tamus communis* and *Humulus lupulus* L. (Cannabaceae) have also been eaten stewed (Tardío et al. 2005). A large number of other plants have been consumed in salads or eaten raw, such as *Foeniculum vulgare* Mill. and *Scandix australis* L. (Apiaceae), *Chondrilla juncea* L., *Mantisalca salmantica* (L.) Briq. & Cavill., *Scorzonera laciniata* L., *Sonchus crassifolius* Pourr. ex Willd. and *Taraxacum* sp. pl. (Asteraceae), *Montia fontana* and *Rorippa nasturtium-aquaticum* (Tardío et al. 2005).

If spring is the season of vegetables, late spring and summer is the season of aromatic plants used for seasoning. The diverse floras of the Mediterranean Basin countries provide numerous highly scented species. Chemical compounds from the essential oils are produced in the secondary metabolism of all these plants and are present in the small uni- or multicellular dots that are transformed hairs located on the epidermis of leaves and flowers. These tiny dots burst, and the essential oil evaporates into the air producing their pleasant *aromas*, even when the plants are dried. One of the biological advantages of the presence of these chemical compounds is that plants are protected from being eaten by large animals or phytophagous insects and from bacterial infections. Taking advantage of these features has certainly been the main reason of why aromatic plants have been used by man since ancient times as medicines, food preservatives and condiments. This was most likely the primary reason for using these plants.

Seasonings feature strongly in the Mediterranean culture as culinary ingredients for cooking, and numerous aromatic plants have been traditionally used as flavouring and food preservatives (Tardío et al. 2006). The essential oils and other components of all these plants also have noteworthy antimicrobial and fungicidal properties.

There is a high biodiversity within this group of species belonging to different families. Lamiaceae, Apiaceae and Asteraceae are the main aromatic plant families in the territories of the Mediterranean Basin.

Most of these species frequently used as seasoning belong to the Lamiaceae family, such as oregano (*Origanum vulgare* L.), thyme (*Thymus* L. sp. pl.), rosemary (*Rosmarinus officinalis* L.) and mint (*Mentha* L. sp. pl.). Oregano is a key ingredient in the home-made products elaborated in the Spanish traditional *matanza* or pig slaughter (Tardío et al. 2006). Some species of the Apiaceae family are also used, such as fennel (*Foeniculum vulgare* s.l.), an excellent seasoning and preservative for olives.

Some of these aromatic plants are also used to make herbal liqueurs, as alcohol is a good dissolvent for the essential oils, the most important chemical compounds in these plants.

Finally, autumn is the season for gathering fruits. Trees are the most important source of berries and nuts, especially those of the Fagaceae and Rosaceae families. In the past, the acorns of *Quercus* sp. pl. species, especially those of *Quercus ilex* subsp. *ballota*, were the most important source of starch (Guarrera 2006; Tardío et al. 2006). Afterwards, the seeds of a cultivated tree, the chestnut *Castanea sativa* (Fagaceae) replaced the acorns as source of starch in some regions. The fleshy fruits of trees in the Rosaceae family, such as *Malus sylvestris* (L.) Mill., *Pyrus cordata* Desv., *Sorbus aria* (L.) Crantz or *Sorbus torminalis* (L.) Crantz, are also consumed. They can be eaten ripe and are also commonly used to elaborate liqueurs. Another important little tree frequently present in the Mediterranean basin is *Arbutus unedo*

L. (Ericaceae), which produces an edible fruit that can be collected and eaten in late autumn (Tardío et al. 2006; Molina et al. 2011).

Shrubs are very common in landscape formations throughout the territories of the Mediterranean biogeographical region (Rivas-Martínez et al. 2007) as potential vegetation or as replacement formations of forests and woodlands after they have been altered. More than 400 different species of shrubs—especially from the Fabaceae, Cistaceae, Ericaceae and Lamiaceae families— are found growing in these diverse plant communities, both as climatic climax vegetation (potential vegetation) in high-mountain altitudes or in extreme xerothermic conditions.

Wild fruits of some species of these shrubs are commonly collected in the Mediterranean region in riparian areas and on forest edges, where a number of shrubs from forest fringe communities usually grow. Fruits of several species of Rosaceae of the genera *Rubus* (such as *R. ulmifolius* Schott), *Rosa* (such as *R. canina* L.) and *Prunus* (such as *Prunus spinosa* L. and *Prunus insititia* L.), and other species such as *Ribes uva-crispa* L. (Grossulariaceae) and *Arctostaphylos uva-ursi* (L.) Spreng (Ericaceae), are collected for being eaten raw or as ingredients in jams, marmalades or plant liqueurs (Tardío et al. 2006).

2.6 Weeds as Wild Vegetables of Croplands

Weeds are one of the main crop problems. Chemical weedkillers are now widely used to clear crops of weeds, whereas in the past, hand-weeding was a routine activity throughout the agricultural year that provided of good wild vegetables for animal and human consumption. One example is *Asparagus acutifolius*, called *espárrago triguero* in Spanish (asparagus from the wheat fields) because in the past it could be found within the cereal crops growing as a weed (Tardío 2010). Besides this species, many other species of weeds were consumed, such as *Chondrilla juncea, Cichorium intybus* L., *Papaver rhoeas* L. and *Sonchus oleraceus*. For example, the basal leaves of *Cichorium intybus*, or chicory, were eaten in salads or stewed, and their toasted roots were widely used in the past as a coffee substitute in the Mediterranean countries. Today with the widespread use of weedkillers, it is sometimes hard to assess whether wild edible plants growing around crops can be eaten or not.

Many of these Mediterranean weeds can be included in the group of thistles. These plant communities are known as *cardales* or *cardizales* in Spanish and sometimes are dominant on disturbed sites and abandoned cultivated lands, but always with a high level of nitrogen in the soil. Almost all the species are from the Asteraceae family, belonging to the genera *Onopordum, Scolymus, Silybum, Carduus, Cirsium, Carlina, Cynara, Cnicus, Centaurea, Galactites, Echinops, Carthamus, Picnomon, Xanthium, Atractylis* and *Notobasis* (Barão and Soveral 2010). Thistles usually have prickles and thorns to protect against animals and humans. They are often sun-loving (heliophilous) plants characterised by their dense flower heads or capitula which are very variable in size and shape but always have the same structure.

As is well known, the artichoke is a cultivated form of the wild artichoke or cardoon (*Cynara cardunculus* L., Asteraceae), an edible thistle whose leaf petioles and immature capitula are traditionally consumed. There are other wild relatives, such as *C. humilis* L. and *C. tournefortii* Boiss. & Reut., whose immature inflorescences are also eaten. Other edible thistles are *Silybum marianum, Onopordum macracanthum* Schousb. and *Scolymus hispanicus* (Tardío 2010). All these plants have edible basal leaves; when peeled they are often used as stewed vegetables, especially in the case of *S. hispanicus*, one of the most highly prized wild vegetables throughout the countries of the Mediterranean Basin.

The flowers of some thistles are used as condiments and as natural food colourings (saffron substitutes), such as *Scolymus hispanicus* and the cultivated *Carthamus tinctorius* known as *azafrán de cardo* (thistle saffron) or *azafrán moruno* and used like saffron. The flowers of *Cynara humilis* and *Cynara cardunculus* are also used to curdle milk for making local varieties of cheese.

Some species from other botanical families are also considered and known as thistles due to their thorny organs. Example includes *Eryngium campestre* L. (Apiaceae), whose basal part of the young shoots and the roots also can be eaten, and where the edible and associated fungus *Pleurotus eryngii* (DC.) Quél (Pleurotaceae) also grows.

2.7 Social Implications: The Pleasure of Collecting Plants. The Countryside as a Mediterranean or Aromatic Garden

Many older people today are interested in collecting wild plants. Some are extremely proficient local experts in the places the plants grow; this knowledge is typically the domain of women, while men are mostly the collectors. For elderly retired people, it can be very pleasant to walk in the countryside during springtime when the sun is shining and temperatures are usually mild, and to gather plants as their ancestors did before them, thus preserving these ancient traditions from a past era when this knowledge was a common place. Collecting wild edible plants is highly recommended for active senior citizens, as it is a healthy pursuit for both body and mind: walking, memorising the names and characteristics of the plants and then consuming high-quality natural plant products.

Also the social importance of gathering plants in Mediterranean environments is evident: good climate conditions, the pleasure and satisfaction of collecting wild edible plants today and a lifeline in times of scarcity, such as during or after wars.

In the countries of the Mediterranean Basin, the landscape is a vast garden with numerous different aromatic plants and a wide range of edible vegetables and fruits in several moments of the year, such as late winter, spring, summer and autumn. For people of limited means, foraging wild plants is a way of supplementing their nutritional need, as their forebears did in the past.

It can also be a good excuse for engaging in local sightseeing activities in some regions by organising excursions for people to collect wild edible plants, learning where they grow, when they can be picked and then how to cook and eat them.

References

Axelrod DI (1973) History of the mediterranean ecosystem in California. In: Di Castri F, Mooney HA (ed) Mediterranean type ecosystems, origin and structure. Springer-Verlag, Berlin, pp 225–277

Barão MJ, Soveral A (2010) The ecology and use of edible thistles in Évora, Alentejo, Southeastern Portugal. In: Pardo-de-Santayana M, Pieroni A, Puri R (eds) Ethnobotany in the New Europe: people, health and wild plant resources. Berghahn Books, Oxford, pp 189–210

Barbour MG, Keeler-Wolf T, Schoenherr AA (2007) Terrestrial vegetation of California, 3rd edn. University of California Press, Berkeley

Blasi C, Carranza ML, Filesi L, Tilia A, Acosta A (1999) Relation between climate and vegetation along a Mediterranean-temperate boundary in central Italy. Global Ecol Biogeogr 8:17–27

Chytrý M (ed) (2009) Vegetace České republiky 2. Ruderální, plevelová, skalní a suťová vegetace. Vegetation of the Czech Republic 2. Ruderal, weed, rock and scree vegetation. Academia, Praha

Dallman PR (1998) Plant life in the world's mediterranean climates. California, Chile, South Africa, Australia, and the Mediterranean Basin (California Native Plant Society). University of California Press, Berkeley

Di Castri F (1981) Mediterranean-type shrublands of the world. In: Di Castri F, Goodall DW, Specht RL (eds) Mediterranean-type shrublands. Elsevier Scientific, Amsterdam, pp 1–52

Emberger L (1954) Une classification biogéographique des climats. Rec Trav Lab Bot Géol Zool Univ Montp Sér Bot 7:3–43

Gavilán R (2005) The use of climatic parameters and indices in vegetation distribution. A case study in the Spanish Sistema Central. Int J Biometeorol 50(2):111–120

Gavilán R, Sánchez Mata D, Vilches B, Entrocassi G (2007) Modelling current distribution of Spanish Quercus pyrenaica forest using climatic parameteres. Phytocoenol 37:561–581

Guarrera PM (2006) Usi e tradizioni della flora italiana. Aracne, Roma

Molina M, Pardo-de-Santayana M, Aceituno L, Morales R, Tardío J (2011) Fruit production of strawberry tree (Arbutus unedo L.) in two Spanish forests. Forestry 84(4):419–429

Morales R, Tardío J, Aceituno L, Molina M, Pardo-de-Santayana M (2011) Biodiversidad y etnobotánica en España. Mem R Soc Esp Hist Nat (2ª época) 9:157–207

Nakamura Y, Krestov PV, Omelko AM (2007) Bioclimate and zonal vegetation in Northeast Asia: first approximation to an integrated study. Phytocoenol 37(3–4):443–470

Pardo-de-Santayana M, Morales R (2010) Chamomiles in Spain. The dynamics of plant nomenclature. In: Pardo-de-Santayana M, Pieroni A, Puri R (eds) Ethnobotany in the New Europe: people, health and wild plant resources. Berghahn Books, Oxford, pp 282–306

Pons A, Quézel P (1985) The history of the flora and vegetation and past and present human disturbance in the Mediterranean. In: Gómez-Campo C (ed) Conservation of Mediterranean plants. Junk, The Hague, pp 25–43

Pott R (1995). Die Pflanzengesellschaften Deutschlands, 2nd edn. Ulmer, Stuttgart

Prentice KC (1990) Bioclimatic distribution of vegetation for general circulation model studies. J Geophys Res 95:11811–11830

Quézel P, Médail F (2003) Écologie et biogeographie des forêts du bassin méditerranéen. Elsevier, Paris

Raven P (1973) Introduction to Section V: plant Biogeography. In: Di Castri F, Mooney HA (ed) Mediterranean type ecosystems, origin and structure: 211–212. Springer, Berlin

Rivas-Martínez S, Fernández-González F, Loidi J, Lousã M, Penas A (2001) Syntaxonomical checklist of vascular plant communities of Spain and Portugal to asssociation level. Itin Geobot 14:5–341
Rivas-Martínez S, Díaz TE, Fernández-González F, Izco J, Loidi J, Lousã M, Penas A (2002) Vascular plant communities of Spain and Portugal. Addenda to the syntaxonomical checklist of 2001. Itin Geobot 15(1–2):5–922
Rivas-Martínez S et al (2007) Mapa de series, geoseries y geopermaseries de vegetación de España (memoria del mapa de vegetación potencial de España) Parte I. Itin Geobot 17:5–436
Rivas-Martínez S, Rivas-Sáenz S, Penas A (2011) Worldwide bioclimatic classification system. Global Geobot 1(1):1–638
Sayre R, Comer P, Harumi H, Cress J (2009) A new map of standardized terrestrial ecosystems of the conterminous United States. US Department of Interior, US Geological Survey Professional Paper 1768. Reston
Sayre, R, Comer P, Hak J, Josse C, Bow J, Warner H, Larwanou M, Kelbessa E, Bekele T, Kehl H, Amena R, Andriamasimanana R, Ba T, Benson L, Boucher T, Brown M, Cress J, Dassering O, Friesen B, Gachathi F, Houcine S, Keita M, Khamala E, Marangu D, Mokua F, Morou B, Mucina L, Mugisha S, Mwavu E, Rutherford M, Sanou P, Syampungani S, Tomor B, Vall A, Vande Weghe J, Wangui E, Waruingi L (2013) A new map of standardized terrestrial ecosystems of Africa. Association of American Geographers. Washington, DC
Tardío J (2010) Spring is coming: the gathering and consumption of wild vegetables in Spain. In: Pardo-de-Santayana M, Pieroni A, Puri R (eds) Ethnobotany in the New Europe: people, health and wild plant resources. Berghahn Books, Oxford-New York, pp 211–238
Tardío J, Pascual H, Morales R (2005) Wild food plants traditionally used in the province of Madrid, Central Spain. Econ Bot 59(2):122–136
Tardío J, Pardo-de-Santayana M, Morales R (2006) Ethnobotanical review of wild edible plants in Spain. Bot J Linn Soc 152:27–71
Tuhkanen S (1980) Climatic parameters and indices in plant geography. Acta Phytogeogr Suec 67:1–105

Chapter 3
Nutritional Ethnobotany in Europe: From Emergency Foods to Healthy Folk Cuisines and Contemporary Foraging Trends

Łukasz Łuczaj and Andrea Pieroni

3.1 Introduction

In several countries and regions of Europe, ethnobotanical studies and reviews give us a picture of traditionally used wild food plants, for example, in Spain (Aceituno-Mata 2010; Benítez 2009; Bonet and Vallès 2002; Criado et al. 2008; Dávila 2010; Gonzalez et al. 2011; Parada et al. 2011; Pardo-de-Santayana et al. 2005, 2007; Pardo-de-Santayana and Morales 2010; Polo et al. 2009; Tardío et al. 2005, 2006; Rivera et al. 2005; Tardío 2010; Velasco et al. 2010; Verde et al. 2003), Portugal (Carvalho and Morales 2010; Pardo-de-Santayana et al. 2007), Italy (di Tizio et al. 2012; Guarrera 2003; Guarrera et al. 2006; Ghirardini et al. 2007; Paoletti et al. 1995; Picchi and Pieroni 2005; Pieroni 1999, 2001; Pieroni et al. 2002, 2005), Greece (Della et al. 2006; Forbes 1976), France (Marco et al. 2003), Bosnia-Herzegovina (Redžić 2006), the whole Mediterranean area (Hadjichambis et al. 2008; Leonti et al. 2006; Rivera et al. 2006a), Austria (Christanell et al. 2010; Schunko and Vogl 2010; Schunko et al. 2012), Switzerland (Abbet et al. in press), Poland (Kujawska and Łuczaj 2010; Łuczaj 2008, 2010a, 2010b, 2011; Łuczaj and Kujawska 2012; Łuczaj and Szymański 2007), Slovakia (Łuczaj 2012a), Hungary (Dénes et al. 2012), Belarus (Łuczaj et al. 2013a), Croatia (Łuczaj et al. 2013b, 2013c), Bulgaria (Nedelcheva 2013), Ukraine (Łuczaj 2012b), Sweden (Svanberg 1998, 2011, 2012), Iceland, and the Faroes (Svanberg and Ægisson 2012). Some of these papers were coming from a special volume of the journal *Acta Societatis Botanicorum Poloniae*, which was devoted to the ethnobotany of wild food plants (Dénes et al. 2012; Kalle and Sõukand 2012; Łuczaj 2012a; Svanberg 2012) and

Ł. Łuczaj (✉)
Department of Botany, Institute of Applied Biotechnology and Basic Sciences, University of Rzeszów, Rzeszów, Poland
e-mail: lukasz.luczaj@interia.pl

A. Pieroni
University of Gastronomic Sciences, Pollenzo, Cuneo, Italy
e-mail: a.pieroni@netcologne.de

M. de C. Sánchez-Mata, J. Tardío (eds.), *Mediterranean Wild Edible Plants*,
DOI 10.1007/978-1-4939-3329-7_3

also contains a separate article summarizing the changing in the use of wild food plants in Europe (Łuczaj et al. 2012). In this chapter, we include the main points drafted in that article extended by looking at the latitudinal and longitudinal aspects of wild food plant use in Europe.

Plant use patterns are usually not constant. Nowadays, in most cases, ethnobotanical studies reveal either a dramatic or gradual loss of traditional knowledge and practices (e.g., Łuczaj 2008; Sõukand and Kalle 2011; Turner et al. 2011). The changes in patterns of wild plant use differ by region and are associated with the general socioeconomic changes in Europe. Moreover, times of famine seem to be in the distant past for industrially developed countries. Food made of cultivated plants and bought from the supermarket appears on the table with relatively little effort, while collecting wild species is more time consuming and season dependent. In spite of that, the importance of wild food plants for food security and in shaping alternative models of consumption is emphasized (Madej et al. 2014; Muller and Almedom 2008). Moreover, in Europe, there are new phenomena associated with plant use appearing in modern societies. Some of them have to do with migration and new ethnic minorities appearing in cities and bringing new traditions with them. For example, Ukrainian migrants throughout Europe tend to collect *Rumex* spp. for their traditional green borsch, and Asian migrants look for bracken fiddleheads in European woods (Pieroni and Gray 2008; Picchi and Pieroni 2005). In Great Britain and Ireland, many members of the large Polish minority search through the woods looking for fungi species traditionally gathered in Poland to the surprise of the local mycophobic population. Other phenomena appear due to new tendencies in nutrition and self-medication facilitated by the rapid spread of information via the Internet. On top of that, not all the traditions are gone; in some areas, old traditions are cultivated for a variety of reasons, while in others, they are lost.

3.2 A Gradual Decrease in the Necessity of Use of Wild Food Plants

The use of wild plants in Europe is often associated with times of famine or food scarcity (although not exclusively). Most of the early studies on the use of wild food plants in Europe—from those coming from the nineteenth century until more or less the 1960s—capture the memory of famine and the use of wild plants for survival, including the consumption of starvation foods that in normal times would be discarded by the community. There were many outbreaks of famine in some parts of Europe in the nineteenth and twentieth centuries. Probably one of the most serious was caused by the potato blight (1844–1849). This affected many potato-dependent countries, from Ireland to Poland (Maurizio 1926, 1927, 1932; Ó Gradá 2009). Locally, famines due to crop failure appeared in some parts of Europe for a few successive decades (Häkkinen 1992; Nelson 1998; Svanberg 2007; Svanberg and Nelson 1992). A severe famine hit Russia in 1892, then there was World War I (1914–1918) and the revolution and the establishment of the Soviet Union which brought famine in 1921–1922 (Sorokin 1975). In 1932–1933, millions of people starved to death in the Ukraine (Conquest 1987; Ó Gradá 2009). The Spanish Civil War (1936–1939) and later World War II (1939–1945) brought another revival of emergency food (Matalas and Grivetti 2007). For most European countries, this was the last episode of serious lack of food, apart from the Balkans when the conflicts, which emerged during the collapse of Yugoslavia, caused a series of food crisis. The best documented and the longest one is the over 3-year siege of Sarajevo, which was captured from inside by the Bosnian botanist Sulejman Redžić, who not only recorded the emergency plants used, but also tried to alleviate the crisis by running media programs explaining the use of wild plants in the besieged city (Redžić 2010; Redžić et al. 2010).

Two kinds of poverty food are dominant in memories or sources from the nineteenth century: plants used to make potherb or soup and those employed as additional ingredients for making bread. The use of potherb/soup wild vegetables composed of many species survived mainly in the Mediterranean, at least in some rural areas, whereas in other parts of Europe, such soups are usually made with few species, for example, nettles (*Urtica dioica* L.) or sorrel (*Rumex* spp.). However, the use of wild plants for making bread has almost completely disappeared from the European diet. The famine bread ingredients consisted of a large variety of products, including even clay, cambium of trees, and wood shavings. Other common additions were acorns, lesser-used seeds of Fabaceae plants (e.g., *Vicia* spp.), dried and milled leaves, rhizomes and roots, sometimes even mosses and lichens. Still, most of the plants used in times of famine were also used when just a shortage of grain occurred in the spring time (Łuczaj 2011; Maurizio 1926; Tardío 2010). Some of the plants used during famine were the species used also at normal circumstances but in small amounts; others were only used in severe famine. Some of the plants commonly used in times of even slight food scarcity are now completely forgotten (Table 3.1).

Although it was probably not perceived as so by the historical users, many of the uses of wild food plants are related to the diversification of a monotonous diet in non-famine times. In many Mediterranean countries (e.g., Portugal, Spain, Italy and Greece), different *Lamiaceae* and *Apiaceae* species have also been used to spice various products. The use of wild food plants for this diversification does not depend so much on the geographical position or the variety of the flora of the users but on access to supplies and the knowledge and creativity of the cook. In northern Europe, the choice of aromatic plants is not so large as in the south, but the available aromatic plants were used in a similar way, particularly juniper (*Juniperus communis* L.) pseudo-fruits and caraway (*Carum carvi* L.) fruits (Kalle and Sõukand 2012; Łuczaj 2011). It is worth noting that local aromatic *Lamiaceae* species, so omnipresent in the Mediterranean cooking, rarely found their way in northern European cuisine. The exception is, for example, the use of *Mentha* spp. (e.g., *M. longifolia* (L.) Huds.) to spice the cheese-and-potato filling of dumplings in parts of southeast Poland (Łuczaj 2008, 2011). The scarcity of aromatic plants in northern Europe sometimes attracted attention to even relatively toxic species such as *Tanacetum*

Table 3.1 Examples of famine plants important in nineteenth century or early twentieth century nutrition in Europe, now nearly forgotten. (Adapted from Łuczaj et al. 2012)

Species (Part used)	Country
Elytrigia repens (L.) Desv. ex Nevski (rhizomes)	Poland
Arum spp. (bulbs)	Croatia
Eryngium spp. (roots)	England, Croatia
Trapa natans (fruits)	Poland, Hungary, Montenegro
Polypodium vulgare (rhizomes)	Poland, Slovakia
Angelica sylvestris L. (young stems)	Estonia
Anthriscus sylvestris (L.) Hoffm. (young stems)	Estonia
Equisetum arvense L. (tubers, spring shoots)	Estonia
Cirsium oleraceum (L.) Scop.(young stems)	Estonia
Cetraria islandica (L.) Ach. (thallus)	Estonia
Crepis vesicaria subsp. *haenseleri* (Boiss. ex DC) P. D. S (basal leaves)	Spain
Quercus ilex subsp. *ballota* (Desf.) Samp. (acorns)	Spain
Crataegus monogyna Jacq. (fruits)	Spain
Rosa canina L. (young shoots)	Spain
Chondrilla juncea L. (young shoots)	Spain
Sonchus asper (L.) Hill (peeled midribs)	Spain
Cichorium intybus L. (basal leaves)	Spain
Scorzonera laciniata L. (tender leaves and stems)	Spain
Silybum marianum (L.) Gaertn. (peeled basal leaves)	Spain
Cynodon dactylon (L.) Pers. (rhizomes)	Spain

vulgare L., which was used to be added to omelettes in England or some dishes in eastern Slovakian (Ruthenian) cuisine (Phillips 1983; Łuczaj 2012a).

The earliest work on children's wild food snacks comes from the Slovakian botanist Jozef Ludovit Holuby (1896). It is believed that the way children approach nature may be a relic of how our ancestors did (Łuczaj and Kujawska 2012; Łuczaj 2012a). Kids often participated in their mother's gathering activities (Łuczaj 2008), but they seem to have had a "folklore" of their own, mainly with plants eaten raw (Kalle and Sõukand 2013; Łuczaj and Kujawska 2012, Tardío et al. 2006). These were often flowers (which contain some sugar in their nectar, such as *Lamium album* L.), mature fleshy fruits, nuts and seeds as well as some tasty and interesting-looking immature fruits, for example, *Capsella bursa-pastoris* (L.) Medik. or *Malva* spp., widely eaten across Europe (Kalle and Sõukand 2013; Łuczaj and Kujawska 2012; Tardío et al. 2006; Tardío 2010). Other snacks are often specific to certain regions of Europe and are reported only from one country.

Children taste everything, sometimes even poisonous plants, but the bitter taste usually warns against future consumption. Thus, early age experimentation may have been a continuing source of introduction or reintroduction of food plants in the diet (Łuczaj and Kujawska 2012). Some of the snacks tasted in childhood were still occasionally "in use" in adulthood, but the majority of modern adults simply do not get many chances to eat them again. Nowadays, even in rural areas, children do not spend as much time in the fields as their parents or grandparents did. They not only do not take part in pastoral and agricultural activities but generally spend little time outdoors.

A large influence on the use of wild foods in the twentieth century was the lowering price of sugar. Sugar was used in cooking centuries before. In Poland, already in the seventeenth and eighteenth centuries, sweets made with candied *Acorus calamus* L. rhizomes were used in large quantities by manor houses (Dumanowski 2010). At this time, in England also, sugar was getting gradually cheaper (Burnett 1989). However, the price of sugar was extremely high. When it became lower in the early twentieth century, rural populations in many countries, following the example of the higher classes, started making preserves using sugar. In northern Europe, in the nineteenth century, the major way of preserving fruits was drying (Łuczaj 2011). However, later, making jams and pasteurized juices and sweetened wines made of both cultivated and wild fruits became popular. This reached its climax in the 1980s during the economic crises after martial law in 1981 and in Estonia in the 1990s when sugar was in short supply. Later, this trend disappeared due to the large choice of products in shops and the bad health reputation of sweetened foods (Łuczaj et al. 2012).

3.3 Reasons for the Contemporary Decrease in the Use of Wild Food Plants

The use of wild food plants in nutrition in many European communities, particularly urban ones, is very low nowadays. In large parts of northern and Eastern Europe, people only collect wild fruits and mushrooms (Łuczaj 2008, 2010a, 2011, 2012b; Dénes et al. 2012; Kalle and Sõukand 2012; Svanberg 2012), whereas in Southern Europe, some wild greens, such as *Asparagus acutifolius* L., *Scolymus hispanicus* L., and *Silene vulgaris* (Moench.) Garcke, are also relatively popular (Tardío et al. 2006, Tardío 2010). In some other areas, only a few species of wild vegetables are collected, for example, *Rumex acetosa* L. in Poland (Łuczaj 2010a) or *Allium ursinum* L. in the Alps (Schunko and Vogl 2010).

The consumption of many wild edible plants in the Western Mediterranean was strongly linked to traditional management activities such as tending livestock, charcoal burning, or bracken harvesting. In some cases, it was also linked to casual walks in the woods, such as walking to school. As most of these activities are not common anymore, people have also abandoned the behaviors associated with them (Tardío et al. 2005). Another important activity in the acquisition and maintenance of knowledge about edible plants was herding (Fig. 3.1). When following cattle or sheep, children and adult herders had a lot of time to observe nature, as they moved through the landscape (Łuczaj et al. 2013c).

Many rural communities in Mediterranean countries still practice the gathering of some wild vegetables, but this knowledge is becoming fragmented and the practice is restricted almost exclusively to older people (e.g., Aceituno-Mata 2010; Hadjichambis et al. 2008; Łuczaj et al. 2013c; Tardío et al. 2006). The loss of ac-

Fig. 3.1 The collection of wild vegetables in the Mediterranean is closely connected with pastoral and agricultural activities and is a by-product of other activities, such as shepherding (as in this photo) or weeding crops. (Vrana (northern Dalmatia, 2013), photo by Łukasz Łuczaj)

cess to nature causes even such a universal plant use category as children's snacks to gradually vanish. Many children have very little access to the rural environment, and the nomenclature of greens growing in urban settings is very limited.

Changes in the availability of species may affect their use. *Chenopodium album* L., once the most widely used wild food plant in Poland, is now difficult to collect in many areas, as herbicide spraying has almost completely eliminated it. Nowadays, the use of nettle (*Urtica dioica*) is more popular as this is a perennial and ruderal species, unaffected (or rather positively affected) by changes in agriculture. In the mid-twentieth century, an agricultural cereal weed, cornflower (*Centaurea cyanus* L.), was extensively used in Poland to make a fermented drink (Łuczaj 2011); however, later this use completely disappeared, probably due to the use of herbicides eliminating the cornflower (Łuczaj 2011). Modern agricultural practices, mainly deep ploughing and the use of herbicides, are also responsible for the lesser abundance of wild vegetables in Spain (Tardío et al. 2005; Polo et al. 2009). Many of them were weeds of cereal crops and usually exploited as human or animal food when crops were hand-weeded. Some of them are currently consigned to roadsides or abandoned agricultural lands.

Generally, access to wild food resources is limited more by a lack of proper habitats (e.g., for urban dwellers) rather than the lack of access to land. Probably only England, which has so-called trespassing laws, is an extreme example of limiting access for nonowners to wild resources—limiting public access to paths, roadsides, seaside, and common lands (see the Land is Ours campaign on the web against this law). In contrast, in Scotland and mainland Europe, public access rights are much broader with the famous *allemansrätt* (i.e., "all people's right") in Scandinavia (allowing even limited camping in someone's empty land). Similarly, in Poland and in the Mediterranean countries, people roam freely through the landscape collecting fungi and medicinal plants in private lands.

In Eastern Europe, for example, in Poland and Estonia, the recent fashion for large short-mown lawns excludes the traditional ruderal flora abounding with species such as *Urtica dioica, Aegopodium podagraria* L., and *Arctium* spp., a potential pool of many edibles (Łuczaj et al. 2012).

In Ukraine, many people stopped collecting birch sap and other wild foods after the nuclear catastrophe in Chernobyl (Łuczaj 2012b). For the few last years in Estonia and Poland, many parents have forbidden their children to eat wild berries for fear of echinococcosis. The parasite is spread by foxes, who became abundant due to mass vaccination (Łuczaj 2012a).

In Spain, the changes in the availability of wild edible plants are caused by the abandonment of grazing, pollution of streams, consolidation of land parcels, and the abandonment of the protection of fruit trees in the landscape (Łuczaj et al. 2012).

The examples of overharvesting of wild food plants in Europe are actually not so common, compared to the overharvesting of medicinal plants. This can be attributed to the fact that food plants must be common to become food plants, whereas some medicinal plants can be a rare expensive produce, susceptible to extermination. The examples of overharvesting in Europe include: *Rorippa nasturtium-aquaticum* in Campoo, Spain; *Artemisia granatensis* for beverages in Spain (Łuczaj et al. 2012); and *Polypodium vulgare* in Poland (Łuczaj 2011). In Poland and Estonia due to the danger of overharvesting, some wild edible species are taken under protection (e.g., *Allium ursinum*—Łuczaj et al. 2012). In Eastern Europe, special combs were traditionally used to collect *Vaccinium myrtillus* L. berries. However, as they also damage the plant's leaves, forest authorities banned them long ago (Łuczaj et al. 2012).

3.4 Latitudinal and Longitudinal Gradients in Food Plant Use in Europe with Special Reference to Wild Green Cuisines

The most visible difference in the present plant use in Europe is the contrast between the Mediterranean and the part of Europe north of the Alps. In the whole European Mediterranean, wild greens used to be commonly eaten in large amount up until very recently, and they are still eaten in many rural areas of Italy, Greece, Turkey, Cyprus, Croatia, Herzegovina, and Spain (e.g., Aceituno-Mata 2010; Benítez 2009; Bonet and Vallès 2002; Carvalho and Morales 2010; Criado et al. 2008; Dávila 2010; Della et al. 2006; Dogan 2012; di Tizio et al. 2012; Forbes 1976; Ghirardini et al. 2007; Guarrera 2003; Guarrera et al. 2006; Gonzalez et al. 2011; Hadjichambis et al. 2008; Łuczaj et al. 2013b, 2013c; Marco et al. 2003; Paoletti et al. 1995; Parada et al. 2011; Pardo-de-Santayana et al. 2005, 2007; Pardo-de-Santayana and

Table 3.2 List of wild greens of the *minestrella* soup, in northwest Tuscany, Italy. (Data Pieroni 1999)

Allium vineale L.	*Picris hieracioides* L.
Apium nodiflorum L.	*Plantago lanceolata* L.
Bellis perennis L.	*Plantago major* L.
Beta vulgaris L. ssp. *maritima* (L.)Thell.	*Primula vulgaris* Hudson
Borago officinalis L.	*Raphanus raphanistrum* L.
Bunias erucago L.	*Ranunculus ficaria* L.
Campanula rapunculus L.	*Reichardia picroides* (L.) Roth
Campanula trachelium L.	*Rumex crispus* L.
Cichorium intybus L.	*Rumex obtusifolium* L.
Cirsium arvense (L.) Scop.	*Salvia pratensis* L.
Crepis leontodontoides All.	*Salvia verbenaca* L.
Crepis sancta (L.) Babcock	*Sanguisorba minor* Scop.
Crepis capillaris (L.) Wallr.	*Silene alba* (Miller) Krause
Daucus carota L.	*Silene vulgaris* (Moench) Garcke
Foeniculum vulgare Miller	*Sisymbrium officinale* (L.) Scop.
Geranium molle L.	*Sonchus asper* L.
Hypochaeris radicata L.	*Sonchus oleraceus* L.
Lapsana communis L.	*Symphytum tuberosum* L.
Leontodon hispidus L.	*Taraxacum officinale* Web.
Lichnis flos-cuculi L.	*Urtica dioca* L.
Malva sylvestris L.	*Urtica urens* L.
Papaver rhoeas L.	*Viola odorata* L.
Picris echioides L.	

Morales 2010; Picchi and Pieroni 2005; Pieroni 1999, 2001; Pieroni et al. 2002, 2005; Polo et al. 2009; Tardío et al. 2005, 2006; Rivera et al. 2005; Tardío 2010; Velasco et al. 2010; Verde et al. 2003; see also Table 3.2 and 3.3). North of the Alps, wild greens are not gathered to such an extent now (e.g., Dénes et al. 2012; Kalle and Sõukand 2012; Łuczaj 2008, 2011, 2012a, 2013a; Svanberg 2012; Svanberg and Ægisson 2012). If they are still used, it is usually one of a few species, mainly used in soups.

Łuczaj (2008, 2010a) introduced the terms *herbophilia* and *herbophobia* to differentiate the opposing cultural attitude towards the use of wild greens: seeking them as a valuable part of the diet and avoiding them/being scared of them. He argued that Polish folk culture can be seen as moderately herbophobous. Katz (2012) showed that Amazonia is a place of total herbophobia; the instances of using green vegetables in general are very rare in Amazonia societies. Łuczaj (2010a) pointed out that the lack of wild greens in modern diets can stem from three different factors:

a. General decrease in the use of wild food
b. "Herbophobia," that is, avoiding any greens due to associations with famine, lack of nutritive properties, or classifying them as animal food
c. The competition between the cultivation of greens, for example, cabbage, other brassicas, spinach and lettuce in particular, and the use of wild greens

Table 3.3 Wild green vegetables used in Dalmatia (southern Croatia) and in the adjacent southwestern Herzegovina. (Adapted from Łuczaj et al. 2013b, 2013c)

Allium ampeloprasum L. (Liliaceae)	*Myagrum perfoliatum* L. (Brassicaceae)
Allium schoenoprasum L. (Liliaceae)	*Ornithogalum umbellatum* L. (Liliaceae)
Allium vineale L. (Liliaceae)	*Papaver rhoeas* L. (Papaveraceae)
Amaranthus retroflexus L. (Amaranthaceae)	*Picris echioides* L. (Asteraceae)
Anchusa arvensis (L.) M. Bieb. and other species of the genus (Boraginaceae)	*Plantago coronopus* L. (Plantaginaceae)
Asparagus spp., mainly *A. acutifolius* L. and *A. officinalis* L. (Asparagaceae)	*Portulaca oleracea* L. (Portulacaceae)
Beta vulgaris L. (Amaranthaceae)- wild	*Ranunculus* cf. *neapolitanus* Ten. (Ranunculaceae)
Borago officinalis L. (Boraginaceae)	*Ranunculus muricatus* L. (Ranunculaceae)
Brassica oleracea L. (Brassicaceae)	*Reichardia picroides* (L.) Roth. (Asteraceae)
Bunias erucago L. (Brassicaceae)	*Rhagadiolus stellatus* (L.) Gaertn. (Asteraceae)
Capparis orientalis Veill. (Capparaceae)	*Rumex* spp., included *R. patientia* (Polygonaceae)
Capsella bursa-pastoris L. (Brassicaceae)	*Ruscus* spp. (Asparagaceae)
Centaurea scabiosa L. (Asteraceae)	*Salicornia herbacea* L. (Amaranthaceae)
Chenopodium album L. (Chenopodiaceae)	*Salvia verbenaca* L. (Lamiaceae)
Chondrilla juncea L. (Asteraceae)	*Scolymus hispanicus* L. (Asteraceae)
Cichorium intybus L. (Asteraceae)	*Scorzonera laciniata* L. (Asteraceae)
Cirsium arvense L. (Asteraceae)	*Scorzonera villosa* Scop. (Asteraceae)
Clematis vitalba L. (Ranunculaceae)	*Silene latifolia* Poir. (Caryophyllaceae)
Crepis sancta (L.) Babc. (Asteraceae)	*Silene vulgaris* (Mch.) Garcke and related species (Caryophyllaceae)
Crepis spp. (Asteraceae)	*Smilax aspera* L. (Smilacaceae)
Crepis zacintha (L.) Babc. (Asteraceae)	*Smyrnium olusatrum* L. (Apiaceae)
Crithmum maritimum L. (Apiaceae)	*Sonchus* spp. (Asteraceae)
Daucus carota L. (Apiaceae)	*Tamus communis* L. (Dioscoreaceae)
Diplotaxis tenuifolia (L.) DC. (Brassicaceae)	*Taraxacum officinale* Weber (Asteraceae) and other species from the genus
Erodium cicutarium (L.) L'Hér. ex Aiton (Geraniaceae)	*Tordylium apulum* L. (Apiaceae)
Eruca sativa Miller (Brassicaceae)	*Torilis nodosa* (L.) Gaertn.
Eryngium maritimum L. and *E. campestre* L. (Apiaceae)	*Tragopogon* spp. (Asteraceae)
Foeniculum vulgare Mill. (Apiaceae)	*Urospermum picroides* (L.) Desf. (Asteraceae)
Geranium molle L. (Geraniaceae)	*Urtica dioica* L. (Urticaceae)
Hirschfeldia incana (L.) Lagr.-Foss. (Brassicaceae)	*Urtica pilulifera* L. (Urticaceae)
Hypochaeris radicata L. (Asteraceae)	*Valerianella locusta* L. (Valerianaceae)
Lactuca serriola L. (Asteraceae)	*Viola arvensis* Murr. (Violaceae)

However still, deconstructing the reasons of this division of Europe in a "herbophilous zone" of the Mediterranean and a moderately herbophobous rest is a complex matter. Both ecological and cultural reasons can be behind it. The Mediterranean region can be characterized by low forest cover, dry summer, mild and wet winters, and it corresponds more or less to the area of olive cultivation. One of the explanations may be the low fertility of many of the degraded maquis habitats, which made people to learn to utilize every possible species in the environment. In contrast, in less populated northern Europe, which also has a surplus of rainfall, large amounts of wild greens can be collected even if one knows only a few commonest species (Łuczaj 2008). That is why only a few species such as *Urtica dioica, Chenopodium album*, or *Rumex* spp. were mainly collected.

In some Italian and Spanish regions, a blend of many different wild species (up to 40 or even 50) has been used in vegetable recipes in particular (Paoletti et al. 1995; Pieroni 1999; Tardío 2010; Turner et al. 2011). In Italy, complex wild vegetable soups are known in the North of the country: in Liguria (*prebuggiun*), in Friuli (*pistic*) and in northwest Tuscany (*minestrella di Gallicano*) (Table 3.2). More often, however, wild greens—especially whorls and other tender aerial parts of several Asteraceae and Brassicaceae (i.e., *Cichorium intybus, Picris echioides, Sonchus oleraceus, Chondrilla juncea, Sinapis arvensis, Hirschfeldia incana, Brassica nigra, Urospermum dalechampii, Brassica fruticulosa, Rahapnus raphanistrum, Bunias erucago*) were stewed, in Southern Italy and especially Sicily, with olive oil and garlic and daily consumed as a side dish. Similar blends of vegetables are made in Dalmatia, called *mišanca, pazija, svakober, divlje zelje* (Łuczaj et al. 2013b, 2013c; Table 3.3), in Spain (Tardío et al. 2006; Tardío 2010) and called *chorta* in Greece. Most of these wild greens have shown a remarkable antioxidant activity and important bio-pharmacological properties, which assign them a primary role in the prevention of ageing-related and other degenerative diseases (Local Food-Nutraceuticals Consortium 2005; Schaffer 2005; Romojaro et al. 2013).

The abovementioned mixes were also known, and even had special names, but to a much lesser extent, in northern Europe. In such countries and Poland, Slovakia, Belarus, and Ukraine, mixed wild greens were usually made into a soup or potherb with other more calorie-rich ingredients (potatoes, flour, milk, cream, etc.). Even in the nineteenth century, however, these mixes usually included only a few species, much fewer than in the Mediterranean. This can probably be attributed to a larger biomass of greens available, which enabled gathering larger amounts of green vegetables even knowing only a few species such as *Chenopodium album, Urtica dioica, Aegopodium podagraria*, or *Heracleum sphondylium*. Nowadays, the use of wild greens in the abovementioned countries has decreased dramatically. In Poland, recent ethnobotanical surveys show that respondents are unable to recall famine plants except for *Urtica dioica* and *Chenopodium album* (Łuczaj 2010a; Łuczaj and Kujawska 2012), thus present ethnobotanical studies, even if the oldest people are interviewed, cannot reveal the whole spectrum of plants used even just at the end of the nineteenth century. That is why it is important to quickly capture the memories of former wild food plants in the countries where few or no ethnobotanical works were carried out as well as to make use of historical sources. Some of the plants used in potherbs and soups in central and northern Europe are the same or closely related species used in the south of Europe. Here we should mention, for example, *Urtica* used throughout Europe. Also, *Chenopodium* species, mainly *Chenopodium album* and *Chenopodium bonus-henricus* have been used for food in many countries of Europe. In northern Europe, *Rumex acetosa* and *R. acetosella* were and sometimes are used for soups. In the Mediterranean, a related species—*Rumex pulcher*—is used, mainly in mixed cooked vegetable recipes.

One of the most reasonable explanations is that actually the exploitation of wild vegetables represented in Southern Europe a by-product of the cereals domestication that took place during the Neolithic Revolution. The so-called *Mediterranean diet* is a food system based on a daily consumption of many grains, pulses, cheese, and especially vegetables; wild greens represented, especially until the very recent past and partially still nowadays, a crucial portion of these plant ingredients especially during the early spring and in the fall, and their role may have been largely underestimated by nutritionists and historians of food systems (Pieroni et al. 2002, 2005; Rivera et al. 2005).

Cultural factors may have played an additional role in enhancing the use of wild greens. Although wild greens were used throughout Europe, in times of food scarcity, their use in the Mediterranean may have been enhanced by the Greek and Roman culture. Ancient Greek writing emphasize the health benefits of wild greens and wild greens mixes or culinarily associated with the use of olive oil, so important for the Roman Empire. These two factors may have additionally increased the intensity of the use of wild greens in the Mediterranean. A similar situation occurs in East Asia (China, Japan) where the wide use of wild greens nowadays is enhanced by the belief of the healthiness of wild products (Kang et al. 2012). The proof that the wide use of many species of wild greens was not always restricted only to the Mediterranean is Belarus. In western Belarus, around 20 species of wild greens were still used at the end of the nineteenth century. Their use was so intense that large amounts of wild greens were even dried for winter (Łuczaj et al. 2013b), a phenomenon not recorded in other European countries but widespread in China (e.g., Kang et al. 2012).

Another factor supporting the existence of wild-green centred cuisines is the maintenance of knowledge of the usefulness of some plants of lesser quality that could be necessary in times of scarcity (Johns 1994). It is a fascinating question, why these multispecies wild green mixes are remembered in some parts of the Near East and the Mediterranean, while they were completely forgotten in northern Europe, where they also existed in the nineteenth century or earlier.

If the north–south, division is very pronounced; clear west–east differences are difficult to see. In the Mediterranean zone, very similar plants, mainly weeds, are used both in the western and eastern Mediterranean. Plants such as *Sonchus oleraceus, Daucus carota, Cichorium intybus, Crepis* spp., *Foeniculum vulgare, Portulaca oleracea, Tragopogon* spp., *Taraxacum* spp., *Silene* spp., *Rumex pulcher, Sinapis* spp. etc. are eaten throughout the Mediterranean, and the small differences in the local choice of, mainly Asteraceae, plants are caused by the availability of particular species (e.g., Aceituno-Mata 2010; Benítez 2009; Bonet and Vallès 2002; Carvalho and Morales 2010; Criado et al. 2008; Dávila 2010; Della et al.

2006; Dogan 2012; di Tizio et al. 2012; Ertuğ 2004; Forbes 1976; Ghirardini et al. 2007; Guarrera 2003; Guarrera et al. 2006; Gonzalez et al. 2011; Hadjichambis et al. 2008; Łuczaj et al. 2013b, 2013c; Marco et al. 2003; Paoletti et al. 1995; Parada et al. 2011; Pardo-de-Santayana et al. 2005, 2005; Pardo-de-Santayana and Morales 2010; Picchi and Pieroni 2005; Pieroni 1999, 2001; Pieroni et al. 2002, 2005; Polo et al. 2009; Tardío et al. 2005, 2006; Rivera et al. 2005; Tardío 2010; Velasco et al. 2010; Verde et al. 2003). Similarly north of the Alps, the commonest wild vegetables, for example, *Chenopodium* spp., *Urtica dioica, Rumex* spp. and *Oxalis* spp., were widely consumed from the Atlantic coast all the way to Russia (e.g., Dénes et al. 2012; Kalle and Soõkand 2012; Łuczaj 2010a; Nedelcheva 2012; Svanberg 2012).

Unfortunately, historical ethnobotanical data from the UK, France, and Germany are rather scattered, so it is difficult to compare the use of the wild greens in those countries compared to central-eastern Europe where recently these data have been well reviewed and summarized. In spite of the extremely small number of wild vegetable species used north of the Alps, some local specialities were still preserved. For example, in parts of Northern England, a special dish called *Easter ledger pudding* is made using the leaves of *Polygonum bistorta* L. (Phillips 1983). In parts of Romania, *Tussilago farfara* L. leaves are commonly used to wrap sarma rolls (Dénes et al. 2012), and in several areas of both Western and Eastern Europe, *Allium ursinum* is used for soups and salads (Łuczaj 2012a).

3.5 New Trends Emerging

More recently, in times of the decreasing quality of supermarket foods, the interest in wild collected foods is gaining a lot of media attention. Numerous field guides are issued, and wild food/foraging workshops are organized. New culinary vogues are promoted by media and health-oriented people (see also Table 3.4).

As a part of this trend, articles such as acorn coffee, *Allium ursinum*-enriched products and birch sap have appeared in health food shops in Poland and many other countries (Łuczaj and Kujawska 2012). In Estonia, health-food shops offer mostly products of non-local origin, although acorn and *chicory*-based coffee have also been reintroduced (in Spain and Poland up until the mid-twentieth century it was poor people's coffee, and now it is a health food), also syrups made of *Juniperus communis* and *Taraxacum* spp. are sold. As a new trend, probably following the example of a similar German product available in health-food shops in Estonia, pasta made with the powder of *Urtica* spp., *Vaccinium myrtillus*, and *Cantharellus cibarius* Fr. is making its way to the customer. In 2010, the company Eesti And (Estonia Gift) started to produce and market pickled and salted forest mushrooms (http://www.eestiand.ee/) in their larger stores—food that was just a few decades ago made and stored in every household regardless of status and distance from the forest but now massively abandoned unless bought from the stores (in other countries, e.g., Poland, pickled mushrooms have been sold in shops for decades).

Table 3.4 Wild plant taxa, traded by a small southern-Swedish foraging enterprise, which is also the main provider of wild food plants at the restaurant NOMA, in Copenhagen (Data from Łuczaj et al. 2012)

Species	Part
Blackthorn (Sloe), *Prunus spinosa* L.	Fruits
Bramble (Blackberry), *Rubus* sp.	Fruits and unripe fruits
Camomile, *Matricaria recutita* L.	Flowers
Chickory, *Cichorium intybus* L.	Leaves and flowers
Chickweed, *Stellaria media* (L.) Vill.	Aerial parts
Chives, *Allium schoenoprasum* L.	Flowers
Cow parsley, *Anthriscus sylvestris* (L.) Hoffm.	Leaves, buds, and flowers
Cuckooflower (Lady's Smock), *Cardamine pratensis* L.	Aerial parts
Daisy, *Bellis perennis* L.	Leaves and flowers
Dandelion, *Taraxacum officinale* Weber s.l.	Leaves, buds, and flowers
Elder, *Sambucus nigra* L.	Flowers and fruits
Garlic mustard, *Alliaria petiolata* (M. Bieb.) Cavara & Grande	Leaves, shoots, and seeds
Grass-Leaved Orache, *Atriplex littoralis* L.	Leaves and unripe fruits
Ground-Elder, *Aegopodium podagraria* L.	Shoots, buds, and flowers
Harebell, *Campanula* sp.	Flowers
Hop, *Humulus lupulus* L.	Shoots
Nettle (Stinging Nettle), *Urtica dioica* L.	Shoots and unripe fruits
Orpine, *Hylotelephium telephium* (L.) H. Ohba	Aerial parts
Ostrich Fern, *Matteuccia struthiopteris* (L.) Tod.	Shoots
Ramsons (Wild Garlic), *Allium ursinum* L.	Shoots, leaves, flowers, unripe fruits, and young seedlings
Raspberry, *Rubus idaeus* L.	Leaves and fruits
Red clover, *Trifolium pratense* L.	Flowers
Purple dead nettle, *Lamium purpureum* L.	Aerial parts
Ribwort plantain, *Plantago lanceolata* L.	Leaves and unripe inflorescences
Rose, *Rosa* spp.	Flowers and fruits
Sand leek, *Allium scorodoprasum* L.	Leaves and seeds
Scurvy grass, *Cochlearia* sp.	Aerial parts
Sea arrowgrass, *Triglochin maritima* L.	Leaves and unripe fruits
Sea aster, *Tripolium vulgare* Nees	Leaves
Sea pea, *Lathyrus japonicus* Willd.	Flowers and shoots
Sea plantain, *Plantago maritima* L.	Aerial parts
Leaf sea rocket, *Cakile maritime* Scop.	Leaves and flowers
Sea sandwort, *Honckenya peploides* (L.) Ehrh.	Aerial parts
Sea-kale, *Crambe maritime* L.	Leaves, flowers, buds, and fruits
Small-flowered winter-cress, *Barbarea stricta* Andrz.	Leaves and flowers
Sorrel (Common Sorrel), *Rumex acetosa* L.	Leaves
Spruce (Norway Spruce), *Picea abies* (L.) H. Karst.	Shoots
Swedish Whitebeam, *Sorbus intermedia* (Ehrh.) Pers.	Fruits
Sweet cicely, *Myrrhis odorata* Scop.	Leaves, flowers, and unripe cones
Tansy, *Tanacetum vulgare* L.	Flowers

Table 3.4 (continued)

Species	Part
Violet, *Viola* sp.	Flowers
Water mint, *Mentha aquatica* L.	Leaves
Water-Cress, *Rorippa nasturtium-aquaticum* (L.) Hayek	Leaves
White deadnettle, *Lamium album* L.	Leaves
Wild marjoram, *Origanum vulgare* L.	Leaves
Wild onion, *Allium vineale* L.	Leaves and fruits
Wild thyme (Creeping Thyme), *Thymus serpyllum* L.	Leaves
Winter-cress, *Barbarea vulgaris* W.T. Aiton	Flowers
Wood sorrel, *Oxalis acetosella* L.	Aerial parts
Woodruff, *Galium odoratum* Scop.	Aerial parts
Wych elm, *Ulmus glabra* Huds.	Unripe fruits
Yarrow, *Achillea millefolium* L.	Leaves and flowers
Yellow archangel, *Lamium galeobdolon* (L.) Crantz	Young shoots and flowers

In Poland, *Dary Natury*, a firm owned by Mirosław Angielczyk, has experienced a tremendous success in selling and promoting natural wild foods and nutraceuticals. It has been very active in the Polish market for a dozen years or so. It mainly sells mixed herbs for herbal infusions and liqueurs, but popular products also include herbal coffee substitutes (acorn, chicory, etc.), birch sap, and even acorn flour. According to the owner, in 2013 they produced a few tons of acorn flour (from *Quercus robur*). Acorns are shelled, leached using pure water, dried and powdered. The flour is available on their Internet site for only 21 PLN/kg (ca. 5 €/kg).

Immigrants from other countries, also outside Europe, are a very little-studied category when it comes to harvesting in the wild. However, observations made in Sweden show that, for instance, Turks, Kurds, Chinese, Koreans, and Thai immigrants are rather widely using the free access to private land and are harvesting wild plants, berries, and mushrooms for their own consumption and also for selling in the markets (Svanberg 2012).

Wild food plants have always been sold in vegetable markets. In the nineteenth-century Poland, these were wild fruits, grains of manna grass (*Glyceria* spp.), and even the rhizomes of *Polypodium vulgare* were sold in a town (Jasło) (Łuczaj 2011, Kujawska and Łuczaj 2010). Apart from that, a variety of mushrooms has been sold in the mycophilous parts of Europe. For example, a few dozen taxa of fungi were sold in the market of Poznań, Poland, in the beginning of the twentieth century (Szulczewski 1996).

Nowadays, in most Eastern European countries, the selling of wild food plants in the market is restricted to wild berries, mushrooms and herbs for making tea, and occasionally also *Rumex acetosa* leaves or horseradish roots—in Estonia, also horseradish leaves (unpublished observations from Poland and Estonia).

Green wild vegetables are rarer than fruits in the markets. However, they are often sold around the Mediterranean, frequently in Italy, Greece, and Croatia (Łuczaj et al. 2012, 2013) and occasionally also in Spain (Tardío 2010).

The twentieth century has seen a decline in the sales of wild food plants not only in Poland, but all over Europe. However, in the twenty-first century, we may witness the reoccurrence of wild products not only in specialist health food stalls but also in ordinary vegetable markets. Such a phenomenon can already be seen in Germany and Austria. Probably the only wild vegetable that has survived from the peasant society in Sweden is *Urtica dioica*, which is still popular among many urban people. However, wild forest berries (*Vaccinium, Rubus*) continue to be very much used among Swedes in general and hold a time-honoured place in both home cooking and restaurant kitchens. Since the mid-twentieth century, there has also been an increasing demand for wild mushrooms. Many people pick their own mushrooms, recognizing everything from 2–3 species to almost 30 edible taxa. Nettles, wild berries, and mushrooms are also available in the weekly street markets during summertime and autumn, but berries and mushrooms can also be found in supermarkets (Svanberg 2012).

Lost traditional knowledge on wild food plants is rediscovered and re-created by individuals particularly interested in the issue. This knowledge is later spread via a variety of workshops, seminars, and particularly media (books and television programs). As far as media is concerned, it teaches edible plants in a new way. Traditionally this knowledge was gained from parents, grandparents, or peers and was a cognitive process not only involving visual, abstract learning but also "rambling" through the countryside, smelling plants and learning their location (Sõukand and Kalle 2010). Maybe that is why edible plant workshops are so popular, being more akin to a traditional way of learning plants. However, usually neither the published guides nor the workshops relate to local practices. They are an amalgamation of proposals regarding how to utilize local floras referring to the traditions of use of these plants in North America, Asia, and other parts of Europe. Thus new species are becoming utilized. For example, in Poland, the use of *Allium ursinum* leaves as food has not ever been recorded in ethnographic sources, but now it is common among many families in the Carpathians due to the media attention this plant has gained (Łuczaj and Kujawska 2012). The decreasing access to wild food plants created a longing for such food, and this gives good ground for all kinds of courses and books. In Estonia, the publishing of books on the use of wild food plants has intensified during the recent decades, as have all publications regarding the use of plants and alternative medicine. Alongside, dozens of courses, local and general, are advertised every year, reintroducing old local uses and introducing new uses of autochtonous and alien plants into the diet of Estonians. Since regaining independence, the Estonian Defence League (voluntary) organizes regular survival courses, which include teaching on the use of wild plants and animals for food. The influence of those books and courses can be evaluated only years later, as people tend to accept teachings selectively, sometimes in a random way (Sõukand and Kalle 2012).

The users of most wild food guides are people who are interested in food independence, survival, or a healthy lifestyle. Probably, the first widely known European guide of this style was Richard Mabey's *Food for Free* sold in Britain in hundreds of thousands of copies (Mabey 1972). Later in the 1990s and 2000s, the

French botanist François Couplan published several similar guides in French and German, for example, Couplan (1989). In the 1980s and 1990s, a wild food guide by the Czech author Dagmar Lánská was sold in large numbers of copies in Eastern Europe, for example, in Czechoslovakia and Poland (Lánská 1992). In the countries of former Yugoslavia, a similarly influential author was the botanist Ljubiša Grlić (2005). Some of the professional field ethnobotanists also published wild food guides for the general public (Łuczaj 2002; Tardío et al. 2002) or even created TV culinary series (*Dziki Obiad Łukasza Łuczaja*, i.e., "Łukasz Łuczaj's Wild Lunch", by Canal Plus). An influential photographic guide by Roger Phillips should also be mentioned (Phillips 1983). Another promoter of wild food was the most known European (British) survival handbook writer Ray Mears (also the author of film series). It should also be noted that Mears authored a book and a TV series with the prominent British archaeobotanist Gordon Hillman (Mears and Hillman 2007).

One very special book should also be mentioned, *L'Ensalada champanèla* (Marco et al. 2003). The new edition of this French guide to wild salad plants also contains a large amount of material on the traditional use of this group of plants in southern France, thus being a guide and a regional monograph in one. A similar guide was published in Albacete, Spain (Rivera et al. 2006a).

In Valchiusella and in the middle Serchio Valley (Gallicano), in northwest Italy, in the last years, local associations have began to organize spring workshops on wild food plant identification, gathering and cooking, lead by local elderly women. Recently, a *Museum of the Wild Herbs* was also born in Liguria (Cosio d'Arroscia, NW Italy), while in diverse Italian areas, especially in Lombardy (Brianza), in the Marche region and in Eastern Sicily, short courses and weekend seminars, mainly animated by local botanists, seem to play a crucial role in re-instilling folk knowledge on locally used wild food plants and this kind of initiative in Italy is becoming extremely popular (e.g. http://www.piantespontaneeincucina.info/struttura/html/header/feste_e_sagre.html, http://www.accademiadelleerbe.it/).

Yet another effort worth mentioning is the *Plants for a Future* database created by Ken Fern from England (PFAF 2015). Though it does not contain data from modern ethnobotanical studies, it is an influential source in spreading knowledge on edible plants.

It is noteworthy that this loss of local knowledge and use of wild gathered plant species is paralleled by an increased interest in such resources by the gastronomic and intellectual elite in the search for new stimuli, culinary experiences, and health food (for example, visit http://www.slowfood.it). The increasing presence of wild food products can also be seen by agritourism farms or local rural restaurants as a part of the local traditional heritage offered by them. In Poland this is, for example, nettle soup and a variety of wild fruit products (Łuczaj 2011). In Spain, herbal teas prepared with species such as *Jasonia glutinosa* (L.) DC. or *Sideritis hyssopifolia* are served in restaurants of tourist areas such as Serranía de Cuenca or Picos de Europa. Moreover, in the case of *Jasonia glutinosa*, new products are appearing and some restaurants offer ice cream made with its infusion. *Sideritis hyssopifolia* L. is also used to aromatise homemade and commercial herb liqueurs, and it is even available on the Internet (Pardo-de-Santayana et al. 2005). Another interesting

Spanish liqueur is *patxaran/pacharán*. It is usually made by macerating *Prunus spinosa* fruits, cinnamon bark, a few coffee seeds, and sugar in anisette and/or liquor. The tradition of preparing liqueurs with its fruits is old, but this recipe is originally from Navarre and now commonly prepared or bought throughout the country. In fact, it is cultivated in the region for the industry of *pacharán*. A similar and popular liqueur based on sloe fruits still exists in Central Europe and Northern Italy.

The use of wild food plants has also been recently promoted by avant-garde restaurants. Here, we should mention above all the pioneering experiences of the chefs Michel Bras and Marc Veyrat in France more than two decades ago and nowadays what is considered one of the best restaurants in the world, NOMA in Copenhagen, run by René Redzepi, offers a cuisine that is largely based on local wild products, including a wide selection of wild food plants (Table 3.4), which are also sometimes foraged by the NOMA staff. Following the aesthetic lines drafted by Redzepi in his *Time and Place in Nordic Cuisine* (Redzepi 2010), wild plants are considered to be a crucial element of a given place and, therefore one of the pillars of a cuisine, which would like to express the "sense of place".

In the meantime, many other top European restaurants tend to use a large number of wild taxa in their kitchen; among them, it is worth mentioning the new generation of the Scandinavian chefs of the "Nordic cuisine" movement, that is, Magnus Nilsson as well as the Argentinian top chef Mauro Colagreco, who owns his restaurant on the French–Italian border. Earlier in the 1990s, a well-known expert on wild foods in the francophone countries, François Couplan (www.couplan.com), worked with leading French chefs incorporating wild plants in their menus. Recently, the Polish top chef, Wojciech Amaro published a book *Natura kuchni polskiej* which incorporates many wild foods into haute-cuisine dishes (Amaro 2006). In Spain, there are also some luxury restaurants, such as the restaurant of the Hotel Alfonso VIII in the city of Plasencia, whose menus offer traditional and re-created dishes with at least six wild plant species, such as *Urtica dioica, Tamus communis, Rorippa nasturtium-aquaticum, Montia fontana, Allium ampeloprasum, Asparagus acutifolius*, and *Scolymus hispanicus* (Recetario Extemeño 2005).

Until recent times, the use of wild food plants in restaurant menus was not practiced in Estonia. But slowly it is becoming an attractive option, for example, in spring 2012 an invitation for the employment of a "gatherer" by a cafe in Tallinn was newsworthy for several news portals (Łuczaj et al. 2012). In England, there is a small rural enterprise called "Forager" engaged in gathering and supplying wild food, mostly to the restaurant trade (http://www.forager.org.uk/), having also written *The Forager Handbook*, a guide to the edible plants that grow in Britain (Irving 2009). Wild berries have a long tradition within Swedish restaurant culture. However, some other wild plants, earlier used only locally by peasants, have become regional specialities. *Allium scorodoprasum* L. was traditionally used in coastal areas as a spring vegetable, especially in stews. On Gotland island, it has been harvested for centuries, used as a remedy against spring fatigue. Nowadays, it is an appreciated early vegetable for the regional speciality, leek soup and is also available in restaurants. Also, berries of *Rubus caesius* L., very little used earlier, are nowadays used as jam and considered a regional specialty of Gotland (Svanberg

1998, 2012). In Finland, products, especially desserts, made with berry juice from *Hippophae rhamnoides*, are seen as regional specialities of Österbotten and Åland Islands and found in many restaurants. In the Faroes and Iceland, and to some extent also in Norway, stalks of *Angelica archangelica* have become a fashionable food made into various products, which can be found in restaurants or bought canned in stores (Svanberg 1998, 2011; Fosså 2006).

In Italy, we (AP and co-workers) recently surveyed ten top chefs, who use flowers in their cuisines. The most interesting finding was that the large parts of the used flowers (including also a few cultivated ornamental plants) do not have any connection to the culinary folk traditions and/or food ethnobotanical literature in Italy.

Changes in plant use are not linear. Some species can become the subject of temporary vogues. In Poland, making a fermented, fizzy, cornflower *Centaurea cyanus* flower lemonade was very common in the mid-twentieth century but has not been reported earlier or later (it probably disappeared mainly due to the decline of *C. cyanus* populations cause by the development of intense agriculture), and *Taraxacum* flower syrup was very popular in the 1990s women's magazines but now seems to be less popular (Kujawska and Łuczaj 2010; Łuczaj 2011). In Poland, the use of dandelion (*Taraxacum*) leaves has had its ups and downs. Dandelion leaves were usually regarded as famine food and there are very few reports of using them (as Polish cuisine avoids bitter tastes). However, vogues for eating dandelion leaves entered Poland a few times, directly from France: first at the end of the nineteenth century among the upper classes, and later in the mid-twentieth century among the families who came back from emigration in France. Both times the trend passed, as it did not withstand the "anti-bitter" attitude of the majority of the Polish population (Łuczaj 2011; Kujawska and Łuczaj 2010). As Kujawska and Łuczaj pointed out (Kujawska and Łuczaj 2010), Polish dishes are now undergoing "Mediterranisation" (e.g., by replacing butter and lard with olive oil and by adding Mediterranean-style aromatic herbs to traditional Polish recipes). One may wonder how long this fashion will last.

3.6 Toxicological Issues Linked to Wild Plants/Fungi Gathering

In many European countries, the sale of fungi in markets is heavily regulated due to the danger of poisoning (Peintner et al. 2013). For example, in Poland, 40 species or genera are legally sold (Łuczaj and Nieroda 2011). Some of the most commonly eaten fungi—*Russula* spp.—cannot be sold to the public due to the similarity of the juvenile fruiting bodies of some species from the genus (e.g., *R. cyanoxantha* (Schaeff.) Fr., *R. virescens* (Schaeff.) Fr., etc.) to the most deadly mushroom, the death cap (*Amanita phalloides* (Fr.) Link). The sale of wild food plants has not seen such regulation, probably for two reasons: They are less present in many countries and also because of the fact that in the large majority of cases the toxic plants are very bitter.

From our own experience, the experience of other people dealing with wild food promotion, and from media coverage, we identified two possibly most dangerous issues (however both of them with only a few cases around Europe):

1. Poisonings with *Convallaria majalis* L. and *Colchicum autumnale* L. leaves by confusing them with the edible *Allium ursinum* (Davanzano et al. 2011; Pilegaard 2012).
2. Confusing edible Apiaceae (e.g., *Pastinaca sativa* L. and *Daucus carota* L.) with the poisonous ones (*Oenanthe crocata* L. and *Conium maculatum* L. respectively)—this issue was extensively discussed by Irving (2009).

In the last few years, some cases of intoxications were caused by the wrong plant identification linked to a new or "re-acquired" knowledge of wild food plants, which was not refined or trained with long-term experience in the field. Informing the public about the possible dangers of wild plants' consumption, particularly when eaten regularly or in large amounts, alongside the benefits, is an increasingly important educational task (Łuczaj et al. 2012).

References

Abbet C, Mayor R, Roguet D, Spichiger R, Hamburger M, Potterat O (in press) Ethnobotanical survey on wild alpine food plants in Lower and Central Valais (Switzerland). J Ethnopharmacol. doi:10.1016/j.jep.2013.11.022

Aceituno-Mata L (2010) Estudio etnobotánico y agroecológico de la Sierra Norte de Madrid. PhD Thesis. Universidad Autónoma de Madrid, Madrid

Amaro W (2006) Natura kuchni polskiej. AmarONE, Warszawa

Benítez G (2009) Etnobotánica y Etnobiología del poniente granadino. PhD thesis. Facultad de Farmacia. Universidad de Granada, Granada

Bonet MA, Vallès J (2002) Use of non-crop food vascular plants in Montseny biosphere reserve (Catalonia, Iberian Peninsula). Int J Food Sci Nutr 53:225–248

Burnett (1989) Plenty and want: a social history of food in England from 1815 to the present day. Routledge, London

Carvalho AM, Morales R (2010) Persistence of wild food and wild medicinal plant knowledge in a Northeastern region of Portugal. In: Pardo-de-Santayana M, Pieroni A, Puri R (eds) Ethnobotany in the New Europe: people, health and wild plant resources. Berghahn Books, Oxford-New York, pp 147–171

Christanell A, Vogl-Lukasser B, Vogl CR, Gütler M (2010) The cultural significance of wild gathered plant species in Kartitsch (Eastern Tyrol, Austria) and the influence of socio-economic changes on local gathering practices. In: Pardo-de-Santayana M, Pieroni A, Puri R (eds) Ethnobotany in the New Europe: people, health and wild plant resources. Berghahn Press, New York-Oxford, pp 51–75

Conquest R (1987) The harvest of sorrow: soviet collectivization and the terror-famine; famine in the Ukraine, 1932–1933. Oxford University Press, Oxford

Couplan F (1989) Le régal végétal. Plantes sauvages comestibles. Encyclopédie des plantes comestibles de l'Europe, vol 1. Equilibres, Flers

Criado J, Fernández López JM, Leocadio G, Núñez Núñez RM, Blanco E (2008) Uso tradicional de las plantas en Toledo. Diputación de Toledo, Toledo

Davanzo F, Miaglia S, Perego S, Assisi F, Bissoli M, Borghini R et al (2011) Plant poisoning (2011) increasing relevance, a problem of public health and education. J Pharm Sci Res 3(7):1338–1343
Dávila P (2010) Estudio etnobotánico comparativo de la colleja [Silene vulgaris (Moench) Garcke] en dos localidades de la Comunidad de Madrid. Proyecto de Fin de Carrera. Universidad Autónoma de Madrid. Facultad de Ciencias, Madrid
Della A, Paraskeva-Hadjichambi D, Hadjichambis AC (2006) An ethnobotanical survey of wild edible plants of Paphos and Larnaca countryside of Cyprus. J Ethnobiol Ethnomed 2:34
Dénes A, Papp N, Babai D, Czúcz B, Molnár Z (2012) Wild plants used for food by Hungarian ethnic groups living in the Carpathian Basin. Acta Soc Bot Pol 81(4):381–396
di Tizio A, Łuczaj Ł, Quave CL, Redžić S, Pieroni A (2012) Traditional food and herbal uses of wild plants in the ancient South-Slavic diaspora of Mundimitar/Montemitro (Southern Italy). J Ethnobiol Ethnomed 8:21
Dogan Y (2012) Traditionally used wild edible greens in the Aegean Region of Turkey. Acta Soc Bot Pol 81(4):329–342
Dumanowski J (2010) Tatarskie ziele w cukrze czyli staropolskie słodycze. Muzeum Pałac w Wilanowie, Warszawa
Ertuğ F (2004) Wild edible plants of the Bodrum Area (Muğla, Turkey). Turk J Bot 28:161–174
Forbes MHC (1976) Gathering in the Argolid: a subsistence subsystem in a Greek agricultural community. In: Dimen M, Friedl E (eds) Regional variation in modern Greece and Cyprus: toward a perspective on the ethnography of Greece. pp 251–264 (Annals of the New York Academy of Science 268) Wiley, New York
Fosså O (2006) Angelica: from Norwegian mountains to the English trifle. In: Hosking R (ed) Wild food: proceedings of the Oxford symposium on food and cookery 2004. Prospect Books, Totnes, pp 131–142
Ghirardini MP, Carli M, Del Vecchio N, Rovati A, Cova O et al (2007) The importance of a taste: a comparative study on wild food plants consumption in twenty-one local communities in Italy. J Ethnobiol Ethnomed 3:22
González JA, García-Barriuso M, Amich F (2011) The consumption of wild and semi-domesticated edible plants in the Arribes del Duero (Salamanca-Zamora, Spain): an analysis of traditional knowledge. Genet Resour Crop Evol 58:991–1006
Grlić L (2005) Enciklopedija samoniklog jestivog bilja. Ex libris, Rijeka
Guarrera PM (2003) Food medicine and minor nourishment in the folk traditions of Central Italy (Marche, Abruzzo and Latium). Fitoterapia 74:515–544
Guarrera PM, Salerno G, Caneva G (2006) Food, flavouring and feed plant traditions in the Tyrrhenian sector of Basilicata, Italy. J Ethnobiol Ethnomed 2:37
Hadjichambis A, Paraskeva-Hadjichambi D, Della A, Giusti ME, De Pasquale C, Lenzarini C et al (2008) Wild and semi-domesticated food plant consumption in seven circum-Mediterranean areas. Int J Food Sci Nutr 59(5):383–414
Häkkinen A (1992) On attitudes and living strategies in the Finnish countryside in the years of famine 1867–1868. In: Häkkinen AJ, (ed). Just a sack of potatoes? Crisis experiences in past and present societies. Societas Historica Finlandiae, Helsinki, pp 149–166
Holuby JL (1896) Aus des Botanik slowakischer Kinder des Trentschiner Komitates in Ungarn. Deutsche Botanische Monatsschrift 19 (8–9):126–131
Irving M (2009) The Forager handbook. Wild food guide to the edible plants of Britain. Ebury Press, London
Johns T (1994) Ambivalence to the palatability factors in wild plants. In: Etkin NL (ed) Eating on the wild side. The University of Arizona Press, Tucson, pp 46–61
Kalle R, Sõukand R (2012) Ethnobotanical review of edible plants of Estonia. Acta Soc Bot Polon 81(4):271–281
Kalle R, Sõukand R (2013) Wild plants eaten in childhood: retrospective of 1970s–1990s Estonia. Bot J Linnean Soc 172:239–253

Kang Y, Łuczaj Ł, Ye S, Zhang S, & Kang J (2012) Wild food plants and wild edible fungi of Heihe valley (Qinling Mountains, Shaanxi, central China): Herbophilia and indifference to fruits and mushrooms. Acta Soc Bot Polon 81(4): 405–413

Katz E, Lopez CL, Fleury M, Miller R, Payê V, Dias T et al (2012) No greens in the forest? Note on the limited consumption of greens in the Amazon. Acta Soc Bot Polon 81(4):283–293

Kujawska M, Łuczaj Ł (2010) Studies of wild food plants in communist and post-communist Poland. Changes in use and in research methodology. In: Pochettino ML, Ladio A, Arenas P (eds) Tradiciones y Transformaciones en Etnobotánica/Traditions and Transformations in Ethnobotany. San Salvador de Jujuy: Edición Cyted (Programa Iberoamericano Ciencia y Técnica para el Desarrollo), pp 539–545

Łuczaj Ł (2002) Dzikie rośliny jadalne Polski, przewodnik survivalowy. Chemigrafia, Krosno

Łuczaj Ł (2008) Archival data on wild food plants eaten in Poland in 1948. J Ethnobiol Ethnomed 4:4

Łuczaj Ł (2010a) Changes in the utilization of wild green vegetables in Poland since the 19th century: a comparison of four ethnobotanical surveys. J Ethnopharmacol 128:395–404

Łuczaj Ł (2010b) Plant identification credibility in ethnobotany: a closer look at Polish ethnographic studies. J Ethnobiol Ethnomed 36:4

Łuczaj Ł (2011) Dziko rosnące rośliny jadalne użytkowane w Polsce od połowy XIX w. do czasów współczesnych. Etnobiologia Polska 1:57–125

Łuczaj Ł (2012a) Ethnobotanical review of edible plants of Slovakia. Acta Soc Bot Polon 81(4):245–255

Łuczaj Ł (2012b) Brzozowy sok, "czeremsza" i zielony barszcz—ankieta etnobotaniczna wśród botaników ukraińskich. Birch sap, ramsons and green borsch—an ethnobotanical survey among Ukrainian botanists. Etnobiologia Polska 2:5–22

Łuczaj Ł, Kujawska M (2012) Botanists and their childhood memories: an under-utilized expert source in ethnobotanical research. Bot J Linnean Soc 168:334–343

Łuczaj Ł, Nieroda Z (2011) Collecting and learning to identify edible fungi in Southeastern Poland: age and gender differences. Ecol Food and Nutr 50:319–336

Łuczaj Ł, Szymański WM (2007) Wild vascular plants gathered for consumption in the Polish countryside: a review. J Ethnobiol Ethnomed 3:17

Łuczaj Ł, Pieroni A, Tardío J, Pardo-de-Santayana M, Sõukand R, Svanberg I, Kalle R (2012) Wild food plant use in 21st century Europe: the disappearance of old traditions and the search for new cuisines involving wild edibles. Acta Soc Bot Pol 81(4):359–370

Łuczaj Ł, Köhler P, Pirożnikow E, Graniszewska M, Pieroni A, Gervasi T (2013a) Wild edible plants of Belarus: from Rostafiński's questionnaire of 1883 to the present. J Ethnobiol Ethnomed 9:21

Łuczaj Ł, Zovko Končić M, Miličević T, Dolina K, Pandža M (2013b) Wild vegetable mixes sold in the markets of Dalmatia (southern Croatia). J Ethnobiol Ethnomed 9:2

Łuczaj Ł, Fressel N, Perković S (2013c) Wild food plants used in the villages of the Lake Vrana Nature Park (northern Dalmatia, Croatia). Acta Soc Bot Pol 82(4):275–281 (in press)

Lánská D (1992) Jadalne rośliny dziko rosnące (edible wild plants). Delta W-Z, Warszawa

Leonti M, Nebel S, Rivera D, Heinrich M (2006) Wild gathered food plants in the European Mediterranean: a comparison analysis. Econ Bot 60:130–142

Local Food-Nutraceuticals Consortium (2005) Understanding local Mediterranean diets: a multidisciplinary pharmacological and ethnobotanical approach. Pharmacol Res 52:353–366

Mabey R (1972) Food for free. Collins, London

Madej T, Pirożnikow E, Dumanowski J, Łuczaj Ł (2014) Juniper beer in Poland: the story of the revival of a traditional beverage. J Ethnobiol, in press

Marco C, Chauvet M, Mathez J, Ubaud J, Passama L, Garrone B et al (2003) Les salades sauvages. L'Ensalada champanèla. Les Ecologistes de L'Euzière, Sant Jean de Cuculles

Matalas A–L, Grivetti LE (2007) Non-food during famine: the Athens Famine Survivor Project. In: MacClancy J, Henry CJ, Macbeth H. Consuming the Inedible: neglected dimensions of food choice. Berghahn, New York, pp 131–139

Maurizio A (1926) Pożywienie roślinne w rozwoju dziejowym (Plant food in human history). Kasa Mianowskiego, Warszawa:
Maurizio A (1927) Geschichte unserer Pflanzennahrung, von den Urzeiten bis zur Gegenwart. Paul Parey, Berlin
Maurizio A (1932) Histoire de l'alimentation depuis la Préhistoire jusqu' à nos jours. Payot, Paris
Mears R, Hillman G (2007) Wild food. Hodder and Stoughton, London
Muller J, Almedom AM (2008) What is "famine food"? Distinguishing between traditional vegetables and special foods for times of hunger/scarcity (Boumba, Niger). Hum Ecol 6(4):599–607
Nedelcheva A (2012) An ethnobotanical study of wild edible plants in Bulgaria. EurAsian J BioSci 77:94
Nedelcheva A (2013) An ethnobotanical study of wild edible plants in Bulgaria. EurAsian J BioSci 7: 77–94
Nelson MC (1998) Bitter bread: the famine in Norrbotten 1867–1868. Almqvist & Wiksell International, Uppsala
Ó Gráda C (2009) Famine: a short history. Princeton University Press, Princeton
Paoletti MG, Dreon AL, Lorenzoni GG (1995) Pistic, traditional food from Western Friuli, N.E. Italy. Econ Bot 49:26–30
Parada M, Carrió E, Vallès J (2011) Ethnobotany of food plants in the Alt Empordà region (Catalonia, Iberian Peninsula). J Appl Bot Food Qual 84(1):11–25
Pardo-de-Santayana M, Morales R (2010) Chamomiles in Spain. The dynamics of plant nomenclature. In: Pardo-de-Santayana M, Pieroni A, Puri R (eds) Ethnobotany in the new Europe: people, health and wild plant resources. Berghahn Press, New York-Oxford, pp 283–307
Pardo-de-Santayana M, Blanco E, Morales R (2005) Plants known as te in Spain: an ethno-pharmaco-botanical review. J Ethnopharmacol 98(1–2):1–19
Pardo-de-Santayana M, Tardío J, Blanco E, Carvalho AM, Lastra JJ, San Miguel E, Morales R (2007) Traditional knowledge of wild edible plants used in the northwest of the Iberian Peninsula (Spain and Portugal). A comparative study. J Ethnobiol Ethnomed 3:27
Peintner U, Schwarz S, Mešić A, Moreau PA, Moreno G, Saviuc P (2013) Mycophilic or mycophobic? Legislation and guidelines on wild mushroom commerce reveal different consumption behaviour in European countries. PloS O N E 8(5):e63926
Phillips R (1983) Wild food. Macmillan, London
Picchi G, Pieroni A (2005) Atlante dei prodotti tipici. Le erbe. RAI, Roma
Pieroni A (1999) Gathered wild food plants in the upper valley of the Serchio river (Garfagnana), Central Italy. Econ Bot 53:327–341
Pieroni A (2001) Evaluation of the cultural significance of wild food botanicals traditionally consumed in Northwestern Tuscany, Italy. J Ethnobiol 21:89–104
Pieroni A, Gray C (2008) Herbal and food folk medicines of the Russlanddeutschen living in Künzelsau/Taläcker, South-Western Germany. Phytother Res 22:889–901
Pieroni A, Nebel S, Quave C, Münz H, Heinrich M (2002) Ethnopharmacology of liakra: traditional weedy vegetables of the Arbëreshë of the Vulture area in southern Italy. J Ethnopharmacol 81:165–185
Pieroni A, Nebel S, Santoro RF, Heinrich M (2005) Food for two seasons: culinary uses of non-cultivated local vegetables and mushrooms in a south Italian village. Int J Food Sci Nutr 56:245–272
Pilegaard K (2012) Ramson confusable with poisonous plants. 2012/2. National Food Institute, Technical University of Denmark, Division of Toxicology and Risk Assessment [Internet] 2012 [cited 2012 July 17]. http://www.food.dtu.dk/upload/f%C3%B8devareinstituttet/food.dtu.dk/publikationer/2012/ramson%20confusable%20with%20poisonous%20plants.pdf
PFAF (2015) Plants for a future database. http://www.pfaf.org. Accessed 10 Oct 2015
Polo S, Tardío J, Vélez-del-Burgo A, Molina M, Pardo-de-Santayana M (2009) Knowledge, use and ecology of golden thistle (*Scolymus hispanicus* L.) in Central Spain. J Ethnobiol Ethnomed 5:42
Recetario extremeño [Internet] 2005 [cited 2012 June 01]. http://www.cacerespatrimonio.com/Gastronomia/index.htm

Redzepi R (2010) Times and place in Nordic Cuisine. Phaidon Press, London

Redžić S (2006) Wild edible plants and their traditional use in the human nutrition in Bosnia-Herzegovina. Ecol Food Nutr 45:189–232

Redžić S (2010) Use of wild and semi-wild edible plants in nutrition and survival of people in 1430 days of siege of Sarajevo during the war in Bosnia and Herzegovina (1992–1995). Coll Antropol 34(2):551–570

Redžić S, Barudanović S, Pilipović S (2010) Wild mushrooms and lichens used as human food for survival in war conditions; Podrinje—Zepa region (Bosnia and Herzegovina, W. Balkan). Hum Ecol Rev 17(2):175–181

Rivera D, Obón C, Inocencio C, Heinrich M, Verde A, Fajardo J, Llorach R (2005) The ethnobotanical study of local Mediterranean food plants as medicinal resources in Southern Spain. J Physiol Pharmacol 56:97–114

Rivera D, Verde A, Fajardo J, Inocencio C, Obón C, Heinrich M (2006a) Guía etnobotánica de los alimentos locales recolectados en la provincia de Albacete. Don Juan Manuel, Albacete

Rivera D, Obón C, Heinrich M, Inocencio C, Verde A, Fajardo J (2006b) Gathered mediterranean food plants—ethnobotanical investigations and historical development. In: Heinrich M, Müller WE, Galli C (eds) Local Mediterranean food plants and nutraceuticals, vol 59. Forum of nutrition. Karger, Basel, pp 18–74

Romojaro A, Botella MÁ, Obón C, Pretel MT (2013) Nutritional and antioxidant properties of wild edible plants and their use as potential ingredients in the modern diet. Int J Food Sci Nutr 64:944–952

Schaffer S, Schmitt-Schillig S, Müller WE, Eckert GP (2005) Antioxidant properties of Mediterranean food plant extracts: geographical differences. J Physiol Pharmacol 1:115–124

Schunko C, Vogl CR (2010) Organic farmers use of wild food plants and fungi in a hilly area in Styria (Austria). J Ethnoiol Ethnomed 6:17

Schunko C, Grasser S, Vogl CR (2012) Intracultural variation of knowledge about wild plant uses in the biosphere reserve grosses walsertal (Austria). J Ethnobiol Ethnomed 8:23

Sorokin P (1975) Hunger as a factor in human affairs. University Presses of Florida, Gainesville

Sõukand R, Kalle R (2010) Plant as object within herbal landscape: different kinds of perception. Biosemiotics 3(3):299–313

Sõukand R, Kalle R (2011) Change in medical plant use in Estonian ethnomedicine: a historical comparison between 1888 and 1994. J Ethnopharmacol 135(2):251–260

Sõukand R, Kalle R (2012) The use of *teetaimed* in Estonia, 1880s–1990s. Appetite 59(2):523–530

Svanberg I (1998) The use of wild plants in the Faroe Islands 1590–1990: a contribution to Scandinavian ethnobotany. Svenska Linnésällskapets Årsskrift 1996–1997:81–130

Svanberg I (2007) Fattigmanskost. In: Bromberg G, Lindell G (eds) Till livs med Linné: om mat hälsa och levnadskonst. Atlantis, Stockholm, pp 185–195

Svanberg I (2011) Folklig botanik. Dialogos, Stockholm

Svanberg I (2012) The use of wild plants as food in pre-industrial Sweden. Acta Soc Bot Pol 81(4):317–327

Svanberg I, Ægisson S (2012) Edible wild plant use in the Faroe Islands and Iceland. Acta Soc Bot Pol 81(4):233–238

Svanberg I, Nelson MC (1992) Bone meal porridge, lichen soup, or mushroom bread: acceptance or rejection of food propaganda 1867–1868. In: Häkkinen AJ (ed) Just a sack of potatoes? Crisis experiences in past and present societies. Societas Historica Finlandiae, Helsinki, pp 119–147

Szulczewski JW (1996) Pieśń bez końca: Zbiór tekstów folkorystyczno-etnograficznych. PSO, Poznań

Tardío J (2010) Spring is coming: the gathering and consumption of wild vegetables in Spain. In: Pardo-de-Santayana M, Pieroni A, Puri R (eds) Ethnobotany in the New Europe: people, health and wild plant resources. Berghahn Books, Oxford-New York, pp 211–238

Tardío J, Pascual H, Morales R (2002) Alimentos silvestres de Madrid. Ediciones La Librería, Madrid

Tardío J, Pascual H, Morales R (2005) Wild food plants traditionally used in the province of Madrid. Econ Bot 59(2):122–136

Tardío J, Pardo-de-Santayana M, Morales R (2006) Ethnobotanical review of wild edible plants in Spain. Bot J Linn Soc 152:27–72

Turner NJ, Łuczaj Ł, Migliorini P, Pieroni A, Dreon AL, Sacchetti L et al (2011) Edible and tended wild plants, traditional ecological knowledge and agroecology. Cr Rev Plant Sci 30:198–225

Velasco JM, Criado J, Blanco E (2010) Usos tradicionales de las plantas en la provincia de Salamanca. Diputación de Salamanca, Salamanca

Verde A, Rivera D, Heinrich M, Fajardo J, Inocencio C, Llorach R, Obón C (2003) Plantas alimenticias recolectadas tradicionalmente en la provincia de Albacete y zonas próximas, su uso tradicional en la medicina popular y su potencial como nutracéuticos. Sabuco Rev Est Albacet 4:35–72

Chapter 4
Ethnobotanical Analysis of Wild Fruits and Vegetables Traditionally Consumed in Spain

Javier Tardío and Manuel Pardo-de-Santayana

4.1 Introduction

Except for a strip of land with Eurosiberian vegetation in the north, the landscape of Iberian Spain can be included in the Mediterranean floristic region. This area has a characteristic Mediterranean climate, with the highest temperatures and a drought period during summer, though with clear differences in terms of annual rainfall, temperature and the length of the drought period throughout the territory. In general, the climate is more arid in the east and south of the Peninsula (the driest areas in the southeast), becoming more continental in the centre. The native potential vegetation is mainly evergreen forest, although some mountain ranges contain deciduous forests and meadows, especially at a certain altitude. These climatic differences, together with a high topographic and geological diversity, have helped confer one of the richest floras of Europe. Following Aedo et al. (2013), the Spanish flora (excluding Canarian plants) contains 6152 species, 53 % of the European flora, and with an important rate (15 %) of plants endemic to this area.

The complex Spanish history, fruit of the crossroads of civilizations in the Mediterranean area, has brought also a great cultural diversity with different languages and traditions. This ethnographical and biological variety has given rise to a rich biocultural heritage and, therefore, a rich ethnobotanical knowledge.

In Spain, as in other Mediterranean countries, wild edible plants have played an important role in complementing and balancing a diet based on agricultural foods, especially during times of shortage. As remarked by Etkin (1994), and contrary to what might be expected, some agricultural populations include significant quanti-

J. Tardío (✉)
Departamento de Investigación Agroalimentaria, Instituto Madrileño de Investigación y Desarrollo Rural, Agrario y Alimentario (IMIDRA), Alcalá de Henares, Madrid, Spain
e-mail: javier.tardio@madrid.org

M. Pardo-de-Santayana
Departamento de Biología (Botánica), Facultad de Ciencias, Universidad Autónoma de Madrid (UAM), Madrid, Spain
e-mail: manuel.pardo@uam.es

M. de C. Sánchez-Mata, J. Tardío (eds.), *Mediterranean Wild Edible Plants*,
DOI 10.1007/978-1-4939-3329-7_4

ties of foraged plants in their diets and exploit a greater variety of plants than some hunters and gatherers do. This is the case of several Mediterranean countries where the consumption of a great number of wild species has been registered (Hadjichambis et al. 2008; Leonti et al. 2006; Nassif and Tanji 2013; Picchi and Pieroni 2005; Rivera et al. 2006; Tardío et al. 2006). These species have been very valuable during certain seasons when fresh agricultural products were scarce, such as winter and spring in the case of wild vegetables or summer and autumn in the case of wild fruits.

However, at present, with the development of modern agriculture and the global supply chains, a great variety of cultivated fruits and vegetables can be easily found in markets throughout the year. Due to these agricultural changes, associated to migration to cities, the use of noncultivated plants has decreased, and nowadays far fewer species are consumed. Furthermore, most of the traditional knowledge of wild food plants is quickly disappearing and, in most cases, survives only with the elderly.

This loss of traditional knowledge about wild useful plants has led to many ethnobotanical studies aiming to register this lore before it disappears. Spain, with more than 30 ethnobotanical doctoral theses and many other ethnobotanical surveys, is one of the European countries with a higher number of studies (Morales et al. 2011). Although most of these works deal with medicinal plants (e.g. González-Tejero 1989; Villar et al. 1987) or useful flora in general (e.g. Aceituno-Mata 2010; Blanco 1998; Criado et al. 2008; Fajardo et al. 2000; Parada 2007; Velasco et al. 2010), some of them, as some of the ones carried out by our research group (Menendez-Baceta et al. 2012; Pardo-de-Santayana et al. 2005, 2007; Tardío et al. 2002, 2005), are focused on wild food plants. During more than 15 years, we have exhaustively compiled in a database the published ethnobotanical information about the wild food plants traditionally used in Spain over the past 100 years. Some unpublished information was also included trying to refill the geographical gaps. In 2006, we published a review with the results obtained at that moment from 46 bibliographical sources (Tardío et al. 2006). For the purposes of our database, we established seven categories of food use to classify wild food plants: vegetables, fruits, beverages, seasoning, preservatives, sweets (including flowers or roots eaten for their sweet flavour) and other food uses, such as oils, flours and pickles (for a detailed description of the categories, see Tardío et al. 2006).

At present, this database contains 6129 records, most of them obtained from 80 primary ethnobotanical sources. It includes data from 42 out of 51 Spanish provinces despite the fact that the information about 6 of them cannot be considered sufficient as these provinces have not been exhaustively surveyed.

In this chapter, we analyse and evaluate the traditional knowledge about the gathering and consumption of wild edible plants in Spain, especially that related to fruits and vegetables, two of the most important food use-categories. For the purpose of a more homogeneous analysis, we only use data from the 80 published references referred to the Iberian territory and the Balearic Islands. Ethnobotanical data referring to the Canary Islands were excluded, since its flora, very different from the rest, belongs to the Macaronesian floristic region.

4.2 Global Figures: Importance of Wild Vegetables and Fruits

For evaluating the relative importance of each of the seven food categories considered, we use two figures: the number of species of each category and the number of use-reports (UR) assigned to each category in the database. Here we will consider a UR as the citation in a bibliographical reference of the use of a species inside a food use-category in a different Spanish province. For instance, one UR is 'the reference 33 mentioned the use of the species *Apium nodiflorum* (L.) Lag. as a vegetable in the province of Segovia', in Central Spain. For permitting comparison with other geographical areas, we have employed a generalization of the cultural importance (CI) index, defined by us (Tardío and Pardo-de-Santayana 2008). Summing up, the CI for a species in an ethnobotanical study was the sum of all the UR for this species divided by the number of informants and, in a similar way, the CI for one use-category is the sum of all the UR for this category divided by the number of informants (Aceituno-Mata 2010). In this review, we treat each study as it was an informant, and so, for calculating each CI we divided the UR by the number of ethnobotanical studies (80).

Table 4.1 shows the global figures of the database for the seven food use-categories considered. At present, 514 species belonging to 74 families have been recorded, accounting for 8.4 % of the 6152 species listed in the Iberian and Balearic Spanish Flora. Figure 4.1 shows the CI index of each of the food use-categories considered for the Spanish wild food plants.

As shown in Table 4.1 and Fig. 4.1, according to the number of UR, the most important categories were green vegetables, beverages and fruits, with 39, 22 and 18 % of the UR, respectively. 'Vegetables' was also the most diverse category, with 51 % of the species followed again by plants used to prepare beverages (31 %). However, most of the species in the category of beverages could not be considered as real food, but as simple condiments, such as those used for making some herb liqueurs. The category of fruits accounted for 79 species being less diverse than sweets (97). The food category of seasoning had 12 % of the UR and 14 % of the species, whereas the category of sweets accounted for a higher number of species (19 %) but a much lower proportion of UR (6 %). Plants used as preservatives and the group of other uses represented 6 and 4 % of species, respectively, though they have almost the same number of UR. As some of the species were included in more than one food use-category, the sum of the percentages of the species is nearly 141, higher than 100. Therefore, there was an overlapping of species in nearly all categories; it was especially higher between seasonings and beverages. Around 65 % of the species used for seasoning were aromatic plants that were also used for making beverages, especially herbal teas (49 %). Also important was the overlapping between wild fruits and beverages. Of the wild fruits, 40 % were also used for beverages, especially for making liqueurs (36 %).

Table 4.1 Global figures about wild food plants mentioned in the Spanish ethnobotanical studies

Food use-category	# of UR[a] (%)		# of species (%)	
Vegetables	1677 (38.9)		263 (51.2)	
Beverages	937 (21.7)		158 (30.7)	
Liqueurs		*377 (8.7)*		*107 (20.8)*
Herbal teas		*504 (11.7)*		*90 (17.5)*
Other beverages		*56 (1.3)*		*25 (4.9)*
Fruits	758 (17.6)		79 (15.4)	
Sweets	268 (6.2)		97 (18.9)	
Flowers		*136 (3.2)*		*58 (11.3)*
Subterranean parts		*120 (2.8)*		*34 (6.6)*
Exudates		*12 (0.3)*		*7 (1.4)*
Seasonings	505 (11.7)		74 (14.4)	
Preservatives	86 (2.0)		32 (6.2)	
Curdling milk		*44 (1.0)*		*14 (2.7)*
Other preservatives		*42 (1.0)*		*19 (3.7)*
Other uses	83 (1.9)		21 (4.1)	
Pickles and brines		*43 (1.0)*		*11 (2.1)*
Flours		*29 (0.7)*		*9 (1.8)*
Oils		*11 (0.3)*		*3 (0.6)*
Total	4314 (100)		514 (140.9[b])	

[a] UR: use-reports, i.e., the citation in a bibliographical reference the use of a species inside a food use-category in a different Spanish province.

[b] As some species are included in more than one use-category, the total number of species is not the sum of the species of each category, and the sum of the percentages is 140.9, higher than 100.

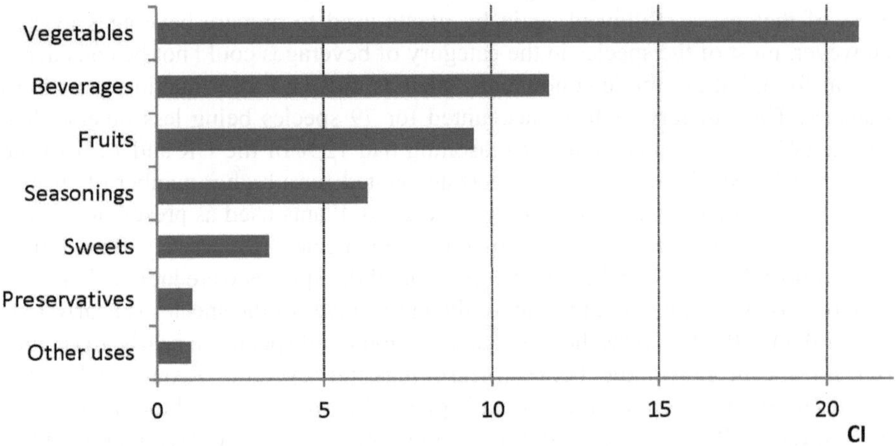

Fig. 4.1 Cultural importance of each of the food use-categories considered for the Spanish wild food plants

In the following sections, we characterize the traditional consumption of wild vegetables and fruits, the most important wild food categories that can be considered as main foods. We try to explain which species people prefer for eating as vegetables or fruits, how and when they were traditionally consumed.

4.3 Wild Vegetables Characterization

Our category of 'vegetables' or 'green vegetables' (*verduras* in Spanish) includes those plants whose vegetative organs (leaves, stems and bulbs) or immature reproductive structures (inflorescences, fruits and seeds) are consumed, either stewed, raw in salads or even raw as a snack. In a previous work (Tardío 2010), with a shorter database, a lower number of species and use-records, the first author carried out an analysis of the wild vegetables traditionally consumed in Spain. In this section, we try to answer some of the questions treated there, with almost 30 % more species and a bigger number of records, and also to complement that work with the cultural importance of the different subjects studied.

4.3.1 Botanical Analysis of Vegetables

4.3.1.1 Botanical Family

The 263 species recorded in our database as wild vegetables belong to 48 different botanical families, though with a diverse contribution of each family that could be explained by the human preference for certain kind of plants.

Regarding the number of species, the family Asteraceae accounts for almost one third of the species ($n=75$, 29 %). Although with much lower figures, other important families are Fabaceae, Brassicaceae, Liliaceae (*sensu lato*), Polygonaceae, Apiaceae and Rosaceae, with between 20 and 11 species each. One point that is often raised is whether the greater number of species selected in a family is only the result of being a diverse family, or whether there are other characteristics that make their species more likely to be used. This analysis can be done supposing a lineal relation between the two variables and studying the residuals of each family (Moerman 1991; Moerman 1994). The residuals, that is to say, the distances from each point and the regression line, indicate how far above or below from the predicted values the actual numbers are. In our case, these two variables measured in each family are the number of wild vegetables species of this family (VEGSP) and the total species per family in the Iberian Flora (FISP). As a result of the lineal regression carried out with our data, we obtained the equation $\text{VEGSP}=0.053\times\text{FISP}-0.367$, which shows the predicted number of species used as vegetables from a given family that might be expected if they were chosen randomly (Fig. 4.2). If so, greater families would have a bigger number of green vegetables. However, this is not true in many cases,

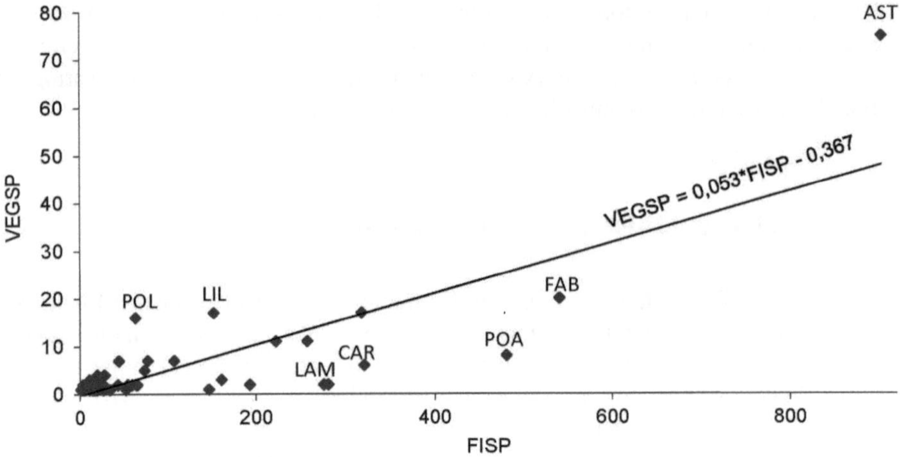

Fig. 4.2 Lineal regression analysis between the number of wild species of a certain family used as vegetables in Spain (VEGSP) and the total number of species of each family in the Iberian Flora (FISP). *AST* Asteraceae; *LIL* Liliaceae (*s.l.*); *POL* Poligonaceae; *FAB* Fabaceae; *POA* Poaceae; *CAR* Caryophyllaceae; *LAM* Lamiaceae

as can be proved analysing the residuals. The families situated over the regression line, with a high and positive residual value, have a number of species higher than predicted. So, Asteraceae (with a residual of 28), Polygonaceae (13) and Liliaceae (9) are families whose species are clearly preferred to be eaten as vegetables. By contrast, other families, such as Poaceae (with a residual of −17), Lamiaceae and Scrophulariaceae (both with −12), Caryophyllaceae (−11) and Fabaceae (−8), are greatly underrepresented in spite of the big size of these families.

What are the characteristics of the species of these families that make them preferred as wild vegetables? In most of the cases, they are abundant species, frequently considered as weeds, with in general favourable nutritional and gastronomic characteristics. They usually have desired biological forms, such as perennials with a rosette of leaves in Asteraceae and Polygonaceae, or bulbs and asparagus in Liliaceae (*s.l.*). In the case of the Asteraceae family, most of the wild vegetables (60 %) are included in the tribe Lactuceae (= Cichorieae) with several genera that account for several species each, such as *Taraxacum* and *Lactuca* (five species each), *Scorzonera, Sonchus* and *Tragopogon* (four species each), and also other genera not so diverse in species but very important in number of UR of one of their species, such as *Scolymus hispanicus* L. and *Silybum marianum* (L.) Gaertn.

In other families there are also genera rich in species, such as *Allium* (6 species) or *Asparagus* (5 species) in the family Liliaceae (*s.l.*) and *Rumex* (12 species) in the family Polygonaceae.

As an index of the cultural importance of each family, we have used the CI for the family as the sum of the CI of all the species belonging to this family (Pardo-de-Santayana et al. 2007). Figure 4.3 shows which families account for a higher

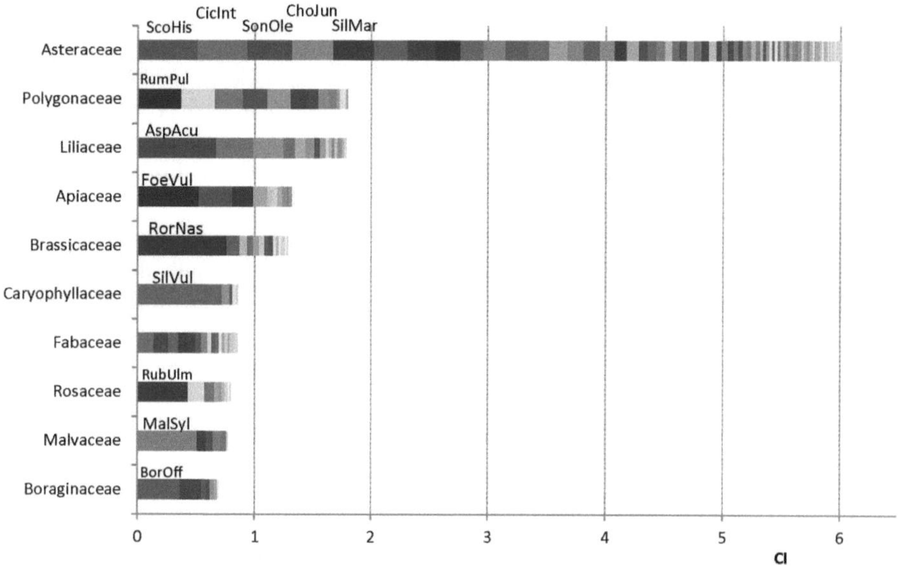

Fig. 4.3 Cultural importance of the ten families with the highest number of use-reports as wild vegetables in Spain, indicating the contribution of each species and labelling only the most important (ScoHis *Scolymus hispanicus;* CicInt *Cichorium intybus;* SonOle *Sonchus oleraceus;* ChoJun *Chondrilla juncea;* SilMar *Silybum marianum;* RumPul *Rumex pulcher;* AspAcu *Asparagus acutifolius;* FoeVul *Foeniculum vulgare;* RorNas *Rorippa nasturtium-aquaticum;* SilVul *Silene vulgaris;* RubUlm *Rubus ulmifolius;* MalSyl *Malva sylvestris;* BorOff *Borago officinalis*)

number of UR, and so a higher CI. As can be seen in the figure, the most important family is also the most diverse family, the Asteraceae (CI: 6.01), though with a very different cultural importance of many of the species. This family also has several species with a high cultural importance, such as *Scolymus hispanicus* (CI: 0.51), *Cichorium intybus* L. (0.42), *Sonchus oleraceus* L. (0.39), *Chondrilla juncea* L. (0.35), *Silybum marianum* (0.35) and *Taraxacum officinale* Weber (0.29).

However, the importance of the second more diverse family, the Fabaceae, is much lower than Polygonaceae or Liliaceae (*s.l.*) that have some species with a bigger cultural importance. The most important species in the Polygonaceae are *Rumex pulcher* L., *Rumex acetosa* L. and *Rumex induratus* Boiss. & Reut., and those in the Liliaceae are *Asparagus acutifolius* L., *Allium ampeloprasum* L. and *Asparagus albus* L.

Other families, such as Caryophyllaceae, 13th in number of species, pass to the sixth place due to the high cultural importance of one of its species, *Silene vulgaris* (Moench) Garcke, which is the second species with more citations (CI: 0.73), only three less than the watercress (*Rorippa nasturtium-aquaticum* (L.) Hayek, CI: 0.76).

4.3.1.2 Life Form

For a simple study of the life form of the species used as wild vegetables in Spain, we have considered four categories: annual herbs, non-annual herbs (both perennial and biennial), shrubs and trees. As shown in Fig. 4.4, the great majority of the wild vegetables are herbs (87 %), with a much lower representation of shrubs (10 %) and trees (3 %).

The less important category of trees includes the consumption of the young shoots of *Chamaerops humilis* L. (a palm, *sensu strictu,* the only autochthonous one of continental Europe) and the immature fruits and seeds of *Ulmus minor* Mill. and *Pinus pinea* L., respectively. Several species of the genera *Asparagus, Rubus* and *Rosa,* whose young shoots are consumed as vegetables, are some of the examples of shrubs.

Nevertheless, as mentioned before, the life form more represented among wild vegetables was herbs and, among them, perennial herbs (47 % in total, see Fig. 4.4), which wither in the summer and sprout again in the next wet season from their rootstocks. Besides some bulbous plants (such as several species of the genus *Allium*) and vines (such as *Tamus communis* L. and *Humulus lupulus* L.), there are a great number of perennial herbaceous plants with a characteristic rosette of leaves, such as *Scolymus hispanicus, Chondrilla juncea, Taraxacum officinale, Anchusa azurea* Mill. and *Rumex pulcher,* among others. Most of them have remarkable powers of regeneration from any part of the rootstock that may be severed—a significant feature that enables them to be used in a sustainable way.

Figure 4.4 shows also that a big percentage of wild vegetables are annual herbs (40 %), mainly winter annuals that germinate in autumn or winter and die after blooming in the spring or early summer, such as *Silybum marianum, Sonchus oleraceus* and *Papaver rhoeas* L. There are also some summer annuals that live in more humid places, such as *Portulaca oleracea* L. and *Chenopodium album* L. Most of these annuals produce a very high number of seeds which have, in general, a great longevity and germinate under favourable environmental situations, sometimes in a discontinuous way.

As statistically shown in a previous work (Tardío 2010), most of these herbaceous plants used as vegetables fall inside the cultural and ecological concept of 'weeds'. They are plants with a high reproductive capacity, rapid growth and ability for adapting to different environmental conditions with a clear preference for markedly human-disturbed habitats (Baker 1974). Our analysis of the proportions of weeds in the wild species selected for vegetables and in the whole flora, by means of a chi-square test, confirmed that there is a significant preference for weeds over other plant forms. This fact seems to be generally applicable to other regions of the world (Ansari et al. 2005; e.g., Díaz-Betancourt et al. 1999; Pieroni et al. 2005; Tanji and Nassif 1995; Vieira-Odilon and Vibrans 2001) and also to medicinal plants, at least in some cases (Stepp 2004; Stepp and Moerman 2001).

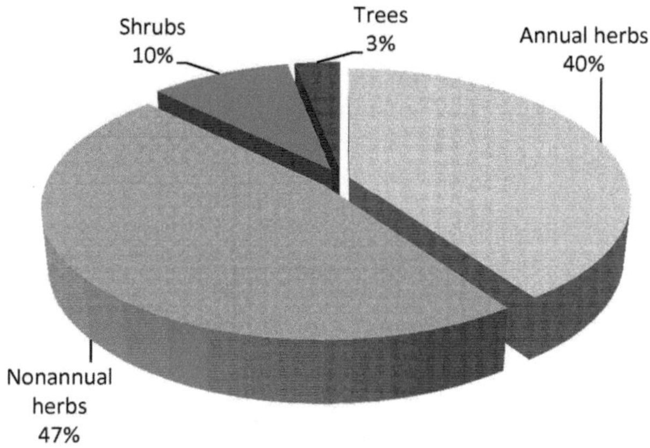

Fig. 4.4 Life form of the wild species used as vegetables in Spain

4.3.1.3 Part of the Plant Consumed

Figure 4.5 shows the ten plant parts categories considered in our analysis, ordered by its CI. As expected, there is a clear predominance of vegetative organs, especially leaves, followed by leaves and stems, and young shoots. The category of leaves is mainly formed by plants with a basal rosette of leaves that are entirely eaten (not only the midribs), many of them of the Asteraceae family. The category of leaves and stems contains the species whose tender leaves and stems are consumed together, such as *Silene vulgaris, Montia fontana* L. and *Rorippa nasturtium-aquaticum*. Young shoots are also relevant, most of them popularly known as 'wild asparaguses', either the sprouts from roots or subterranean stems from genera such as *Asparagus, Tamus* and *Ruscus,* or the tips of some climbing plants, such as *Bryonia, Humulus* and *Clematis,* or the peeled sucker of some bushes, such as *Rubus* and *Rosa.*

Other important used parts are the midribs of the basal leaves, typically used in case of thistles of several genera such as *Scolymus, Silybum, Cynara, Onopordum* and *Arctium,* but especially for the species *Scolymus hispanicus,* which is one of the most widely consumed thistles in Spain. This category could be integrated in 'Leaves'; however, it has the particularity that the prickly leaves are peeled leaving only the principal veins.

The category of fruits and seeds includes species whose immature reproductive structures are consumed, such as those of the genera *Malva, Vicia* and *Lathyrus.* For more detail and examples of the other plant part categories, see Tardío (2010).

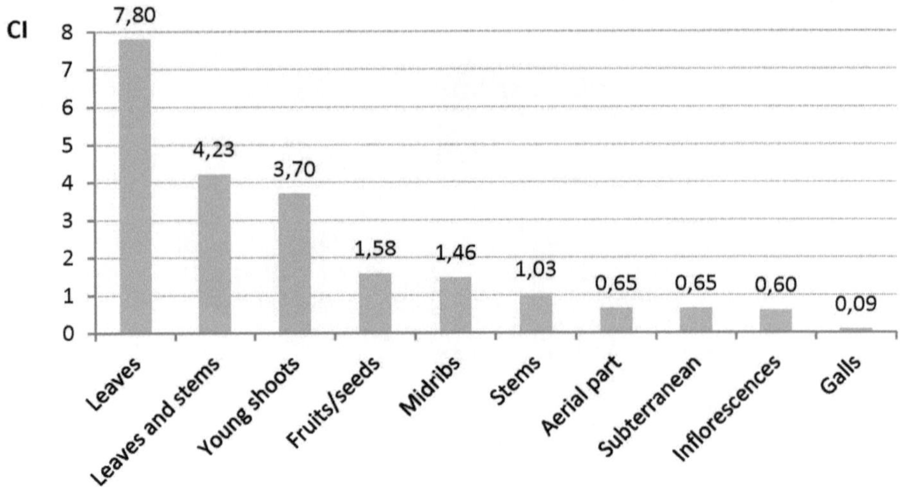

Fig. 4.5 Cultural importance (CI) of the different plant parts consumed as wild vegetables

4.3.2 The Traditional Consumption of Wild Vegetables

4.3.2.1 Mode of Consumption

The analysis of the mode of consumption of the species used as vegetables is very interesting, especially for the possible relation with their sensorial characteristics. We have considered three different but nonexclusive categories: raw as a snack, raw in salads and cooked. Wild vegetables are eaten 'raw as a snack' when they are consumed directly in the field. In other occasions, the plants are taken home to be prepared in a salad, maybe a sign of a greater level of appreciation. Some species are eaten cooked, both alone and mixed with other vegetables in a great variety of stews.

The Venn diagram of Fig. 4.6 shows the number of species of vegetables consumed in each category and the overlaps among them. For better understanding what kind of species are in each section, we have also included its CI, that is the sum of CI of the species that are integrated in it. We can hypothesize that the cultural importance of a widely distributed species is directly related to its gastronomic quality. Therefore, a wild vegetable that has been reported in many ethnobotanical sources has a high CI, probably because of its favourable sensorial and nutritive characteristics.

As can be seen in Fig. 4.6, 186 species (71 %) are consumed cooked, a bit less than half (79) exclusively, while the other species are also consumed raw, either as a snack or in salads, the latter being a somewhat larger category. If we look at the figure, we can also notice that, besides a high number of species, the highest CI

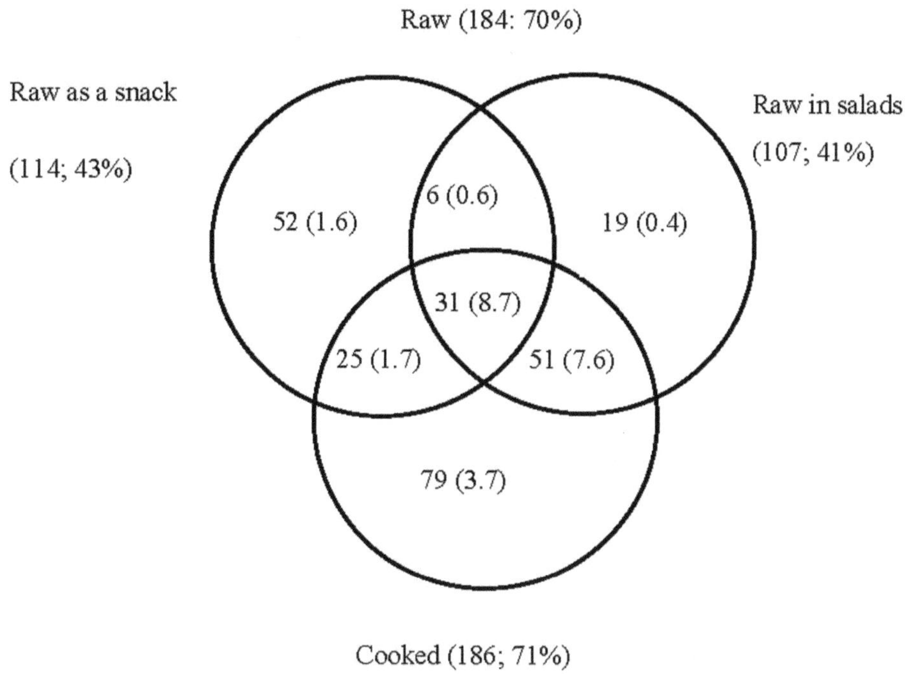

Fig. 4.6 Venn diagram showing the number of species of vegetables consumed in each category and the overlaps among them. In parentheses is the CI of that category (sum of CI of their species)

values are also found in these overlapping areas. The greatest cultural importance is reached in the intersection of the three ways of consumption, with 31 species including some with the highest CI, such as *Foeniculum vulgare* Mill., *Scolymus hispanicus, Malva sylvestris* L., *Cichorium intybus, Silybum marianum, Rubus ulmifolius* Schott, *Chondrilla juncea* and *Taraxacum officinale.* The second highest CI is found in the group of species that are eaten both cooked and raw in salads, including very important species, such as *Rorippa nasturtium-aquaticum, Silene vulgaris, Portulaca oleracea* and *Sonchus oleraceus.* Within these two groups, many of the most extensively consumed plants can be identified. Some of these species are appreciated for their quality and, because of their good flavour, they can be eaten stewed or raw. That is the case of *Silene vulgaris* (*colleja*), whose tender leaves and stems are eaten mostly stewed in many parts of Spain, in omelettes, with scrambled eggs, and also as a garnish for *potaje,* a typical Spanish dish often consumed during Lent. The peeled basal leaves of *Scolymus hispanicus* are traditionally boiled and then fried lightly in olive oil with garlic and cured ham and sometimes with scrambled eggs, often also served as a garnish for *cocido,* another traditional Spanish dish. Finally, the tender leaves and stems of *Rorippa nasturtium-aquaticum* (watercress) are consumed mainly raw in salads, but sometimes also in stews and soups.

The numerous groups of species that are only consumed cooked include a few plants with great cultural importance, such as *Asparagus acutifolius* (CI: 0.68),

Rumex pulcher (CI: 0.38) and *Bryonia dioica* Jacq. (CI: 0.29) and many others with a much lower CI. Most of these species are only consumed in this way because of their taste, which is not so pleasant when eaten raw. Many of them are bitter or strongly flavoured plants. This unpleasant taste is often reduced through repeated boiling and discarding the boiling water, as reported for *Ruscus aculeatus* L., *Cichorium intybus* L., *Rumex pulcher* L. and *Asparagus albus* L. Following Johns (1996), boiling may be a good method for food detoxification that combines the application of heat and also a means of increasing the rate and solubility of some compounds and leaching them out of a material. Other species are unpleasant when raw because of their rough external texture, for instance *Anchusa azurea, Echium plantagineum* L. and *Picris echioides* L., though that can be mitigated when boiled, too. As in other parts of the world (e.g. Johns 1994; Nebel et al. 2006; Pieroni 2003), some of these less palatable species have been usually consumed in mixed recipes with up to 15 wild vegetables (Tardío 2010).

Figure 4.6 also shows that 184 species (70 % of the wild vegetables) are consumed raw. Those species considered of better quality because of their pleasant flavour are used in salads. Besides some aforementioned species, the small leaves and stems of *Montia fontana,* the young shoots of *Chamaerops humilis,* the basal leaves of *Reichardia picroides* (L.) Roth and the leaves and roots of *Campanula rapunculus* L. are some of the species most widely consumed raw in salads. Among the numerous species that are only eaten raw as a snack, we can remark the consumption of the basal part of the stems of *Scirpoides holoschoenus* (L.) Soják, the leaves of *Oxalis acetosella* L. and other species of the same genus, the peeled young shoots of *Rosa canina* L., the unripe fruits of several species of *Erodium* and the immature seeds of *Vicia villosa* Roth and *V. lutea* L.

4.3.2.2 Harvesting and Consumption Time: Preservation Techniques

As mentioned before, the plant parts utilized as vegetables vary from vegetative organs to immature reproductive structures. Since most Spanish plants bloom in spring or early summer, the immature reproductive organs have to be collected in late spring. In the case of vegetative organs, in general, humans search for the maximum development of these edible parts being still tender, what usually happens before blooming, from the beginning to middle spring. However, there are some additional periods when the gathering of wild vegetables is also possible, depending on the species, on the region, and sometimes, due to the irregularity of the Mediterranean climate, on the year. The development attained by some perennial herbs in rainy and mild autumns can make their use interesting as a vegetable, especially in warmer areas such as the lowlands of the south and east of Spain, where some species are also collected in winter. Nevertheless, in colder and mountainous areas of inland Spain, harvesting can last until the beginning of summer.

This strong seasonality of wild vegetables is precisely, in our opinion, one of the reasons for their maintenance in agricultural societies, since it complements the harvesting period for the majority of cultivated plants. Nevertheless, at present, many kinds of cultivated vegetables can be found in the markets, at almost any time of the year. So, wild vegetables are no longer necessary and their consumption today is much less than it was in the past.

Regarding the consumption time of wild vegetables, most plants were traditionally consumed fresh; therefore, harvesting and consumption times coincide. Nevertheless, sometimes simple preservation techniques were used to make this food available throughout the year. One of the oldest methods, documented in historical sources, is the conservation in pickles and brines. This preservation method has been used in Spain according to the ethnobotanical sources consulted for *Crithmum maritimum* L., *Inula crithmoides* L., *Sedum sediforme* (Jacq.) Pau, *Pinus pinea, Capparis spinosa* L., *Allium ampeloprasum* and *Portulaca oleracea,* the first one and the last one already mentioned in the first century AD by the Spanish-Roman Columella in his agricultural work *De Re Rustica* (Columela 1824).

Other modern preservation processes used for some wild vegetables are home canning (bottling) and freezing that help provide the product outside of the harvest season. Bottling has been used at least for *Scolymus hispanicus* and *Asparagus acutifolius,* whereas the use of freezing has been mentioned for *Silene vulgaris, Tamus communis* and *Atriplex hortensis* L.

4.4 Wild Fruits Characterization

Our category of wild 'fruits' includes those plants whose fruits (or seeds) are consumed when fully ripe. They are usually eaten raw directly in the field, but sometimes they are brought home and even used in the elaboration of different sweet desserts, such as homemade jam or cakes.

4.4.1 Botanical Analysis of Fruits

4.4.1.1 Botanical Family

The 79 species recorded in our database as wild fruits are included in 27 botanical families. Almost half (44 %, n=35) are from the Rosaceae family. Other important families are Fagaceae (11 %, n=9), Ericaceae and Grossulariaceae (5 %, n=4, each).

The regression analysis defined the equation FRUSP=0.003 × FISP+2.569 (Fig. 4.7) for the predicted number of species used as fruits from a given family that might be expected if they were chosen randomly. The residual analysis clearly shows that the species from the Rosaceae and Fagaceae families (with residuals of 32 and 6, respectively), especially of the first one, are clearly selected for their edible fruits, with the number of species much higher than those that could be expected if randomly chosen.

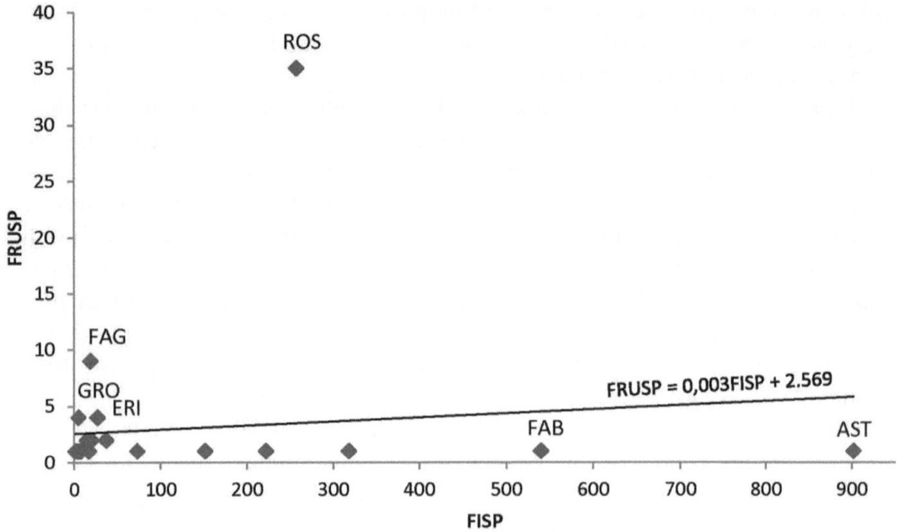

Fig. 4.7 Lineal regression analysis between number of wild species of a certain family used as edible fruits in Spain (FRUSP) and total number of species of each family in the Iberian Flora (FISP). *ROS* Rosaceae; *FAG* Fagaceae; *ERI* Ericaceae; *GRO* Grossulariaceae; *AST* Asteraceae; *FAB* Fabaceae

Inside the Rosaceae family, the use of some very diverse genera can be highlighted, such as *Prunus, Sorbus* and *Rosa* (six species each), *Rubus* (five species), *Pyrus* and *Crataegus* (three species). In the family Fagaceae, the very diverse genus *Quercus* stands out with seven species.

Regarding the cultural significance of each family, Fig. 4.8 shows that the two most important families are again the most diverse. The family Rosaceae (CI: 4.59), besides its diversity of species, accounts for several species with a high CI, such as *Rubus ulmifolius* (0.70), the most important, *Crataegus monogyna* Jacq. (0.50), *Prunus spinosa* L. (0.40) and *Fragaria vesca* L. (0.34). In the family Fagaceae (CI: 1.25), the importance of *Quercus ilex* L. (0.55), *Castanea sativa* Mill. (0.39) and *Fagus sylvativa* L. (0.14) must be remarked.

The third family Ericaceae includes *Arbutus unedo* L. (CI: 0.55), the second most culturally important species in the category of fruits tied with *Quercus ilex*.

4.4.1.2 Life Form

As can be seen in Fig. 4.9, the great majority of the species whose mature fruits or seeds are traditionally eaten in Spain are trees (46%) and shrubs (42%), with a much lower percentage of perennial and annual herbs (10 and 2%, respectively). There are lots of examples of trees, such as *Arbutus unedo, Quercus ilex* and *Pinus*

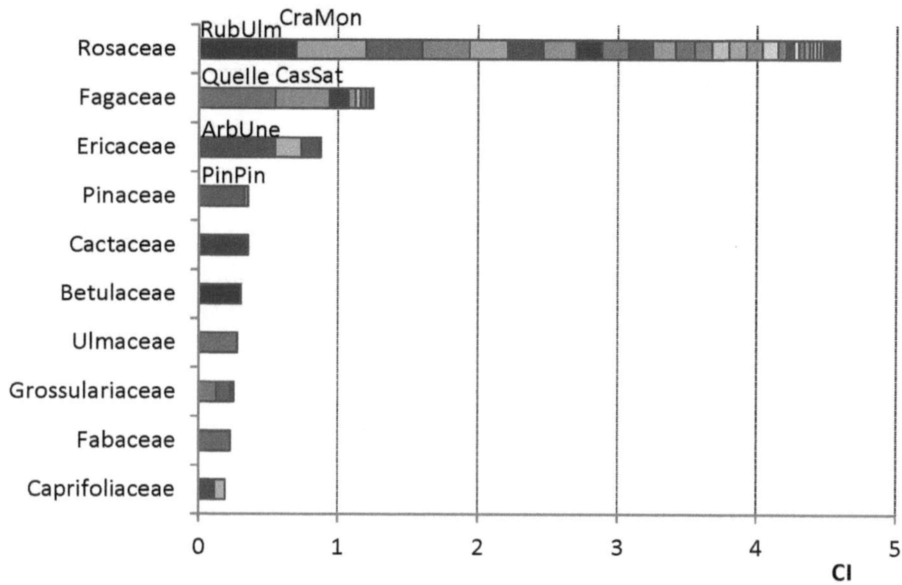

Fig. 4.8 Cultural importance of the ten families that include species with the highest number of use-reports as wild fruits in Spain, indicating the contribution of some of the most important species (RubUlm *Rubus ulmifolius;* CraMon *Crataegus monogyna;* PruSpi *Prunus spinosa;* QueIle *Quercus ilex;* CasSat *Castanea sativa;* ArbUne *Arbutus unedo;* PinPin *Pinus pinea*)

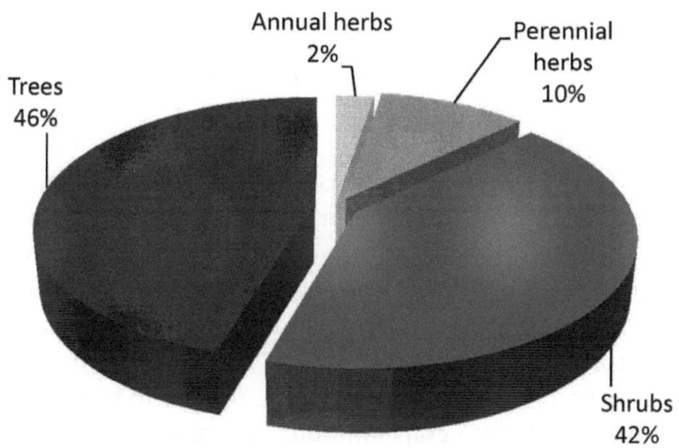

Fig. 4.9 Life form of the wild species used as fruits in Spain

pinea, and shrubs, such as *Amelanchier ovalis* Medik., *Berberis vulgaris* L. and *Rubus ulmifolius,* but there are a few perennial and annual herbs, such as *Fragaria vesca* and *Silybum marianum.*

4.4.1.3 Part of the Plant Consumed

As mentioned before, our category of wild 'fruits' includes those plants whose fruits (or seeds) are consumed when fully ripe. However, if we differentiated them by organs, it could be remarked that the majority of the species (91 %) are used for their fruits, while a few are used only for their seeds (9 %), such as *Pinus pinea* and *Cistus ladanifer* L. The divergence is even higher if we consider also the UR (95 and 5 %, respectively).

4.4.2 The Traditional Consumption of Wild Fruits

4.4.2.1 Mode of Consumption

Regarding the mode of consumption of fruits, three categories have been considered. Two of them are consumed raw: the first one directly in the field, as a snack, and the second at home, sometimes after being stored. The third category implied the cooking of the fruits, both for the elaboration of jam or other dishes and simply for consumption through boiling or roasting. Most of the species are consumed in more than one category, producing overlapping among them, as shown in the Venn diagram of Fig. 4.10. This figure shows the number of species exclusive or shared in each section of the categories and also their cultural importance (CI), as the sum of the CI of the species included.

The fruits of most of the species (97 %) were eaten raw, the majority of them directly in the field (91 %), traditionally consumed by children or by shepherds. This mode of consumption was predominant or exclusive in some of the species, such as *Crataegus monogyna, Rosa* sp. pl., *Arctostaphylos uva-ursi* (L.) Spreng., *Celtis australis* L., *Fragaria vesca, Sorbus aria* (L.) Crantz and *S. torminalis* (L.) Crantz. The fruits of some of the species were also taken for being consumed at home, sometimes even stored. Some of them, such as those of *Prunus spinosa* L., *Pyrus bourgaeana* Decne. and *Mespilus germanica* L., needed to be stored for some time (usually in hay or grain) before consumption. This process of overripening, which increases the sugar content of the fruits and therefore improves considerably their palatability, was also used with *Malus sylvestris* (L.) Mill. and *Viburnum lantana* L.

The fruits of a few species were cooked, used both for the elaboration of jam (*Arbutus unedo* and several species of the genera *Rubus, Rosa* and *Vaccinium*) and simply for the purpose of consumption through boiling or roasting (*Castanea sativa* and several species of *Quercus*). As can be seen in the figure, the section with a higher CI is that integrated by the species that are consumed in the three ways, such

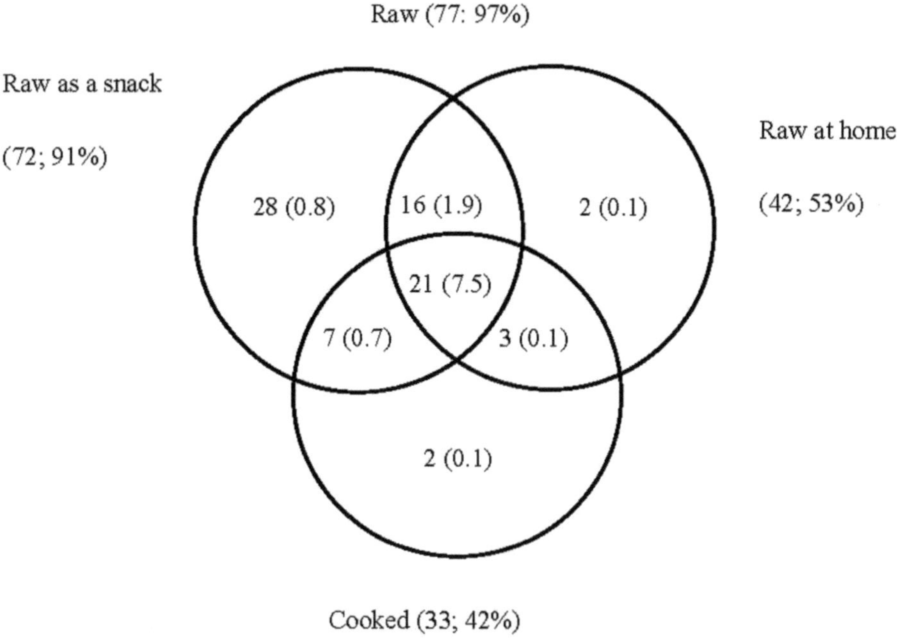

Fig. 4.10 Venn diagram showing the number of species of fruits consumed in each category and the overlapping among them. In parentheses is the CI of that category (sum of CI of their species)

as *Rubus ulmifolius, Quercus ilex, Castanea sativa, Arbutus unedo* and *Vaccinium myrtillus* L. Although the acorns of *Quercus ilex* subsp. *ballota* (Desf.) Samp., especially those of some sweet specimens, can be directly consumed, those of the subspecies *ilex* and other species of this genus (e.g. *Q. faginea* Lam. and *Q. suber* L.) required a cooking treatment that removes the bitterness. In the case of the fruits of *Sambucus nigra* L., besides their use for making beverages, all the reports referred to their consumption after cooking, used for elaborating a delicious jam. This is due to the slight toxicity of its raw fruits that completely disappears with the cooking process (Couplan 1990; Velasco et al. 2010).

4.4.2.2 Harvesting and Consumption Time: Preservation Techniques

Even though there are important regional differences resulting from the range of latitude and altitude, the fruits of the majority of the species registered in the database (n=71; 90% of the species) are collected in summer and autumn (Fig. 4.11). It must be noted that the harvesting time of some species lasts for several seasons, both because the maturity period extends to more than one season (e.g. the end of summer and the autumn) and for the climatic variability among regions.

Some species from more humid regions, such as wild strawberries *(Fragaria vesca),* cherries (*Prunus avium* L.) and blueberries *(Vaccinium myrtillus),* are

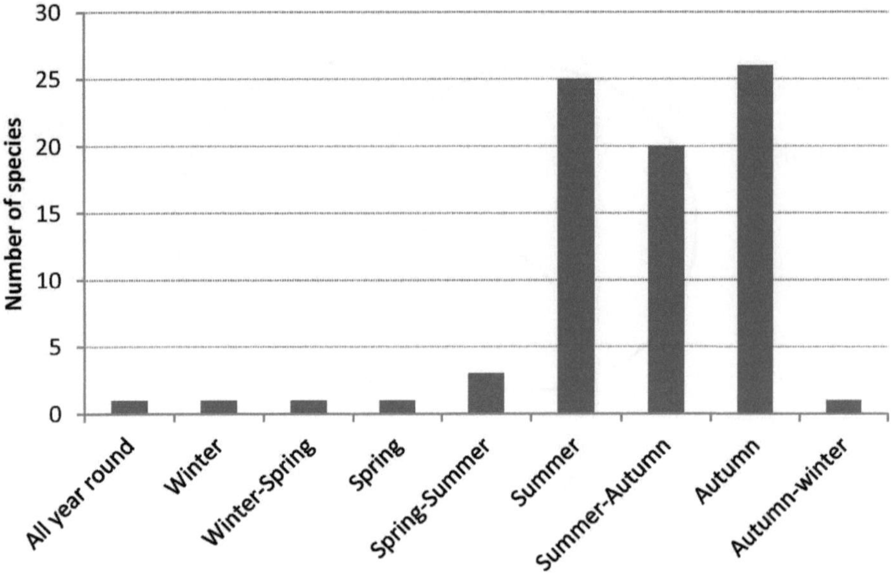

Fig. 4.11 Number of wild species whose fruits (or seeds) are harvested in each season

typically collected in summer (or final spring). However, there are also some species of drier Mediterranean areas that were also harvested in summer, such as the seeds of *Cistus ladanifer* and the fruits of *Corema album* (L.) D. Don, *Silybum marianum* and *Chamaerops humilis*.

There is also an important number of species whose fruits are collected in summer and autumn, such as *Rubus ulmifolius, Berberis vulgaris, Celtis australis* and *Ceratonia siliqua* L.

The acorns of several species of *Quercus* genus (especially *Q. ilex*), the chestnuts (*Castanea sativa*) or the fruits of *Fagus sylvatica, Malus sylvestris* or *Sambucus nigra* are naturally collected in autumn.

There are a few species whose fruits' harvesting time is out of this period, such as *Arbutus unedo,* from the middle-end of autumn to the beginning of winter, *Viscum album* L. in winter and *Pinus halepensis* Mill. in spring. The pine nuts of *Pinus pinea* reach their maturity in the autumn of the third year after flowering and are collected all year round.

As for preservation techniques, some simple methods were sometimes used to make fruits available throughout the year. Some wild fruits, such as hazelnuts (*Corylus avellana* L.), were dried as they are today as a mode of preservation. However, in the fruits of other species, such as *Prunus spinosa* and *Arctostaphylos uva-ursi,* this process also made them more palatable by increasing sugar concentration. Without being completely dried, the flavours of other species were also improved by a process of overmaturing in hay or straw. That was the case with *Malus sylvestris, Pyrus bourgaeana* or *Viburnum lantana.* Finally, some species have been used to make jam, such as *Rubus ulmifolius, Prunus insititia* L. and *Arbutus unedo.*

4.5 Regional Differences in Traditional Wild Plant Consumption

As analyzed in a previous work (Tardío and Pardo-de-Santayana 2014), there are important regional differences in traditional wild food plants consumption in Spain, both in overall number of species and in the relative importance of the different food use-categories. Northern regions show, in general, a lower number of edible species and a greater importance of the 'wild fruits' category. People from Mediterranean regions, from the centre, east and south of Spain, have traditionally consumed a much higher number of species, most of them vegetables. Ethnobotanical studies carried out in other European countries seem to confirm the same general trend. Northern countries, such as Poland (Łuczaj 2008; Łuczaj and Szymański 2007), account for a lower number of wild edibles consumed, both in overall figures and in the proportion of vegetables and condiments used. However, southern and Mediterranean regions such as the south of Italy (Ghirardini et al. 2007; Pieroni et al. 2005) and Bosnia-Herzegovina (Redžić 2006) are characterized both by a considerably much higher number of wild species consumed and especially wild vegetables, being known in terms of Łuczaj (2008), as herbophilous regions opposed to the herbophobous regions of northern Europe. This phenomenon is also discussed by Łuczaj and Pieroni in Chap. 3 of this book.

4.6 Evolution of Wild Plant Consumption in Spain

It should be noted that wild species registered in our database are those that have been traditionally consumed in Spain until about 50 years ago. Most of them are only a reminder of the wild plant consumption in the past, and they are rarely collected today. Apart from the reasons for this change already mentioned in Sect. 4.1, such as social and agricultural changes and the present availability of cultivated fruit and vegetables, some of these wild foods are considered famine food and others were simply children or shepherd snacks that today appear as unfashionable.

However, there are a number of species (e.g. *Asparagus acutifolius, Scolymus hispanicus, Silene vulgaris, Tamus communis* and *Montia fontana*) that are still collected or even marketed on a small scale, but in a much lower proportion than in the past. In order to assess the level of this decrease, our research group has carried out some recent ethnobotanical surveys in rural areas of Central Spain (Andrés 2012; Dávila 2010; García-Cervigón 2013; Molina-Bustamante 2014; Polo et al. 2009). Table 4.2 shows the results of these five works referring to past and present consumption of five wild plants in seven municipalities. It can be seen that even in species still consumed (24–53 % of the informants), such as the wild vegetables *Scolymus hispanicus* and *Silene vulgaris* or the fruits of *Rubus ulmifolius,* the mean decreasing rate ranges between 24 and 58 %. The table also shows two species whose consumption has been almost abandoned (mean decreasing rate 89–97 %), such as *Rumex papillaris,* which was consumed as a raw vegetable by almost 50 %

Table 4.2 Mean percentage of interviewees who mentioned having consumed in the past and at present different wild species, and the corresponding decreasing rate in five recent ethnobotanical surveys carried out in seven municipalities of central Spain

Species	Past (%)	Present (%)	Decreasing rate (%)	References
Scolymus hispanicus	57	26	58	Polo et al. 2009; Andrés 2012; García-Cervigón 2013; Molina-Bustamante 2014
Silene vulgaris	31	24	24	Dávila 2010; Andrés 2012
Rubus ulmifolius	95	53	45	García-Cervigón 2013; Molina-Bustamante 2014
Rumex papillaris	49	7	89	García-Cervigón 2013; Molina-Bustamante 2014
Crataegus monogyna	31	1	97	García-Cervigón 2013; Molina-Bustamante 2014

of the informants in the past, and *Crataegus monogyna,* whose fruits were eaten in the past by 30% of the informants.

These studies also indicate that older people clearly consume and gather more species than the younger generations, but they do not find important differences in the gendered distribution of ethnobotanical knowledge about wild edible plants (Andrés 2012; García-Cervigón 2013; Polo et al. 2009).

Therefore, there is a general decline in the gathering and consumption of wild edible plants though it is maintained for some selected species that are especially appreciated for their flavour in addition to the recreational motivation of gathering.

At the same time, both in Spain and in other Western countries, there is a growing interest in wild food plants attributable to several reasons. On one hand, it is because of their gastronomic interest that has caused them to be even included in the haute-cuisine restaurant menus (see also Chap. 2). On the other hand, as amply discussed in many other chapters of this book, there is also great interest for their nutritional value and their role in maintaining human health (e.g. Bharucha and Pretty 2010; Johns and Eyzaguirre 2006; Simopoulos 2004) and as possible sources of nutraceuticals (Heinrich et al. 2006).

Finally, as it is happening in other European countries (Łuczaj et al. 2012), we have detected in Spain an increasing interest in wild food gathering courses, books and other outreach activities carried out by our research team among urban young people concerned with rural traditional knowledge. We hope that through all these efforts, the ethnobotanical knowledge about edible wild plants that has been formed over thousands of years shall not disappear and will continue to be used in the future.

References

Aceituno-Mata L (2010) Estudio etnobotánico y agroecológico de la Sierra Norte de Madrid. PhD Thesis. Universidad Autónoma de Madrid

Aedo C, Medina L, Fernández-Albert M (2013) Species richness and endemicity in the Spanish vascular flora. Nord J Bot 31(4):478–488

Andrés A (2012) Plantas silvestres de uso alimentario y medicinal en La Mancha conquense. Proyecto de Fin de Carrera. Universidad Autónoma de Madrid. Facultad de Ciencias. Madrid

Ansari NM, Houlihan L, Hussain B, Pieroni A (2005) Antioxidant activity of five vegetables traditionally consumed by South-Asian Migrants in Bradford, Yorkshire, UK. Phytother Res 19:907–911

Baker HG (1974) The evolution of weeds. Annu Rev Ecol Syst 5:1–24

Bharucha Z, Pretty J (2010) The roles and values of wild foods in agricultural systems. Phil Trans R Soc B 365(1554):2913–2926

Blanco E (1998) Diccionario de etnobotánica segoviana. Ayuntamiento de Segovia, Segovia

Columela LJM (1824) Los doce libros de Agricultura (De Re Rustica). Trad. J.M. Álvarez de Sotomayor. Imprenta de D. Miguel de Burgos, Madrid

Couplan F (1990) Les belles veneneuses. Encyclopedie des plantes comestibles de l'Europe vol. 3. Equilibres Aujourd'houi, Flers

Criado J, Fernández López JM, Leocadio G, Núñez RM, Blanco E (2008) Uso tradicional de las plantas en Toledo. Diputación de Toledo, Toledo

Dávila P (2010) Estudio etnobotánico comparativo de la colleja [*Silene vulgaris* (Moench) Garcke] en dos localidades de la Comunidad de Madrid. Proyecto de Fin de Carrera. Universidad Autónoma de Madrid. Facultad de Ciencias. Madrid

Díaz-Betancourt M, Ghermandi L, Ladio AH, López-Moreno IR, Raffaele E, Rapoport EH (1999) Weeds as a source for human consumption. A comparison between tropical and temperature Latin America. Rev Biol Trop 47(3):329–338

Etkin NL (1994) The cull of the wild. In: Etkin NL (ed) Eating on the wild side. The University of Arizona Press, Tucson, pp 1–21

Fajardo J, Verde A, Rivera D, Obón C (2000) Las plantas en la cultura popular de la provincia de Albacete. Instituto de Estudios Albacetenses, Albacete

García-Cervigón LC (2013) Evaluación temporal del conocimiento y práctica etnobotánicos en Buitrago del Lozoya (Sierra Norte de Madrid). Proyecto de Fin de Carrera. Universidad Autónoma de Madrid. Facultad de Ciencias. Madrid

Ghirardini M, Carli M, del Vecchio N, Rovati A, Cova O, Valigi F, Agnetti G, Macconi M, Adamo D, Traina M, Laudini F, Marcheselli I, Caruso N, Gedda T, Donati F, Marzadro A, Russi P, Spaggiari C, Bianco M, Binda R, Barattieri E, Tognacci A, Girardo M, Vaschetti L, Caprino P, Sesti E, Andreozzi G, Coletto E, Belzer G, Pieroni A (2007) The importance of a taste. A comparative study on wild food plant consumption in twenty-one local communities in Italy. J Ethnobiol Ethnomed 3(1):22

González-Tejero MR (1989) Investigaciones etnobotánicas en la provincia de Granada. PhD Thesis. Universidad de Granada

Hadjichambis AC, Paraskeva-Hadjichambi D, Della A, Giusti ME, De Pasquale C, Lenzarini C, Censorii E, González-Tejero MR, Sanchez-Rojas CP, Ramiro-Gutiérrez JM, Skoula M, Johnson C, Sarpaki A, Hmamouchi M, Jorhi S, El-Demerdash M, El-Zayat M, Pieroni A (2008) Wild and semi-domesticated food plant consumption in seven circum-Mediterranean areas. Int J Food Sci Nutr 59(5):383–414

Heinrich M, Müller WE, Galli C (2006) Local Mediterranean food plants and nutraceuticals. Karger, Basel

Johns T (1994) Ambivalence to the palatability factors in wild plants. In: Etkin NL (ed) Eating on the wild side. The University of Arizona Press, Tucson, pp 46–61

Johns T (1996) The origins of human diet and medicine. Chemical ecology. The University of Arizona Press, Tucson

Johns T, Eyzaguirre PB (2006) Linking biodiversity, diet and health in policy and practice. Proc Nutr Soc 65(2):182–189

Leonti M, Nebel S, Rivera D, Heinrich M (2006) Wild gathered food plants in the European Mediterranean: a comparison analysis. Econ Bot 60(2):130–142

Łuczaj L (2008) Archival data on wild food plants used in Poland in 1948. J Ethnobiol Ethnomed 4(1):4

Łuczaj L, Szymański W (2007) Wild vascular plants gathered for consumption in the polish countryside: a review. J Ethnobiol Ethnomed 3(1):17

Łuczaj L, Pieroni A, Tardío J, Pardo-de-Santayana M, Soukand R, Svanberg I, Kalle R (2012) Wild food plant use in 21st century Europe: the disappearance of old traditions and the search for new cuisines involving wild edibles. Acta Soc Bot Pol 81(4):359–370

Menendez-Baceta G, Aceituno-Mata L, Tardío J, Reyes-García V, Pardo-de-Santayana M (2012) Wild edible plants traditionally gathered in Gorbeialdea (Biscay, Basque Country). Genet Resour Crop Evol 59(7):1329–1347

Moerman DE (1991) The medicinal flora of native North American: an analysis. J Ethnopharmacol 31:1–42

Moerman DE (1994) North American food and drug plants. In: Etkin NL (ed) Eating on the wild side. The University of Arizona Press, Tucson, pp 1–21

Molina-Bustamante M, (2014) Tendencias actuales en el uso de plantas silvestres comestibles en la Sierra del Rincón, Madrid. Proyecto de Fin de Carrera. Universidad Autónoma de Madrid. Facultad de Ciencias. Madrid

Morales R, Tardío J, Aceituno L, Molina M, Pardo-de-Santayana M (2011) Biodiversidad y Etnobotánica en España. In: Viejo-Montesinos JL (ed) Biodiversidad. Aproximación a la diversidad botánica y zoológica de España. Real Sociedad Española de Historia Natural, Madrid. pp 157–207

Nassif F, Tanji A (2013) Gathered food plants in Morocco: the long forgotten species in ethnobotanical research. Life Sci Leafl 3(3):17–54

Nebel S, Pieroni A, Heinrich M (2006) *Ta chòrta:* wild edible greens used in the Graecanic area in Calabria, Southern Italy. Appetite 47(3):333–342

Parada M (2007) Estudi etnobotànic de L'Alt Empordà. PhD Thesis. Facultat de Farmàcia. Universitat de Barcelona

Pardo-de-Santayana M, Tardío J, Morales R (2005) The gathering and consumption of wild edible plants in the Campoo (Cantabria, Spain). Int J Food Sci Nutr 56(7):529–542

Pardo-de-Santayana M, Tardío J, Blanco E, Carvalho AM, Lastra JJ, San Miguel E, Morales R (2007) Traditional knowledge on wild edible plants in the northwest of the Iberian Peninsula (Spain and Portugal): a comparative study. J Ethnobiol Ethnomed 3:27

Picchi G, Pieroni A (2005) Atlante dei prodotti tipici: Le erbe. Agra, RAI-Eri, Roma

Pieroni A (2003) Wild food plants and Arbëresh women in Lucania, Southern Italy. In: Howard PL (ed) Women and plants. Gender relations in biodiversity, management and conservation. Zed Books, London, pp 66–82

Pieroni A, Nebel S, Santoro RF, Heinrich M (2005) Food for two seasons: culinary uses of non-cultivated local vegetables and mushrooms in a south Italian village. Int J Food Sci Nutr 56(4):245–272

Polo S, Tardío J, Vélez-del-Burgo A, Molina M, Pardo-de-Santayana M (2009) Knowledge, use and ecology of golden thistle (*Scolymus hispanicus* L.) in Central Spain. J Ethnobiol Ethnomed 5(1):42

Redžić S (2006) Wild edible plants and their traditional use in the human nutrition in Bosnia-Herzegovina. Ecol Food Nutr 45(3):189–232

Rivera D, Obon C, Heinrich M, Inocencio C, Verde A, Fajardo J (2006) Gathered Mediterranean food plants—Ethnobotanical investigations and historical development. Forum Nutr 59:18–74

Simopoulos AP (2004) Omega-3 fatty acids and antioxidants in edible wild plants. Biol Res 37:263–277

Stepp JR (2004) The role of weeds as sources of pharmaceuticals. J Ethnopharmacol 92:163–166

Stepp JR, Moerman DE (2001) The importance of weeds in ethnopharmacology. J Ethnopharmacol 75:19–23

Tanji A, Nassif F (1995) Edible weeds in Morocco. Weed Technol 9:617–620

Tardío J (2010) Spring is coming: the gathering and consumption of wild vegetables in Spain. In: Pardo-de-Santayana M, Pieroni A, Puri R (eds) Ethnobotany in the New Europe: people, health and wild plant resources. Berghahn Books, Oxford–New York, pp 211–238

Tardío J, Pardo-de-Santayana M (2008) Cultural importance indices: a comparative analysis based on the useful wild plants of southern Cantabria (Northern Spain). Econ Bot 62(1):24–39

Tardío J, Pardo-de-Santayana M (2014) Wild food plants traditionally used in Spain: regional analysis. In: Chevalier A, Marinova E, Peña-Chocarro L (eds) Plants and people: choices and diversity through time. Oxbow Books, Oxford.

Tardío J, Pascual H, Morales R (2002) Alimentos silvestres de Madrid. Guía de plantas y setas de uso alimentario tradicional en la Comunidad de Madrid. Ediciones La Librería, Madrid

Tardío J, Pascual H, Morales R (2005) Wild food plants traditionally used in the province of Madrid. Econ Bot 59(2):122–136

Tardío J, Pardo-de-Santayana M, Morales R (2006) Ethnobotanical review of wild edible plants in Spain. Bot J Linn Soc 152(1):27–72

Velasco JM, Criado J, Blanco E (2010) Usos tradicionales de las plantas en la provincia de Salamanca. Diputación de Salamanca, Salamanca

Vieira-Odilon L, Vibrans H (2001) Weeds as crops: the value of maize field weeds in the Valley of Toluca, Mexico. Econ Bot 55(3):426–443

Villar L, Palacín JM, Calvo C, Gómez García D, Montserrat G (1987) Plantas medicinales del Pirineo Aragonés y demás tierras oscenses. CSIC y Diputación de Huesca, Huesca

Chapter 5
Natural Production and Cultivation of Mediterranean Wild Edibles

María Molina, Manuel Pardo-de-Santayana and Javier Tardío

5.1 Introduction

As discussed in previous chapters, foraging wild plants for food is still a popular activity in rural contexts of the Mediterranean region (Hadjichambis et al. 2008). Although it is no longer a subsistence practice, recreational harvesting for domestic consumption currently attains traditional and new collectors such as retired people and Sunday excursionists from urban areas, and a renewed interest for commercial harvesting is arising (Molina et al. 2012). Green wild vegetables are frequently sold in Italy, Greece, and Croatia (D'Antuono and Lovato 2003; Łuczaj et al. 2012). For instance, the vegetable mix called *mišanca/pazija* in Croatia, containing several wild species (Fig. 5.1), is currently sold in every market of the Dalmatian coast (di Tizio et al. 2012; Łuczaj et al. 2013). In Spain, gourmet liqueurs and marmalades made from wild fruits are sold in some street markets and shops (Pardo-de-Santayana et al. 2010), and wild vegetables can be occasionally found in local markets and restaurants (Parada et al. 2011; Tardío 2010). Even avant-garde restaurants, such as the Danish restaurant "Noma" in Copenhagen, considered one of the best restaurants in the world nowadays, offers a very wide selection of wild food plants (See Chap. 3). As in the case of mushrooms (de Frutos et al. 2009; Martínez de Aragón et al. 2007), the demand for some wild plants seems to be increasing, becoming an economically profitable activity (Łuczaj et al. 2012).

M. Molina (✉) · M. Pardo-de-Santayana
Departamento de Biología (Botánica), Facultad de Ciencias, Universidad Autónoma de Madrid (UAM), Madrid, Spain
e-mails: m.molina.simon@gmail.com; manuel.pardo@uam.es

J. Tardío
Departamento de Investigación Agroalimentaria, Instituto Madrileño de Investigación y Desarrollo Rural, Agrario y Alimentario (IMIDRA), Alcalá de Henares, Madrid, Spain
e-mail: javier.tardio@madrid.org

M. de C. Sánchez-Mata, J. Tardío (eds.), *Mediterranean Wild Edible Plants*,
DOI 10.1007/978-1-4939-3329-7_5

Fig. 5.1 *Left and right,* vegetable mixes called *mišanca/pazija,* with several species collected from the wild, are currently sold in some markets of Dalmatia, Croatia. (Photographs by Łukasz Łuczaj; with permission)

The present tendency towards the recovery of food traditions and the need for product diversification may also offer opportunities for new crop domestication (D'Antuono and Lovato 2003; Egea-Gilabert et al. 2013). Native species of wild fruits and vegetables often represent unexploited resources that can promote a healthier and more diverse diet (Sánchez-Mata et al. 2012). In this way, there is a large set of wild-growing species used in Mediterranean traditional cuisines that deserves more attention. The cultivation of plants already known and traditionally used for food may represent alternative crops for niche markets (Benincasa et al. 2007; D'Antuono et al. 2009).

Some wild edibles are either obtained from the wild or from cultivation. For instance, the golden thistle (*Scolymus hispanicus* L.; Fig. 5.2) has been subject to cultivation in the past (Hernández Bermejo and León 1994) but is currently a minor crop only cultivated in a few areas of southern Spain (Soriano 2010) and southern Italy (Laghetti 2009). It is also a culturally important species gathered from the wild in several Mediterranean countries such as Greece, Italy, Morocco, and Spain (Hadjichambis et al. 2008; Leonti et al. 2006; Nassif and Tanji 2013). Similarly, bladder campion (*Silene vulgaris* (Moench) Garcke) growing in agricultural areas is tolerated and gathered for domestic consumption, and it is cultivated in some home gardens in Spain (Alarcón 2013) and Italy (Laghetti et al. 1994).

We can also find examples of recent domestication processes in other traditional wild vegetables such as the rocket salads (*Eruca vesicaria* (L.) Cav., and *Diplotaxis tenuifolia* (L.) DC.) and watercress (*Rorippa nasturtium-aquaticum* Hayek). In these "new" vegetables, the favorable combination of positive experience (sensory component of acceptance) and information (local gastronomy, health promotion) has contributed to successfully spreading their use (D'Antuono et al. 2009).

Despite the economic and nutritional interest of wild edible plants, there is still a poor knowledge of their natural production and its agronomic potential (Molina et al. 2014). In this chapter, we firstly present a review of worldwide studies on availability and production of wild edible plants, both on their natural environments and also some cultivation assays with Mediterranean plants. Afterwards, we resume our investigations on wild food plants production in their natural habitats and under cultivation in central Spain, taking into account the edible parts that are traditionally collected and consumed.

Fig. 5.2 Commercial cultivation of golden thistle in southern Spain by a local agricultural cooperative. Seedbed (**a**), crop plants before harvesting (**b**), harvested and peeled midribs, being cleaned in water tanks, where they remain turgid (**c**), edible plant material ready for commercialization (**d**). (Photographs courtesy of Centro de Conservación de Recursos Fitogenéticos del Instituto Nacional de Investigaciones Agrarias (CRF-INIA); with permission)

5.2 Studies on Availability and Production of Wild Edible Plants

5.2.1 Overview of Natural Production Studies

According to the great heterogeneity of wild edibles regarding life and growth forms, natural habitats, and edible parts, a broad range of plant yield rates and natural supplies of these species should be expected. Wild edible plants include fruit-tree species, tubers, edible young shoots, and leafy greens, among others. They can be found in a wide diversity of habitats, ranging from forests to human-disturbed areas, such as those from the agricultural landscape. Local supply and harvesting impact may also differ depending on life and growth forms, distribution areas, or parts used.

There are some studies that aim to evaluate quantitatively the natural availability of wild edible plants. The results of some of them are shown in Table 5.1. We can find examples from Mexico (Farfán et al. 2007; González-Amaro et al. 2009; González-Insuasti et al. 2008; Pérez-Negrón and Casas 2007), Argentina (Díaz-Betancourt et al. 1999; Ladio and Rapoport 2005; Rapoport 1995), North America (Kerns et al. 2004; Lepofsky et al. 1985; Murray et al. 2005), Finland (Ihalainen et al. 2003; Miina et al. 2009; Turtiainen et al. 2011), or South Africa (Youngblood 2004).

Table 5.1 Natural production of some wild vegetables and fruits in non-Mediterranean territories

Species	Production ($kg\ ha^{-1}$)[a]	Site	Source
Wild vegetables			
Amaranthus hybridus L.	45	Santiago Quiotepec, Mexico	Pérez-Negrón and Casas 2007
Amaranthus hybridus L.	3000	Tlaxcala, Mexico	González-Amaro et al. 2009
Brassica rapa L.	600	Tlaxcala, Mexico	González-Amaro et al. 2009
Calandrinia micrantha Schltdl.	400	Tlaxcala, Mexico	González-Amaro et al. 2009
Chenopodium berlandieri Moq.	700	Tlaxcala, Mexico	González-Amaro et al. 2009
Claytonia perfoliata Donn ex Willd.[b]	11000	Bariloche, Argentina	Díaz-Betancourt et al. 1999
Malva parviflora L.	400	Tlaxcala, Mexico	González-Amaro et al. 2009
Portulaca oleracea L.	4.6	Santiago Quiotepec, Mexico	Pérez-Negrón and Casas 2007
Wild fruits			
Berberis buxifolia Lam.	280	Bariloche, Argentina	Ladio and Rapoport 2005
Empetrum nigrum L.	6.5–12.4	Mackenzie River Delta Region, Canada	Murray et al. 2005
Prunus serotina Ehrh.	1436	Monarch Butterfly Biosphere Reserve, Mexico	Farfán et al. 2007
Rosa rubiginosa L.	2000	Bariloche, Argentina	Ladio and Rapoport 2005
Rubus chamaemorus L.	0.9	Central Finland	Raatikainen et al. 1984
Rubus chamaemorus L.	0-12.3	Mackenzie River Delta Region, Canada	Murray et al. 2005
Rubus idaeus L.	0.2	Central Finland	Raatikainen et al. 1984
Vaccinium microcarpon L.	1.3	Central Finland	Raatikainen et al. 1984
Vaccinium myrtillus L.	22.3	Finland	Turtiainen et al. 2011
Vaccinium oxycoccos L.	2	Central Finland	Raatikainen et al. 1984
Vaccinium uliginosum L.	0.7	Central Finland	Raatikainen et al. 1984
Vaccinium uliginosum L.	0.8–7.5	Mackenzie River Delta Region, Canada	Murray et al. 2005
Vaccinium vitis-idaea L.	22.7	Finland	Turtiainen et al. 2011

[a] Some original data were expressed as $t\ ha^{-1}$ or as $g\ m^{-2}$

[b] Syn. *Montia perfoliata*. Estimations were obtained from three consecutive cuts in 0.25-m^2 plots

Some of these studies assess the local supply and current demand of wild food resources. For instance, Farfán et al. (2007) documented an extraction of 7.47, 4.40, and 1.82 t of fruits per year of *Prunus serotina* Ehrh., *Rubus liebmanii* Focke, and *Crataegus mexicana* Moc. & Sessé ex DC. by the Mazahua indigenous community of Mexico. Bearing in mind the local availability of these species (Farfán 2001), the extraction rates were of 2.4%, 73.3%, and 5.4%, respectively. Among wild leafy greens, these authors documented some less variable extraction rates of 18.2%, 19.6%, and 13.2% in *Brassica campestris* L., *Amaranthus hybridus* L., and *Rorippa nasturtium-aquaticum* (L.) Hayek, respectively. Other studies conducted in the Tehuacán–Cuicatlán Valley in central Mexico also report important amounts of wild greens and cactus fruits (Pérez-Negrón and Casas 2007). According to these authors, the local supply of wild food plants widely meets their current demand and apparently does not endanger their natural populations.

The importance of the spontaneous weed vegetation in traditional agroecosystems is also an interesting example of the economic potential of weedy vegetables. Some studies demonstrate that the potential benefits from the weed vegetation are sometimes higher than those derived from the main crop (González-Amaro et al. 2009). For instance, useful spontaneous plants growing in a maize field and its margins at Tlaxcala, Mexico, produce 14.8 t ha^{-1} (fresh weight), being the forage species the major contributors (9.7 t ha^{-1}) together with wild edible herbs such as *A. hybridus, Chenopodium berlandieri* Moq., *Brassica rapa* L., *Malva parviflora* L., and *Calandrinia micrantha* Schltdl. The first two species are widely available in the supermarkets of Mexico city as well as in traditional weekly markets (Vieyra-Odilon and Vibrans 2001). However, maize grain (1.5 t ha^{-1}), although supposedly the main purpose of maize cultivation, is a minor contributor to total productivity and potential net return. Maize straw is generally overlooked but is worth considerably more (3.7 t ha^{-1} of dry weight) and gives a modest profit. This research suggests that the non-crop production can be economically highly important in traditional cropping systems as well as a risk mitigator to compensate crop failures.

Comparative studies were also performed between tropical and temperate areas of America in order to assess the potential amount of wild food provided by common weeds in urban and agricultural environments. Edible fresh biomass varies between 1277–3582 kg ha^{-1} in Coatepec, Mexico, and 287–2939 kg ha^{-1} in Bariloche, Argentina, with average values of 2.1 and 1.3 t ha^{-1}, respectively (Díaz-Betancourt et al. 1999). It suggests that tropical weeds are more productive than temperate weeds. Among the most profitable wild edible plants growing in Bariloche, the authors indicate that *Claytonia perfoliata* Donn ex Willd. (syn. *Montia perfoliata* (Donn ex Willd.) Howell), a North American invader of Patagonian urban forests, shows clear capabilities to recover after harvesting the aerial parts (Díaz-Betancourt et al. 1999; Rapoport et al. 1998). Overall, exotic plants were the major contributors to wild edible biomass in disturbed areas, both in terms of species richness and coverage. Among fruits, yields of 0.28 t ha^{-1} of the native species *Berberis buxifolia* Lam. and 2 t ha^{-1} of the exotic species *Rosa rubiginosa* L. were reported in the surroundings of Bariloche, covering 33 and 1.2% of the study area, respectively. They are currently consumed by 10 and 20%, respectively, of the population at this site (Ladio and Rapoport 2005).

Regarding commercial wild berries, such as bilberry (*Vaccinium myrtillus* L.) and cowberry (*V. vitis-idaea* L.), many studies on fruit production have been carried out for decades (e.g., Raatikainen et al. 1984; Rossi et al. 1984). Mathematical models were also developed for predicting berry yields on national scales by means of forest stands data (Ihalainen et al. 2003; Miina et al. 2009). Bilberry is one of the economically most important wild fruit species in Finland, Sweden, and Norway (Turtiainen et al. 2011). Picking wild berries in the Nordic countries has been a popular traditional household and recreational activity, providing important additional income in some areas. Nowadays, approximately 60% of the Finnish population participates in berry picking every year (Turtiainen et al. 2011). According to recent studies, Finnish annual berry production varies from 92 to 312 million kg (22.3 kg ha^{-1} on average) of bilberry and from 129 to 386 million kg (22.7 kg ha^{-1}) of cowberry (Turtiainen et al. 2011). Estimates from 1997 to 1999 indicate that approximately 5–6% and 8–10% of the total production of bilberries and cowberries, respectively, were collected in Finland, although it can be presumed that commercial wild berry picking after the phenomenon of foreign pickers has so far affected current utilisation rates of wild berries (Turtiainen et al. 2011).

Other studies also estimate annual yields of wild fruits, such as those carried out in North America (Murray et al. 2005) on blueberry (*Vaccinium uliginosum* L.), cloudberry (*Rubus chamaemorus* L.), and crowberry (*Empetrum nigrum* L.) and the works conducted in central Finland (Raatikainen et al. 1984) on black cowberry (*Empetrum* spp.), blueberry (*V. uliginosum*), raspberry (*Rubus idaeus* L.), cranberry (*V. oxycoccos* L. and *V. microcarpon* L.), and cloudberry (*R. chamaemorus*), among others.

Research on yield rates of edible wild bulbs and tubers of *Cyperus usitatus* Burch. ex Roem. & Schult., *Albuca canadensis* (L.) F.M.Leight., *Pelargonium sidoides* DC., and *Talinum caffrum* (Thunb.) Eckl. & Zeyh. were performed in South Africa (Youngblood 2004).

In the Mediterranean region, some studies on the availability and yield of wild food species have recently been conducted by our research group in central Spain (Dávila 2010; Molina et al. 2011, 2012, 2014; Polo et al. 2009; Tardío et al. 2011). However, as far as we know, research on this topic has been poorly addressed regarding Mediterranean wild edibles, despite its interest to design sustainable strategies of resource management and to promote environmentally friendly extraction practices for the use and commercialization of wild vegetables and fruits.

5.2.2 Some Cultivation Assays on Mediterranean Wild Plants

Cultivation assays of traditional wild edible plants are generally scarce, and some local experiences are only available on national publications, not accessible to the international scientific community. However, some morphological, agronomical, and/or biochemical analysis of several Mediterranean wild food plants have been recently conducted in Italy, Spain, Turkey, and Tunisia, including wild vegetables such as bladder campion, purslane, wild asparagus, and fleshy fruited species such as strawberry tree. These studies aim to assess the agronomic potential of traditional wild edible plants, identifying germplasm accessions with the most interesting phenotypic and nutritional qualities suited for breeding programs to develop commercial cultivars for the agricultural market (Egea-Gilabert et al. 2013). According to these studies, some wild species are considered promising new foods with possibilities for marketing as high-quality, minimally processed products.

Bladder campion

Bladder campion (*Silene vulgaris*) is one of the most appreciated leafy vegetables in the traditional gastronomy of many Mediterranean countries. This species also exhibits a good nutritional potential according to nutritional analysis performed on plant material growing wild (Morales et al. 2012a, b; Sánchez-Mata et al. 2012) and cultivated (Alarcón et al. 2006; Egea-Gilabert et al. 2013). For these reasons, it is considered an attractive candidate for cultivation and commercialization as a ready-to-eat product.

Cultivation experiments of this species were conducted in Spain, including outdoor field trials (Fernández and López 2005; García and Alarcón 2007) and indoor greenhouse cultivation, both in soil-based culture (Alarcón et al. 2006; Arreola et al. 2006; Franco et al. 2008) and in a hydroponic floating system (Egea-Gilabert et al. 2013). Agronomical, morphological, and/or nutritional parameters were evaluated, in some cases under organic production systems (Alarcón et al. 2006). Average crop yields of 3400 g m^{-2} were obtained at low labor requirements, including hand weeding and irrigation (Fernández and López 2005). It is considered a seasonal crop in which several harvests can be performed throughout the year, with a pick of production in spring. Harvest takes place before flowering, when the plants have four to six leaves. The tender aerial parts are cut, leaving the roots intact and allowing plant regeneration (Fernández and López 2005).

Since bladder campion shows high inter- and intra-population genetic variation, accessions with desirable agronomical traits for marketing, such as a large leaf blade width/leaf length ratio, short internode length, late flowering, intense green color of the leaves, and of course, high yield, could be selected to meet commercial size and quality standards (Egea-Gilabert et al. 2013; García and Alarcón 2007). In addition, accessions with a high level of nutritional compounds (glutathione, total phenols, and antioxidant capacity) and low concentrations of antinutritional compounds are desired (Alarcón et al. 2006; Conesa et al. 2009; Egea-Gilabert et al. 2013). The use of vegetative reproduction can be the starting point to the cultivation of the most interesting genotypes (Alarcón 2013). Other issues of interest for its cultivation were studied, such as the influence of fertilization strategies on yield (Arreola et al. 2004) and the influence of nursery irrigation regimes on vegetative growth and root development, assessing its potential as a crop for semiarid conditions (Arreola et al. 2006; Franco et al. 2008).

Purslane

Purslane (*Portulaca oleracea*) also has a long history of use for human food. Some varieties such as Golden Gerber, Garden (The Netherlands), and Golden (England) are commercially grown. It is considered a minor crop in the USA, and it is also cultivated on a small scale in France and Holland, whereas in other parts of the world, it is regarded as a weed (Cros et al. 2007). The interest in cultivating this annual species has grown in the last decades since it is considered an exceptionally rich source of bioprotective nutrients, particularly of ω-3 fatty acids and antioxidants (Dkhil et al. 2011; Mortley et al. 2012; Simopoulos et al. 1995).

In the Mediterranean region, some cultivation experiments have been conducted in Spain (Cros et al. 2007; Franco et al. 2011) and Italy (Gonnella et al. 2005) to grow common purslane for its interest as a baby-leaf vegetable or as a component of mixed ready-to-use vegetables. Plants grown in a hydroponic floating system and harvested at the five-leaf-pair stage produced yields of 1800–2200 g m^{-2} of fresh weight (Cros et al. 2007). Different substrates were tested, and peat was found to be the best one for floating system culture according to yield and fatty acid content parameters (Cros et al. 2007). Considerably higher yields of 9–15 kg m^{-2} were recorded by Gonnella et al. (2005) for purslane under floating system cultivation. In this case, a considerable portion of shoots, made up of leaves and succulent stems, was harvested.

As happens in other leafy vegetables, acceptable contents of nitrate and oxalates are required to meet quality standards. This topic has been extensively studied in purslane grown in soilless culture systems (Fontana et al. 2006; Palaniswamy et al. 2002, 2004) as well as the influence of harvest intervals on fatty acid content (Mortley et al. 2012). Seed germination methods were also analysed (Fernández et al. 2008). Since this species is moderately tolerant to salinity and drought, it is considered a promising vegetable for agriculture in dry areas or areas where the irrigation water contains high salt concentrations (Franco et al. 2011; Teixeira and Carvalho 2009; Yazici et al. 2007).

Wild Asparagus

Another traditional vegetable with a great potential to become a new crop is wild asparagus (*Asparagus acutifolius*). In this case, cultivation experiments (Fig. 5.3) were conducted in Italy to provide suitable field techniques that support recent attempts of farmers for producing spears from this perennial Mediterranean shrub (Benincasa et al. 2007; Rosati et al. 2005). Since the wild asparagus market already exists in several Mediterranean countries, cultivation may increase the limited availability from the wild. Additionally, the frugality of the wild asparagus allows for a crop virtually free of pests and diseases, perfectly suited for organic or any other natural farming techniques (Aliotta et al. 2004; Benincasa et al. 2007).

According to Benincasa et al. (2007), average spear weight varied from 5.6 to 6 g in mature plants that are 3 years old, and 4–8 spears per plant were collected at regular intervals from March-April to May, reaching average crop yields of 1000–1400 kg ha^{-1} and 2000 kg ha^{-1} from the best plots (crop density of 33,000 plants ha^{-1}). Similar results were obtained by Rosati et al. (2005). In contrast to cultivated asparagus (*A. officinalis* L.), the evergreen and prickly vegetation of wild asparagus poses an obstacle to fast and comfortable harvesting. According to these authors, cutting the vegetation to ease harvest reduces harvest labor by approximately half but may reduce plant vigor and yield and compromise plant

Fig. 5.3 a–d Cultivation experiments of *Asparagus acutifolius* at the Instituto de Horticultura Pontacagnano of Salerno, Italy. Nursery production (**a**) and adult asparagus plants (**b**). Wild asparagus cultivated under olive trees: initial stage (**c**) and adult plants (**d**). (Photographs by Adolfo Rosati; with permission)

longevity (Rosati et al. 2005). However, they hypothesize that better irrigation and nutrition as well as interrupting harvest earlier should allow the plant to recover from the stress of having to replace the evergreen vegetation. The spears could also be directly harvested by the consumers in pick-your-own operations associated with tourism in order to reduce harvest costs (Benincasa et al. 2007).

Since wild asparagus has similar ecological requirements to olive trees, and it grows under both sun and shade expositions, its cultivation as an understory crop could provide economic sustainability in traditional olive cultivation, with no detriment on the yield of either crop (Rosati 2001; Rosati et al. 2009). It would ease pressure on forests and help control soil erosion on agricultural lands.

Strawberry Tree

Some studies have also been conducted in fruit-tree species, although to a lesser extent. For instance, morphological and genetic studies of strawberry tree (*Arbutus unedo*) and breeding programs have been carried out in Italy (Mulas et al. 1998; Mulas and Deidda 1998), Turkey (Celikel et al. 2008), and Tunisia (Takrouni and Boussaid 2010) with the aim of promoting extensive cultivation and preventing deforestation and over-collecting of natural stands. Other candidates for cultivation include *Myrtus communis* for the liqueur industry (Mulas et al. 1998).

5.3 Natural Production of Wild Edibles in Central Spain

Research on natural production and availability of wild edible plants was performed by our research group in central Spain (Molina et al. 2011, 2012, 2014; Molina 2014; Tardío et al. 2011). Field work was conducted during the years 2007–2009 in several outlying villages around Madrid city. We estimated the production per plant (gram of edible fresh matter) and the potential food supplies (production per hectare) of some wild vegetables and fruits traditionally consumed in Spain and other Mediterranean countries. For each species, the sampling areas were limited to the specific places where these plants spontaneously grow. Since plant production is widely influenced by ecological and climatic factors, yield average values were calculated from different data sets obtained during 2 or 3 consecutive years and from two different sites. A detailed description of methodological procedures can be found in the aforementioned references.

5.3.1 Wild Vegetables Growing in Human-Disturbed Areas

Wild vegetables constitute the most important group of wild food plants harvested in the Mediterranean countries according to several ethnobotanical studies (Ghirardini et al. 2007; Nassif and Tanji 2013; Tardío et al. 2006; see also Chap. 4). A remarkable percentage of these leafy vegetables are weeds or ruderal species growing in human-disturbed areas such as agricultural fields, orchards, pastures, fallow lands, vacant lots, and roadsides (Fig. 5.4).

As shown in Table 5.2, some weedy vegetables are very productive, reaching average yields of 260–280 g per plant of sea beet (*Beta maritima* L.) and fennel (*Foeniculum vulgare* Mill.) and 120–130 g per plant of chicory (*Cichorium intybus* L.) and bugloss (*Anchusa azurea* Mill.). Other species show individual plant yields lower than 60 g per plant, such as those of skeleton weed (*Chondrilla juncea* L.), field poppy (*Papaver rhoeas* L.), sow thistle (*Sonchus oleraceus* L.), and dandelion (*Taraxacum obovatum* (Willd.) DC.). Production varied from 50 to 100 g per plant in golden thistle and milk thistle (*Silybum marianum* (L.) Gaertn.), respectively, out of which only the midribs of the basal leaves are eaten. In wild leek (*Allium ampeloprasum* L.), the edible portion formed by the bulb and the pseudostem of overlapping leaves yields approximately 14 g per plant.

Most of the aforementioned species were grouped as non-clonal plants since the aerial parts can be assumed to have developed from a single "rooted unit". Other species that have branching rhizomes, such as sorrel (*Rumex papillaris* Bois. & Reut.) or fiddle dock (*Rumex pulcher* L.), and stoloniferous stems, such as bladder campion, were considered clonal species. They usually grow forming clumps; thus, dense rosettes or patches may have originated from one or more "rooted unit". Yields of clonal species presented in Table 5.2 were referred as gram per 20×20-cm quadrat (0.04 m^2). It represents a plant–unit surface comparable to the surface occupied by the basal rosette of the non-clonal species. Among clonal species, yields

Table 5.2 Natural productions of wild vegetables growing in human-disturbed areas (mean ± standard error). (Data source: Molina et al. 2014)

Species	Parts used	Total yield	Units
Non-clonal species			
Allium ampeloprasum	Bulb and pseudostem	13.8 ± 0.6	g/plant
Anchusa azurea	Basal leaves	117.1 ± 8.4	g/plant
Beta maritima	Basal leaves	284.4 ± 24.3	g/plant
Chondrilla juncea	Basal leaves	30.1 ± 2.7	g/plant
Cichorium intybus	Basal leaves	130.7 ± 11.2	g/plant
Foeniculum vulgare	Young leaves and stems	261.7 ± 21.1	g/plant
Papaver rhoeas	Young leaves and stems	58.4 ± 6.1	g/plant
Scolymus hispanicus	Midribs of basal leaves	52.7 ± 4.9	g/plant
Silybum marianum	Midribs of basal leaves	117.1 ± 8.4	g/plant
Sonchus oleraceus	Young leaves and stems	28.3 ± 2.5	g/plant
Taraxacum obovatum	Basal leaves	12.7 ± 0.8	g/plant
Clonal species[a]			
Rumex papillaris	Basal leaves	98.3 ± 4.0	g/quadrat
Rumex pulcher[b]	Basal leaves	86.1 ± 5.1	g/quadrat
Silene vulgaris[c]	Young leaves and stems	20.4 ± 0.8	g/quadrat

[a] In clonal species, yields are referred as gram per 20 × 20-cm quadrat that represent a plant–unit surface comparable to the surface occupied by the basal rosette of the non-clonal species

[b] *Rumex pulcher* subsp. *pulcher*

[c] *Silene vulgaris* subsp. *vulgaris*

Fig. 5.4 Collecting *Allium ampeloprasum* (**a, b**) and *Silybum marianum* (**c, d**) in the borders of pathways (**a**) and in fallow lands (**c**), respectively. (Photographs by María Molina)

Fig. 5.5 Collecting *Apium nodiflorum* (**a, b**) and *Montia fontana* (**c, d**) in irrigation channels nearby farmlands (**a**) and in swampy areas (**c**), respectively. (Photographs by María Molina)

of 85–100 g per quadrat of sorrel and fiddle dock could be obtained, whereas lower production rates were found in bladder campion (20 g per quadrat).

5.3.2 Wild Vegetables Growing in Aquatic Environments

Some wild vegetables occur in damp places, frequently in water, that is, in springs, streams, moist pastures, irrigation channels nearby farmlands, and/or swampy areas. This is the case of fool's water-cress (*Apium nodiflorum* (L.) Lag.) and water blinks (*Montia fontana* L.), as shown in Fig. 5.5. Both species develop branching stems, sometimes prostrate and rooting, and they usually grow forming clumps. The aerial parts, including young leaves and stems, are consumed. As can be seen in Table 5.3, natural production rates oscillate between 100–150 g per quadrat.

Table 5.3 Natural productions of wild vegetables growing in aquatic environments (mean ± standard error). (Data source: Tardío et al. 2011; Molina et al. 2014)

Species	Parts used	Total yield[a]	Units
Apium nodiflorum	Young leaves and stems	152.5 ± 11.5	g/quadrat
Montia fontana[b]	Young leaves and stems	105.7 ± 5.1	g/quadrat

[a] Yields are referred as gram per 20 × 20-cm quadrat

[b] *Montia fontana* subsp. *amporitana* Sennen.

Fig. 5.6 Collecting young sprouts of *Asparagus acutifolius* (**a, b**) and *Tamus communis* (**c, d**) in Mediterranean holm oak forests (**a**) and in the wooded edge margins of pasturelands (**c**), respectively. (Photographs by María Molina)

5.3.3 Edible Young Sprouts Growing in Forestlands

Apart from the so-known wild asparagus (*Asparagus acutifolius* and other species of the same genus), there are other climbing plants whose young sprouts are traditionally collected in the Mediterranean region, such as red bryony (*Bryonia dioica* Jacq.), black bryony (*Tamus communis* L.), and hop (*Humulus lupulus* L.). They can be found in forested areas including Mediterranean sclerophyllous and deciduous forests, riverbank forests, and/or wooded edge margins of croplands and pastures (Fig. 5.6). The apical parts of their young stems (spears) are generally consumed cooked.

As shown in Table 5.4, average spear weight barely varied from 1.5 to 2.8 g in these species. However, total yields depend not only on spear weight but on the spear number as well. Apart from the spears that progressively emerge from the subterranean organs during the growing season, the spears that originated from lateral buds when the apical bud is removed can also be collected in red bryony and hop, increasing its productivity. Taking into account the total number of spears and their weight, individual plant yields of 8 g per plant were obtained in the wild asparagus, whereas the production of red bryony was five times higher (40 g per plant). Individual plant yields could not be measured in black bryony. Nevertheless, according to personal observations, the spears of black bryony that originated from lateral buds when the apical one is cut are very thin and they do not achieve harvestable sizes. Thus, harvestable new spears could only be originated from the tuber.

Table 5.4 Natural productions of edible wild sprouts collected in forested areas (mean ± standard error). (Data source: Molina et al. 2012; Molina 2014)

Species	Spear weight (g)	Spear number	Units	Total yield	Units
Asparagus acutifolius	2.76 ± 0.09	2.96 ± 0.25	Spears/plant	8.15 ± 0.76	g/plant
Bryonia dioica	1.76 ± 0.02	23.60 ± 2.25	Spears/plant	40.67 ± 4.74	g/plant
Humulus lupulus	1.69 ± 0.05	63.83 ± 8.31	Spears/m^2 of plant	79.3 ± 14.7	g/m^2 of plant
Tamus communis	1.58 ± 0.07	N/A[a]	–	N/A	–

[a] *N/A* not available

Finally, in the clonal species *H. lupulus,* which usually propagates by vegetative growth by means of branching rhizomes, approximately 79.3 g m^{-2} of spears could be obtained in the natural patches of vegetation formed by hop.

5.3.4 *Wild Fruits Growing in Forestlands*

Wild fruits are also an interesting food resource of Mediterranean forestlands. Perennial species with edible fruits can be found in sclerophyllous and deciduous forests, riverbank forests, and/or wooded edge margins of croplands and pastures. Apart from dry fruits and seeds, such as acorns, chestnuts, hazelnuts, and pine nuts, there is a large set of fleshy fruits traditionally consumed in the Mediterranean region (Fig. 5.7).

The natural production rates of some of them are presented in Table 5.5. For instance, average fruit weight of strawberry tree and hawthorn (*Crataegus monogyna* Jacq.) was 3.69 g and 0.36 g, respectively. A total of 4–4.4 kg per tree could be obtained in these species. Other plants, such as blackberry (*Rubus ulmifolius* Schott),

Fig. 5.7 Wild fleshy fruits of *Arbutus unedo* (left) and *Crataegus monogyna* (right) growing as understory species in Mediterranean forests. (Photographs by María Molina)

Table 5.5 Natural productions of wild fruits growing in forested areas (mean ± standard error). (Data source: Molina et al. 2011; Molina 2014)

Species	Fruit weight (g)	Number of fruits	Units	Total yield	Units
Arbutus unedo	3.69 ± 0.08	1136 ± 150	fruits/tree	4.39 ± 0.63	kg/tree
Crataegus monogyna	0.36 ± 0.01	11,109 ± 1572	fruits/tree	4.00 ± 0.56	kg/tree
Rubus ulmifolius	1.01 ± 0.02	504 ± 51	fruits/m^2 of plant	0.51 ± 0.05	kg/m^2 of plant

produce 500 fruits m^{-2} of approximately 1 g each. Potential yields of 0.51 kg m^{-2} are expected in the natural patches of vegetation covered by this clonal species with branching rhizomes.

5.3.5 Potential Food Supplies of Wild Edibles

Taking into account individual plant yields (Tables 5.1–5.4) and plant density estimates (Molina 2014), the approximate figures of the potential food supplies that these species could offer in their natural habitats are presented in Fig. 5.8. Among leafy vegetables and sprouts, yields lower than 100 kg ha^{-1} were observed in *Scolymus hispanicus* (28 kg ha^{-1}) and *Sonchus oleraceus* (86 kg ha^{-1}) and *Asparagus acutifolius* (6 kg ha^{-1}), *Tamus communis* (13 kg ha^{-1}), and *Bryonia dioica* (91 kg ha^{-1}), respectively. Yields varied between 100 and 500 kg ha^{-1} in leafy vegetables such as *Allium ampeloprasum, Anchusa azurea, Chondrilla juncea, Cichorium intybum, Rumex papillaris,* and *Silene vulgaris;* sprouts such as those of *Humulus lupulus;* and fruit species such as *Arbutus unedo* and *Crataegus monogyna*. Other leafy vegetables such as *Beta maritima, Papaver rhoeas, Rumex pulcher*, and *Silybum marianum* obtained yields from 500 to 1000 kg ha^{-1}. The species which reached the highest yields were *Foeniculum vulgare* (1760 kg ha^{-1}), *Montia fontana* (2140 kg ha^{-1}), and *Rubus ulmifolius* (2416 kg ha^{-1}).

5.4 Cultivation Experiments on Wild Leafy Vegetables at IMIDRA (Madrid, Central Spain)

Cultivation experiments aimed to explore the agronomic feasibility of some traditional leafy vegetables were also conducted by our research group at the Madrid Institute for Research in Food and Agriculture (IMIDRA) in Spain. Crop yields from five culturally important wild species were assessed: skeleton weed (*Chondrilla juncea*), chicory (*Cichorium intybus*), fiddle dock (*Rumex pulcher*), golden thistle (*Scolymus hispanicus*), and bladder campion (*Silene vulgaris*).

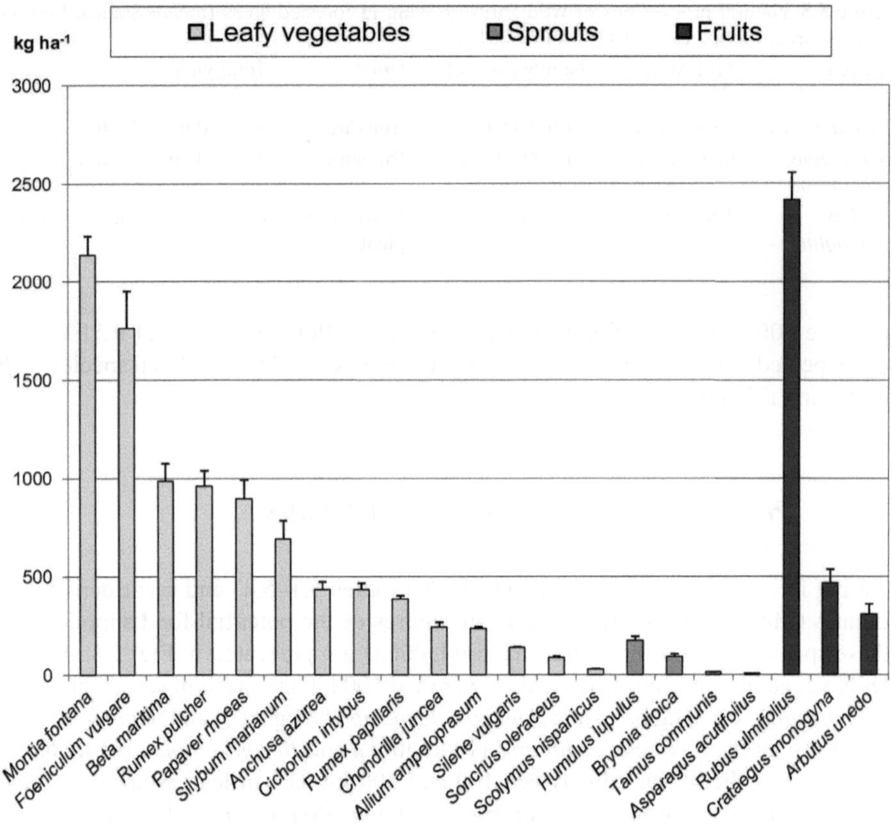

Fig. 5.8 Potential food supplies estimations of some traditional wild edible plants in the Mediterranean region. Yield rates are exclusively referred to the specific natural habitats where each plant occurs spontaneously (kg ha^{-1}; mean ± standard error). (Data source: Molina et al. 2011, 2012, 2014; Molina 2014; Tardío et al. 2011)

5.4.1 Cultivating Wild Vegetables

Seeds from two different sites (two accessions per species) were collected in summer 2009 and sown at the end of winter 2010 in a nursery. Jiffy pots located in plastic trays were sown with several seeds and thinned, leaving one plant per pot for being transplanted to the field in spring at the end of March (Fig. 5.9). We used a randomized design with four replications. The 2.40 × 0.8-m (1.92 m^2) elementary plot included 24 plants, placed in two rows separated 40 cm and 20 cm within them, with corridors of 2.6 m width among the four blocks (Fig. 5.9). Labor requirements were limited to occasional drip irrigation at the driest periods and hand weeding.

Due to the late nursery and transplant in 2010, crop yields of these five perennial species were measured during the second and third year after the plantation (2011–2012). In general, 5 plants per plot, that is, 40 plants per species (5 plants × 4 blocks

Fig. 5.9 Cultivation experiments at Madrid Institute for Research in Food and Agriculture (IMIDRA), Spain. Seedling of *Silene vulgaris* ready for being transplanted to the field (**a**). Randomized block design with four replicates was employed (**b**). Individual plot of *Cichorium intybus* (**c**). Harvesting 20 × 20 cm quadrats of *Silene vulgaris* (**d**). (Photographs by Javier Tardío)

× 2 accessions), were harvested, and the fresh edible part was immediately weighed. In the clonal species *Silene vulgaris,* we collected the edible plant material of five 20 × 20-cm quadrats. The number of harvest episodes per year varied depending on the capacity of the species for regrowth after harvesting and the annual weather conditions. Harvest began in spring, and in some species, such as *Cichorium intybus, Rumex pulcher,* and *Silene vulgaris,* a maximum of two episodes of harvest in springtime and one in autumn were performed. In *Chondrilla juncea* and *Scolymus hispanicus,* only one harvest in spring was conducted.

5.4.2 Production Under Cultivation

As shown in Table 5.6, most of the selected species showed crop yields around 5000–7000 kg ha^{-1} $year^{-1}$, except *Chondrilla juncea* (1674 kg ha^{-1} $year^{-1}$). *Scolymus hispanicus* stood out for its high yield per plant (279 g on average), whereas *Silene vulgaris* showed the highest degree of tolerance to be harvested several times per year (2.5 harvest episodes per year on average). Overall, *Cichorium intybus* obtained the highest yields because of the combination of high rates of production per plant (159 g on average) and high tolerance to be harvested several times throughout the year (two harvest episodes per year on average).

Table 5.6 Crop yields of five wild leafy vegetables under experimental culture conditions during 2011–2012 (mean ± standard error)

Species	g/plant-quadrat	kg/ha[a] per harvest	Number of harvests	Total yield (kg/ha[a])
Chondrilla juncea	75.9±7.54	1674±333	1.0±0	1674±333
Cichorium intybus	159.0±6.15	3508±271	2.0±0.6	7016±543
Rumex pulcher[b]	127.5±5.51	2813±243	1.8±0.3	4923±425
Scolymus hispanicus	279.0±17.73	6155±782	1.0±0	6155±782
Silene vulgaris[c]	60.5±2.03	2668±90	2.5±0.3	6670±225

[a] Considering the whole surface of the experiment

[b] *Rumex pulcher* subsp. *pulcher*

[c] *Silene vulgaris* subsp. *vulgaris*. In this species, yield refers to a 20 × 20-cm quadrat

According to other cultivation experiments previously mentioned (Fernández and López 2005), *Silene vulgaris* can hold a greater number of harvest episodes per year, and consequently, crop yields could be almost duplicated. It could also be possible in other species such as *Cichorium intybus,* although our experience indicates that an excessive harvest could degenerate the culture prematurely since the plants become exhausted due to a repeated collection of the same individuals. In our opinion, the yields of *Chondrilla juncea* could be considerably increased in better environmental conditions. Our experiments were performed on clay loam soil, developing small plants, but higher yields would presumably be obtained on sandy soils according to the specific soil preferences of this species.

5.4.3 Comparison with Natural Production

Crop yields of these five traditional wild vegetables can be compared with those obtained in their natural populations and previously presented in Table 5.3. Plant yields of cultivated versus growing wild plants are represented in Fig. 5.10. As shown in the figure, slightly higher figures of production per plant were obtained for *Cichorium intybus* and *Rumex pulcher* under culture in comparison with yield rates of these species growing wild. However, yields of cultivated *Scolymus hispanicus, Chondrilla juncea,* and *Silene vulgaris* duplicate or even triplicate its natural yield rates. As shown in Table 5.6, between 1.8 and 2.5 harvest episodes on average were performed in cultivated *C. intybus, R. pulcher,* and *S. vulgaris,* reaching total crop yields around 5000–7000 kg ha^{-1} year^{-1}. Overall, we can conclude that even under non-intensive culture conditions, comparable to those of organic farming, the yield rates of these wild vegetables can be considerably increased.

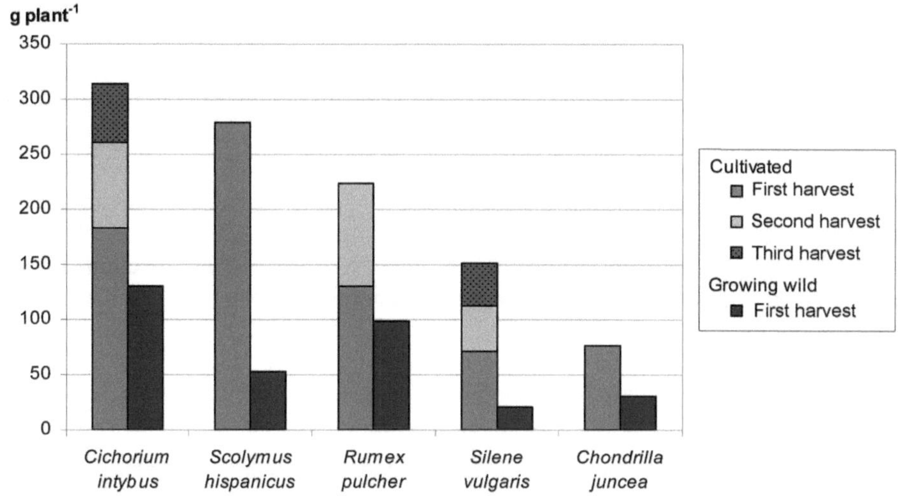

Fig. 5.10 Average plant production of five wild vegetables under culture and growing wild in g/plant, except for *Silene vulgaris,* in gram per 20 × 20-cm quadrat

5.5 Opportunities and Current Challenges for Gathering and Cultivating Wild Food Plants

5.5.1 Gathering in the Wild

The gathering of wild edible plants is a complementary food resource that also offers social benefits to local communities, such as contact with nature; reinforcement of social relationships and family ties; entertainment; physical, emotional, and spiritual well-being; revitalization of local identities and traditions; etc. (Emery et al. 2006; Menendez-Baceta et al. 2012). Therefore, the sustainable harvesting of wild food plants might be encouraged as a multifunctional use of biodiversity.

Most of the wild food plants traditionally consumed in the Mediterranean region are non-endangered species commonly found in disturbed areas and forestlands. Since the aerial parts of wild vegetables are collected before flowering and fruiting, intense harvesting might impact the abundance of their natural populations, especially in annual herbs. Nevertheless, as shown in Chap. 4 of this book, many of the wild vegetables are perennial herbs, in which their harvesting usually leaves the subterranean organs intact and prevents the plants from being exhausted due to a repeated collection of the same individuals. Even in the case of the bulbous plant *Allium ampeloprasum,* whose collection might be considered destructive, our experience shows that many small bulbs produced around the central bulb remain in the collecting place (Molina et al. 2014). Many other weedy vegetables are annual and fast-growing species that are never completely collected, allowing the self-regeneration of their populations. Neither does the extraction of wild fruits represent a significant threat to ensure the long-term maintenance of natural stands in widely distributed species.

According to our estimations, the yield rates of wild food species growing in their natural environments were considerably high in most cases. It suggests that the impact of harvesting might be relatively low and that it would be possible to increase this practice to develop alternative commercial products, as proposed by Farfán et al. (2007). As commented in Sect. 5.2.1, the extraction rates found by these authors in Mexico were in most cases lower than 20 % of the local availability.

Our results may indicate that the production of Mediterranean wild edible plants showed a great heterogeneity of yield rates depending on life and growth forms, distribution areas, or parts used. Our yield rates of wild vegetables (30–2140 kg ha^{-1}), edible sprouts (6–170 kg ha^{-1}), and fruit species (300–2400 kg ha^{-1}) were slightly lower than those reported in the bibliography of other non-Mediterranean territories (Table 5.1). These differences may be due to several reasons. One of them would be the specific morphological characteristics of each species. For instance, the edible leaves of *Montia fontana* are smaller (0.03–0.2 cm long) than those of the related species *Claytonia perfoliata* (0.5–4 cm long), which consequently obtained higher yields. However, we can also find remarkable intraspecific variation in several studies, even in the same country. As shown in Table 5.1, the production of *Amaranthus hybridus* varied from 45 kg ha^{-1} in the report of Pérez-Negrón and Casas (2007) to 3000 kg ha^{-1} in that of González-Amaro et al. (2009), both in Mexico. These differences might be due to ecological and climatic factors that can influence the local availability of wild species but also due to the application of different methodological procedures that should be taken into account for comparison purposes.

The potential food supplies obtained in our field studies were calculated from a limited number of localities of central Spain. Therefore, they obviously are neither representative of Spain nor of the Mediterranean region. We should also note that the yield rates shown in Fig. 5.8 exclusively refer to the habitats and places where each species occurs spontaneously. Although some species could be locally abundant, they are not necessarily common on a country basis. For instance, *Montia fontana* was one of the most productive species according to Fig. 5.8, but the aquatic environments where it naturally occurs do not cover large areas, and additionally, its distribution is mainly limited to noncalcareous soils. Other wild vegetables growing in human-disturbed areas and with no specific soil preferences such as *Cichorium intybus* or *Papaver rhoeas* are more widely distributed on a country basis.

The changes in land uses and management practices could affect the local availability of these species. According to local perceptions, the abundance of some wild edible plants have diminished in the past decades due to deforestation, urban spreading, the abandonment of traditional agricultural practices, and the use of modern agricultural practices such as deep ploughing and pesticide spraying (Celikel et al. 2008; Laghetti 2009; Polo et al. 2009; Takrouni and Boussaid 2010). Since the risk of contamination by car exhausts or pesticide spraying could compromise the safe consumption of wild edibles in some cases, weedy vegetables could be collected in organic farms in order to guarantee food safety and to increase crop profitability (Molina et al. 2014).

5.5.2 Cultivation

As we have previously pointed out, the cultivation of traditional wild food species could represent an interesting alternative for the most culturally appreciated species, particularly of those that are scarce on a country basis. Rather than forage them, some species could be intentionally grown in order to satisfy their demand. It would contribute to ensure food quality and to avoid the contamination risks that wild vegetables growing on roadsides and in agricultural areas are exposed to (Molina et al. 2014). Cultivation would also help to control the risk of parasites in some aquatic plants growing in areas frequented by cattle.

However, there are some problems that could limit the success of growing traditional wild vegetables. Firstly, one of the limitations that can be found is the low seed germination rates or even seed dormancy that appears in some species (Benincasa et al. 2007; Casco 2000). This is not a general problem in wild vegetables since most of them are weeds. However, there are some cases in which it is an important problem, such as in *Asparagus acutifolius,* whose seeds have a strong dormancy and do not germinate easily. In this species, pre-germination treatments and a suitable technique for production of transplants have been proposed (Conversa and Elia 2009; Rosati and Falavigna 2000). Vegetative propagation could also be appropriate for avoiding seed germination problems, besides maintaining favorable morphological features. This vegetative propagation can be carried out by means of stolons as in *Silene vulgaris* (Alarcón 2013), rhizomes as in *Ruscus aculeatus* and *Smilax aspera,* or tubers as in *Tamus communis* (D'Antuono and Lovato 2003). Since requirements of seeds and seedlings of many wild edible plants are still not well known, further research is needed.

Secondly, available information on cultural systems and techniques for commercial production of traditional wild vegetables is also scarce (Benincasa et al. 2007; Casco 2000). More research is needed to know which are the most suitable cultural systems, field techniques, substrates, irrigation systems, and labor requirements for growing these species according to their yield potential and their agronomical characteristics.

Thirdly, labor costs derived from cultivation of traditional wild edibles have been poorly assessed and deserve more attention. It should be taken into account that the cost of labor is today one of the main drivers of land use change in Europe since labor costs are sometimes higher than potential profits, and it has lead to the abandonment of agricultural lands in some territories (Rosati et al. 2009). In this way, some authors indicate that traditional wild vegetables could supply a specialized market in which high quality and product differentiation represent a competitive strategy that allows for a higher price. For instance, the spears of *Asparagus acutifolius* collected from the wild are sold at US$ 9–32 kg^{-1} in Italy, depending on season and market (Benincasa et al. 2007). Bunches of wild asparagus of about 250 g are sold at 5 € in some Spanish regions (Molina et al. 2012). According to Benincasa et al. (2007), it represents two to four times the price of the cultivated asparagus (*A. officinalis*). Agronomic trials conducted by these authors showed that harvest efficiency of *A. acutifolius* was approximately of 1.2 kg of spears per h of labor, and it increases to 3 kg per h when the prickly evergreen vegetation is previously cut and removed. In the latter case, harvest would cost approximately one third of the gross income of the crop, suggesting that the crop could easily be economically viable. Fernández and López (2005) estimated the labor costs, including hand weeding and harvesting costs, for *Silene vulgaris* as 18 € m^{-2}, and concluded that a minimum sale price of 7.5 € kg^{-1} would make the crop profitable. According to these authors, the young leaves and stems of *S. vulgaris* collected from the wild are sold in Spanish local markets at 6–8 € kg^{-1}. Apart from these and other limited examples, the economic viability of growing traditional wild edibles for the agricultural market has been poorly addressed.

5.6 Concluding Remarks

Wild edible plants are a significant food resource that has been widely underestimated, despite their interest in promoting food security and rural development. Particularly, the gathering of wild vegetables and fruits is deeply rooted in the Mediterranean food traditions, and, as some authors have suggested, the consumption of wild edibles have probably contributed to the so-called health benefits of the Mediterranean diet (Trichopoulou and Vasilopoulou 2000).

Although information on natural yield rates of Mediterranean wild edible plants is scarce, the available data presented in this chapter suggest that some of the wild food plants traditionally eaten in the Mediterranean region are abundant and productive species that could contribute to enhance food diversity in contemporary diets. More research is needed to have a wider characterization of their natural production and abundance. Besides, the main factors that could determine local supplies should also be addressed. For instance, urban spreading, changes in land uses and management practices, and deforestation could affect the local availability of these plants.

Apart from traditional harvesting for domestic consumption, some culturally important species could also be collected for commercial purposes, given the renewed interest for traditional food products. Although there are no available data on the quantities of wild food plants harvested in the Mediterranean region, results suggest that the impact of harvesting might be relatively low, and they could possibly tolerate higher extraction rates. Sustainable levels of harvest should be established, taking into account the edible parts used and their production rates as well as the specific habitats where they naturally occur. Moreover, safety issues may be addressed to ensure food quality, such as contamination risks derived from car exhausts or pesticide spraying.

Wild edible plants are a good reservoir of potential new crops too. Cultivation experiments could contribute to assess the agronomic potential of culturally appreciated wild species, especially of those that may be more prone to overexploitation. It would help to ease pressure on natural stands, providing new options for employment and income generation. Particularly, cultivation under organic conditions of wild vegetables might be an interesting alternative and a complementary economic resource for Mediterranean farmers. Since wild species are well adapted to local environments, they can ensure steady productions under adverse environmental conditions. Some of them are tolerant to drought and salinity, and they are considered promising vegetables for agriculture in marginal areas. Nowadays, health claims and product differentiation may represent successful commercial strategies for the cultivation of traditional wild edible plants (D'Antuono et al. 2009). Thus, domestication opportunities should also be considered, taking into account nutritional characteristics, cultural traits, and consumer acceptance.

References

Alarcón R (2013) Entre "malas hierbas", criptocultivos y plantas cultivadas: la colleja (*Silene vulgaris*). Ambienta 102:80–88

Alarcón R, Ortiz LT, García P (2006) Nutrient and fatty acid composition of wild edible bladder campion populations [*Silene vulgaris* (Moench.) Garcke]. Int J Food Sci Technol 41:1239–1242

Aliotta G, Aceto S, Farina A, Gaudio L, Rosati A, Sica M, Parente A (2004) Natural history, cultivation and biodiversity assessment of *Asparagus*. In: Research Advances in Agricultural and Food Chemistry. Global Research-Network, pp 1–12

Arreola J, Franco JA, Martínez-Sánchez JJ (2004) Fertilization strategies for *Silene vulgaris* (Caryophyllaceae) production, a wild species with alimentary use. Hortscience 39:796

Arreola J, Franco JA, Vicente MJ, Martínez S (2006) Effect of nursery irrigation regimes on vegetative growth and root development of *Silene vulgaris* after transplantation into semi-arid conditions. J Hortic Sci Biotechnol 81(4):583–592

Benincasa P, Tei F, Rosati A (2007) Plant density and genotype effects on wild asparagus (*Asparagus acutifolius* L.) spear yield and quality. Hortscience 42(5):1163–1166

Casco JC (2000) Trabajando juntos inventamos nuevas oportunidades de futuro. Ayuntamiento de Plasenzuela, Plasenzuela, Cáceres

Celikel G, Demirsoy L, Demirsoy H (2008) The strawberry tree (*Arbutus unedo* L.) selection in Turkey. Sci Hortic 118:115–119

Conesa E, Niñirola D, Vicente MJ, Ochoa J, Bañón S, Fernández JA (2009) The influence of nitrate/ammonium ratio on yield quality and nitrate, oxalate and vitamin C content of baby leaf spinach and bladder campion plants grown in a floating system. Acta Hortic 843:269–274

Conversa G, Elia A (2009) Effect of seed age, stratification, and soaking on germination of wild asparagus (*Asparagus acutifolius* L.). Sci Hortic 119(3):241–245

Cros V, Martínez-Sánchez JJ, Franco JA (2007) Good yields of common purslane with a high fatty acid content can be obtained in a peat-based floating system. HortTechnology 17(1):14–20

D'Antuono LF, Lovato A (2003) Germination trials and domestication potential of three native species with edible sprouts: *Ruscus aculeatus* L., *Tamus communis* L. and *Smilax aspera* L. Acta Hort (ISHS) 598:211–218

D'Antuono LF, Elementi S, Neri R (2009) Exploring new potential health-promoting vegetables: glucosinolates and sensory attributes of rocket salads and related *Diplotaxis* and *Eruca* species. J Sci Food Agric 89(4):713–722

Dávila P (2010) Estudio etnobotánico comparativo de la colleja [Silene vulgaris (Moench) Garcke] en dos localidades de la Comunidad de Madrid. Proyecto de Fin de Carrera. Universidad Autónoma de Madrid

de Frutos P, Martínez-Peña F, Ortega-Martínez P, Esteban S (2009) Estimating the social benefits of recreational harvesting of edible wild mushrooms using travel cost methods. Inv Agrar-Sist Rec F 18(3):235–246 di Tizio A, Łuczaj ŁJ, Quave CL, Redžic S, Pieroni A (2012) Traditional food and herbal uses of wild plants in the ancient South-Slavic diaspora of Mundimitar/Montemitro (Southern Italy). J Ethnobiol Ethnomed 8:21

Díaz-Betancourt M, Ghermandi L, Ladio AH, López-Moreno IR, Raffaele E, Rapoport EH (1999) Weeds as a source for human consumption. A comparison between tropical and temperature Latin America. Rev Biol Trop 47(3):329–338

Dkhil MA, Moniem AEA, Al-Quraishy S, Saleh RA (2011) Antioxidant effect of purslane (*Portulaca oleracea*) and its mechanism of action. J Med Plants Res 5(9):1589–1563

Egea-Gilabert C, Niñirola D, Conesa E, Candela ME, Fernández JA (2013) Agronomical use as baby leaf salad of *Silene vulgaris* based on morphological, biochemical and molecular traits. Sci Hortic 152:35–43

Emery M, Martin S, Dyke A (2006) Wild harvests from Scottish woodlands: social, cultural and economic values of contemporary nontimber forest products. Scotland Forestry Commission, Edinburgh

Farfán B (2001) Aspectos ecológicos y etnobotánicos de los recursos vegetales de la comunidad mazahua Francisco Serrato, municipio de Zitácuaro, Michoacán, México. BSc Dissertation. Facultad de Biología, Universidad Michoacana de San Nicolás de Hidalgo, Morelia. México

Farfán B, Casas A, Ibarra-Manriquez G, Pérez-Negrón E (2007) Mazahua ethnobotany and subsistence in the Monarch Butterfly Biosphere Reserve, México. Econ Bot 61(2):173–191

Fernández J, López JA (2005) La colleja, el cultivo de una verdura silvestre tradicional. Agricultura: Revista agropecuaria 876:548–550

Fernández JA, Navarro A, Vicente MJ, Peñapareja D, Plana V (2008) Effect of seed germination methods on seedling emergence and earliness of purslane (*Portulaca oleracea* L.) cultivars in a hydroponic floating system. Acta Hortic 782:207–212

Fontana E, Hoeberechts J, Nicola S, Cros V, Palmegiano GB, Peiretti PG (2006) Nitrogen concentration and nitrate/ammonium ratio affect yield and change the oxalic acid concentration and fatty acid profile of purslane (*Portulaca oleracea* L.) grown in a soilless culture system. J Sci Food Agric 86(14):2417–2424

Franco JA, Arreola J, Vicente MJ, Martínez-Sánchez JJ (2008) Nursery irrigation regimes affect the seedling characteristics of *Silene vulgaris* as they relate to potential performance following transplanting into semi-arid conditions. J Hortic Sci Biotech 83(1):15–22

Franco F, Cros V, Vicente MJ, Martínez-Sánchez JJ (2011) Effects of salinity on the germination, growth, and nitrate contents of purslane (*Portulaca oleracea* L.) cultivated under different climatic conditions. J Hortic Sci Biotech 86(1):1–6

García P, Alarcón R (2007) La colleja (*Silene vulgaris*) una verdura silvestre de calidad. Agricultura 901:802–805

Ghirardini M, Carli M, del Vecchio N, Rovati A, Cova O, Valigi F, Agnetti G, Macconi M, Adamo D, Traina M, Laudini F, Marcheselli I, Caruso N, Gedda T, Donati F, Marzadro A, Russi P, Spaggiari C, Bianco M, Binda R, Barattieri E, Tognacci A, Girardo M, Vaschetti L, Caprino P, Sesti E, Andreozzi G, Coletto E, Belzer G, Pieroni A (2007) The importance of a taste. A comparative study on wild food plant consumption in twenty-one local communities in Italy. J Ethnobiol Ethnomed 3(1):22

Gonnella M, Charfeddine M, Conversa G, Santamaria P (2005) Portulaca: da infestante ad alimento funzionale? Colture Protette 34(3):49–55

González-Amaro RM, Martínez-Bernal A, Basurto-Peña F, Vibrans H (2009) Crop and non-crop productivity in a traditional maize agroecosystem of the highland of Mexico. J Ethnobiol Ethnomed 5:38

González-Insuasti MS, Martorell C, Caballero J (2008) Factors that influence the intensity of non-agricultural management of plant resources. Agroforest Syst 74:1–15

Hadjichambis AC, Paraskeva-Hadjichambi D, Della A, Giusti ME, De Pasquale C, Lenzarini C, Censorii E, González-Tejero MR, Sánchez-Rojas CP, Ramiro-Gutiérrez JM, Skoula M, Johnson C, Sarpaki A, Hmamouchi M, Jorhi S, El-Demerdash M, El-Zayat M, Pieroni A (2008) Wild and semi-domesticated food plant consumption in seven circum-Mediterranean areas. Int J Food Sci Nutr 59(5):383–414

Hernández Bermejo JE, León J (1994) (eds) Neglected crops: 1492 from a different perspective. (FAO Plant Production and Protection Series, No. 26). Botanical Garden of Cordoba, Spain

Ihalainen M, Salo K, Pukkala T (2003) Empirical prediction models for *Vaccinium myrtillus* and *V. vitis-idaea* berry yields in North Karelia, Finland. Silva Fennica 37(1):95–108

Kerns BK, Alexander SJ, Bailey JD (2004) Huckleberry abundance, stand conditions, and use in Western Oregon: evaluating the role of forest management. Econ Bot 58(4):668–678

Ladio A, Rapoport EH (2005) La variación estacional de las plantas silvestres comestibles en baldíos suburbanos de Bariloche, Parque Nacional Nahuel Huapi, Patagonia, Argentina. Vida Silvestre Neotropical 11(1–2):33–41

Laghetti G (2009) Microevolution of *Scolymus hispanicus* L. (Compositae) in South Italy: from gathering of wild plants to some attempts of cultivation. In: Buerkert A, Gebauer J (eds) Agrobiodiversity and genetic erosion. Contributions in Honor of Prof. Dr. Karl Hammer. Supplement No. 92 to the Journal of Agriculture and Rural Development in the Tropics and Subtropics. Kassel University Press GmbH, pp 119–126

Laghetti G, Perrino P, Hammer K (1994) Utilization of *Silene vulgaris* (Moench) Garcke in Italy. Econ Bot 48(3):337–339

Leonti M, Nebel S, Rivera D, Heinrich M (2006) Wild gathered food plants in the European Mediterranean: a comparison analysis. Econ Bot 60(2):130–142

Lepofsky D, Turner NJ, Kuhnlein HV (1985) Determining the availability of traditional wild plant foods: an example of Nuxalk foods, Bella Coola, British Columbia. Ecol Food Nutr 16:223–241

Łuczaj Ł, Pieroni A, Tardío J, Pardo-de-Santayana M, Sõukand R, Svanberg I, Kalle R (2012) Wild food plant use in 21st century Europe: the disappearance of old traditions and the search for new cuisines involving wild edibles. Acta Societatis Botanicorum Poloniae 81(4):359–370

Łuczaj Ł, Končić MZ, Miličević T, Dolina K, Pandža M (2013) Wild vegetable mixes sold in the markets of Dalmatia (southern Croatia). J Ethnobiol Ethnomed 9:2

Martínez de Aragón J, Bonet JA, Fischer CR, Colinas C (2007) Productivity of ectomycorrhizal and selected edible saprotrophic fungi in pine forests of the pre-Pyrenees mountains, Spain: predictive equations for forest management of mycological resources. Forest Ecol Manag 252(1–3):239–256

Menendez-Baceta G, Aceituno-Mata L, Tardío J, Reyes-García V, Pardo-de-Santayana M (2012) Wild edible plants traditionally gathered in Gorbeialdea (Biscay, Basque Country). Genet Resour Crop Evol 59(7): 1329–1347

Miina J, Hotanen JP, Salo K (2009) Modelling the abundance and temporal variation in the production of bilberry (*Vaccinium myrtillus* L.) in Finnish mineral soil forests. Silva Fennica 43(4):577–593

Molina M (2014) Plantas silvestres comestibles: estudio de la producción y abundancia natural de verduras de hoja, espárragos y frutos carnosos de uso tradicional en España. PhD Thesis. Universidad Autónoma de Madrid

Molina M, Pardo-de-Santayana M, Aceituno L, Morales R, Tardío J (2011) Fruit production of strawberry-tree (*Arbutus unedo* L.) in two Spanish forests. Forestry 84:419–429

Molina M, Pardo-de-Santayana M, García E, Aceituno-Mata L, Morales R, Tardío J (2012) Exploring the potential of wild food resources in the Mediterranean region: natural yield and gathering pressure of the wild asparagus (*Asparagus acutifolius* L.). Spanish J Agric Res 10(4):1090–1100

Molina M, Tardío J, Aceituno-Mata L, Morales R, Reyes-García V, Pardo-de-Santayana M (2014) Weeds and food diversity: natural yield assessment and future alternatives for traditionally consumed wild vegetables. J Ethnobiol 34(1):44–67

Morales P, Carvalho AM, Sánchez-Mata MC, Cámara M, Molina M, Ferreira ICFR (2012a) Tocopherol composition and antioxidant activity of Spanish wild vegetables. Genet Resour Crop Evol 59:851–863

Morales P, Ferreira ICFR, Carvalho AM, Sánchez-Mata MC, Cámara M, Tardío J (2012b) Fatty acids characterization of twenty Spanish wild vegetables. Food Sci Technol Int 18:281–290

Mortley DG, Oh JH, Johnson DS, Bonsi CK, Hill WA (2012) Influence of harvest intervals on growth responses and fatty acid content of purslane (*Portulaca oleracea*). Hortscience 47(3):437–439

Mulas M, Deidda P (1998) Domestication of woody plants from Mediterranean maquis to promote new crops for mountain lands. Acta Hort (ISHS) 457:295–301

Mulas M, Cani MR, Brigaglia N, Deidda P (1998) Selezione varietale da popolazioni spontanee per la coltivazione di mirto e coberzzolo in Sardegna. Riv Fruttic 3:45–50

Murray G, Boxall PC, Wein RW (2005) Distribution, abundance, and utilization of wild berries by the Gwich'in people in the Mackenzie River Delta Region. Econ Bot 59(2):174–184

Nassif F, Tanji A (2013) Gathered food plants in Morocco: the long forgotten species in ethnobotanical research. Life Sci Leafl 3:17–54

Palaniswamy UR, Bible BB, McAvoy RJ (2002) Effect of nitrate: ammonium nitrogen ratio on oxalate levels of purslane. In: Janick J, Whipkey A (eds) Trends in new crops and new uses. ASHS Press, Alexandria, pp 453–455

Palaniswamy UR, Bible BB, McAvoy RJ (2004) Oxalic acid concentrations in Purslane (*Portulaca oleracea* L.) is altered by the stage of harvest and the nitrate to ammonium ratios in hydroponics. Sci Hortic 102:267–275

Parada M, Carrió E, Vallès J (2011) Ethnobotany of food plants in the Alt Empordà region (Catalonia, Iberian Peninsula). J Appl Bot Food Qual 84:11–25

Pardo-de-Santayana M, Pieroni A, Puri R (2010) The ethnobotany of Europe, past and present. In: Pardo-de-Santayana M, Pieroni A, Puri R (eds) Ethnobotany in the New Europe: people, health and wild plant resources. Berghahn Press, New York, Oxford, pp 1–15

Pérez-Negrón E, Casas A (2007) Use, extraction rates and spatial availability of plant resources in the Tehuacán-Cuicatlán Valley, Mexico: the case of Santiago Quiotepec, Oaxaca. J Arid Environ 70:356–379

Polo S, Tardío J, Vélez-del-Burgo A, Molina M, Pardo-de-Santayana M (2009) Knowledge, use and ecology of golden thistle (*Scolymus hispanicus* L.) in Central Spain. J Ethnobiol Ethnomed 5(1):42

Raatikainen M, Rossi E, Huovinen J, Koskela M-L, Niemelä M, Raatikainen T (1984) The yields of the edible wild berries in central Finland. (In Finnish with English summary). Silva Fennica 18(3):199–219

Rapoport EH (1995) Edible weeds: a scarcely used resource. Bull Ecol Soc Am 76(3):163–166

Rapoport EH, Ladio A, Raffaele E, Ghermandi L, Sanz EH (1998) Malezas comestibles. Hay yuyos y yuyos… Ciencia Hoy 9(49)

Rosati A (2001) Un possibile futuro per l'asparago selvatico. L'Informatore Agrario 7:89–92

Rosati A, Falavigna A (2000) Germinazione dei semi di asparago selvatico. L'Informatore Agrario 56:53–55

Rosati A, Pepe R, Senatore A, Perrone D, Falavigna A (2005) Produttività dell'asparago selvatico. L'Informatore Agrario 8:75–77

Rosati A, Paoletti A, Caporali S (2009) Biodiversita nell'oliveto: una risorsa economica per il recupero del reddito e del paesaggio. In: Alvino A, Cervelli C, Sarli G (eds.) Atti del IV Convegno Nazionale sulle Piante Mediterranee

Rossi E, Raatikainen M, Huovinen J, Koskela M-L, Niemelä M (1984) The picking and use of edible wild berries in Central Finland. (In Finnish with English summary). Silva Fennica 18:221–236

Sánchez-Mata MC, Cabrera-Loera RD, Morales P, Fernández-Ruiz V, Cámara M, Díez-Marqués C, Pardo-de-Santayana M, Tardío J (2012) Wild vegetables of the Mediterranean area as valuable sources of bioactive compounds. Genet Resour Crop Evol 59(3):431–443

Simopoulos AP, Norman HA, Gillapsy JE (1995) Purslane in human nutrition and its potential for world agriculture. World Rev Nutr Diet 77:7–74

Soriano JJ (2010) Los recursos genéticos entre la soberanía alimentaria y la interdependencia global. In: Soler M, Guerrero C (eds) Patrimonio cultural en la nueva ruralidad andaluza. PH Cuadernos 26. Instituto Andaluz del Patrimonio Histórico, Junta de Andalucía, pp 41–61

Takrouni MM, Boussaid M (2010) Genetic diversity and population's structure in Tunisian strawberry tree (*Arbutus unedo* L.). Sci Hortic 126:330–337

Tardío J (2010) Spring is coming: the gathering and consumption of wild vegetables in Spain. In: Pardo-de-Santayana M, Pieroni A, Puri R (eds) Ethnobotany in the New Europe: people, health and wild plant resources. Berghahn Books, Oxford-New York, pp 211–238

Tardío J, Pardo-de-Santayana M, Morales R (2006) Ethnobotanical review of wild edible plants in Spain. Bot J Linn Soc 152(1):27–72

Tardío J, Molina M, Aceituno-Mata L, Pardo-de-Santayana M, Morales R, Fernández-Ruiz V, Morales P, García P, Cámara M, Sánchez-Mata MC (2011) *Montia fontana* L. (Portulacaceae), an interesting wild vegetable traditionally consumed in the Iberian Peninsula. Genet Resour Crop Evol 58(7):1105–1118

Teixeira M, Carvalho IS (2009) Effects of salt stress on purslane (*Portulaca oleracea*) nutrition. Ann Appl Biol 154:77–86

Trichopoulou A, Vasilopoulou E (2000) Mediterranean diet and longevity. Br J Nutr 84(2):S205–S209

Turtiainen M, Salo K, Saastamoinen O (2011) Variations of yield and utilisation of bilberries (*Vaccinium myrtillus* L.) and cowberries (*V. vitis-idaea* L.) in Finland. Silva Fennica 45(2):237–251

Vieyra-Odilon L, Vibrans H (2001) Weeds as crops: the value of maize field weeds in the Valley of Toluca, Mexico. Econ Bot 55(3):426–443

Yazici I, Türkan I, Sekmen AH, Demiral T (2007) Salinity tolerance of purslane (*Portulaca oleracea* L.) is achieved by enhanced antioxidative system, lower level of lipid peroxidation and proline accumulation. Env Exp Bot 61(1):49–57

Youngblood D (2004) Identification and quantification of edible plant foods in the Upper (Nama) Karoo, South Africa. Econ Bot 58:43–65

Part II
Nutrients and Bioactive Compounds

Chapter 6
The Contribution of Wild Plants to Dietary Intakes of Micronutrients (I): Vitamins

María de Cortes Sánchez-Mata, María Cruz Matallana-González and Patricia Morales

6.1 Introduction

Among the wide variety of food components, micronutrients (vitamins and minerals) are needed only in small amounts, being essential for body functions. The importance of micronutrients for health status is known from ancient times, as deficiency diseases such as scurvy or rachitism were frequent in populations with deficient intake of certain vitamins or minerals. Dietary recommendations were often based on observations such as those of James Lind, a surgeon in the British Navy during the eighteenth century who demonstrated that limes and oranges cured sailor scurvy. Forty years after the publication of his findings, all the ships in the British Navy were ordered to carry lime juice, and scurvy on board ships was virtually eradicated (Hughes 1975). During the nineteenth and twentieth centuries, many studies conducted on the knowledge of the essential functions of micronutrients led to the establishment of reference intakes, in order to avoid diseases related to deficiencies or excess of vitamins or minerals.

The first tables of recommended intake of nutrients were published in 1938 in Canada (daily recommended nutrient intakes (DRNIs)) and, in the same year, in UK (Technical Commission on Nutrition of the League of Nations), with the purpose of preventing deficiency diseases of nutritional origin in a population suffering shortages during the two World Wars. In 1941, the Food and Nutrition Board (FNB) of the American Institute of Medicine of the National Academies (formerly National Academy of Sciences) established the so-called recommended dietary allowances

P. Morales (✉) · M. C. Sánchez-Mata · M. C. Matallana-González
Departamento de Nutrición y Bromatología II, Bromatología, Facultad de Farmacia, Universidad Complutense de Madrid, Madrid, Spain
e-mails: patricia.morales@farm.ucm.es; cortesm@ucm.es; mcmatall@ucm.es

M. de C. Sánchez-Mata, J. Tardío (eds.), *Mediterranean Wild Edible Plants*,
DOI 10.1007/978-1-4939-3329-7_6

(RDAs), published in 1943 (FNB 1943). In the 1950s, FAO/WHO started working on the evaluation of scientific evidences to provide recommendations on nutrient intake (FAO 1957). Since then, many revisions have been made, and many countries, institutions, and scientific societies have created expert groups to establish their own reference values and guidelines (Cuervo et al. 2009; FAO/WHO 2001; FNB 2000), taking into account the recent scientific advances, as well as the particular characteristics of each population.

However, the objective nowadays is not only the avoidance of deficiency diseases but also the prevention of chronic and degenerative diseases related to incorrect nutrition habits, focusing on achieving an equilibrated diet that covers the physiological and metabolic human needs for an optimal health status. Nowadays, nutritional recommendations have become quite variable, since they may consider gender, age, and physiological variability, as well as genetic factors, geographical implications (e.g., different sun exposure), and food habits. Throughout Europe, we can find reference values for individual countries such as Spain (Moreiras et al. 2011), Italy (Società Italiana di Nutrizione Umana 1998), France (Martin 2001), Belgium (Conseil Supérieur de la Santé 2009), UK (Department of Health of UK 1991), and Ireland (Nutrition Sub-committee of the Food Safety Authority of Ireland 1999). There are also reference values for groups of countries with sociocultural similarities, such as German-speaking countries, including Germany, Austria, and Switzerland (Deutsche Gesellschaft für Ernährung et al. 2000), and Scandinavian countries (Becker et al. 2004).

Thus, recommendations may vary widely, and so in Europe, harmonizing and aligning the nutrition policy and public health strategies has become an important goal. The Scientific Committee on Food of the European Union is making efforts in this direction, and the European Commission funded the EURopean micronutrient RECommendations Aligned (EURRECA) Network of Excellence with the aim of producing Europe-wide scientific consensus on the evidence for developing micronutrient recommendations (EURRECA 2013; Doets et al. 2008; Cavelaars et al. 2010). In this context, equilibrium between the diversification (to take into account differences) and the harmonization (to avoid inconsistencies or error sources) should be reached.

In this complex scenery, and considering also the individual variability, each institution may use different terms or parameters to establish the needs of energy and nutrients for each population group. Sometimes, the recommendations are based on the amount of nutrient estimated to be suitable to cover the needs of 50 % healthy individuals (of a particular age and gender), but this would leave 50 % of the population with inadequate intake. Other times, the reference values are those covering the needs of 97–98 % healthy individuals (of a particular age and gender), which could induce an excessive intake in a group. The former is often used for reference energy values and the latter is used for most nutrients. Often, an upper or maximum level should also be used in the case of nutrients that may cause disorder related to an excessive intake. For that reason, the clarification between the different terms and concepts used when referring to recommended intakes is necessary. Figure 6.1 graphically shows a summary of these concepts and their equivalences in the different recommendations.

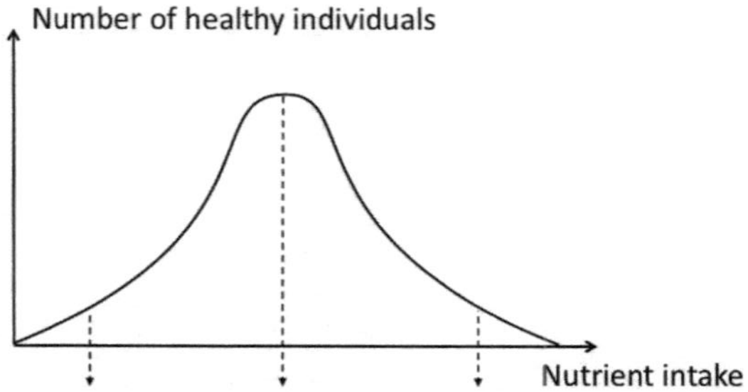

Country/ institution	Average daily level of intake estimated to meet the requirements of: 2.5 % healthy individuals	50 % healthy individuals	97-98 % healthy individuals	Maximum daily intake likely to pose no adverse health effects	Other parameters		
Spain			IR				
Italy			LRA				
France			ANC				
Belgium		BM	ANR				
United Kingdom (DRV)	LRNI	EAR	RNI		SF		
Ireland	LTI	AR	RDA				
Germany-Austria-Switzerland			Recommendations		Estimated values		
Scandinavian Countries		RV	RI	UL			
Scientific Committee on Foods-EU	LTI	AR	PRI		ARI	AMDR	RLV
Food Nutrition Board-US (DRI)		EAR	RDA	UL	AI		
FAO/WHO		EAR	RNI	UL	AI	RSI	PNI

Fig. 6.1 Summary and equivalences of parameters for different nutrients (based on Cuervo et al. 2009). *AI* adequate intake (average daily level of intake estimated to meet the requirements of 50 % of healthy individuals when there is not enough evidence to calculate RDA), *AMDR* acceptable macronutrient distribution range (interval for a low risk of chronic diseases), *ANC* apports nutritionnels conseillés, *ANR* apports nutritionnels (journelieur) recommandés, *AR* average requirement, *ARI* acceptable range of intake, *BM* besoin moyen, *DRI* dietary reference intake, *DRV* dietary reference value, *EAR* estimated average requirement, *IR* ingestas recomendadas, *LRA* livelli raccomandati di assunzione, *LRNI* lower reference nutrient intake, *LTI* lowest threshold intake, *PNI* protective nutrient intake (for certain nutrients: supplementary amount for special needs), *PRI* population reference intake, *RDA* recommended dietary allowance, *RDI* recommended daily intake, *RLV* reference labeling value (in the EU, nutrient reference value), *RNI* recommended nutrient intake, *RSI* recommended safe intake (for certain nutrients when data are lacking, levels to prevent deficiency signs), *RV* reference value, *SF* safe intake, *UL* upper tolerable level

Food-based dietary guidelines (FBDG) have also been established in Europe to provide nutrient advice in a way consumers can understand, avoiding the use of numerical recommended intake of nutrients, but providing a practical way of interpreting these into dietary advice for individuals within a population. In 1996, FAO/WHO published guidelines for the development of FBDG, describing their characteristics as the expression of the principles of nutrition education, mostly as foods (instead of nutrients) intended for use by individual members of the general public; and if not expressed entirely as foods, written in a language that avoids, as far as possible, the technical terms of nutritional science (FAO/WHO 1996; WHO 2003).

The European International Life Sciences Institute (ILSI Europe) and the European Food Safety Authority (EFSA) have organized several workshops focusing on FBDG. Nowadays, nearly all the European countries have developed FBDG. They are all based on the principle to provide guidance for a healthy balanced diet that will help prevent noncommunicable diseases such as heart diseases and cancer. Common recommendations include eating plenty of fruits, vegetables, and complex carbohydrates, and choosing foods which are low in saturated fat, salt, and sugar (European Food Information Council 2009; FAO/WHO 1996).

With the purpose of giving more complete information to consumers, in the last years, nutrition labeling for foodstuffs has included a lot of mentions regarding nutrients and bioactive compounds present in foods. Sometimes too much information on food labels may be confusing for consumers who are not always familiar with the terminology. The EU is making great efforts in this regard to establish the reference labeling values (RLVs) based on population reference intake (PRI), recommended dietary allowance (RDA), and recommended nutrient intake (RNI) of several European countries (see Fig. 6.1), as well as US and FAO/WHO recommendations. As a result, the Regulation (EU) No. 1169/2011 of the European Parliament and of the Council of 25 October 2011 on the provision of food information to consumers has been published, recording several rules for nutrition labeling with the purpose of achieving a high level of health protection of consumers and guaranteeing their right to information about the food they consume. The Annex XIII of this last Regulation includes data about harmonized nutrient reference values (NRVs) for vitamins and minerals (adults), which should be used for the expression of the calculated percentages of the reference intakes covered by a portion of a given food product, as they should appear on its label.

With the purpose of examining how wild edible plants may contribute to the intake of micronutrients in the human diet, through the present and the next chapter, the reference values selected have been those equivalent to RDAs (average daily level of intake estimated to meet the requirements of 97–98 % healthy individuals to be sure of covering the needs of almost all of them). Intervals made of different accepted values of recommendations for adults have been considered as reference values, mainly population reference intake (PRI) given by the Scientific Committee on Foods of the European Union (1992) and recommended nutrient intake (RNI) from FAO/WHO (2001). Also, some of those Mediterranean countries that have published specific reference values, such as Spain (*ingestas recomendadas* (IR); from Moreiras et al. 2011), Italy (*livelli raccomandati di assunzione* (LRA); from

Società Italiana di Nutrizione Umana 1998), and France (*apports nutritionnels conseillés* (ANC), from *Agence Française de Securité Sanitaire des Aliments*; Martin 2001), have been taken into account.

6.2 Vitamins: Classification and Dietary Requirements

The concept of "vitamin," suggested in 1880 by Nikolai Lunin (1844–1920), and coined in 1911 by Casimir Funk (1884–1967), refers to a group of organic compounds that are, in very small amounts, essential for the normal function of the human body, since most of them cannot be synthesized in the body tissues. They are involved in a wide variety of chemical and physiological functions, and are broadly distributed in natural food sources. Vitamin research reached a high degree of development during the first years of twentieth century, leading to the discovery of many chemical compounds. They possess a wide diversity of chemical structures and activities through different mechanisms, and have been named using capital letters. Different vitamers (structurally related compounds belonging to a particular vitamin display) often have similar qualitative biological properties to one another, but they can exhibit varying degrees of potency, due to the subtle differences in their chemical structures. Provitamins or vitamin precursors are naturally occurring substances which are not vitamins themselves but can be converted into vitamins by normal body metabolism (Ball 2006).

Classically, vitamins have been classified into two groups, depending on their solubility:

- Hydrophilic or hydrosoluble vitamins, including vitamin B group (B_1, B_2, B_3, B_5, B_6, biotin, B_9, and B_{12}) and vitamin C. They are quite abundant in plant and animal tissues, vegetables being one of their main sources in the human diet.
- Lipophilic or liposoluble vitamins, including vitamins A, D, E, and K, which are distributed in nature associated to lipid fraction of plant or animal tissues. Many of these compounds can be found in low amounts in fat-poor foods, such as fruits and vegetables, acting as pigments inside chromoplasts, or as antioxidants in low amounts in plant tissues. In this way, fruits and vegetables may contribute to some lipophilic vitamins intake, such as provitamin A, vitamin E, or K.

As previously mentioned, these vitamins have been subject to the establishment of reference vitamin intake values for nutritional studies as well as for food labeling purposes in different countries or organisms (Table 6.1).

Table 6.1 Recommended intake of some vitamins (milligrams or micrograms per day): most widely accepted values for adults in Mediterranean countries

	International recommendations			National recommendations			
	FAO/WHO, RNI[a]	EU, PRI[b]	EU, NRV[c]	Spain, IR[d]	Italy, LRA[e]	France, ANC[f]	USA, DRI[g]
Vitamin C (mg/day)							
Male	45	45	80	60	60	110–120	90
Female							75
Vitamin B_9 (μg/day)							
Male	400	200–400[h]	200	400	200	330–400	400
Female						300–400	400
Vitamin A (μg/day)							
Male	600	700	800	1000	700	700–800	900
Female	500–600	600		800	600	600	700
Vitamin E (mg/day)							
Male	10[i]	–	12	12	4	12[j]	15
Female	7.5[i]				3		
Vitamin K (μg/day)							
Male	65	–	75	–	–	45–70	120
Female	90						90

[a] *RNI* recommended nutrient intake (FAO/WHO 2001)
[b] *PRI* population reference intake (SCF 1992)
[c] *NRV* nutrient reference values, EU Regulation 1169/2011
[d] *IR* Ingestas recomendadas (Moreiras et al. 2011)
[e] *LRA* Livelli raccomandati di assunzione (SINU, Società Italiana di Nutrizione Umana 1998)
[f] *ANC* Apports nutritionnels conseillés (Martin 2001)
[g] *DRI* Dietary reference intakes (Recommended Dietary Allowances or Adequate Intakes), (Food and Nutrition Board; Trumbo et al. 2002)
[h] Levels of 400 μg/day are recommended for women before conception
[i] "Acceptable intakes" (no sufficient data to establish recommended intake)
[j] According to French ANC, the intake of vitamin E should be increased to 20–50 mg/day for adults over 75 years old

6.3 Hydrosoluble Vitamins

6.3.1 B-Complex Vitamins

This group comprises different compounds functioning as coenzymes in many reactions of the human body, such as thiamin (vitamin B_1), riboflavin (vitamin B_2), niacin (vitamin B_3), pantothenic acid (vitamin B_5), pyridoxine (vitamin B_6), biotin, and folic acid (vitamin B_9). They may be involved in several deficiency diseases such as beriberi and pellagra, among others.

These wide variety of compounds are present in either plant or animal foods. For example, whole grains are generally a good source of these vitamins. Data about B-complex vitamins in wild plant foods are very scarce. Wetherilt (1992) reported some values of vitamins B_1, B_2, B_3, and B_6 in nettle leaves (*Urtica dioica* L.), vitamin B_3 being the most abundant (620 µg/100 g), but vitamin B_2 the most relevant if compared with recommended intake of these vitamins (100 g would provide about 15–25 % of the daily requirements of riboflavin, according to different recommendations). Data of the contents of some B-complex vitamins have also been reported by Souci et al. (2008) for the leaves of wild *Eruca vesicaria* (L.) Cav., as well as by Boudraa et al. (2010) for the fruits of *Ziziphus lotus* (L.) Lam. (see Chap. 13).

However, the most important contribution of wild edible foods to vitamin B-complex in the human diet is due to their folic acid content.

6.3.1.1 Vitamin B_9 (Folic Acid)

The denomination vitamin B_9 comprises a wide group of water-soluble compounds belonging to the B-complex group and known as folic acid and its derivatives (folates). They were first isolated from spinach leaves (Latin term *folium* = leaf) by Herschel K. Mitchell (1913–2000) in 1941, after the suggestion of their vitamin activity by Lucy Wills (1844–1920). Folic acid (5-methyltetrahydrofolic acid, 5-CH_3-H_4folic acid) consists of a pteridine-derived ring, linked by a methylene bridge to a residue of p-amino benzoic acid, which is linked by an amide bond to a glutamic acid residue (Fig. 6.2).

As folates are synthesized only by microorganisms and plants, humans depend on a variety of dietary sources for its intake. Liver (around 240 µg/100 g), fresh green leafy vegetables (e.g., spinach, 140 µg/100 g; endive, 267 µg/100 g), some legumes (red beans, 394 µg/100 g), wheat germ, and yeast are considered as good sources of folic acid (Lucock 2000; Souci et al. 2008). In food, folates are naturally present as polyglutamates ($PteGlu_n$), mainly as mono-, penta-, and hexaglutamates (Scott 1994).

Natural folates cannot be directly absorbed by the intestinal mucosa (Fig. 6.3); therefore, they should be enzymatically hydrolyzed in the intestinal lumen, along with the reduction of the pteridine ring, and the methylation to 5-methyltetrahydrofolate (5-CH_3-THF) in the enterocyte. 5-CH3-THF is considered the only vitamer

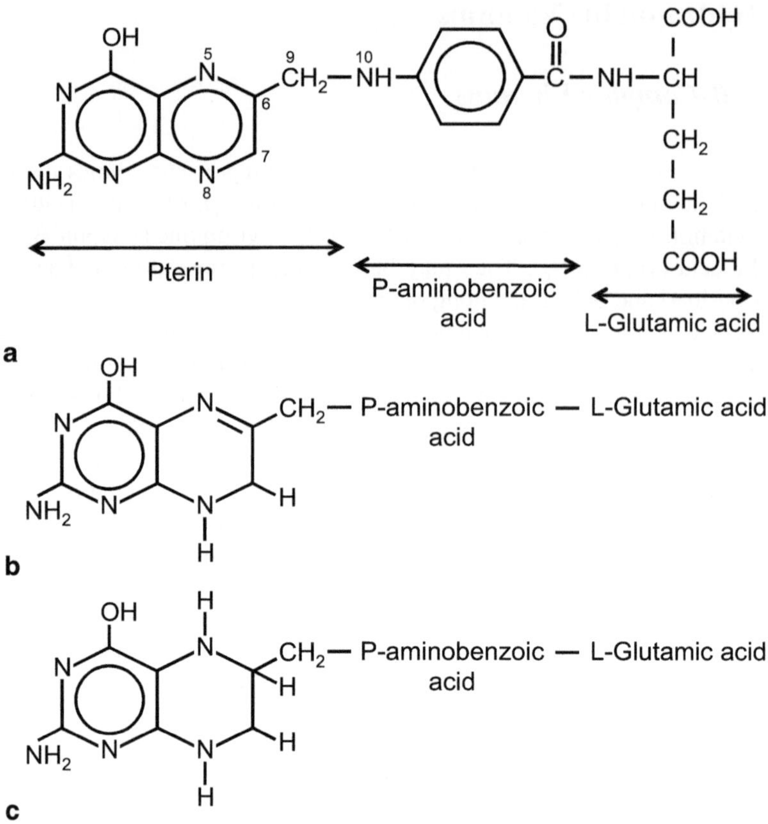

Fig. 6.2 Chemical structures of folic acid and natural folates. **a** Folic acid. **b** Dihydrofolic acid. **c** Tetrahydrofolic acid

with physiological activity, and it circulates in plasma in the free form (10%) or bound to specific proteins (Cuskelly et al 1996; FAO/WHO 2001; Lucock 2000; Scott 1994).

Since folic acid is crucial in tissues formation during embryo development, low maternal folate levels are closely related to the development of neural tube formation defects such as spina bifida and anencephaly (Steegers-Theunissen 1995; Wright et al. 2010). Low folate levels are also correlated with the hyperhomocysteinemia, which is considered as an independent risk factor for atherosclerotic and atherothrombotic damage (Shai et al 2004; Vanizor Kural 2003). Other functions of folates include preventive effects against different tumor types and some degenerative diseases such as Alzheimer's (Gerber 2001).

To avoid folic acid deficiency, a daily intake of 200 (Italian LRA and EU-PRI) to 400 μg (Spanish IR together with FNB-RDA, FAO/WHO-RNI), mainly during

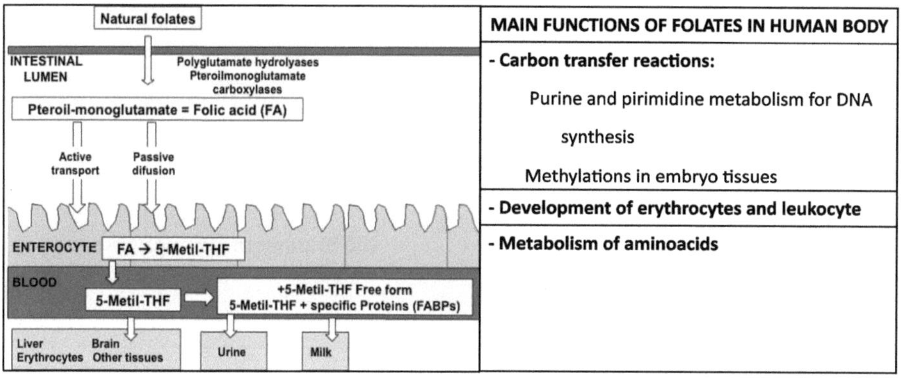

Fig. 6.3 Scheme of absorption and functions of folates absorption. (Based on Kim 1999; Mahan et al. 2012)

the preconceptional period, is recommended (Table 6.1). In all cases, additional 100–200 μg/day should be ingested during pregnancy (especially in the first months to reduce the risk of neural tube formation defects) and lactation.

There are very few studies on folate content in wild vegetables. Some of them are conducted on African or Canadian plants (Kruger et al. 1998; Kuhnlein 1990), but data about folate content in Mediterranean wild vegetables are very scarce. Morales et al. (2015) reported very high folate levels in many wild Mediterranean vegetables, most of them over 100 μg/100 g of fresh edible product. To evaluate if the content of a given nutrient in a food product is enough to be considered as a source of that nutrient, Regulation 1169/2011 on the provision of food information for consumers can be taken as a reference. A food product can be claimed as "source of a vitamin/mineral" if a 100-g portion contains 15 % or more of Nutrient Reference Values (NRVs) given by this Regulation for food labelling. It can also be considered as "high content of a vitamin/mineral" if a 100-g portion contains 30 % of NRV or more.

Considering the published values for folate content in Mediterranean wild vegetables, many of them could be considered as very good folate sources, providing more than 15 % or even 30 % of NRV recorded in European Regulation 1169/2011 (30 and 60 μg/100 g, respectively). Special mention should be made of *Rumex pulcher* L., with average total folate content of around 500 μg/100 g, and also many other wild plants such as *Beta maritima* L., *Anchusa azurea* Mill., *Foeniculum vulgare* Mill., *Silene vulgaris* (Moench) Garcke, *Cichorium intybus* L., and *Asparagus acutifolius* L., which can provide the total amount of folate daily needed by eating a 100-g portion (Fig. 6.4).

As it happens to other vitamins some folates are lost during the cooking process, due to lixiviation in boiling liquid and thermal degradation. Studies on folate content in boiled wild plants are very scarce and show a reduction of about 30–80 % of total folate content in the boiled leaves of *S. vulgaris* and *R. pulcher*, and a lesser

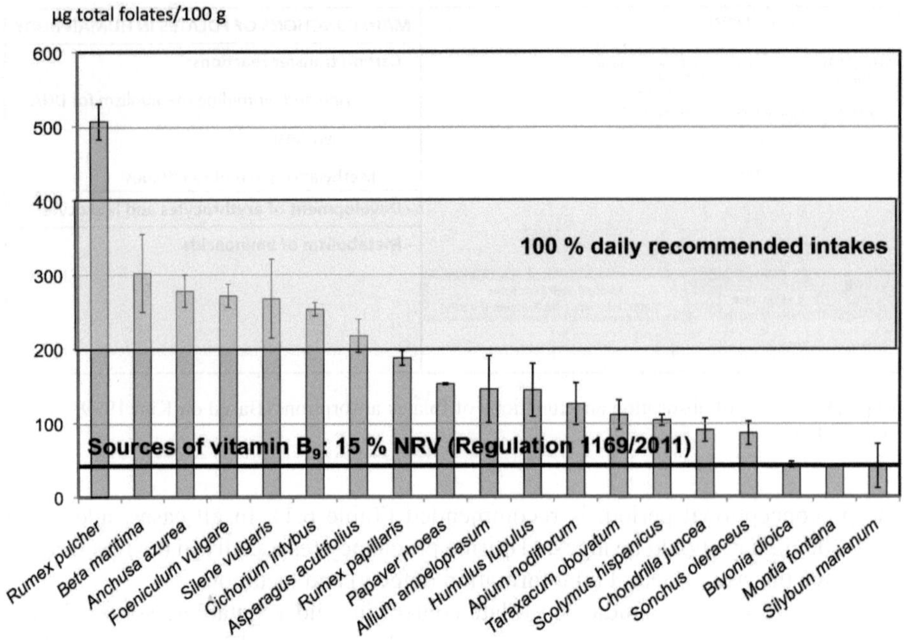

Fig. 6.4 Selected wild vegetables standing out for their high vitamin B_9 levels. (Data from Morales et al. 2015)

reduction (30–53 %) in wild *Asparagus* shoots. The higher loss of folate in leaves than in stems could be related to the different structures of the tissues of both plant organs; the former being thinner and softer could make the loss of nutrients easier. However, due to the high levels of folate found in these species, in spite of cooking losses, they may be good folate sources for the diet (150–200 µg/100 g).

6.3.2 *Vitamin C*

As shown in Fig. 6.5, two chemical species have been shown to provide vitamin C activity: L-ascorbic acid (AA) and the first product of its oxidation, L-dehydroascorbic acid (DHA). Ascorbic acid is very quickly oxidized, especially at high temperatures, alkaline medium, and in the presence of oxygen (Brubacher et al.1985).

Following Souci et al. (2008), fresh fruits, especially kiwi and citrus fruits (with values around 36–59 mg/100 g or more), and green vegetables, such as pepper (131 mg/100 g) and parsley (100 mg/100 g), constitute rich sources of vitamin C, mainly in the reduced form (AA), although many factors such as genetic, maturity, climate, and storage may affect the levels of this vitamin (Ball 2006).

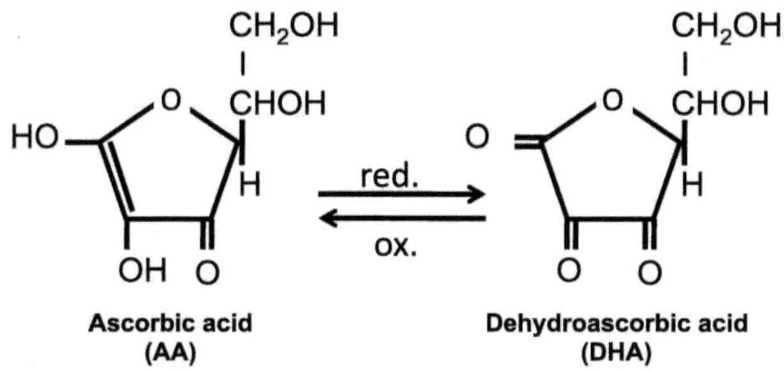

Fig. 6.5 Chemical structures of ascorbic and dehydroascorbic acids

Vitamin C is ingested in reduced (most of the dietary vitamin C) and oxidized forms, which are absorbed across the brush-border membrane, as it is schematically shown in Fig. 6.6. This vitamin circulates mainly as unbound ascorbic acid, available in blood and interstitial fluids. When oxidative stress is high, ascorbic acid may be oxidized to dehydroascorbic acid, which may be later recycled to ascorbic acid. Dehydroascorbic acid has a short half-life and is metabolized to excretory products (mainly oxalic acid).

The nutritional importance of vitamin C as an essential water-soluble vitamin has been well established. As shown in Fig. 6.6, ascorbic acid is a cofactor in numerous physiological reactions (Bender 2003; Phillips et al. 2005). Vitamin C is a powerful antioxidant, and it is the first line of antioxidative defense in water-soluble compartments (Marchioli 1999). It also has synergistic effects in the regeneration of fat-soluble vitamin E, playing an important role in lipid peroxidation inhibition (Halliwell 2001; Traber 2007). Other functions of vitamin C include the inhibition

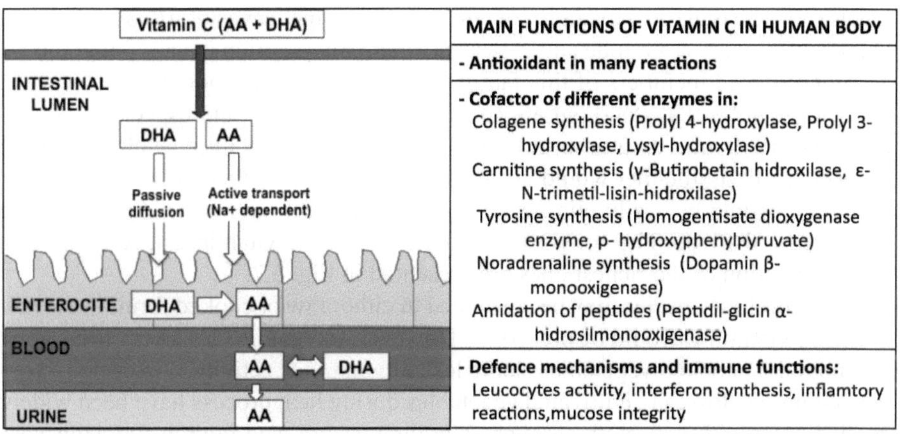

Fig. 6.6 Scheme of the intestinal absorption and functions of vitamin C. (Based on Malo and Wilson 2000; Paddayatty and Levine 2001)

of the synthesis of mutagenic compounds such as nitrosamines in stomach and intestine (Correa 1992).

The deficiency of vitamin C results in scurvy, whose primary symptoms are hemorrhages in the gums, skin, bones, and joints, and the failure of wound healing. This disease is not frequent nowadays in occidental countries due to a better access of the population to fresh fruits and vegetables. However, nowadays it is known that vitamin C is involved in many other mechanisms of the human body, including antioxidant mechanisms of protection against oxidative stress which causes several chronic diseases of high prevalence in developed countries. Some vitamin C dietary recommendations have increased, being established nowadays from 45 mg/day (FAO/WHO-RNI and EU-PRI) to 120 mg/day (French ANC) for adults (Table 6.1).

Fresh fruits and vegetables are the best sources of vitamin C, and wild edible species are not an exception. Most studies measure total vitamin C contents, but others differentiate between AA and DHA contents, which add information about its antioxidant potential.

Considering the NRV recorded in the European Regulation 1169/2011, a food could be considered as a source of vitamin C if a 100-g portion contains at least 12 mg of vitamin C per 100 g; and the mention "high content of vitamin C" could be possible over 24 mg/100 g. These levels could be achieved through many fruits and vegetables, either conventional or unusual. A great number of wild vegetables are valuable sources of vitamin C for the human diet, since most of them contain over 12 mg/100 g fresh matter. Although many Asteraceae species usually present low levels of vitamin C, around 5 mg/100 g or less, there are some exceptions, such as *Sonchus asper* (L.) Hill, whose leaves have been reported to contain more than 50 mg/100 g of ascorbic acid (see Fig. 6.7).

On the contrary, many Cruciferae species, such as *Sisymbrium officinale* (L.) Scop., *E. vesicaria* (L.) Cav., *Cardaria draba* (L.) Desv., *Diplotaxis tenuifolia* (L.) DC., and *Capsella bursa-pastoris* (L.) Medik., stand out for their high vitamin C levels (50–150 mg/100 g average content), as well as the leaves of *Alliaria petiolata* (M. Bieb.) Cavara & Grande and *Cakile maritima* Scop. (often over 150 mg/100 g). Many cultivated members of this family are also very good sources of vitamin C. Other vegetables that stand out are *Chenopodium album* L. and especially *U. dioica* leaves (200 mg/100 g or more). The intake of these very high levels of vitamin C sporadically in the diet would not present any toxicity effect since the excess of vitamin C would be easily eliminated from the body through urine. Among all the wild greens that could be considered as good sources of vitamin C (see Chap. 13), only a selection of the richest ones are presented in Fig. 6.7.

Many of these species may be consumed in either raw or cooked form. In the raw material, vitamin C would be preserved. However, as vitamin C is a very heat-labile compound, in cooked vegetables, some loss of vitamin C activity is expected to occur. Vitamin C losses in cultivated vegetables during heat process have been widely studied using different methods such as boiling, steam, and microwave. However, the data about the effect of thermal process on the nutritive value of wild vegetables

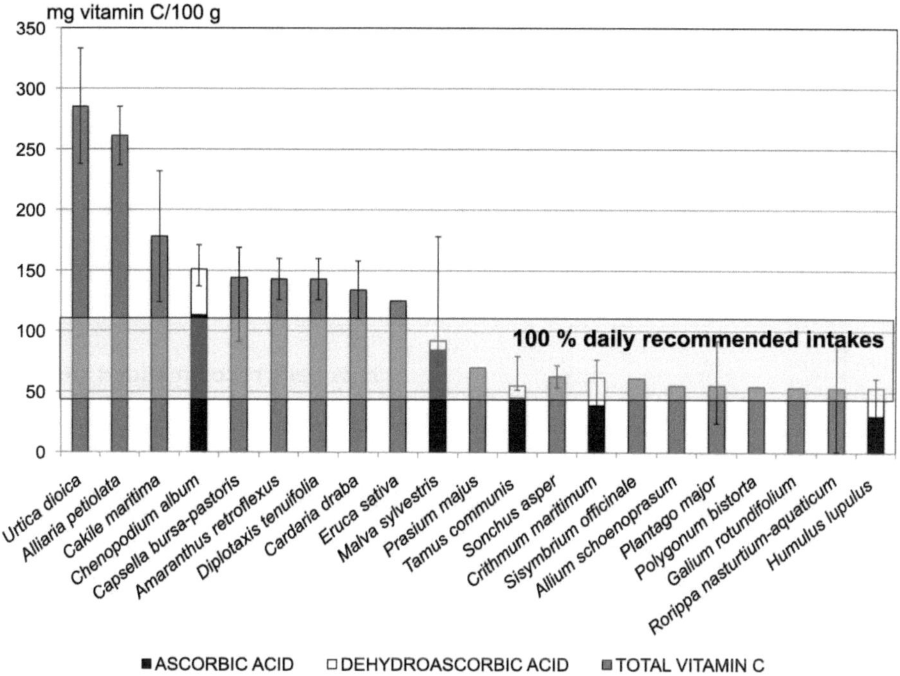

Fig. 6.7 Best vitamin C sources among Mediterranean wild vegetables. (Data from Aliotta and Pollio 1981; Franke and Kensbock 1981; Guil-Guerrero and Torija Isasa 1997; Guil-Guerrero et al. 1999; Kiliç and Coskun 2007; Morales 2011; Sánchez-Mata et al. 2012; Vardavas et al. 2006; Yildrim et al. 2001)

are not sufficient. In the study of Morales (2011), vitamin C was reduced to about 40–50 % in the boiled leaves of *S. vulgaris* and *R. pulcher*, and to a higher extent in other plant tissues such as wild *Asparagus* shoots (14–75 %). Other studies on plants from different origins report values up to 95 % losses of vitamin C, which minimize in pressure cooking compared to conventional cooking (Somsub et al. 2008; Yadav and Sehgal 1997). Vitamin C losses in these matrices should be attributed to either lixiviation in boiling liquid or thermal degradation, dehydroascorbic acid being often more unstable than ascorbic acid. Nevertheless, due to the significant presence of ascorbic acid in these species, some of them are still good sources of vitamin C. For example, a 100-g portion of cooked shoots of *A. acutifolius* could provide more than 50 % of daily recommendations of vitamin C.

Like vegetables, many fruits can be considered as good sources of vitamin C (Fig. 6.8). The fruits of *Arbutus unedo* L. and *Rosa canina* L. may present average levels of around 100 mg/100 g or more. For this reason, these fruits could be very good alternatives to conventional fruits for their vitamin C contribution. The habitual fresh consumption of most fruits is an advantage to preserve all this nutritional potential.

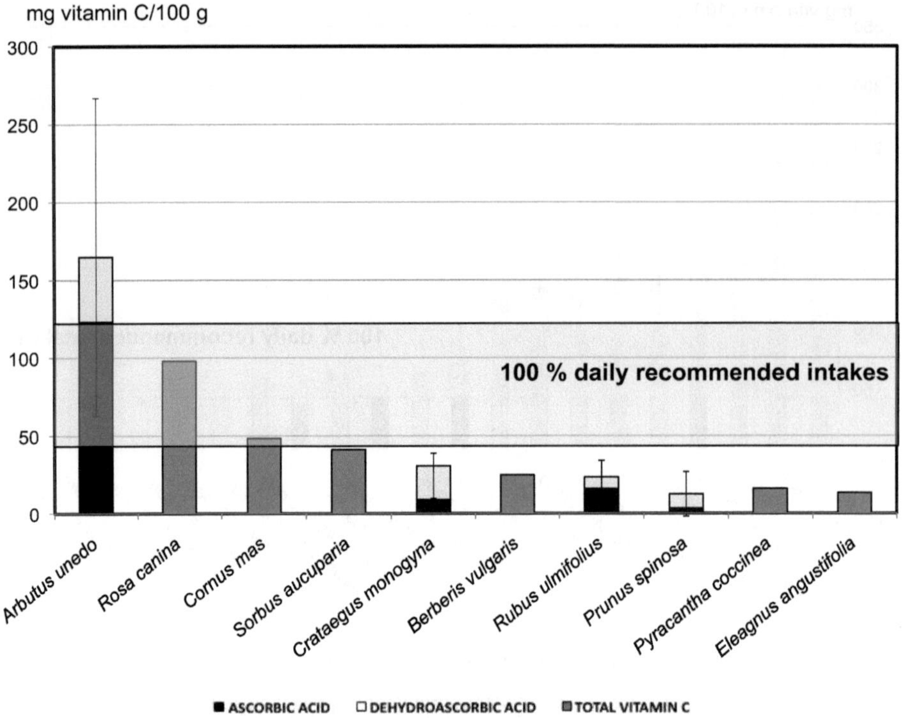

Fig. 6.8 Selected wild fruits standing out for their high vitamin C levels. (Data from Alarcão-E-Silva et al. 2001; Boudraa et al. 2010; Barros et al. 2010b, 2011; Doležal et al. 2001; Egea et al. 2010; Jabłońska-Ryś et al. 2009; Morales et al. 2013; Pallauf et al. 2008; Pereira et al. 2013; Ruiz-Rodríguez et al. 2011, 2014a, b)

6.4 Liposoluble Vitamins

6.4.1 Vitamin and Provitamin A

Vitamin A is a generic term that includes retinol as well as some of its esters and isomers. The international standard for vitamin A is all-*trans*-retinol, for which the international unit (IU) was defined as 0.3 μg of this form of retinol, as stated by FAO/WHO (2001). Retinol is only naturally present in animal tissues, but plants usually contain carotenoids in their chloroplasts, many of which have demonstrated provitamin A activity after in vivo conversion to retinol.

Since nineteenth century, when W. H. Wackenroder (1798–1854) isolated carotenoids for the first time from carrot roots (from which they took their name), around 600 carotenoids and their isomers have been identified (Bauernfeind 1972). They are polienic isoprenoids, with long conjugated bonds leading to yellow to red colors, acting as natural pigments in many plant tissues. They are derived from the precursor lycopene (a completely linear compound), by the way of circulation, isomeriza-

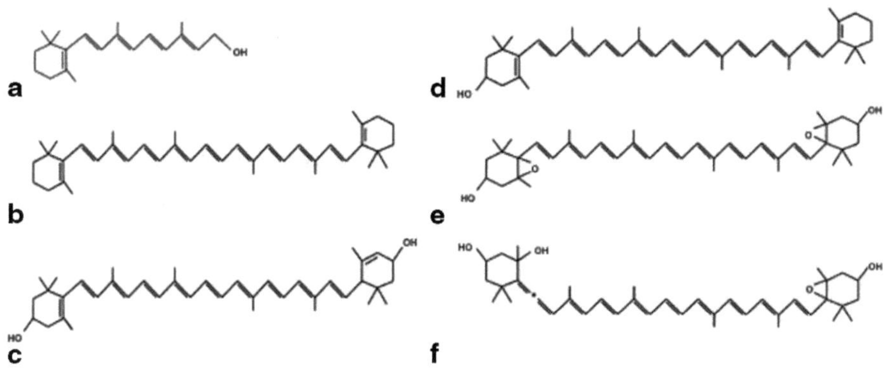

Fig. 6.9 Chemical structures of vitamin A (retinol) and carotenoids. **a** Retinol. **b** β-carotene. **c** Lutein. **d** β-cryptoxanthin. **e** Violaxanthin. **f** Neoxanthin

tion, oxidation, or a combination of these processes. Carotenoids are divided into two groups: carotenes (with only carbon and hydrogen atoms, including lycopene) and xanthophylls (with oxygen atoms). Some of the most important carotenoids are β-carotene, γ-carotene, α-carotene, β-cryptoxanthin, lutein, and lycopene (Fig. 6.9), exhibiting different functions in the human body.

As a storage organ, liver is a rich source of vitamin A in the form of retinol. Whole milk (12–40 μg retinol/100 g) and dairy products, such as butter or cheese, as well as eggs (800 μg retinol/100 g), are important dietary sources. Among vegetables, carrots and sweet potatoes (up to 93 mg β-carotene/100 g), as well as green leafy vegetables, such as spinach (3–8 mg β-carotene/100 g), are major contributors of provitamin A in the diet. Green leafy vegetables are usually rich in lutein, as well as some β-carotene, and other compounds such as neoxanthin and violaxanthin. In these foods, green chlorophylls often mask the color of carotenoids. Fruits also contain a great variety of carotenoids, responsible for their different colors, such as orange β-carotene in peaches, apricots, and some variety of melons, or rose-red colors due to lycopene in tomato, watermelon, and grapefruit (Ball 2006; Souci et al. 2008).

Vitamin A can be produced within the body from certain carotenoids such as α-carotene, β-carotene, and β-cryptoxanthin (Britton et al. 1995; Ibrahim et al. 1991; Patton et al. 1990). Their vitamin A activity is measured as retinol equivalents (RE), or more frequently nowadays as retinol activity equivalents (RAE); 1 μg RAE = 1 μg of retinol, equivalent to 12 μg β-carotene, or to 24 μg other provitamin A carotenoids, as α-carotene or β-cryptoxanthin (Mahan et al. 2012).

Figure 6.10 schematically presents vitamin and provitamin A absorption and their functions in the human body. Critical factors for efficient retinol solubilization are the presence of a lipid medium (presence of fatty acids, monoglycerides, cholesterol, and phospholipids) and the secretion of bile salts and hydrolytic enzymes (Blomhoff 1991; Dew and Ong 1994; Erdman et al. 1993; Parker 1996,1997).

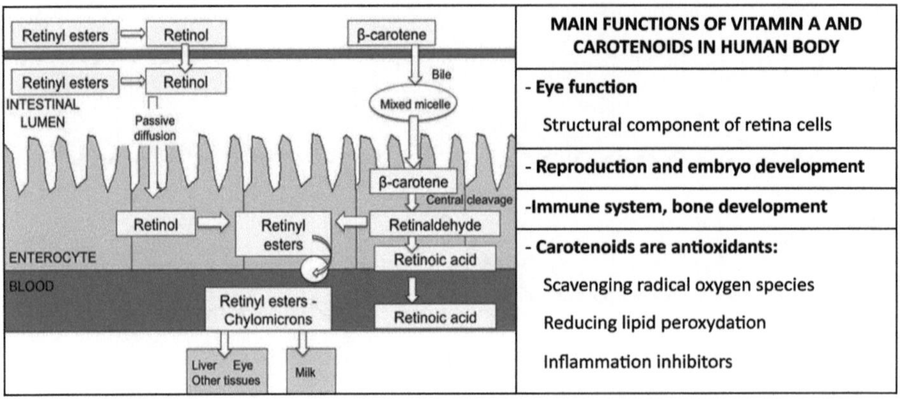

Fig. 6.10 Scheme of the intestinal absorption and functions of retinyl esters and carotenoids. (Based on Zeb and Mehmood 2004; Mahan et al. 2012; Salvatore et al. 2005; Maiani et al. 2009)

As some carotenoids, such as lutein and zeaxanthin, are stored in the retina, acting as powerful antioxidants for a normal eye functioning, their role in prevention of macular degeneration is being studied. Other carotenoids have shown a preventive role against other chronic diseases such as prostate cancer and cardiovascular diseases (Cámara et al. 2013; Zeb and Mehmood 2004).

On the one hand, deficiency of vitamin A is a major cause of premature death in developing countries, particularly among children. In addition to the specific signs and symptoms of xerophthalmia and the risk of irreversible blindness, nonspecific symptoms include increased morbidity and mortality, poor reproductive health, increased risk of anemia, and contributions to slowed growth and development (FAO/WHO 2001). On the other hand, an excessive dietary intake of this vitamin may be related to chronic toxicity, including teratogenicity in developing fetuses. Normally, vitamin A toxicity results from the indiscriminate use of pharmaceutical supplements, and not in usual diet.

The daily intake of vitamin A for adults should be in the range of 0.5–0.8 mg/day for woman (> 18 years) and 0.6–1 mg/day for man, Spanish IR being one of the highest ones (Table 6.1). Vitamin A activity in a given food can only be calculated if the contents of provitamin A carotenoids are known. In this respect, not many studies on wild edible plants apply analytical techniques that allow analysis of individual carotenoids, and most of them report just total amount of carotenoids expressed as β-carotene, which reflects a global amount of these bioactive antioxidant compounds, but is not indicative of vitamin A activity. Some of these references indicate interesting total carotenoid contents of over 6 mg/100 g in some wild vegetables, such as some *Chenopodium* species among others (see Chap. 9, according to several authors (Aliotta and Pollio 1981; Guil-Guerrero 2001; Guil-Guerrero and Torija-Isasa 1997; Guil-Guerrero et al. 1998). However, in leafy vegetables, non-provitamin A carotenoids, such as xanthophylls (lutein, neoxanthin, and violaxanthin), are usually the major carotenoids, and in many cases, they are even present in higher levels than β-carotene (García-Herrera 2014; Guil-Guerrero et al. 2003; Salvatore et al. 2005; Šircelj et al. 2010).

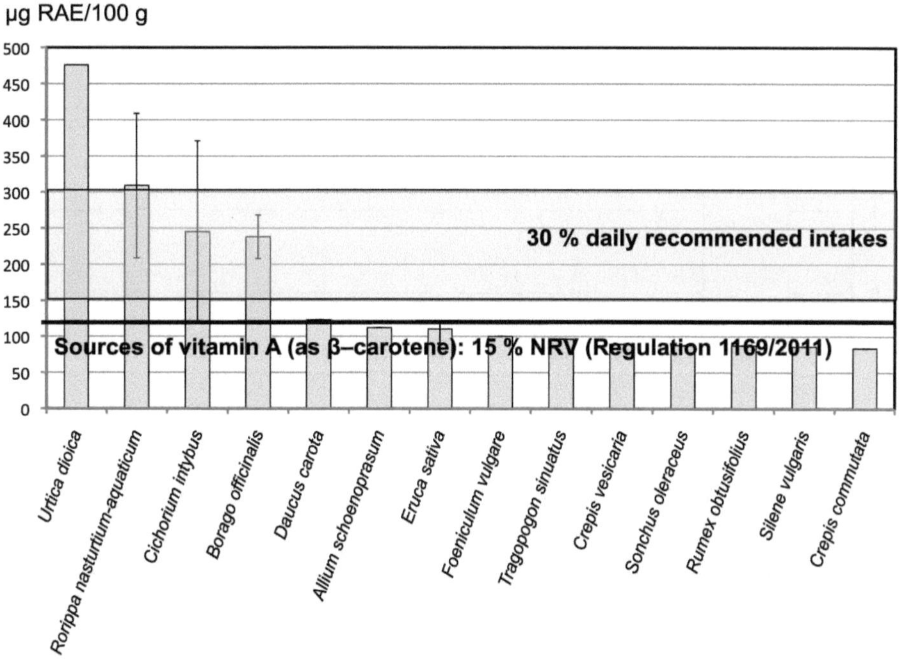

Fig. 6.11 Selected wild vegetables standing out as better sources of vitamin A (as RAE calculated from individually quantified provitamin A carotenoids contents). (Data from Guil-Guerrero et al. 2003; Kudritsata et al. 1987; Salvatore et al. 2005; Vardavas 2006; Wetherilt 1992)

From scientific literature data about individual contents of provitamin A in wild edible plants, the amount of vitamin A has been estimated, as RAE, and some species have been selected for providing higher amounts of vitamin A activity among wild vegetables (Fig. 6.11) and wild fruits (Fig. 6.12).

Based on these data, and according to previously mentioned European Regulation 1169/2011, some wild vegetables can be considered as sources of vitamin A activity (more than 0.12 mg RAE/100 g). Special mention should be made of *U. dioica* leaves, providing a very high contribution to daily recommended intake of vitamin A, as well as those of *Rorippa nasturtium-aquaticum* (L.) Hayek, *C. intybus* L., and *Borago officinalis* L., reaching 30% of daily recommended intake of vitamin A for adults. As previously indicated, *C. album* and *Rumex acetosella* L. are also probably rich sources of provitamin A, given their carotenoid contents, but available data of RAE of these species come from analysis of plants gathered in non-Mediterranean areas, such as India. As many of these species are eaten after thermal treatment, some losses after processing could be expected. Yadav and Sehgal (1997) reported losses up to 24% (lower than for other nutrients, such as vitamin C) on different Indian vegetables after thermal process.

Many wild fruits are richer sources of provitamin A than wild vegetables. In Fig. 6.12, a few wild fruits that can be considered as good sources of vitamin A activity (according to European Regulation 1169/2011) are compiled. As can be

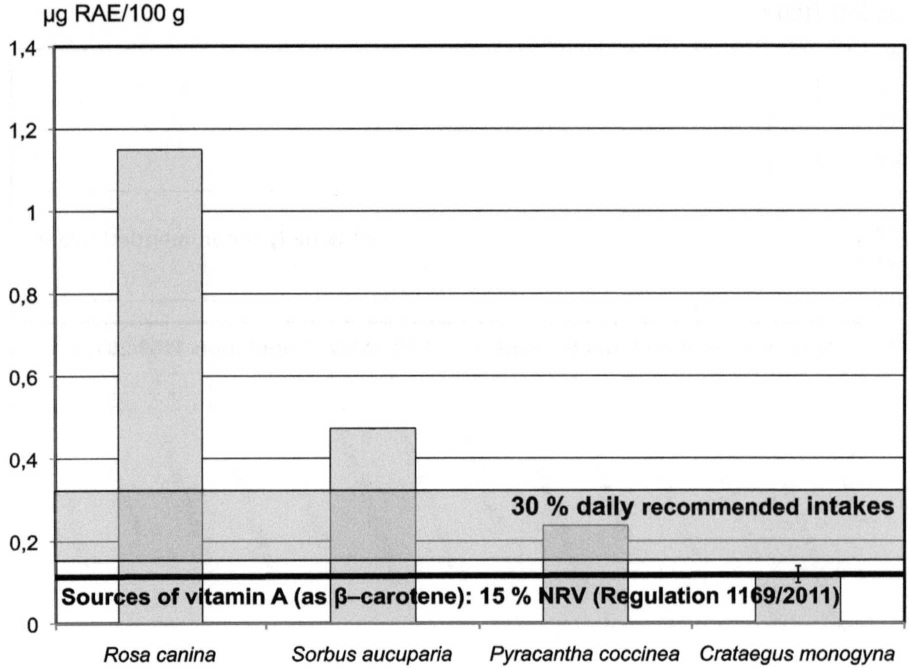

Fig. 6.12 Some vitamin A sources among wild edible fruits (as RAE calculated from individually quantified provitamin A carotenoids contents). (Data from Boudraa et al. 2010; Doležal et al. 2001; Egea et al. 2010; Ruiz-Rodríguez 2014)

seen, rose hips stand out, since a 100-g portion could provide the whole amount of RAE (as β-carotene) daily needed in the human diet. For this reason, the consumption of these fresh fruits would be an excellent strategy to improve the nutritional quality of the human diet.

6.4.2 *Vitamin E*

Vitamin E is the term used to designate a family of chemically related compounds, namely tocopherols and tocotrienols, with a chromanol head and an isoprene side chain (Fig. 6.13). They differ in both the number of methyl substituents and their positions on the phenolic ring (Burton and Traber 1990; Kamal-Eldin et al. 2000).

The most important food sources of vitamin E are fish, meat, eggs, dairy products, and some plant foods such as nuts, beans, and seeds, which are also rich in polyunsaturated lipids. Green leafy vegetables also provide significant amounts ranging between 0.1 and 4.8 mg/100 g, with parsley and cabbage as the richest sources (Souci et al. 2008).

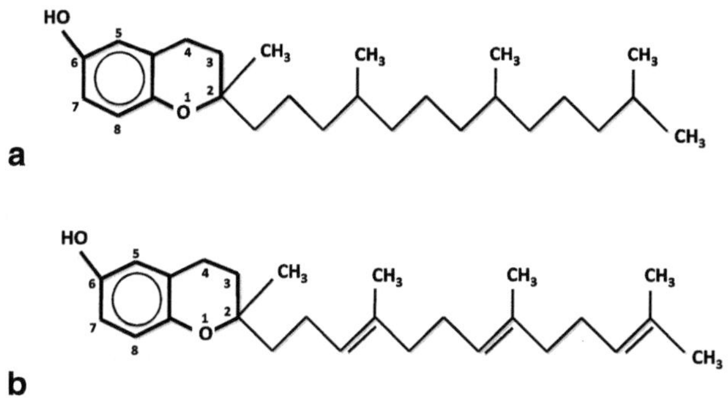

Tocopherol or Tocotrienol	
5,7,8-Trimethyl	α
5,8-Dimethyl	β
7,8-dimethyl	γ
8-Methyl	δ

Fig. 6.13 Chemical structures of vitamin E compounds. **a** Tocol (basic structures of tocopherols). **b** Tocotrienol

Absorption of vitamin E in the intestine depends on adequate pancreatic function, biliary secretion, and micelle formation, and it is schematically represented in Fig. 6.14. Vitamin E is transported in the blood by the plasma lipoproteins and erythrocytes and enters the cells of peripheral tissues within the intact lipoprotein through the low-density lipoprotein (LDL) receptor pathway (Brigelius-Flohe and Traber 1999; Drevon 1991; Traber 2007).

The major form of vitamin E present in plant tissues is α-tocopherol. Moreover, it has been considered the most active form in humans due to the prefer-

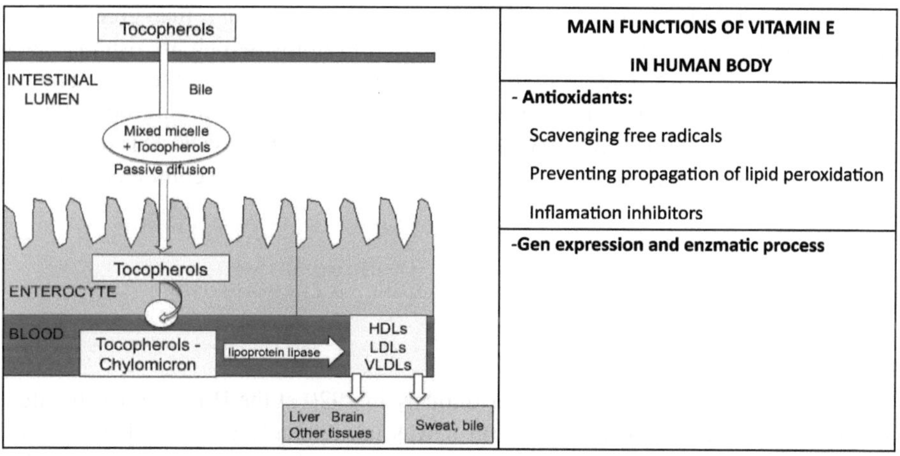

Fig. 6.14 Scheme of the intestinal absorption and functions of vitamin E. (Based on Flórez 2003; Traber 2007)

ential absorption and distribution of this compound in the human body (Caretto et al. 2009). Relative vitamin E activity of tocopherols and tocotrienols has been reported by FAO/WHO (2001). Compared to D-α-tocopherol (100% vitamin E activity), β-tocopherol would provide 50% of vitamin E activity, γ-tocopherol, 10%, δ-tocopherol, 3%, D-α-tocotrienol, 30%, and D-β-tocotrienol, 5%.

Because of its role as a free radical scavenger, vitamin E is also believed to protect against degenerative processes, and other disease states involving oxidative stress, being under intensive investigation nowadays (Burton and Traber 1990; Kamal-Eldin and Schwenke 2002)

A deficiency of vitamin E results in a variety of pathological conditions that affect the muscular, cardiovascular (atherosclerosis, ischemic heart disease), reproductive, and central nervous systems (cerebella ataxia), as well as the liver, kidney, and red blood cells, and development of different types of cancers (Azzi 2004). To avoid these disorders, a daily intake between 7.5–10 mg/day (FAO/WHO) and 12 mg/day (Spanish IR and French ANC) is advisable in the diet, although Trumbo et al. (2002) recommend 15 mg/day as RDA (Table 6.1).

Wild vegetables have been reported to present different tocopherols, α-tocopherol being often the major compound. As this compound is the most active like vitamin E, and taking into account the equivalences of tocopherols activity given by FAO/WHO (2001), the potential vitamin E activity of several wild plants reported in scientific literature has been estimated, as shown in Fig. 6.15. Some wild leafy vegetables have shown enough vitamin E activity to be considered as its sources (according to European Regulation 1169/2011, more than 1.8 mg/100 g). Even if wide natural variability is often found, the richest ones are several species of wild shoots such as *Asparagus acutifolius* and *Humulus lupulus* L., as well as the young stems with leaves of *Montia fontana* L. or the basal leaves of dandelion. Besides the species shown in Fig. 6.15, the tender leaves of *Malva sylvestris* L. have also been reported to contain very high levels of tocopherols (83 mg/100 g on dry basis; in Barros et al. 2010a).

However, wild fruits may be much better sources of vitamin E than greens, probably due to the fact that they contain seeds, rich in lipids, which are usually eaten together with the fruit. The estimation of vitamin E activity from their tocopherols content (Fig. 6.16) shows average contents of α-tocopherol ranging between 2.8 and 5.2 mg/100 g, with strawberry tree fruits being the richest ones. This means that 100 g of these fruits could easily cover around 30% of daily recommended intake of vitamin E.

6.4.3 Vitamin K

The discovery of vitamin K arose fortuitously in 1929 at the Biochemical Institute of the University of Copenhagen, and it was immediately associated with blood coagulation. In the following decade, the principal K vitamers, phylloquinone (K_1),

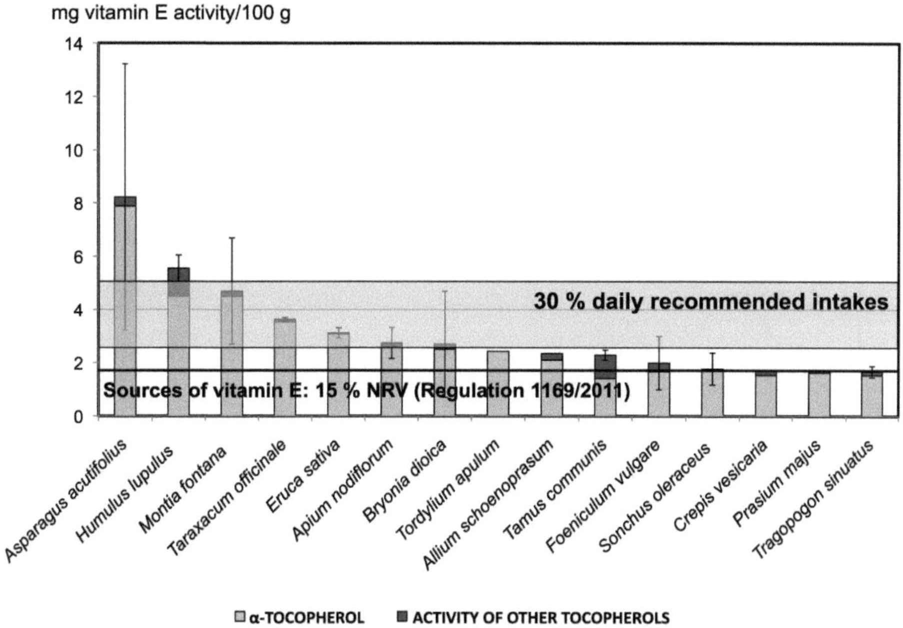

Fig. 6.15 Some vitamin E sources among wild vegetables (vitamin E activity has been calculated taking into account the equivalences of tocopherols activity given by FAO/WHO 2001). (Data from Barros et al. 2010a; Dias et al. 2009; Martins et al. 2011; Morales et al. 2012, 2014; Vardavas et al. 2006; Zeghichi et al. 2005)

synthesized by green vegetables (Fig. 6.17), and the menaquinones (K_2), synthesized by microorganisms, were isolated and fully characterized. Later, a synthetic isomer (menadione) was designed as vitamin K_3 (Dam 1943; FAO/WHO 2001).

Phylloquinone is distributed ubiquitously throughout the diet, and the range of concentrations in different food categories is very wide. In general, the relative values in green leafy vegetables (400–700 μg/100 g) confirm the known association of phylloquinone with photosynthetic tissues. Also, certain legumes and some vegetable oils, such as rapeseed and soybean oils (50–200 μg/100 g), contain phylloquinone (Shearer et al. 1996; Souci et al. 2008).

After the absorption of vitamin K in the intestine (Fig. 6.18), it enters into systemic circulation, chemically unchanged in association with chylomicrons, from which it is taken to the liver where the vitamin can be utilized for the synthesis of clotting factors (Dam 1943; Groenen-van Dooren et al. 1995; Lamon-Fava et al. 1998; Shearer et al. 1974).

Vitamin K deficiency is very uncommon in humans, aside from a small percentage of infants who suffer from hemorrhagic disease of the newborn, a potentially fatal disorder. In adult humans, a prolonged blood clotting time is the predominant, if not sole, clinical sign of vitamin K deficiency. The deficiency is more likely to arise from secondary causes such as malabsorption syndromes or biliary obstruction

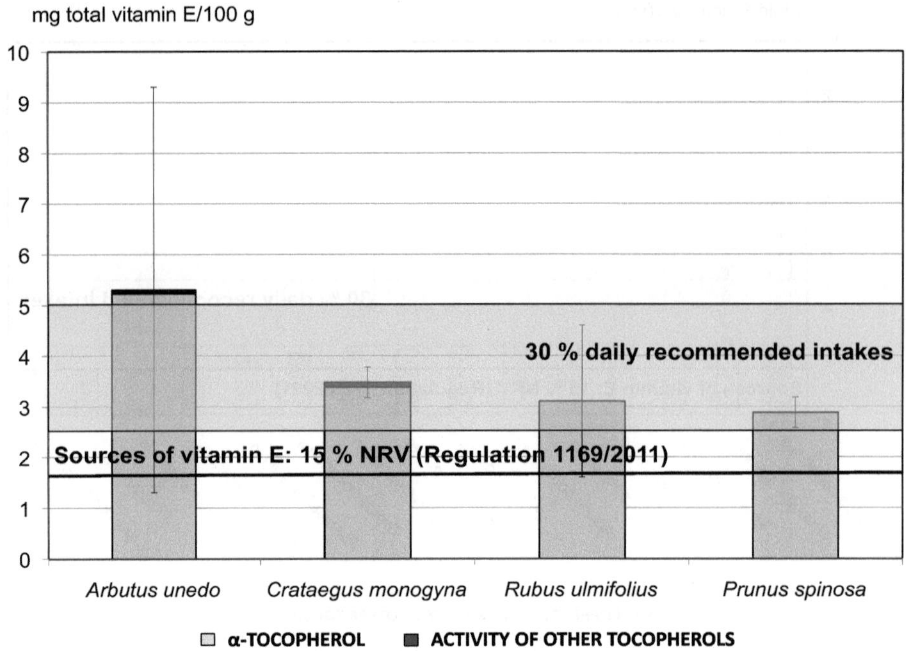

Fig. 6.16 Some vitamin E sources among the edible fruits. (Data from Morales et al. 2013)

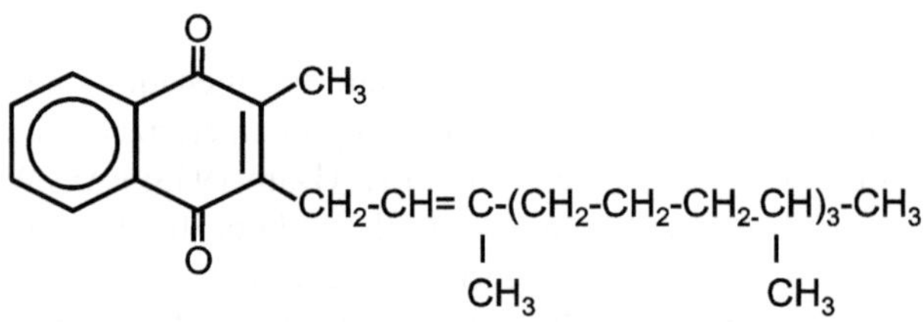

Fig. 6.17 Chemical structure of phylloquinone (vitamin K1)

than from a dietary inadequacy of vitamin K (FAO/WHO 2001). To avoid this, different organisms recommend a daily intake of 45–90 μg of vitamin K, while FNB advises a higher intake of 90–120 μg/day (Table 6.1). Mediterranean countries such as Spain and Italy, as well as the European Commission have not established a specific recommendation for this vitamin.

To the best of our knowledge, there is only one study on vitamin K levels in Mediterranean wild vegetables (Vardavas et al. 2006), showing that most of them can be considered as very good sources of this nutrient (according to European Regulation 1169/2011, providing more than 11.3 μg/100 g) and even more than three times the daily requirements for adults can be provided with a single portion of 100 g, as it can be seen in Fig. 6.19. No toxicity has been associated with this very

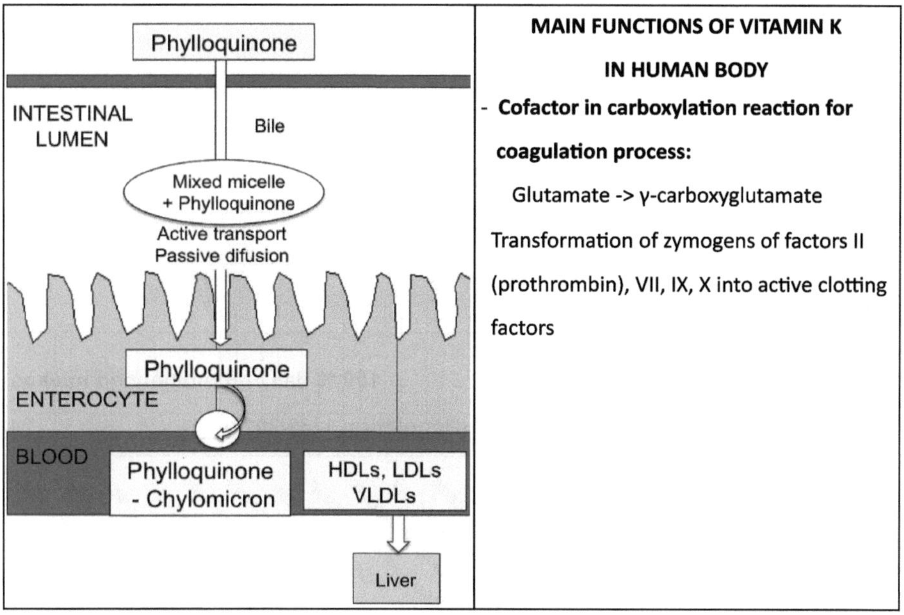

Fig. 6.18 Scheme of the intestinal absorption and functions of vitamin K. (Based on Suttie 1985; Davie 1995)

high intake of phyllochinone (Mahan et al. 2012). *Prasium majus* L., *Rumex obtusifolius* L., *Daucus carota* L., *Petroselinum sativum* Hoffm., *Rorippa nasturtium-aquaticum* (L.) Hayek, *Cichorium spinosum* L., and *Foeniculum vulgare* are just a few examples of the wild species containing very high levels of vitamin K. These contents are in the range of some of the highest values found in many cultivated vegetables such as broccoli, cauliflower, and parsley (240–421 μg/100 g).

6.5 Conclusions

Even considering the natural variability, which is expected to occur in biological tissues, scientific literature presents sufficient evidence to consider many Mediterranean wild edible plants as interesting vitamin sources. A portion of 100 g of them may provide, in many cases, the total amount of vitamins B_9 (folates), C, and/or K daily needed by adults. In other cases, 30 % of daily recommendations of vitamin A and E could also be provided.

If some wild vegetables have to stand out, mention could be made of nettle (*U. dioica*) leaves as a source of vitamin C and provitamin A; *F. vulgare* and *S. vulgaris*, both as sources of vitamins B_9 and K; *C. intybus* as a source of vitamins B_9, K and provitamin A; *A. acutifolius* as a source of vitamins B_9 and E; and many wild crucifer vegetables as good sources of vitamin C.

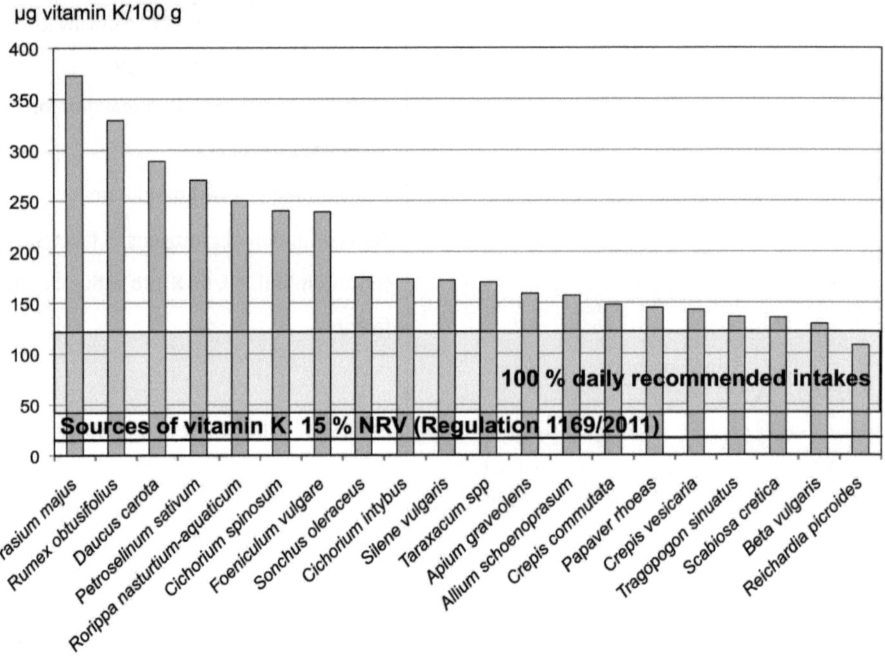

Fig. 6.19 Vitamin K sources among the wild vegetables. (Data from: Vardavas et al. 2006)

Wild fruits are usually also good sources of vitamin C, and especially liposoluble vitamins, such as vitamins E and provitamin A, due to the frequent presence of seeds, carrying a lipid fraction. Strawberry tree (*A. unedo*) fruits are remarkable for being a source of vitamin E, and rose hips (*R. canina*) of vitamin A. Both species are also noticeable for their very high levels of vitamin C.

Although many vegetables are frequently eaten cooked, and some vitamins are highly heat labile, the cooked product may still retain a remarkable portion of the initial contents of vitamins. Fruits are often eaten fresh, so their contribution to vitamin intake in the diet is quite interesting to improve the nutritional status of the Mediterranean populations. Anyway, for fruits or vegetables, the consumption of fresh products, for example, salads, would be advisable, if possible, to retain the great nutritional potential of these wild edible plant foods, which have been traditionally a part of our diet and deserve a renewed interest in modern diets.

References

Alarcão-E-Silva MLCMM, Leitão AEB, Azinheira HG, Leitão MCA (2001) The Arbutus berry: studies on its color and chemical characteristics at two mature stages. J Food Compos Anal 14:27–35

Aliotta G, Pollio A (1981) Vitamin A and C content in some edible wild plants in Italy. Riv Ital EPPOS 63:47–48

Azzi A (2004) The role of α-tocopherol in preventing disease. Eur J Nutr 43(1):18–25
Ball GFM (2006) Vitamins in foods. Analysis, bioavailability and stability. Taylor & Francis Group, Florida
Barros L, Carvalho AM, Ferreira ICFR (2010a) Leaves, flowers, immature fruits and leafy flowered stems of *Malva sylvestris*: a comparative study of the nutraceutical potential and composition. Food Chem Toxicol 48(6):1466–1472
Barros L, Carvalho AM, Morais JS, Ferreira ICFR (2010b) Strawberry-tree, blackthorn and rose fruits: detailed characterisation in nutrients and phytochemicals with antioxidant properties. Food Chem 120(1):247–254
Barros L, Carvalho AM, Ferreira ICFR (2011) Comparing the composition and bioactivity of *Crataegus monogyna* flowers and fruits used in folk medicine. Phytochem Anal 22:181–188
Bauernfeind JC (1972) Carotenoid vitamin A precursors and analogs in foods and feeds. J Agric Food Chem 20:456–473
Becker W, Lyhne N, Pedersen AN, Aro A, Fogelholm M, Phórsdottir I, Alexander J, Anderssen SA, Meltzer SM, Pedersen JI (2004) Nordic Nutrition Recommendations 2004—integrating nutrition and physical activity. Scand J Food Nutr 48:178–187
Bender DA (2003) Vitamin C (ascorbic acid). In: Bender DA (ed) Nutritional biochemistry of the vitamins, 2nd edn. Cambridge University Press, Cambridge, pp 357–384
Blomhoff R (1991) Vitamin A metabolism: new perspectives on absorption, transport, and storage. Physiol Rev 71:951–990
Boudraa S, Hambaba L, Zidani S, Boudraa H (2010) Composition minérale et vitaminique des fruits de cinq espèces sous exploitées en Algérie: *Celtis australis* L., *Crataegus azarolus* L., *Crataegus monogyna* Jacq., *Elaeagnus angustifolia* L. et *Zizyphus lotus* L. Fruits 65(2):75–84
Brigelius-Flohe R, Traber MG (1999) Vitamin E: function and metabolism. FASEB J 13:1145–1155
Britton G, Liaaen-Jensen S, Pfander H (1995) Carotenoids today and challenges for the future. In: Britton G, Liaaen-Jensen S, Pfander H (eds) Carotenoids vol. 1 A: isolation and analysis. Birkhäuser, Basel, p. 13–26
Brubacher G, Müller-Mulot W, Southgate DAT (1985) Methods for determination of vitamins in foods. Elsevier Applied Science Publishers, London
Burton GW, Traber MG (1990) Vitamin E: antioxidant activity, biokinetics, and bioavailability. Annu Rev Nutr 10:357–382
Cámara M, Sánchez-Mata MC, Fernández-Ruiz V, Cámara RM, Manzoor S, Cáceres JO (2013) Lycopene. A review of chemical and biological activity related to beneficial health effects. In: Rahman A (ed) Studies in natural products chemistry, vol 40. Elsevier, pp 383–426
Caretto S, Nisi R, Paradiso A, De Gara L (2009) Tocopherol production in plant cell cultures. Mol Nutr Food Res 54:726–730
Cavelaars AEJM, Kadvan A, Doets EL, Tepsić J, Novaković R, Dhonukshe-Rutten R, Renkema M, Glibetic M, Bucchini L, Matthys C, Smith R, van't Veer P, de Groot LC, Gurinovic M (2010) Nutri-RecQuest: a web-based search engine on current micronutrient recommendations. Eur J Clin Nutr 64(Suppl 2):S43–S47
Conseil Supérieur de la Santé (2009) Recommandations nutritionnelles pour la Belgique (Révision 2006). N° 8309, Bruxelles. http://www.health.fgov.be/css_HgR/brochures. Accessed 15 Jan 2015
Correa P (1992) Human gastric carcinogenesis. Cancer Res 52:6735–6740
Cuervo M, Corbalán M, Baladía E, Cabrerizo L, Formiguera X, Iglesias C, Lorenzo H, Polanco I, Quiles J, Romero de Ávila MD, Russolillo G, Villarino A, Alfredo Martínez J (2009) Comparison of dietary reference intakes (DRI) between different countries of the European Union, The United States and the World Health Organization. Nutr Hosp 24(4):384–414
Cuskelly G, McNulty H, Scott J (1996) Effect of increasing dietary folate on redcell folate: implications for prevention of neural tube defects. Lancet 347:657–659
Dam H (1943) The discovery of vitamin K, its biological functions and therapeutical applications. Nobel lecture. http://www.nobelprize.org/nobel_prizes/medicine/laureates/1943/dam-lecture.pdf. Accessed 15 Jan 2015

Davie EW (1995) Biochemical and molecular aspects of the coagulation cascade. Thromb Haemost 74:1–6

Department of Health of UK (1991) Dietary reference values for food energy and nutrients in the United Kingdom. HMSO, London

Deutsche Gesellschaft für Ernährung, Österreichische Gesellschaft für Ernährung, Schweizerische Gesellschaft für Ernährungsforschung, Schweizerische Vereinigung für Ernährung (2000) D-A-CH Referenzwerte: Referenzwerte für die Nährstoffzufuhr. Umschau/Braus Verlag, Frankfurt

Dew S, Ong DE (1994) Specificity of the retinol transporter of the rat small intestine brush border. BioChemistry 33:12340–12345

Dias MG, Camoes MFGFC, Oliveira L (2009) Carotenoids in traditional Portuguese fruits and vegetables. Food Chem 113(3):808–815

Doets EL, de Wit LS, Dhonukshe-Rutten RA, Cavelaars AE, Raats MM, Timotijevic L, Brzozowska A, Wijnhoven TM, Pavlovic M, Totland TH, Andersen LF, Ruprich J, Pijls LT, Ashwell M, Lambert JP, van't Veer P, de Groot LC (2008) Current micronutrient recommendations in Europe: towards understanding their differences and similarities. Eur J Nutr 47(Suppl 1):17–40

Doležal M, Velíšek J, Famfulíková P (2001) Chemical composition of less known wild fruits. In: Pfannhauser W, Fenwick GR, Khokhar S (eds) Biologically-active phytochemicals in food analysis, metabolism, bioavailability and function. Royal Society of Chemistry, London, pp 241–244

Drevon CA (1991) Absorption, transport, and metabolism of vitamin E. Free Rad Res Commun 14:229–246

Egea I, Sánchez-Bel P, Romojaro F, Pretel MT (2010) Six edible wild fruits as potential antioxidant additives or nutritional supplements. Plant Foods Hum Nutr 65:121–129

Erdman JW Jr, Bierer TL, Gugger ET (1993) Absorption and transport of carotenoids. Ann NY Acad Sci 691:76–85

European Food Information Council (2009). EUFIC Review 10/2009: Food-based dietary guidelines in Europe. www.eufic.org/article/en/expid/food-based-dietary-guidelines-in-europe/. Accessed 15 Jan 2015

European Parliament and Council (2011) Regulation (EU) No 1169/2011 of the European Parliament and of the Council of 25 October 2011 on the provision of food information to consumers. Official Journal of the European Union, 22.11.2011: L304/18-L304/63

EURRECA Network of Excellence (2013) European commission contract n°FP6 036196. http://www.eurreca.org. Accessed 15 Jan 2015

FAO (1957) Calorie requirements: report of the Second Committee on calorie requirements. FAO Nutritional Studies N 15. FAO, Rome

FAO/WHO (1996). Preparation and use of food-based dietary guidelines. Report of a joint FAO/WHO consultation Nicosia, Cyprus. http://www.fao.org/docrep/X0243E/x0243e00.htm. Accessed 15 Jan 2015

FAO/WHO (2001) Human vitamin and mineral requirements. Report of a joint FAO/WHO expert consultation, Bangkok, Thailand. Food and Nutrition Division. FAO, Rome

Flórez J (2003) Farmacología Humana. Ed. Masson, Barcelona

FNB (Food and Nutrition Board) of the National Academy of Sciences (1943) Recommended dietary allowances. National Research Council Reprint and Circular Series. Washington DC

FNB (Food and Nutrition Board) of the Institute of Medicine (2000) Dietary reference intakes: applications in dietary assessment. National Academy Press, Washington DC

Franke W, Kensbock A (1981) Vitamin C content of native wild-growing vegetables and greens. Ernährungs Umschau 28:187–191

García-Herrera P (2014) Plantas silvestres de consumo tradicional. Caracterización de su valor nutricional y estimación de su actividad antifúngica. PhD Thesis. Universidad Complutense de Madrid, Facultad de Farmacia, Departamento de Nutrición y Bromatología II

Gerber M (2001) Protective vegetal micronutrients and microcomponents for breast cancer. Bull Cancer 88(10):943–953

Groenen-van Dooren MMCL, Ronden JE, Soute BAM, Vermeer C (1995) Bioavailability of phylloquinone and menaquinones after oral and colorectal administration in vitamin K-deficient rats. Biochem Pharmacol 50:797–801

Guil-Guerrero JL (2001) Nutritional composition of *Plantago* species (*P. major* L., *P. lanceolata* L., and *P. media* L.). Ecol Food Nutr 40(5):481–495

Guil-Guerrero JL, Torija Isasa ME (1997) Nutritional composition of leaves of *Chenopodium* species (*C. album* L., *C. murale* L. and *C. opulifolium* Shraeder). Int J Food Sci Nutr 48:321–327

Guil-Guerrero JL, Giménez-Giménez A, Rodriguez-Garcia I, Torija-Isasa ME (1998) Nutritional composition of *Sonchus* species (*S. asper* L., *S. oleraceus* L. and *S. tenerrimus* L.). Sci Food Agric 76:628–632

Guil-Guerrero JL, Giménez-Martínez JJ, Torija-Isasa ME (1999) Nutritional composition of wild edible crucifer species. J Food Biochem 23:283–294

Guil-Guerrero JL, Rebolloso–Fuentes MM, Torija-Isasa ME (2003) Fatty acids and carotenoids from stinging nettle (*Urtica dioica* L.). J Food Compos Anal 16:111–119

Halliwell B (2001) Vitamin C and genomic stability. Mutat Res 475:29–35

Hughes RE (1975) James Lind and the cure of scurvy: an experimental approach. Med Hist 19(4):342–351

Ibrahim K, Hassan TJ, Jafarey SN (1991) Plasma vitamin A and carotene in maternal and cord blood. Asia Oceania J Obstet Gynaecol 17:159–164

Jabłońska-Ryś E, Zalewska-Korona M, Kalbarczyk J (2009) Antioxidant capacity, ascorbic acid and phenolics content in wild edible fruits. J Fruit Ornam Plant Res 17(2):115–120

Kamal-Eldin A, Stefanie G, Jan P, Anna-Maija L (2000) Normal-phase high-performance liquid chromatography of tocopherols and tocotrienols. Comparison of different chromatographic columns. J Chromatogr A 881:217–227

Kiliç CS, Coskun M (2007) *Capsella bursa-pastoris* (l.) Medik. (Cruciferae) askorbik asit Içerigi üzerinde karsilastirmali bir çalisma (A comparative study on the ascorbic acid content of *Capsella bursa pastoris* (L.) Medik. (Cruciferae). J Fac Pharm Ankara 36(3):153–160

Kim Y (1999) Folate and cancer prevention: a new medical application of folate beyond hyperhomocysteinemia and neural tube defects. Nutr Rev 57(10):314–321

Kruger M, Sayed N, Langenhoven M, Holing F (1998) Composition of South African foods: vegetables and fruit. Medical Research Council, Cape Town

Kudritsata SE, Filman GM, Zagorodskaya LM, Chikovanii DM (1987) Carotenoids of *Urtica dioica*. Khimiya Priridnykh Soedinenii 6(5):640–641

Kuhnlein HV (1990) Nutrient values of indigenous wild plant greens and roots used by the Nuxalk people of Bella Coole, British Columbia. J Food Compos Anal 3:38–46

Lamon-Fava S, Sadowski JA, Davidson KW, O'Brien ME, McNamara JR, Schaefer EJ (1998) Plasma lipoproteins as carriers of phylloquinone (vitamin K1) in humans. Am J Clin Nutr 67:1226–1231

Lucock M (2000) Folic acid: nutritional biochemistry, molecular biology, and role in disease processes. Mol Genet Metab 71:121–138

Mahan LK, Escott-Stump S, Raymond JL (2012) Krause Dietoterapia, 13th edn. Elsevier, Madrid

Maiani G, Castón MJ, Catasta G, Toti E, Cambrodón IG, Bysted A, Granado-Lorencio F, Olmedilla-Alonso B, Knuthsen P, Valoti M, Böhm V, Mayer-Miebach E, Behsnilian D, Schlemmer U (2009) Carotenoids: actual knowledge on food sources, intakes, stability and bioavailability and their protective role in humans. Mol Nutr Food Res 53:S194–S218

Malo C, Wilson JX (2000) Glucose modulates vitamin C transport in adult human small intestinal brush border membrane vesicles. J Nutr 130:63–69

Marchioli R (1999) Antioxidant vitamins and prevention of cardiovascular disease: laboratory, epidemiological and clinical trial data. Pharmacol Res 40(3):227–238

Martin A (2001) Apports nutritionnels conseilles pour la population française, 3rd edn. Agence Française de Securité Sanitaire des Aliments (AFSSA). Tec et Doc Lavoisier, París

Martins D, Barros L, Carvalho AM, Ferreira ICFR (2011) Nutritional and in vitro antioxidant properties of edible wild greens in Iberian Peninsula traditional diet. Food Chem 125(2):488–494

Morales P (2011) Vegetales silvestres de uso alimentario. Determinación de compuestos bioactivos y valoración de la capacidad antioxidante. Thesis, Universidad Complutense de Madrid. Madrid, Spain

Morales P, Carvalho AM, Sánchez-Mata MC, Cámara M, Molina M, Ferreira ICFR (2012) Tocopherol composition and antioxidant activity of Spanish wild vegetables. Genet Resour Crop Evol 59:851–863

Morales P, Ferreira IC, Carvalho AM, Fernández-Ruiz V, Sánchez-Mata MC, Cámara M, Morales R, Tardío J (2013) Wild edible fruits as a potential source of phytochemicals with capacity to inhibit lipid peroxidation. Eur J Lipid Sci 115(2):176–185

Morales P, Ferreira ICFR, Carvalho AM, Sánchez-Mata MC, Cámara M, Fernández-Ruiz V, Pardo-de-Santayana M, Tardío J (2014) Mediterranean non-cultivated vegetables as dietary sources of compounds with antioxidant and biological activity. LWT-Food Sci Technol 55:389–396

Morales P, Fernández-Ruiz V, Sánchez-Mata MC, Cámara M, Tardío J (2015) Optimization and application of FL-HPLC for folates analysis in 20 species of Mediterranean wild vegetables. Food Anal Method 8:302–3011

Moreiras O, Carbajal A, Cabrera L, Cuadrado C (2011) Tablas de composición de alimentos, 14th edn. Pirámide, Madrid

Nutrition Sub-committee of the Food Safety Authority of Ireland (1999) Recommended dietary allowances for Ireland. Food Safety Authority of Ireland, Dublin

Paddayatty SJ, Levine M (2001) New insights into the physiology and pharmacology of vitamin C. CMAJ 164:353–355

Pallauf K, Rivas-Gonzalo JC, del Castillo MD, Cano MP, Pascual-Teresa S (2008) Characterization of the antioxidant composition of strawberry tree (*Arbutus unedo* L.) fruits. J Food Compos Anal 21(4):273–281

Parker RS (1996) Absorption, metabolism, and transport of carotenoids. FASEB J 10:542–551

Parker RS (1997) Bioavailability of carotenoids. Eur J Clin Nutr 51(Suppl 1):S86–S90

Patton S, Canfield LM, Huston GE, Ferris AM, Jensen RG (1990) Carotenoids of human colostrum. Lipids 25:159–165

Pereira C, Barros L, Carvalho AM, Ferreira IC (2013) Use of UFLC-PDA for the analysis of organic acids in thirty-five species of food and medicinal plants. Food Anal Method 6(5):1337–1344

Phillips KM, Wunderlich KM, Holden JM, Exler J, Gebhardt SE, Haytowitz DB (2005) Stability of 5-methyltetrahydrofolate in frozen fresh fruits and vegetables. Food Chem 92:587–595

Ruiz-Rodríguez BM (2014) Frutos silvestres de uso tradicional en la alimentación. Evaluación de su valor nutricional, compuestos bioactivos y capacidad antioxidante. PhD Thesis. Universidad Complutense de Madrid, Facultad de Farmacia, Departamento de Nutrición y Bromatología II

Ruiz-Rodríguez BM, Morales P, Fernández-Ruiz V, Sánchez-Mata MC, Cámara M, Díez-Marqués C, Pardo-de-Santayana M, Molina M, Tardío J (2011) Valorization of wild strawberry-tree fruits (*Arbutus unedo* L.) through nutritional assessment and natural production data. Food Res Int 44(5):1244–1253

Ruiz-Rodríguez BM, De Ancos B, Sánchez-Moreno C, Fernández-Ruiz V, Sánchez-Mata MC, Cámara M, Tardío J (2014a) Wild blackthorn (*Prunus spinosa* L.) and hawthorn (*Crataegus monogyna* Jacq.) fruits as valuable sources of antioxidants. Fruits 69(1):61–73

Ruiz-Rodríguez BM, Sánchez-Moreno C, De Ancos B, Sánchez-Mata MC, Fernández-Ruiz V, Cámara M, Tardío J (2014b) Wild *Arbutus unedo* L. and *Rubus ulmifolius* Schott fruits are underutilized sources of valuable bioactive compounds with antioxidant capacity. Fruits 69(6):435–448

Salvatore S, Pellegrini N, Brenna OV, Del Rio D, Frasca G, Brighenti F, Tumino R (2005) Antioxidant characterization of some Sicilian edible wild greens. J Agric Food Chem 53:9465–9471

Sánchez-Mata MC, Cabrera-Loera RD, Morales P, Fernández-Ruiz V, Cámara M, Díez-Marqués C, Pardo-de-Santayana M, Tardío J (2012) Wild vegetables of the Mediterranean area as valuable sources of bioactive compounds. Genet Resour Crop Evol 59(3):431–443

Schwenke DC (2002) Does the lack of tocopherols and tocotrienols put women at increased risk of breast cancer? J Nutr Biochem 13:2–20

Scientific Committee on Foods of the European Union (1992) Nutrient and energy intakes for the European community. Opinion adopted by the SCF on 11 December 1992. In: Reports of the SCF (ed.), 31st series. European Commission, Luxembourg

Scott JM, Weir DG, Molloy A, MacPartlin J, Daly L, Kirke P (1994) Folic acid metabolism and mechanisms of neural tube defects. CIBA Foundation Symposium 181:180–191

Shai I, Stampfer MJ, Mab J, Manson JE, Hankinson SE, Cannuscio C, Selhub J, Curhan G, Rimn EB (2004) Homocysteine as a risk factor for coronary heart diseases and its association with inflammatory biomarkers, lipids and dietary factors. Atherosclerosis 177:375–381

Shearer MJ, McBurney A, Barkhan P (1974) Studies on the absorption and metabolism of phylloquinone (vitamin K1) in man. Vitam Horm 32:513–542

Shearer MJ, Bach A, Kohlmeier M (1996) Chemistry, nutritional sources, tissue distribution and metabolism of vitamin K with special reference to bone health. J Nutr 126:1181S–1186S

Šircelj H, Mikulic-Petkovšek M, Batic F (2010) Antioxidants in spring leaves of *Oxalis acetosella* L. Food Chem 123:351–357

Società Italiana di Nutrizione Umana (1998) Livelli di assunzione raccomandati di energia e nutrienti per la poplazione Italiani, revisione 1996. EDRA, Milano

Somsub W, Kongkachuichai R, Sungpuag P, Charoensiri R (2008) Effects of three conventional cooking methods on vitamin C, tannin, myoinositol phosphates contents in selected Thai vegetables. J Food comps Anal 21:187–197

Souci SW, Fachmann W, Kraut H (2008) Food composition and nutrition tables. Medpharm, Sttutgart

Steegers-Theunissen RP (1995) Folate metabolism and neural tube defects: a review. Eur J Obstet Gynecol Reprod Biol 61:39–48

Suttie JW (1985) Vitamin K. In: Diplock AD (ed) Fat-soluble vitamins: their biochemistry and applications. Heinemann, London, pp 225–311

Traber MG (2007) Vitamin E regulatory mechanisms. Annu Rev Nutr 27:347–362

Trumbo P, Schlicker S, Yates AA, Poos M (2002) Dietary reference intakes for energy, carbohydrate, fiber, fat, fatty acids, cholesterol, protein and amino acids. J Am Diet Assoc 102:1621–1630

Vanizor Kural B (2003) Plasma homocysteine and its relationships with atherotrombotic markers in psoriatic patients. Clin Chim Acta 332:23–30

Vardavas CI, Majchrzak D, Wagner KH, Elmadfa I, Kafatos A (2006) The antioxidant and phylloquinone content of wildly grown greens in Crete. Food Chem 99:813–821

Wetherilt H (1992) Evaluation of *Urtica* species as potential sources of important nutrients. Food Sci Hum Nutr 29:15–26

WHO European Region (2003) Food based dietary guidelines in the WHO European Region. http://www.euro.who.int/__data/assets/pdf_file/0017/150083/E79832.pdf. Accessed 15 Jan 2015

Wright AJA, King MJ, Finglas PM (2010) Folate-supplemented oral contraceptives: does 6 S-5-methyltetrahydrofolic acid (Metafolin) offer advantages over folic acid? Gynaecol Forum 15:29–32

Yadav SK, Sehgal S (1997) Effect of home processing and storage on ascorbic acid and, B-carotene content of Bathua *(Chenopodium album)* and fenugreek *(Trigonellafoenum graecum)* leaves. Plant Foods Hum Nutr 50:239–247

Yildrim E, Dursuna A, Turan M (2001) Determination of the nutrition contents of the wild plants used as vegetables in Upper Çoruh Valley. Turk J Bot 25:367–371

Zeb A, Mehmood S (2004) Carotenoids contents from various sources and their potential health applications. Pak J Nutr 3(3):199–204

Zeghichi S, Kallithraka S, Simopoulos AP, Kypriotakis Z (2005) Nutritional composition of selected wild plants in the diet of Crete. In: Simopoulos AP, Gopalan C (eds) Plants in human health and nutrition policy. Karger, Basel, pp 22–40

Chapter 7
The Contribution of Wild Plants to Dietary Intakes of Micronutrients (II): Mineral Elements

Patricia García-Herrera and María de Cortes Sánchez-Mata

7.1 Introduction: Classification and Dietary Requirements of Minerals

Many different mineral elements are spread in biological tissues, as either soluble ions (cations or anions) or part of organic molecules, such as hemoglobin (Fe) and cobalamin (Co) in animal or human tissues or chlorophyll (Mg) in plant tissues. Many of them are well known for their physiological functions and requirements. From a nutritional point of view, mineral elements have been classified into the following groups:

- *Macroelements*, needed in amounts of about 100 mg/day or higher, such as potassium, sodium, calcium, magnesium or phosphorus
- *Microelements* or *oligoelements*, needed in lower amounts but may be essential to maintain body functions, such as iron, zinc or manganese. Those needed in very low amounts are also called oligoelements or trace elements (e.g., selenium or chromium)

In general, minerals are involved in several functions such as the regulation of enzymes activities, the preservation of osmotic and acid–base equilibrium, membrane transport mechanisms, and muscular and neural transmissions. They may also have a structural role, being constituents of bones or other tissues (Mahan et al. 2012).

As previously explained (see Sect. 6.1), many national and international organisms, institutions, and committees and scientific societies, such as Food and Agricultural Organisation/World Health Organization (FAO/WHO), Food and Nutrition Board (FNB) of the American Institute of Medicine, Scientific Committee on Foods (SCF) of the European Union (EU), among others, have established nutritional

M. C. Sánchez-Mata (✉) · P. García-Herrera
Departamento de Nutrición y Bromatología II, Bromatología, Facultad de Farmacia, Universidad Complutense de Madrid, Madrid, Spain
e-mail: cortesm@ucm.es; patrigarcia31@hotmail.com

M. de C. Sánchez-Mata, J. Tardío (eds.), *Mediterranean Wild Edible Plants*,
DOI 10.1007/978-1-4939-3329-7_7

Table 7.1 Recommended intakes of mineral elements (mg/day) for adults, most widely accepted for Mediterranean countries

	International recommendations			National recommendations			
	FAO/WHO, RNI[a]	EU, PRI[b]	EU, NRV[c]	Spain, IR[d]	Italy, LRA[e]	France, ANC[f]	USA, DRI[g]
K							
Males		3100	2000	3500	3100		
Females							
P							
Males		550	700	700	800–1000	750	700
Females						750–800	
Ca							
Males	1000–1300	700	800	800	800–1000	900–1200	1000–1200[i]
Females					800–1500[h]		
Mg							
Males	230–260		375	350	150–500	420	400–420
Females	190–220			300–330		360	310–320
Fe							
Males	9–27[j]	9	14	10	10	9	8
Females	8–59[j,k]	20		10–18[k]	10–18[k]	9–16[k]	8–18[k]
Mn							
Males			2				2.3[i]
Females							1.8[i]
Zn							
Males	4.2–14	9.5	10	15	10	10–13	11
Females	3–9.8	7			7		0.9
Cu							
Males		1.1	1		1.2	1.5–2	
Females							

[a] *RNI* Recommended nutrient intake (FAO/WHO 2002)
[b] *PRI* Population reference intake (SCF 1992)
[c] *NRV* Nutrient reference values, *EU* European Union Regulation 1169/2011
[d] *IR* Ingestas Recomendadas (Moreiras et al. 2011)
[e] *LRA* Livelli Raccomandati di Assunzione (*SINU* Società Italiana di Nutrizione Umana 1998)
[f] *ANC* Apports Nutritionnels Conseillés (Martin 2001)
[g] *DRI* Dietary reference intakes (Recommended Dietary Allowances or Adequate Intakes), *FNB* Food and Nutrition Board (Trumbo et al. 2002)
[h] 1500 mg is advisable for females over 50 years old due to lack of estrogenic protection against osteoporosis
[i] Adequate intakes (no sufficient evidence to establish RDA)
[j] Depending on bioavailability
[k] Higher values are recommended for fertile age women; after menopause, requirements are similar for males and females

recommendations for daily intake of minerals (Cuervo et al. 2009). The European Parliament and Council (2011), through Regulation 1169/2011, have also established nutrient reference values (NRVs) for food-labelling purposes. Some of the most accepted recommendations and reference values for mineral elements intake are recorded in Table 7.1.

7.2 Macroelements

7.2.1 Sodium

Sodium is an important element to maintain electrolyte equilibrium in an organism. Its absorption and main functions in the human body are summarized in Fig. 7.1. Sodium deficit causes hyponatremia, which is not frequent. However, an excessive intake of Na in the diet may lead to arterial hypertension, pulmonary edema, cardiovascular and renal damage as well as an increased urinary excretion of Ca, being a potential risk factor for osteoporosis (Badui 2006; Mahan et al. 2012; Sacks et al. 2001). For that reason, nowadays, the limitation of the presence of Na in the diet is a key point in many public health programs promoted by WHO, FAO, or EU, which focus on the prevention of chronic diseases such as hypertension, cardiovascular disease, osteoporosis, and renal problems.

In this way, the recommendations of SCF, FNB, or UK Department of Health range a reference intake of 0.5–3.4 g Na/day. Nowadays, most countries' food-based dietary guidelines advice on lower salt intake as it is the form in which Na is added to foods. WHO recommends 5 g salt per day as a maximum (equivalent to 2 g Na/day), and the European Regulation 1169/2011 establishes a reference intake of 6 g salt per day (2.4 g Na/day) for food nutrition labelling purposes. Recent studies on the actual Na dietary intake of the European population found an intake of 8–11 g salt (3–5 g Na) per day, which is much higher than the recommendations. It should be taken into account that Na is naturally present in most natural or manufactured foods (10–15% and 70–75% of total Na intake, respectively), and so, the real Na intake will be always higher than that provided by salt. For that reason, besides the limitation in salt use, also the selection of preferably low-Na foods in the diet is

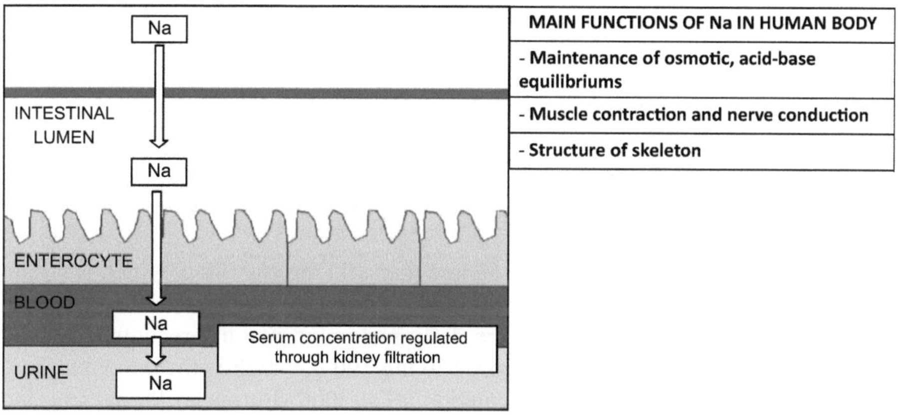

Fig. 7.1 Scheme of the intestinal absorption and main functions of Na. (Data from Mahan et al. 2012)

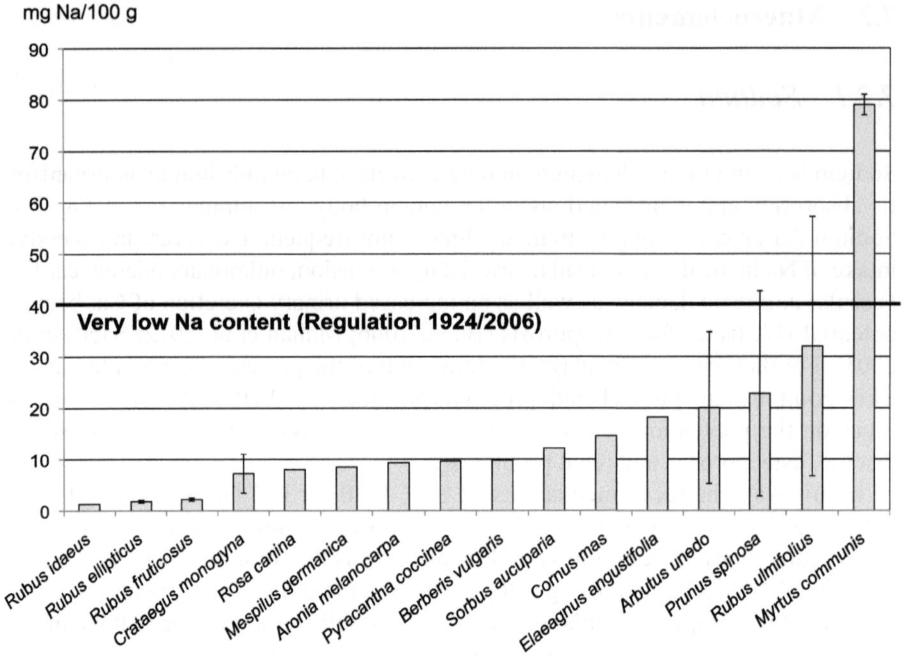

Fig. 7.2 Selected wild fruits standing out for their low Na levels. (Data from Dolezal et al. 2001; Hacıseferoğulları et al. 2012; Marakoglu et al. 2005; Özcan and Hacıseferoğulları 2007; Ruiz-Rodríguez et al. 2011; Saklani et al. 2012; Ruiz-Rodríguez 2014)

necessary to keep Na intake at the minimum, and thus contribute to a better health status (European Food Safety Authority 2005, 2009).

In this context, foods from vegetal origin should be preferred, among other reasons, for their relatively low Na content, compared with animal-origin foods. Fruits and vegetables are a very good choice for people who want to reduce Na intake, and wild plants provide foods quite recommendable from this point of view. Na seems to be preferably accumulated in stems or leaves, and thus, wild fruits have usually shown lower Na levels than wild vegetables (Figs. 7.2 and 7.3); the fruits of *Rubus* spp. stand out for their very low Na levels. A wide natural variability is found in Na levels in plants due to many factors; one of the most important is soil composition, which may influence to a high extent the amount of minerals in plant tissues (Sun et al. 2013). Some examples of Na content in plants growing on Na-rich soils are mentioned below.

Wild vegetables contain, in general, less than 100 mg Na/100 g, which is within the normal levels in most vegetables (either wild or cultivated, according to Souci et al. 2008), and represent 5% of the daily maximum amount recommended by WHO. Most wild vegetables are below 120 mg/100 g, which is the maximum level established by the European Regulation 1924/2006 (European Parliament and Council 2006) to consider a food as "low Na content", and even many of them are below 40 mg/100 g ("very low Na content"). Considering samples from a different

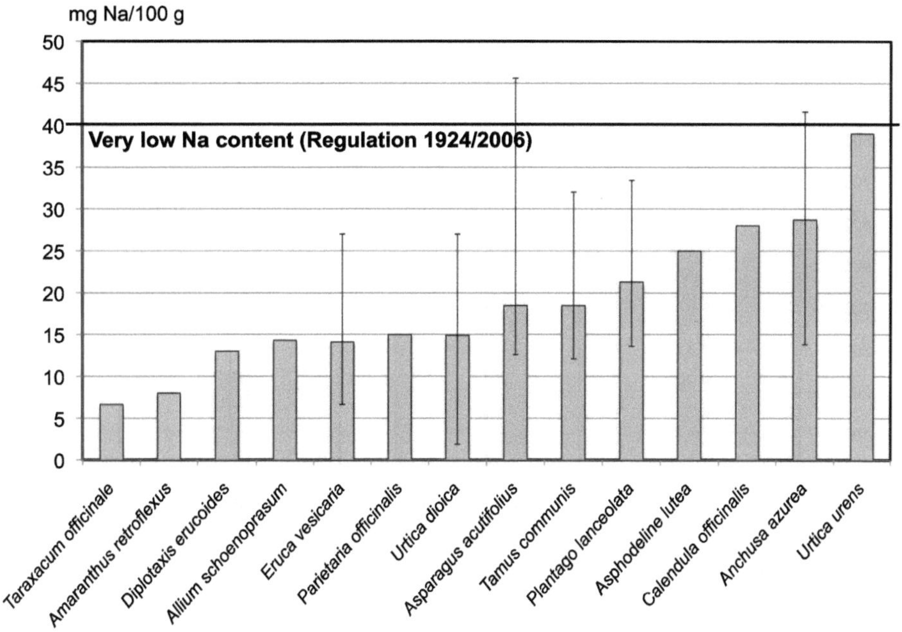

Fig. 7.3 Selected wild vegetables standing out for their low Na levels. (Data from Martínez-Para et al. 1979; Adamski and Bieganska 1980; Ayan et al. 2006; Barlas et al. 2011; Bianco et al. 1998; Bockholt and Schnittke 1996; García-Herrera 2014; Guil-Guerrero 2001; Krstic-Pavlovic and Dzamic 1985; Queralt et al. 2005; Trichopoulou et al. 2000; Villatoro-Pulido et al. 2012; Wetherilt 1992; Zeghichi et al. 2005)

origin, which include a wide natural variability, many wild edible fruits and greens can be considered as presenting very low Na contents such as for example the leaves of *Taraxacum officinale* (L.) Weber ex F.H.Wigg., some *Urtica* species and two members of Cruciferae family, *Diplotaxis erucoides* (L.) DC., and *Eruca vesicaria* (L.) Cav. (Figs. 7.2 and 7.3). Although quite variable, the wild shoots of *Asparagus acutifolius* L. as well as the leaves of *Anchusa azurea* Mill. can also be considered as poor-Na foods.

Moreover, attention should be paid to some species that have shown high levels of Na, which are unusual in plant foods. Figure 7.4 shows some wild greens containing more than 180 mg Na/100 g. This is the case of many halophytes, such as *Salicornia europaea* L., *Parietaria judaica* L., *Crithmum maritimum* L., *Chenopodium spp.* and *Cakile maritima* Scop. Some of them reach up to 900 mg/100 g (Guil-Guerrero et al. 1996a, 1999a), being equivalent to more than 2.25 g of salt in the non-salted raw product, which means a contribution of up to 45 % of the maximum daily amount recommended by WHO. In a lesser extent, some *Sonchus* species have shown average levels of around 150 mg/100 g (see Chap. 13). People who must reduce Na in their diet should avoid these plants or boil them in a high water/plant proportion since this mineral is easily dissolved in the cooking liquid. For example, García-Herrera (2014) found losses of up to 50 % of the initial Na

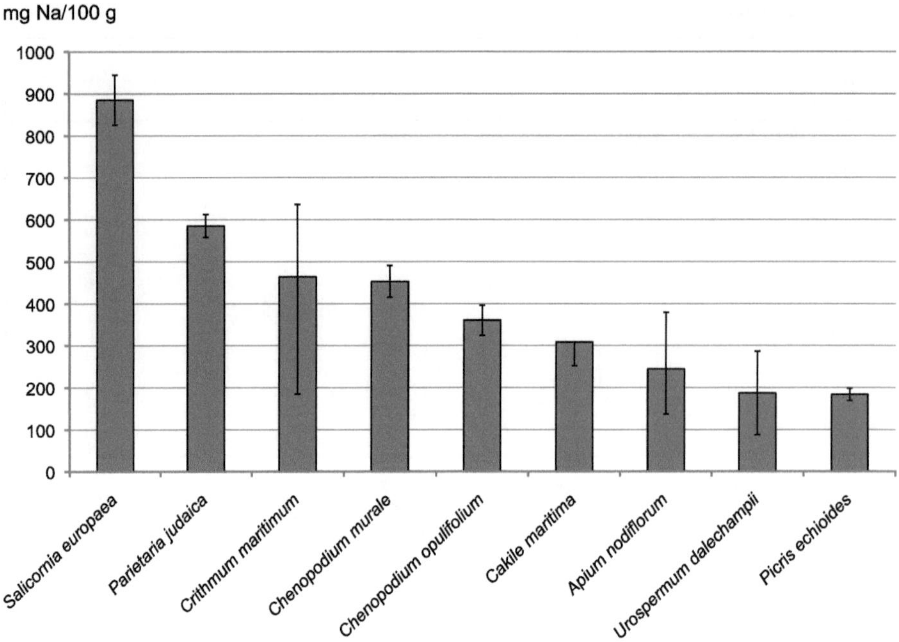

Fig. 7.4 Selected wild vegetables containing more than 150 mg Na/100 g. (Data from Bianco et al. 1998; García-Herrera et al. 2014; Guil-Guerrero and Torija-Isasa 1997; Guil-Guerrero et al. 1996a, 1998b, 1999a, b; Romojaro et al. 2013)

content in some boiled wild leafy vegetables, and this can be corroborated in the composition data of some boiled wild vegetables recorded in Chap. 13.

7.2.2 Potassium

The main sources of potassium in the human diet are plant foods as well as milk and meat. Potassium is widespread in nature, being present in many foods. It is easily absorbed, acting in different functions in the human body (Fig. 7.5). Its deficiency is rare; however, diets poor in fruits and vegetables are related to inadequate potassium intake, which could result in muscle weakness and atrial fibrillation (Mahan et al. 2012). The most widely accepted potassium requirements for adults are in the range of 3100–3500 mg/day (Table 7.1).

From previously published scientific data, most wild edible plants usually contain more than 300 mg K/100 g (minimum level to be considered as K sources according to Regulation 1169/2011), and many of them range into 500–1000 mg/100 g as an average (see Chap. 13), which are habitual values for vegetables (up to 500 mg/100 g in stems or bulbs and up to 1000 mg/100 g in leaves, according to Souci et al. 2008). Some wild greens can be stood out for containing unusually high potassium levels,

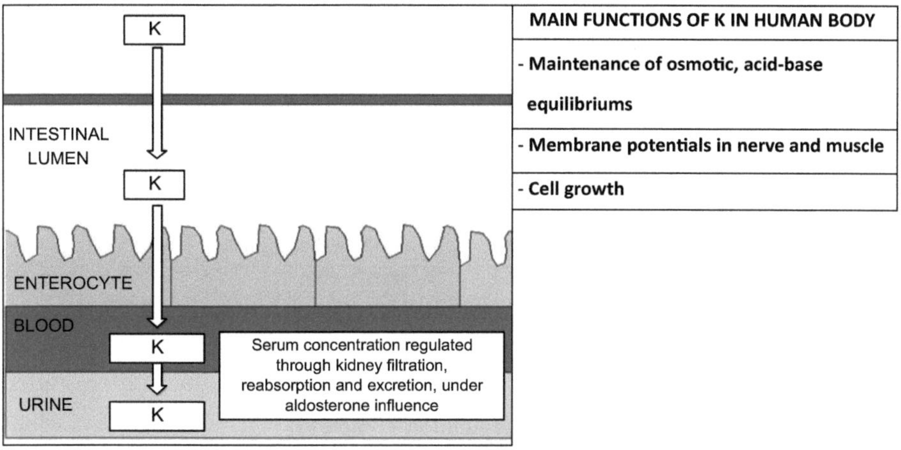

Fig. 7.5 Scheme of the intestinal absorption and main functions of K. (Data from Mahan et al. 2012)

such as *Beta maritima* L., *Chondrilla juncea, Camelina rumelica* Velen., *Scolymus hispanicus* L., *and Chenopodium album* L., reaching contents of 1 g/100 g or more (Fig. 7.6), which means that with a 100-g portion, an adult could get around one third of the daily amount of potassium needed according to different recommendations mentioned above. Other vegetables, such as *Bellis perennis* L. or *Bunias erucago* L., have also been reported to contain high K amounts by Ranfa et al. (2014). Wild fruits may reach also high values in some cases (Fig. 7.7); however, they usually contain less K than vegetables, with average values up to 800 mg/100 g, standing out are rose hips as well as *Elaeagnus angustifolia* L. and *Crataegus monogyna* Jacq. fruits, among others.

Given the role of K in the human body and the relationships with other elements, foods with high K/Na ratio should be advisable. All the abovementioned species, among others, such as the leaves of *Anchusa azurea, Taraxacum Taraxacum officinale, Chondrilla juncea* as well as wild shoots of *Asparagus acutifolius* and *Tamus communis* L. present a high K/Na ratio, above 8; special mention should be made for the leaves of *Chenopodium album*, with a K/Na ratio of 100 or more, due to its extremely low Na levels (Bianco et al. 1998; García-Herrera 2014; Guil-Guerrero and Torija-Isasa 1997; Yildirim et al. 2001). On the other hand, the leaves of *Sonchus* species show a low K/Na ratio, below 3.8; this parameter has been related to saline tolerance in some plants (Asch et al. 2000; Guil-Guerrero et al. 1998b).

As it has been previously mentioned, many wild plants are consumed after cooking; during this process, losses of minerals may take place by lixiviation to the cooking liquid. According to García-Herrera (2014), boiled *Silene vulgaris* (Moench) Garcke young stems with leaves and *Rumex pulcher* L. leaves still retain levels over 200 mg K/100 g.

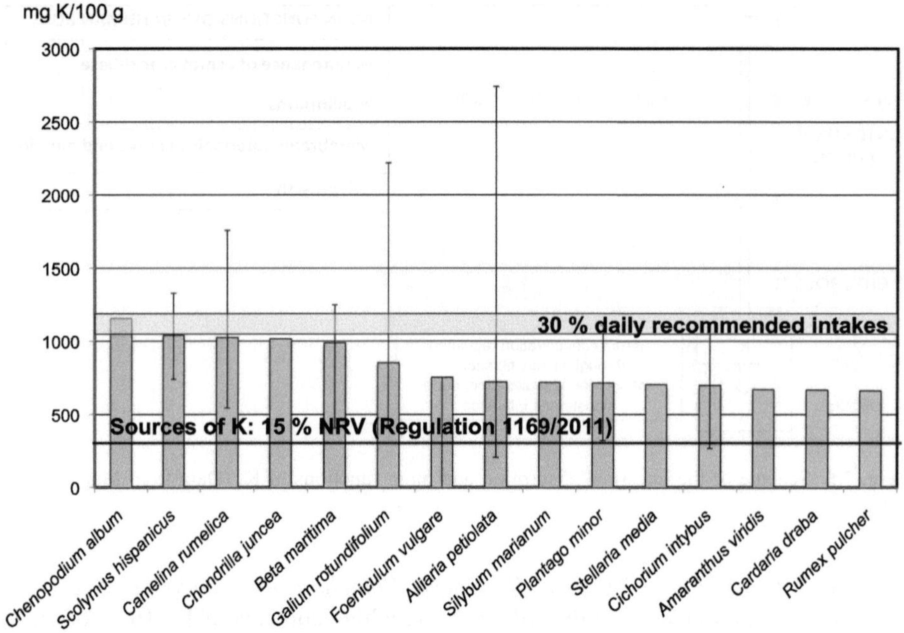

Fig. 7.6 Selected wild vegetables standing out as K sources. (Data from Bianco et al. 1998; Guil-Guerrero and Torija-Isasa 1997; García-Herrera 2014; García-Herrera et al. 2014; Guil-Guerrero et al. 1999a, b; Özcan et al. 2008; Ranfa et al. 2014; Romojaro et al. 2013; Trichopoulou et al. 2000; Yildirim et al. 2001; Zeghichi et al. 2005)

7.2.3 Phosphorus

Phosphorus is present in foods in organic molecules, from which it should be hydrolyzed. The ratios of Ca/P must be close to unity to achieve good calcium and phosphorous intestinal absorption. This mineral is involved in a great number of functions in the human body, including formation of DNA, RNA, and membrane phospholipids; regulation process of intracellular fluids; and being part of hydroxiapatite, the main inorganic molecule of bones and teeth. (Belitz and Grosh 1997; Mahan et al. 2012).

Nutritional requirement of P for healthy adults is in the range of 550–1000 mg/day (Table 7.1). P deficiency is very infrequent in occidental countries because it is widely distributed in meat, fish, eggs, milk and dairy products as well as nuts and grains in levels up to 480 mg/100 g (Souci et al. 2008).

Leafy vegetables, either wild or cultivated, do not stand out by their P dietary contribution, being around 20–70 mg/100 g fresh product, which hardly reaches 13 % of dietary recommendations, and thus, vegetables cannot be considered as good P sources (see Chap. 13). However, among wild greens traditionally eaten in Mediterranean areas, *Malva sylvestris* L., *Urtica dioica* L., and *Alliaria petiolata* (M.Bieb.) Cavara & Grande show remarkable levels compared to other species, up

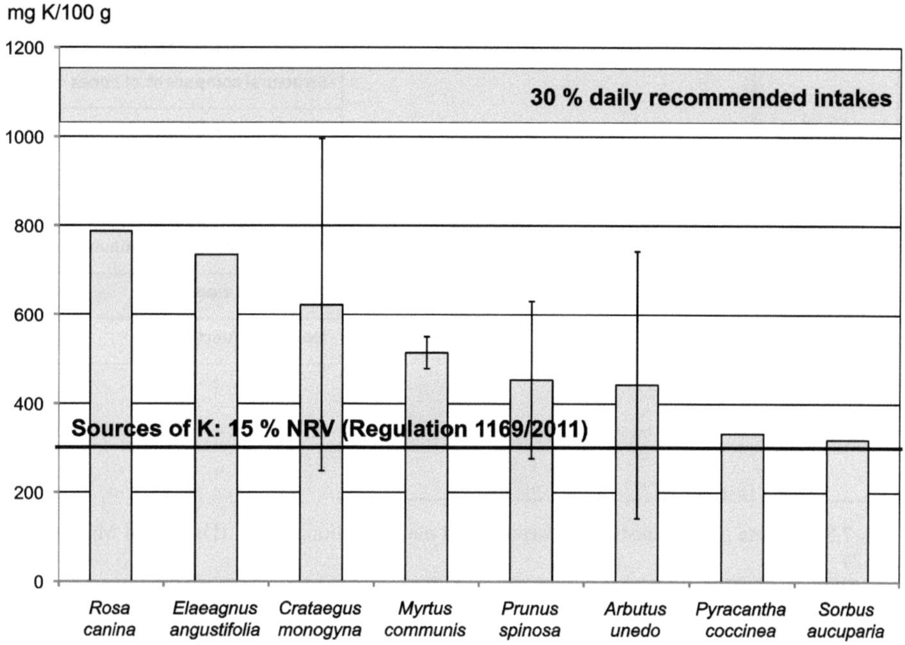

Fig. 7.7 Selected wild fruits standing out as K sources. (Data from Boudraa et al. 2010; Dolezal et al. 2001; Hacıseferoğulları et al. 2012; Özcan and Hacıseferoğulları 2007; Marakoglu et al. 2005; Ruiz-Rodríguez et al. 2011; Ruiz-Rodríguez 2014; Souci et al. 2008)

to 177 mg/100 g (Guil-Guerrero et al. 1999a, b; Krstic-Pavlovic and Dzamic 1985; Wilman and Riley 1993).

7.2.4 Calcium

The most abundant element in human body is calcium, accounting for up to 2% of the body weight (almost half of the total mineral content in the human body), with 99% of calcium distributed in bones (Badui 2006). Figure 7.8 briefly illustrates Ca absorption and main functions in the human body.

According to FNB (2010), humans may absorb about 10–60% of dietary calcium, but this varies depending on many factors, such as:

- *Amount consumed*: The efficiency of absorption decreases as calcium intake increases
- *Age and life stage*: Calcium absorption is as high as 60% in infants and young children, who need substantial amounts of the mineral to build bone; absorption decreases to 15–20% in adulthood (though it is increased during pregnancy, after menopause, or other circumstances of increased requirements) and continues to decrease as people age

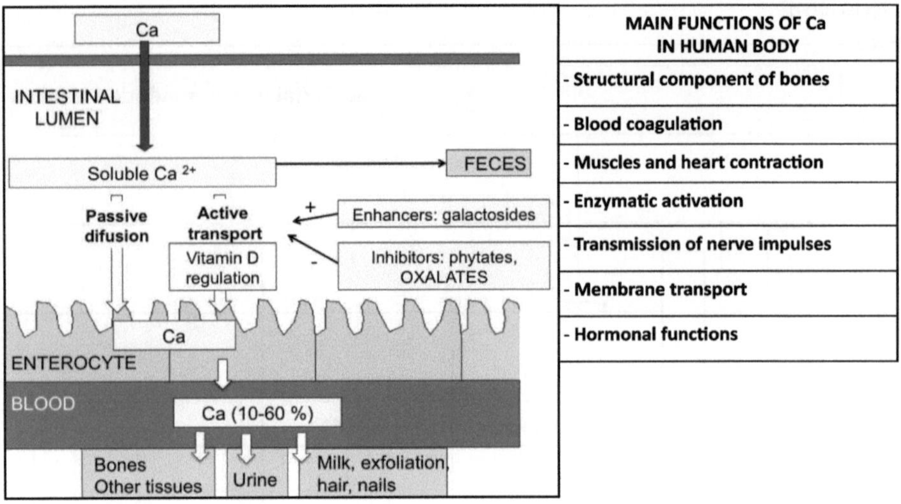

Fig. 7.8 Scheme of the intestinal absorption and main functions of Ca. (Data from Mahan et al. 2012)

- *Vitamin D intake*: as 1,25-dihydroxycholecalciferol stimulates the production of a calcium-binding protein and intestinal alkaline phosphatase
- *Other components in food*:
  - Lactose (hydrolyzed or non-hydrolyzed) or some related galactooligosaccharides in the gut may also have an enhancer effect on calcium absorption (Gibson and Robertfroid 2008; Kwak et al. 2012)
  - Calcium absorption may be limited by formation of complexes with phytates or oxalates, which is the main limitation for the bioavailability of Ca from plant-origin food. Oxalates in plants can be classified into insoluble (mostly calcium-bound) and water-soluble (potassium-bound, sodium-bound, or free) oxalates. Soluble oxalates are assumed to have a negative impact on mineral absorption due to the ability to bind free minerals in the small intestine, thus reducing their absorption (Weaver and Heaney 1991). Wide variability has been found in the proportions of soluble and insoluble oxalates in leafy vegetables (Judprasong et al. 2006)

Higher calcium intake should be achieved during the development of bone mass in the earlier stages of life. A deficiency in calcium intake in infants may induce a form of rachitism; in the elderly, it leads to the development of osteoporosis and tetany in skeletal muscles. A low calcium intake has been also suggested to be a risk factor in chronic diseases such as colon cancer or hypertension. Adequate intake of dietary calcium is associated with decreased prevalence of overweight and obesity. However, intakes higher than 2 g/day may cause hypercalcemia, leading to excessive calcification of soft tissues, including kidney calculus. The excessive calcium intake may induce constipation and also interfere with the absorption of other divalent cations such as iron, zinc and manganese (Mahan et al. 2012; Schrager 2005).

For those reasons, the recommended calcium intake for adults and children over 9 years old is around 700–1000 mg/day, depending on the source, while infants and younger children require 210–800 mg/day and aged women may need up to 1500 mg/day due to the lack of estrogen protection against osteoporosis (Table 7.1). These levels take into account factors affecting calcium bioavailability, such as individual conditions, as well as the form present in the food, and the presence of components enhancing or decreasing its absorption. Leafy vegetables are one of the main sources of oxalates in the diet, and daily dietary oxalate intake in industrialized countries varies depending vegetable consumption, usually ranging between 70 and 2000 mg/day (Bohn 2008).

Besides the reduction of the availability of dietary Ca, oxalic acid can also form kidney calculus. The minimal lethal dose of oxalic acid for adults has been established at 5 g, which is a level very difficult to reach in the diet. However, particularly for the case of some plant foods, which may have high levels of oxalic acid, some authors have established that a ratio of oxalic acid/Ca below 2.5 should be desirable to avoid impairing dietary calcium absorption (Guil-Guerrero et al. 1996b).

In this way, wild edible plant foods have been found to be often good sources of calcium, from a point of view of the high levels that many wild leafy vegetables contain, even higher than many foods widely accepted as good calcium sources, such as dairy products. According to a variety of scientific literature data and Regulation 1169/2011, many wild leafy vegetables can be considered as calcium sources (providing more than 15 % of NRV in 100 g, which means > 120 mg/100 g) as can be seen in Chap. 13. Conventional cultivated vegetables usually present up to 130 mg/100 g (stems and bulbs) or 400 mg/100 g of calcium (leaves), according to Souci et al. (2008). Some wild vegetables may even reach 800 mg of calcium per 100 g of fresh vegetable, as the leaves of some species of *Parietaria* or nettle, although, in this case, a wide variability has been found. This means that a 100-g portion of these vegetables may reach 50–100 % of daily Ca recommendations. Figure 7.9 shows a selection of the richest calcium sources among wild vegetables, according to literature data.

Although many wild leafy vegetables may be considered as calcium sources, attending to their calcium contents, it has to be taken into account that many of them may present high oxalic acid levels. Generally, oxalic acid may reduce calcium absorption by about one sixth, so foods with a ratio of oxalic acid/Ca lower than 2.5 are preferably for human diet (Concon 1988; Derache 1990; Mahan et al. 2012).

There are not always analytical data available about both Ca and oxalic acid content in edible wild species. From all the selected species that can be considered as Ca sources (> 120 mg Ca/100 g, according to European Regulation 1169/2011), the ratio of oxalic acid/Ca has been calculated when data are available, and those species with a suitable ratio below 2.5 have been represented in Fig. 7.10.

Table 7.2 classifies different wild leafy vegetables exceeding the data on calcium contribution in relation to their oxalic acid/Ca ratio. The edible parts of some species, such as *Foeniculum vulgare, Malva sylvestris, Capsella bursa-pastoris* (L.) Medik, *Eruca vesicaria*, or some *Plantago* species, stand out because of their high amount of calcium (35–50 % of daily reference values in a 100-g portion), as well as low oxalic acid/Ca ratio; this would mean a potential better absorption of calcium

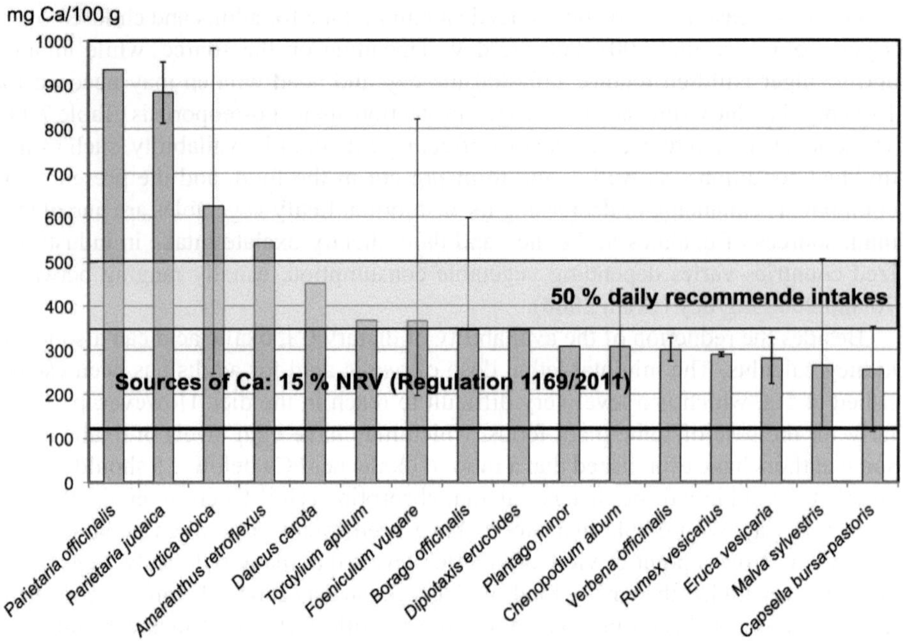

Fig. 7.9 Selected wild vegetables standing out as Ca sources. (Data from Adamski and Bieganska 1980; Alfawaz 2006; Ayan et al. 2006; Barlas et al. 2011; Bianco et al. 1998; Krstic-Pavlovic and Dzamic 1985; García-Herrera 2014; García-Herrera et al. 2014; Guil-Guerrero and Torija-Isasa 1997; Guil-Guerrero et al. 1998b, 1999a, b; Martínez-Para et al. 1979; Özcan et al. 2008; Ranfa et al. 2014; Romojaro et al. 2013; Trichopoulou et al. 2000; Villatoro-Pulido et al. 2012; Wetherilt 1992; Yildirim et al. 2001; Zeghichi et al. 2005)

than from other plant sources (Morales 2011; García-Herrera 2014; Guil-Guerrero et al. 1996b, 1998b; 1999a, b). These species would be a good choice for people with low calcium intake in their diet, for example, those with restriction of dairy products, or children and women with increased calcium needs.

Other wild edible plants, though representing a lower contribution to dietary calcium, may also have interest. However, apart from impairing calcium absorption effect and the wide environmental variability often found in oxalic acid levels within the same species, care should be taken with species with high oxalic acid levels due to the risk of formation of renal calculus. According to literature, wild greens such as *Portulaca oleracea* L., *Sonchus oleraceus* (L.) L., *Beta maritima, Stellaria media* (L.) Vill., *Apium nodiflorum* (L.) Lag., *Borago officinalis* L., *Silene vulgaris, Papaver rhoeas* L., some *Rumex* species, *Salicornia europaea* L., *Scolymus hispanicus* L., or *Amaranthus viridis* L. (with 400 mg/100 g of oxalic acid as an average, or more), or especially *Silybum marianum* (L.) Gaertn., *Chenopodium album* or *C. murale* L. leaves (with 1 g/100 g or more of oxalic acid as an average) should be avoided by people with the tendency to form renal calculus, even if some of them may be sources of calcium (Morales 2011; Guil-Guerrero and Torija-Isasa 1997; Guil-Guerrero et al. 1996b, 1998a). Attention should be paid to dandelion leaves

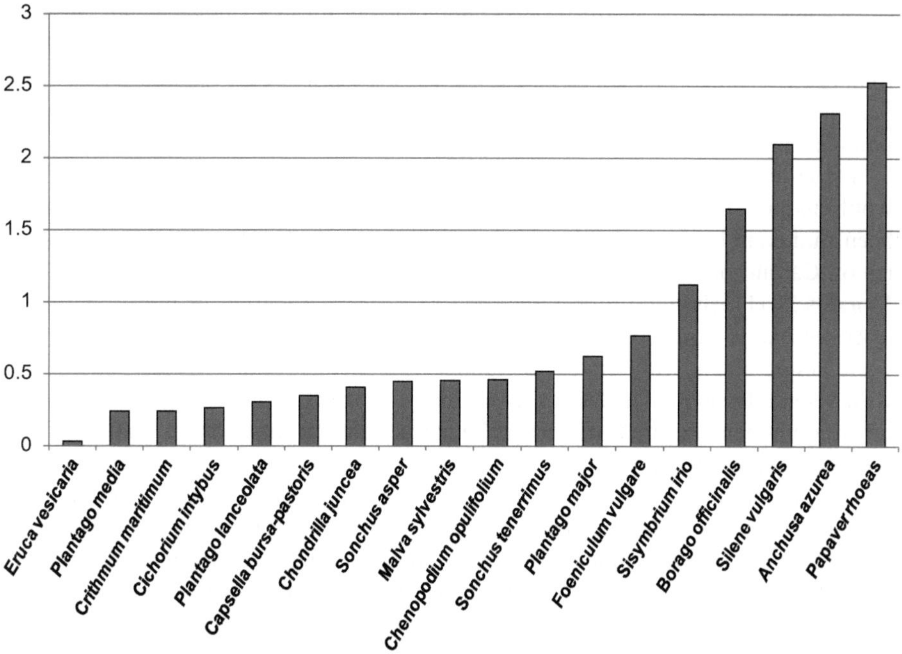

Fig. 7.10 Profile of oxalic acid/Ca ratio of some wild species that could be considered as Ca sources. (Data from: Ayan et al. 2006; Barlas et al. 2011; Bianco et al. 1998; García-Herrera 2014; García-Herrera et al. 2014; Guil et al., 1997; Guil-Guerrero et al. 1996a, 1998a, b; Queralt et al. 2005; Morales et al. 2014; Pereira et al. 2013; Romojaro et al. 2013; Ranfa et al. 2014; Sánchez-Mata et al. 2012; Trichopoulou et al. 2000; Villatoro-Pulido et al. 2012; Zeghichi et al. 2005)

Table 7.2 Selected leafy vegetables for their Ca content (higher % nutrient reference value) and oxalate-dependent availability of Ca (oxalic acid/Ca ratio), according to literature. In bold, the potentially best calcium contributors

	Oxalic acid/Ca ratio		
% Ca NRV	<1	1–2.5	>2.5
>35 % NRV (>280 mg Ca/100 g)	***Foeniculum vulgare, Malva sylvestris, Capsella bursa-pastoris, Eruca vesicaria, Plantago lanceolata, Plantago media***	*Borago officinalis*	*Chenopodium album*
>15 % NRV (>120 mg Ca/100 g)	*Chenopodium opulifolium, Sonchus tenerrimus, Sonchus asper, Chondrilla juncea, Cichorium intybus, Scolymus hispanicus, Crithmum maritimum*	*Sisymbrium irio, Anchusa azurea, Silene vulgaris, Papaver rhoeas*	*Stellaria media, Apium nodiflorum, Silybum marianum*

since some vegetables as *Taraxacum obovatum* (Willd.) DC. have shown very low oxalic acid content (below 40 mg/100 g), while *T. officinale* sect. *Ruderalia* has been found to contain up to 1 g/100 g of oxalates (Dias et al. 2014; Morales et al. 2014; Sánchez-Mata et al. 2012), so as identification of species is not always easy, dandelion leaves should not be recommended when oxalates are to be avoided in the diet.

These oxalates-rich plants, if consumed, should be cooked thoroughly, which would induce some dissolution of soluble oxalates into the cooking liquid as it has been studied in leafy vegetables (Yadav and Sehgal 2003). There are very few studies on Ca and oxalic acid losses in boiled wild vegetables (García-Herrera 2014; Morales 2011), showing high retention of Ca (90% retention or more) and some losses of oxalic acid (30–40% losses) in some vegetables after boiling (*Silene vulgaris* young stems with leaves, and *Rumex pulcher* leaves). Calcium retention in wild vegetables is higher than other minerals, probably related to some binding to other components in the food that make extraction into the cooking liquid difficult (García-Herrera 2014).

On the contrary, *Plantago media, Crithmum maritimum*, or *Cichorium intybus* leaves have shown low oxalates levels (less than 60 mg/100 g), and so, they would be the best choice for people with renal alterations (Guil et al. 1997; Guil-Guerrero 2001; Guil-Guerrero et al. 1996a, b, 1998a; Morales 2011). According to Morales (2011), many wild young shoots such as *Asparagus acutifolius* and *Humulus lupulus* L. as well as *Allium ampeloprasum* L. bulbs could also be also advisable (less than 200 mg/100 g, average, being less in the boiled product).

Regarding wild fruits, not many of them could be considered as good Ca sources except for the fruits of *Celtis australis* L., with very high calcium levels as reported by Boudraa et al. (2010). Other species, such as *Ziziphus lotus* (L.) Lam., *Rosa canina* L., and *Elaeagnus angustifolia*, and hawthorn fruits, could also be stood out for their calcium contents (Fig. 7.11). It should also be taken into account that oxalic acid levels are usually low in fruits (57–98 mg/100 g). The scarce data about oxalic acid/Ca ratio in wild fruits show values below 2 (Morales et al. 2013; Ruiz-Rodríguez 2014), so Ca, although less abundant, would be expected to be better absorbed from these wild fruits than from some vegetables.

7.2.5 Magnesium

Magnesium is the second most abundant intracellular cation in the human body, 60% being found in bones, 26% in muscles and other soft tissues and body fluids (Mahan et al. 2012).

Magnesium absorption and human functions are summarized in Fig. 7.12. Although Mg deficiency is rare, high magnesium depletion would lead to a delay in bone growth in young patients and osteoporosis in the elderly; a severe deficiency of Mg may cause tremors, spasms, neurologic changes, anorexia, nausea and vomiting as well as sodium retention. On the contrary, too high intake of magnesium are associated with higher bone density but also with central nervous system depression and even paralysis, especially in patients with renal insufficiency (Mahan et al. 2012).

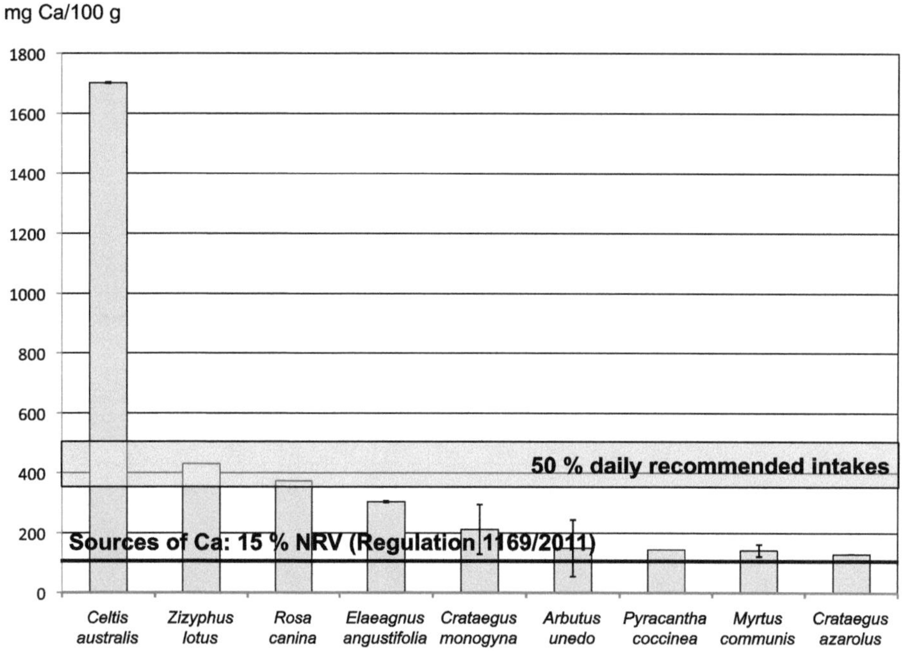

Fig. 7.11 Selected wild fruits standing out as Ca sources. (Data from Boudraa et al. 2010; Dolezal et al. 2001; Hacıseferoğulları et al. 2012; Özcan and Hacıseferoğulları 2007; Ruiz-Rodríguez et al. 2011; Ruiz-Rodríguez 2014)

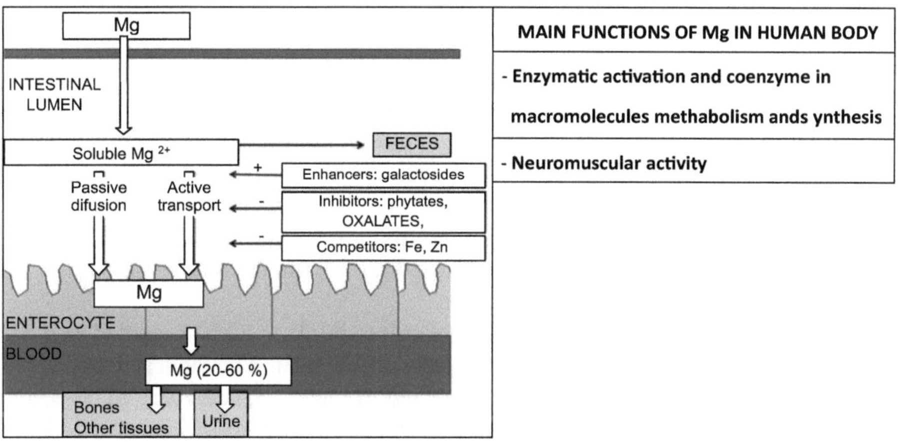

Fig. 7.12 Scheme of the intestinal absorption and main functions of Mg. (Data from Bohn 2008; Brink et al. 1992; Mahan et al. 2012)

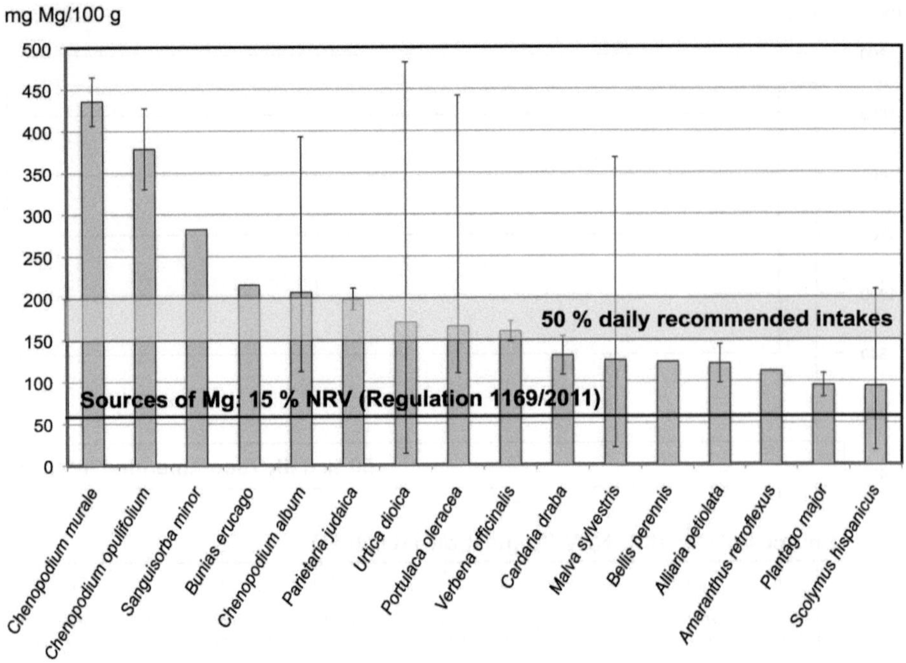

Fig. 7.13 Selected wild vegetables standing out as Mg sources. (Data from Adamski and Bieganska 1980; Bianco et al. 1998; Bockholt and Schnittke 1996; Krstic-Pavlovic and Dzamic 1985; García-Herrera 2014; García-Herrera et al. 2014; Gjorgieva et al. 2011; Guil-Guerrero and Torija-Isasa 1997; Guil-Guerrero et al. 1998b, a; Hiçsönmez et al. 2009; Martínez-Para et al. 1979; Özcan et al. 2008; Wetherilt 1992; Wilman and Riley 1993; Ranfa et al. 2014; Romojaro et al. 2013; Trichopoulou et al. 2000; Yildirim et al. 2001; Zeghichi et al. 2005)

Milk is a moderate source of magnesium, but the main sources of this mineral are whole grains, nuts, legumes, and dark green vegetables because it is an essential constituent of chlorophyll. Magnesium requirements are quite variable. As shown in Table 7.1, FAO/WHO recommends 190–220 mg/day for females and 230–260 mg/day for males. Recommendations of EU as well as those of the Mediterranean countries are higher (300–360 mg/day for females and 350–420 mg/day for males). Regulation 1169/2011 established an intermediate reference value for nutrition-labelling purposes of 375 mg/day.

Leafy vegetables, due to the presence of chlorophyll, could be regarded as moderate contributors to cover Mg requirements in the diet. The richest magnesium sources among wild vegetables are represented in Fig. 7.13, according to compiled literature data. The leaves of different *Chenopodium* species are remarkable from this point of view (providing more than 300 mg/100 g, reaching almost the whole Mg daily recommendations), as well as the edible parts of *Sanguisorba minor* Scop., *Bunias erucago*, and *Parietaria judaica* L., which often reach half of the daily magnesium recommendations.

Some of these species are consumed after boiling and a part of magnesium contents will be lost in the cooking liquid, up to 60% in *Silene vulgaris* young stems with leaves and *Rumex pulcher* leaves, while up to 40% in tender shoots (García-Herrera 2014). As in the case of other nutrients, the thinner and softer structure of leaves may influence higher mineral extraction than in grosser and harder plant tissues, as was found by this author.

Apart from these species, very variable magnesium levels have been found by different authors in the leaves of *Malva sylvestris, Portulaca oleracea*, and *Urtica dioica*, which may obey to environmental factors of the analyzed samples. Thus, these species could not always, but in many cases, be considered as good magnesium sources for the diet. Furthermore, considering the low oxalic acid/(Ca+Mg) ratio of *Malva sylvestris* obtained from average contents found for this species (0.3), a more efficient absorption of both minerals should be expected. As a result, it could be concluded that these wild edible plants, with more than 120 mg Mg/100 g on average, may be a good alternative to other conventional vegetables in the diet, which usually present oxalic acid/(Ca+Mg) ratios of 3 or more (Souci et al. 2008).

With respect to wild fruits, only some examples as *Celtis australis* or *Zizyphus lotus* fruits could be considered as Mg sources, as can be seen in Fig. 7.14.

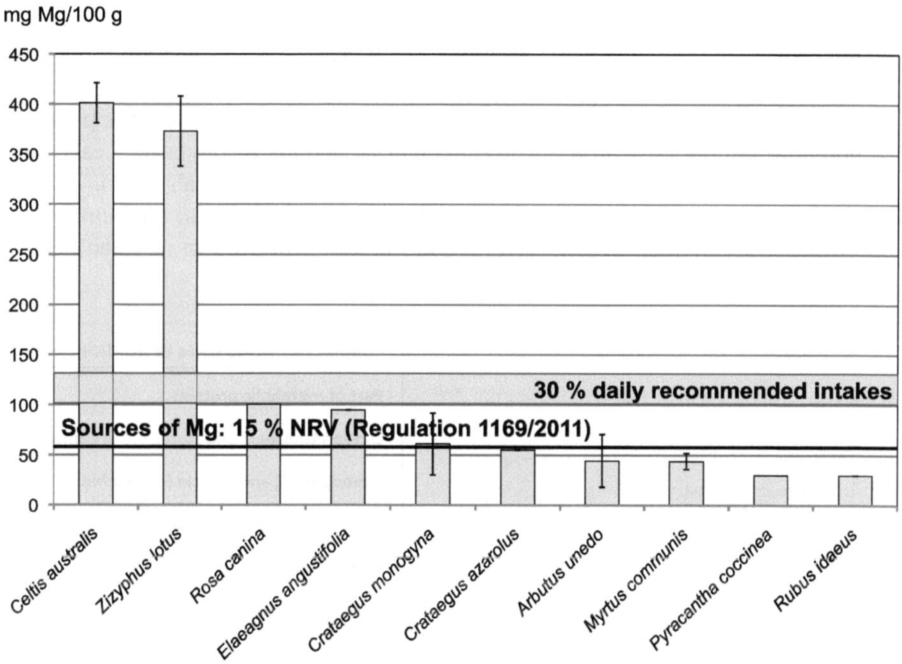

Fig. 7.14 Selected wild fruits standing out as Mg sources. (Data from Abdeddaim et al. 2014; Boudraa et al. 2010; Dolezal et al. 2001; Ruiz-Rodríguez 2014)

7.3 Microelements

7.3.1 Iron

In the body, iron is found in metaloproteins, such as hemoglobin or myoglobin, being transported and stored into the structure of transferrin, ferritin, and hemosiderin. In this way, in food it is present in two forms, with two different mechanisms of absorption (Fig. 7.15): heme Fe in animal tissues and non-heme Fe (inorganic) in plant tissues, such as legumes and vegetables. The former is more easily absorbed (bioavailability of 20–30 %), while only 2–10 % of inorganic Fe is absorbed (Bothwell et al. 1989).

Iron loss through menstrual hemorrhage is estimated on about 1.5–2.4 mg per day in fertile age women (compared to 1 mg per day in males). Most reference values recommended as daily Fe intake are about 8–10 mg/day for men and elderly women and about 16–20 mg/day for women below 50–55 years old (Table 7.1). High iron losses or low iron intakes cause anemia in infants below 2 years old, adolescent girls, or pregnant women, some of the main groups of risk. On the contrary, diseases associated with iron excess, hemosiderosis and hemochromatosis are usually associated with a gene acting as a promoter of excessive iron absorption (Mahan et al. 2012).

The main Fe sources in the human diet are meat and other animal products, which provide more bioavailable iron than plants. However, some plant foods may moderately contribute to human iron intake. Legumes are a good example, providing approximately 4–7 mg/100 g to the diet, even taking into account the presence of potential inhibitors of iron absorption such as tannins or phytates. Leafy vegetables, such as spinach, may contribute 2–4 mg/100 g (Souci et al. 2008).

Many wild vegetables can also be considered as iron sources to the human diet, according to Regulation 1169/2011 (>2.1 mg in 100 g of product as eaten). Some

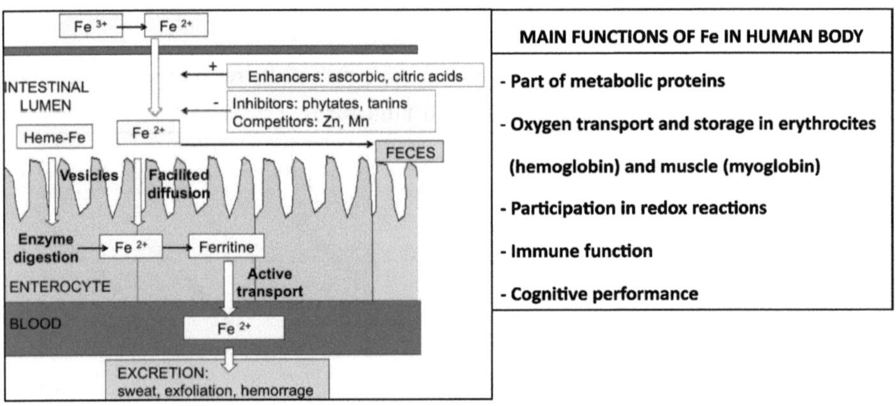

Fig. 7.15 Scheme of the intestinal absorption and main functions of Fe. (Data from Mahan et al. 2012)

wild greens could be regarded as interesting iron contributors to the human diet, for example *Tragopogon sinuatus* Avé-Lall. or *Cichorium pumilum* Jacq., with more than 150 mg/100 g on dry basis, according to Zeghichi et al. (2005) and Ayan et al. (2006). Other authors have reported interesting values for iron content in wild vegetables, the most relevant recorded in Fig. 7.16. Taking into account both the natural variability in iron contents and the differences in the recommendations given by different organizations, we could conclude that 100 g of some wild leafy vegetables such as dandelion leaves could provide more than 2–3 mg of Fe (about 15 % of the most generally accepted recommended daily intake of iron in Europe for females and 30 % of the recommendation for males; see Table 7.1 and Fig. 7.16). Many other species, such as *Cakile maritima* Scop., *Rumex vesicarius* L., *Portulaca oleracea, Verbena officinalis* L., *Sonchus tenerrimus* L., *Sisymbrium irio* L., or *Picris echioides* L. leaves, have shown higher iron values, around 4 mg/100 g. Conventional vegetables, including those traditionally considered as Fe sources (for example, spinach) are usually within this range, with some exceptions such as parsley, which may provide up to 6 mg/100 g of Fe (Souci et al. 2008). In relation to this, *Polygonum bistorta* L. or *Amaranthus viridis* edible parts can stand out as some of the best iron sources known among the wild vegetables traditionally eaten in the Mediterranean area, reaching values close to 6 mg per 100 g of fresh leaves (which represents about 30 % of daily recommended intake of iron for females and 60 % of the recommendation for males);

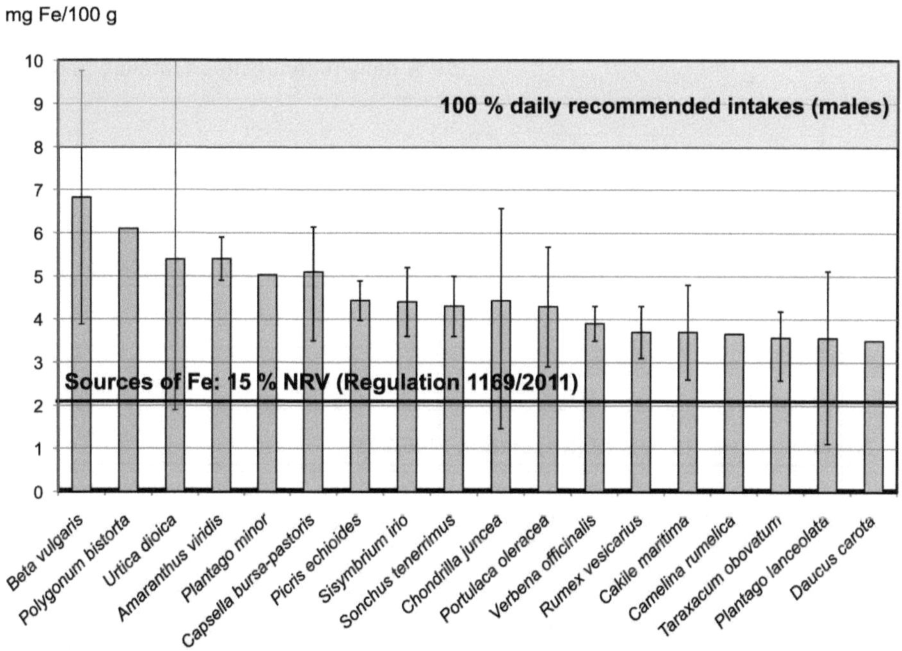

Fig. 7.16 Selected wild vegetables standing out as Fe sources. (Data from Adamski and Bieganska 1980; Alfawaz 2006; Krstic-Pavlovic and Dzamic 1985; Guil-Guerrero et al. 1998a, b, 1999b; Martínez-Para et al. 1979; Wetherilt 1992; Yildirim et al. 2001)

also, *Beta vulgaris* and *Urtica dioica* leaves have presented average values within 5–6 mg/100 g, although wide variability has been found in these species, reaching in some cases 100 % of the reference values for males in a 100-g portion.

It should be taken into account that these values refer to fresh samples of wild edible plant; if cooked, these species could be subject to some losses due to dissolution of Fe into the cooking liquid. In some studies done on different wild edible species, the losses of Fe in the boiled plant are in most cases slight (up to 12 %). This high retention of iron in plant tissues may be attributable to some retention to plant components that could impair the extraction into the cooking liquid (García-Herrera 2014). As for other nutrients, the loss of iron due to cooking process can be avoided with the consumption of cooking liquids (for example as soups) or eating the raw plants in salads when possible.

Fruits, either from cultivated or wild origin, do not usually contain high levels of Fe, with some exceptions, for example, the wild fruits of *Elaeagnus angustifolia*, and sometimes, *Crataegus monogyna* fruits are in the range of the iron sources among wild vegetables (Fig. 7.17).

All the abovementioned species could be a good alternative to increase the iron presence in the human daily diet, and especially in cases of ferropenic anemia, which may be often subclinical, especially in women. Although iron is less bioavailable in these foods, compared to animal iron sources, the animal-origin foods

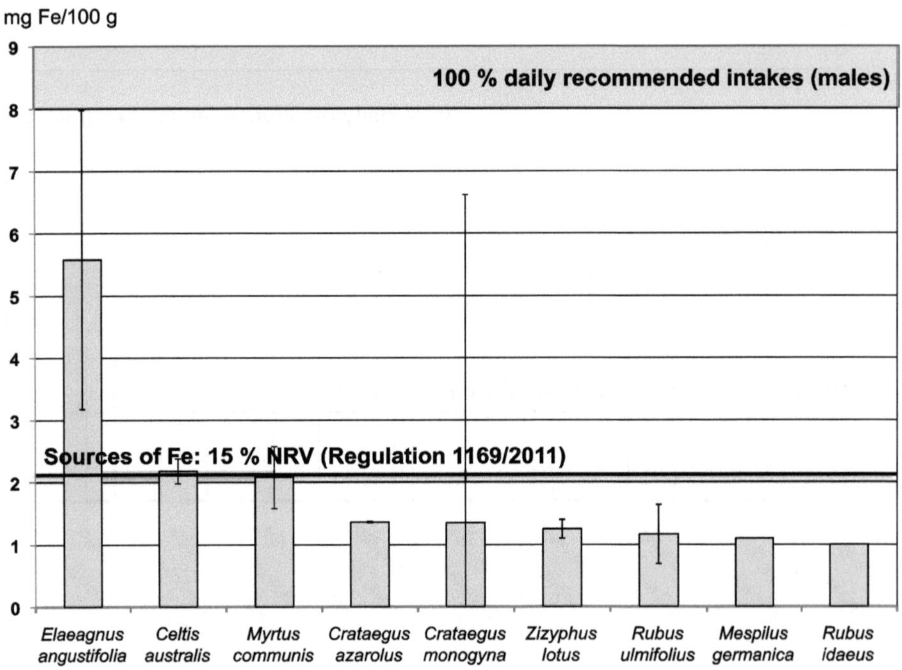

Fig. 7.17 Selected wild fruits for their high Fe levels. (Data from Abdeddaim et al. 2014; Boudraa et al. 2010; Dolezal et al. 2001; Hacıseferoğulları et al. 2012; Marakoglu et al. 2005; Ruiz-Rodríguez 2014)

have the disadvantage of being usually rich in saturated lipids, and for that reason, the inclusion of these wild plants in the diet should be encouraged as contributors to prevent or help iron deficiency treatment.

7.3.2 Copper

Copper is a trace element located in blood, liver, brain, heart, and kidneys, as well as muscular tissues, acting as a cofactor of different enzymes in the human body. The absorption of Cu is produced in the short intestine by facilitated diffusion, and the exit is by active transport to the bloodstream where it is bound to ceruloplasmin. Fiber, phytates, zinc, and ascorbic acid can decrease copper absorption. Excretion is by urine, sweat, and menstrual blood. Copper deficiency is rare, and symptoms can be attributed to enzyme deficiencies, being mainly anemia, neutropenia, and skeletal abnormalities, particularly demineralization; also, some hair or skin alterations may appear, as well as neurologic changes. Copper toxicity due to ingestion of food is considered almost impossible, but an excessive supplementation may lead to liver cirrhosis and alterations in the formation of erythrocytes (Mahan et al. 2012).

The generally accepted recommendations of copper can vary between 0.9–2 mg/day, although the most commons values are 1–1.5 (Table 7.1). The main sources are meat, seafood, nuts, and seeds (Fennema 2000). Among wild vegetables, only a few species could be considered as a copper source for the human diet, according to Regulation 1169/2011 (content of 0.15 mg/100 g or more), as is recorded in Fig. 7.18. For example, wild shoots of *Asparagus acutifolius*, as well as leaves of some *Chenopodium* and *Sonchus* species, *Verbena officinalis* L., *Salicornia europaea*, or *Portulaca oleracea* could be regarded as copper sources for the diet, despite their natural variability. Other species, such as *Urtica dioica, Chondrilla juncea*, or *Papaver rhoeas* (not represented in Fig. 7.18), present average contents over 0.4 mg/100 g, sometimes reaching one third of daily Cu recommendations, but as a wide variability was found, it is not possible to assure that they are always natural sources of this nutrient.

As some species are consumed after boiling, copper levels could be reduced in a range that may vary between up to 20 % in *Rumex pulcher* and *Silene vulgaris* leaves and up to 63 % in wild asparagus shoots, according to García-Herrera (2014).

Wild fruits may also be copper sources for the diet, providing interesting values often over 0.15 mg/100 g, as for example, *Myrtus communis* L., *Rubus ulmifolius* Schott, *Pyracantha coccinea* M. Roem., and *Crataegus monogyna* fruits (Fig. 7.19).

7.3.3 Manganese

Manganese is an essential nutrient acting as a cofactor of enzymes, such as pyruvate carboxylase and mitochondrial superoxide dismutase (Fig. 7.20). Manganese deficiency is extremely rare, being first described in 1972, with symptoms such as

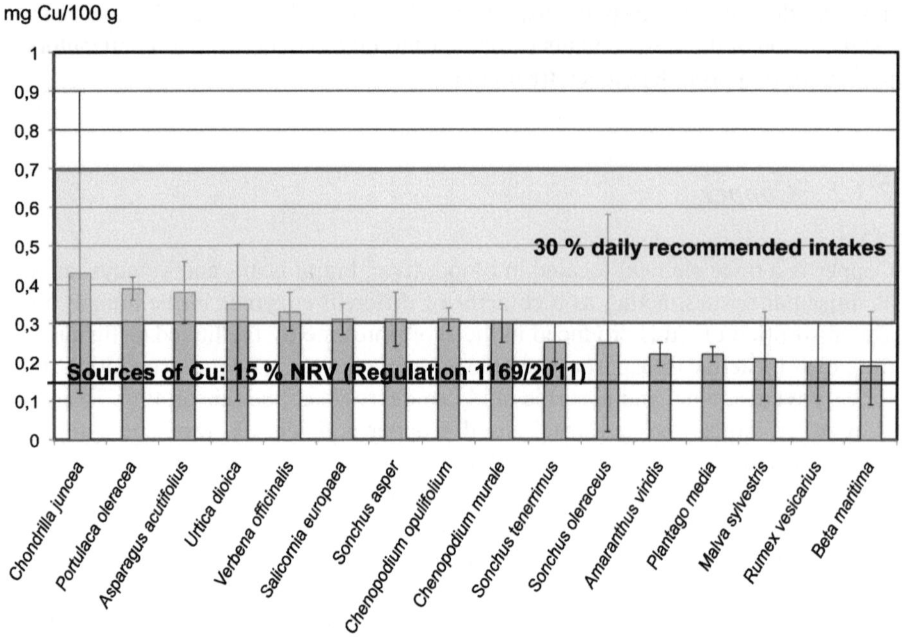

Fig. 7.18 Selected wild vegetables standing out as Cu sources. (Data from Adamski and Bieganska 1980; Alfawaz 2006; García-Herrera et al. 2014; Guil-Guerrero 2001; Guil-Guerrero and Torija-Isasa 1997; Guil-Guerrero et al. 1998a, b; 1999a; Hiçsönmez et al. 2009; Krstic-Pavlovic and Dzamic 1985; Martínez-Para et al. 1979; Wetherilt 1992; García-Herrera 2014; Zeghichi et al. 2005)

weight loss, gastrointestinal disorders as well as skin and hair alterations. An excess of manganese may be accumulated in the liver and central nervous system, producing Parkinson-type symptoms (Mahan et al. 2012).

The reference values for manganese intake range between 1.8 and 2.3 mg/day for adults (Table 7.1). These amounts may be achieved through the intake of whole grains, nuts, tea, fruits, and vegetables as the main sources of this mineral in the diet (Fennema 2000). According to Regulation 1169/2011, many wild vegetables could be good manganese sources (providing more than 0.3 mg/100 g), the most important are shown in Fig. 7.21. Despite natural variability, very wide in the case of *Urtica dioica*, many wild vegetables, such as *Chenopodium* spp., *Chondrilla juncea*, and *Montia fontana* L., among others, may reach or surpass 1 mg/100 g (50 % of daily recommended intakes of Mn), values rarely found in conventional cultivated vegetables (Souci et al. 2008). Furthermore, manganese is one of the microelements showing the highest retention in vegetables after boiling, being higher than 65 % in wild asparagus shoots and above 91 % in *Rumex pulcher* leaves and *Silene vulgaris* young stems with leaves (García-Herrera 2014). For that, they can be considered as a very good source of this mineral, not only in raw form but also after boiling.

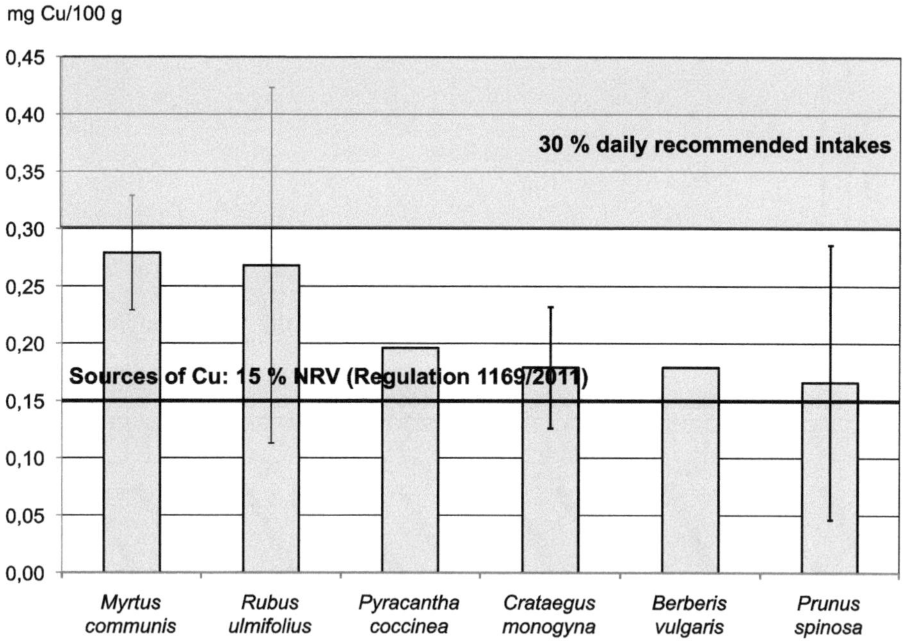

Fig. 7.19 Selected wild fruits standing out as Cu sources. (Data from Boudraa et al. 2010; Dolezal et al. 2001; Hacıseferoğulları et al. 2012; Ruiz-Rodríguez 2014)

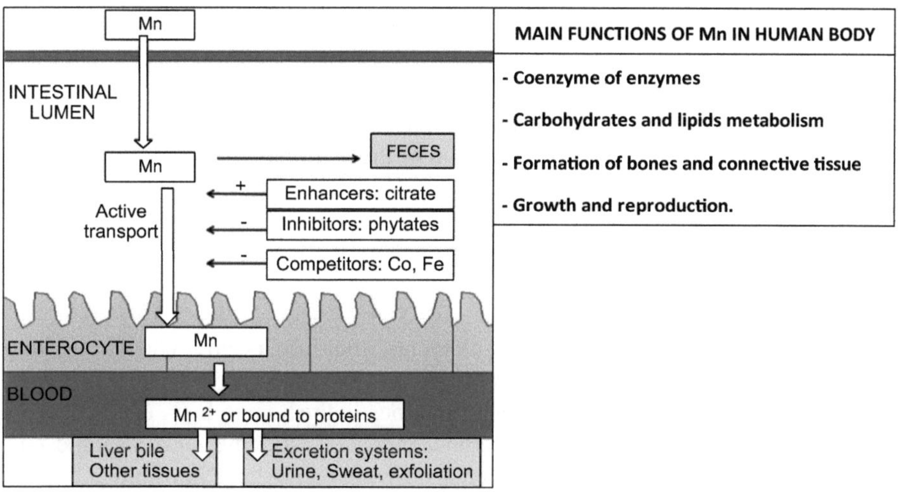

Fig. 7.20 Scheme of the intestinal absorption and main functions of Mn. (Data from Gropper et al. 2008; Mahan et al. 2012)

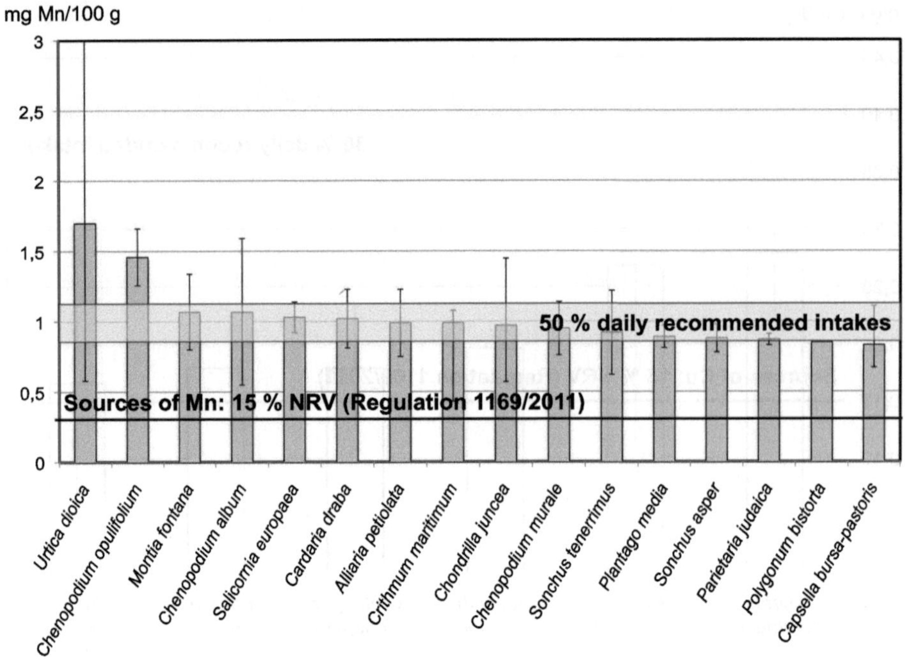

Fig. 7.21 Selected wild vegetables standing out as Mn sources. (Data from Ayan et al. 2006; García-Herrera et al. 2014; Guil-Guerrero 2001; Guil-Guerrero and Torija-Isasa 1997; Guil-Guerrero et al 1996a, 1998a, b, 1999a, b; Yildirim et al. 2001; Tardío et al. 2011)

Also, different fruits species could be considered as Mn sources, such as *Ziziphus lotus, Rubus ulmifolius*, or *Celtis australis*, providing often more than 1 mg/100 g (Fig. 7.22) and, in some cases, reaching even 2 mg/100 g in *R. ulmifolius* fruits (Ruiz-Rodriguez 2014), which means a contribution of the whole amount of manganese needed daily in every 100-g portion of these fruits.

7.3.4 Zinc

The second in abundance among trace elements, after iron, is zinc. Widely distributed in the human body, with high concentrations in the pancreas, liver, kidney, bones, and muscle, it serves several important functions in the human body (Fig. 7.23).

Most institutions recommend a reference intake of 7–9.5 mg Zn/day (Table 7.1). The main sources are meat, fish, and cereals. However, phytates in whole grains can limit zinc absorption in some populations, causing slow growth problems in young people, anemia, hypogeusia, loss of appetite, healing problems, skin and hair alterations, and immune defects. By contrast, excess of zinc (intake of 100–300 mg/day) interferes with the absorption of copper; however, zinc intoxication usually takes place by mechanisms different from ingestion, such as inhalation or parenterally (Badui 2006; Mahan et al. 2012).

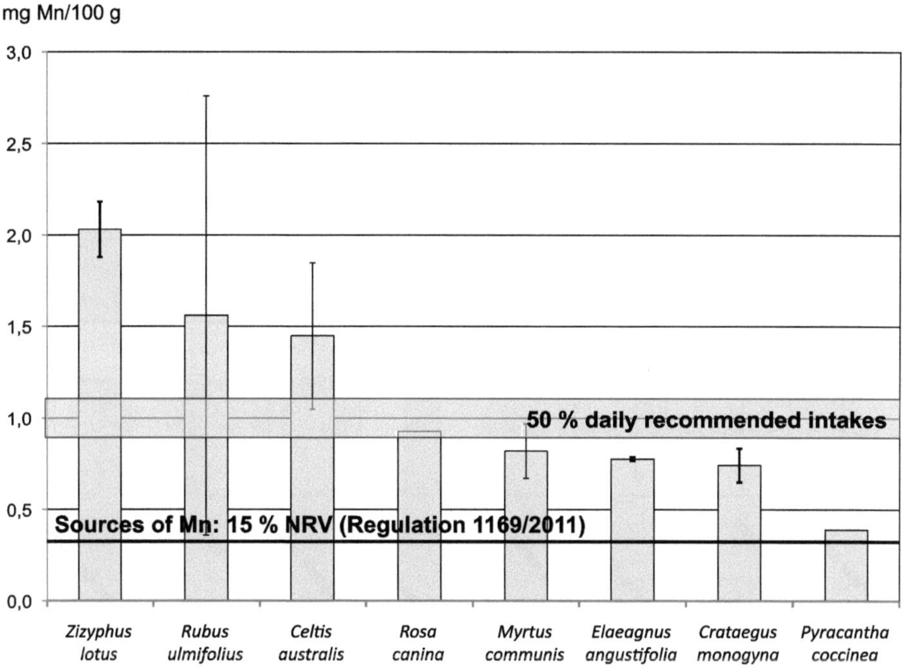

Fig. 7.22 Selected wild fruits standing out as Mn sources. (Data from Abdeddaim et al. 2014; Boudraa et al. 2010; Dolezal et al. 2001; Ruiz-Rodríguez 2014)

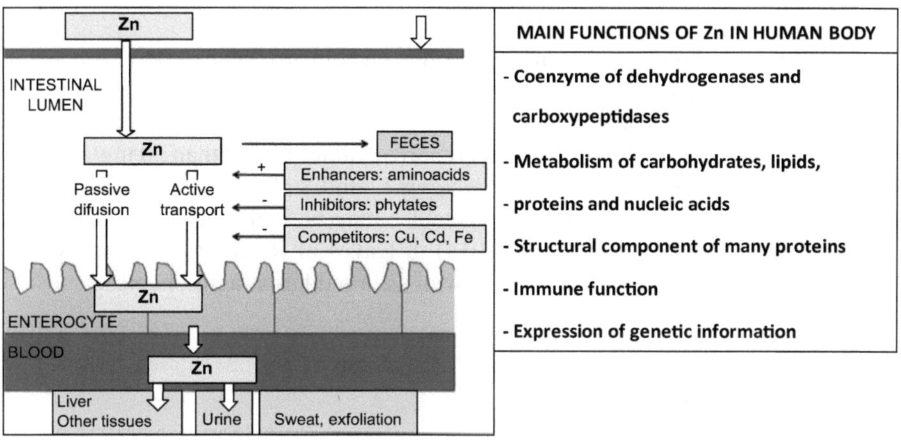

Fig. 7.23 Scheme of the intestinal absorption and main functions of Zn. (Data from Mahan et al. 2012)

Wild edible plants are not generally recognized as good Zn sources, exhibiting Zn levels usually below 1 mg/100 g, which are also the levels usually found in conventional vegetables (Souci et al. 2008); traditionally eaten wild fruits are often below 0.6 mg Zn/100 g (Ruiz-Rodríguez 2014). However, there are some wild plants

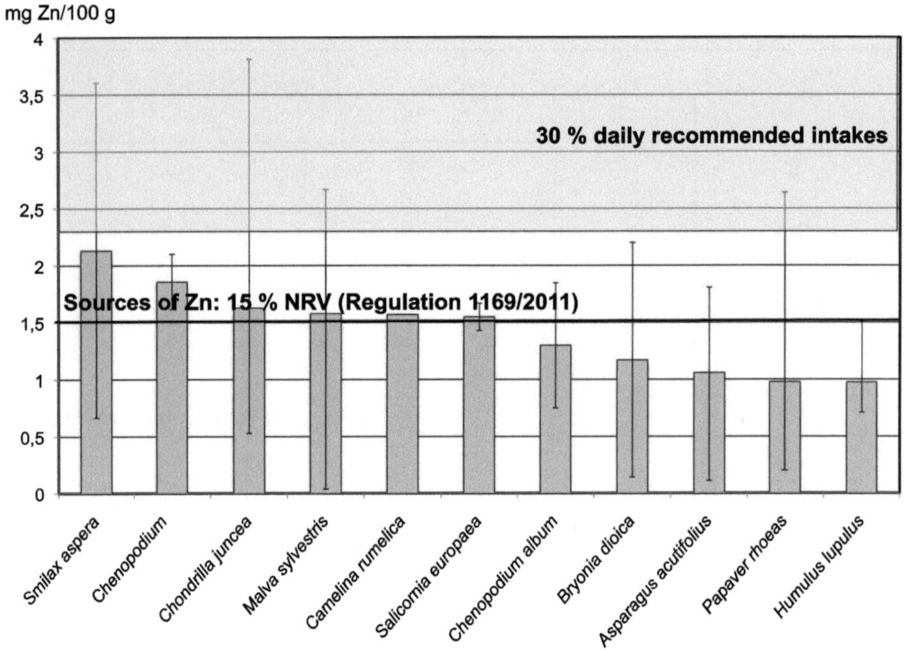

Fig. 7.24 Selected wild vegetables standing out as Zn sources. (Data compiled from Guil-Guerrero and Torija-Isasa 1997; García-Herrera 2014; García-Herrera et al. 2014; Guil-Guerrero et al. 1999a; Hiçsönmez et al. 2009; Poschenrieder et al. 2012; Romojaro et al. 2013; Trichopoulou et al. 2000; Yildirim et al. 2001; Zeghichi et al. 2005)

that must be stood out because of their higher Zn contents, such as *Myrtus communis* fruits (Hacıseferoğulları et al. 2012) and several greens such as *Chenopodium opulifolium* Schrad. ex W.D.J.Koch & Ziz or *Salicornia europaea*, among others (Fig. 7.24). Ayan et al. (2006) have also reported values higher than 50 mg/100 g on dry basis in edible parts of *Geranium lucidum* L. and *Stellaria holostea* L.

Retention of zinc in boiled wild vegetables may be in the range of 40–68 % as has been found by García-Herrera (2014) in leaves of *Rumex pulcher* and young stems with leaves of *Silene vulgaris*, as well as in wild asparagus shoots.

7.4 Conclusions

From the studies available in scientific literature, many wild greens and fruits are valuable contributors to dietary macro and microelements intakes. They may provide Ca, Mg, Fe, and Mn amounts that in some cases may reach 50 % of the daily requirements recommended by different institutions and K and Cu levels reaching 30 % of the recommendations. In general, wild vegetables may contribute to these requirements to a higher extent than fruits, Among fruits, *Crataegus monogyna* usually provides a source of K, Ca, Mg, and Cu; rose hips are usually rich in K, Ca, Mg, and Mn; *Celtis australis* and *Ziziphus lotus* could be stood out as sources of Mg, and together with some *Rubus* species, also as sources of Mn.

All biological tissues are subject to variability in their composition due to genetic and environmental factors. This phenomenon is especially important in mineral levels in plants since these levels are often highly dependent on soil characteristics. In some cases, the absence of a wide variability for a given species may obey the fact that just one sample was analyzed, and there are no enough data to reflect this variability. Therefore, sampling is a key point to assure the reliability of the results in nutrient analysis studies.

However, despite this natural variability, some features may be common factors in wild green composition as mineral contributors to human diet. For example, the leaves of some *Chenopodium* species often provide relevant K, Mg, Cu, Mn, and Zn amounts. *Chondrilla juncea* leaves are usually rich in K, Fe, Cu, and Mn. *Portulaca oleracea* stands out for its Mg, Fe and Cu levels, and the leaves of some *Urtica* species may be often Mg and Fe sources.

In the case of calcium contribution, other factors such as the presence of oxalates in leafy vegetables, lowering gut availability of this mineral, should be also taken into account. In this way, *Foeniculum vulgare, Malva sylvestris, Capsella bursa-pastoris, Eruca vesicaria, Plantago lanceolata,* or *Plantago media*, among others, could be considered as valuable Ca contributors, with better potential availability than other plants, due to their relation to oxalic acid levels. Also, in terms of avoiding potential harmful effects of oxalate due to renal calculus formation, *Plantago media, Crithmum maritimum*, or *Cichorium intybus* leaves are good alternatives for people who have to limit oxalate intake in their diet.

Most of them are also foods with very low Na amounts, with some exceptions, mainly those species that usually grow on saline soils. Therefore, wild plant foods should be regarded as good sources of many minerals in the human diet, and their consumption should be encouraged not only with the purpose of preserving traditional food habits as a valuable cultural heritage but also as a useful tool to improve the nutritional quality of current human diet.

References

Abdeddaim M, Lombarkia O, Bacha A, Fahloul D, Abdeddaim D, Farhat R, Saadoudi M, Noui Y, Lekbir A (2014) Biochemcial characterization and nutritional properties of *Zizyphus lotus* L. fruits in Aures region, northeastern of Algeria. Ann Food Sci Technol 15(1):75–81

Adamski R, Bieganska J (1980) Studies of chemical substances present in *Urtica dioica* L. leaves. Part 1. Trace elements. Herba Pol 26:177–180

Alfawaz MA (2006) Chemical composition of hummayd (*Rumex vesicarius*) grown in Saudi Arabia. J Food Compos Anal 19:552–555

Asch F, Dingkuhn M, Dörffling K, Miezan K (2000) Leaf K/Na ratio predicts salinity induced yield loss in irrigated rice. Euphytica 113(2):109–118

Ayan I, Acar Z, Mut H, Basaran U, Asci O (2006) Morphological, chemical and nutritional properties of forage plants in a natural rangeland in Turkey. Bangladesh J Bot 35(2):133–142

Badui S (2006) Química de los alimentos, 4th edn. Pearson Educación, Mexico
Barlas NT, Irget ME, Tepecik M (2011) Mineral content of the rocket plant (*Eruca sativa*). Afr J Biotechnol 10(64):14080–14082
Belitz HD, Grosch W (1997) Química de los alimentos, 2nd edn. Acribia, Zaragoza
Bianco VV, Santamaria P, Elia A (1998) Nutritional value and nitrate content in edible wild species used in southern Italy. Acta Hortic 467:71–87
Bockholt R, Schnittke C (1996) Contents of nutrients and minerals in herb species of intensive managed peat soil meadows. Wirtsch Futter 42(3):209–216
Bohn T (2008) Dietary factors influencing magnesium absorption in humans. Curr Nutr Food Sci 4(1):53–72
Bothwell TH, Baynes RD, MacFarlane BJ, MacPhail AP (1989) Nutritional iron requirements and food iron absorption. J Intern Med 226(5):357–365
Boudraa S, Hambaba L, Zidani S, Boudraa H (2010) Composition minérale et vitaminique des fruits de cinq espèces sous exploitées en Algérie: *Celtis australis* L., *Crataegus azarolus* L., *Crataegus monogyna* Jacq., *Elaeagnus angustifolia* L. et *Zizyphus lotus* L. Fruits 65(2):75–84
Brink EJ, Beynen AC, Dekker PR, van Beresteijn EC, van der Meer R (1992) Interaction of calcium and phosphate decreases ileal magnesium solubility and apparent magnesium absorption in rats. J Nutr 122:580–586
Concon JM (1988) Toxicology. Principles and concepts. Marcel Dekker, New York
Cuervo M, Corbalán M, Baladía E, Cabrerizo L, Formiguera X, Iglesias C, Lorenzo H, Polanco I, Quiles J, Romero de Ávila MD, Russolillo G, Villarino A, Alfredo Martínez J (2009) Comparison of dietary reference intakes (DRI) between different countries of the European Union, the United States and the World Health Organization. Nutr Hosp 24(4):384–414
Derache R (1990) Toxicología y seguridad de los alimentos. Omega, Barcelona
Dias MI, Barros L, Alves RC, Oliveira M, Santos-Buelga C, Ferreira IC (2014) Nutritional composition, antioxidant activity and phenolic compounds of wild Taraxacum sect. Ruderalia. Food Res Int 56:266–271
Dolezal M, Velísek J, Famfulíková P (2001) Chemical composition of less known wild fruits. In: Pfannhauser W, Fenwick GR, Khokhar S (eds) Biologically-active phytochemicals in food. Analysis, metabolism, bioavailability and function. Royal Society of Chemistry, London, pp 241–244
European Food Safety Authority (2005) Opinion of the scientific panel on dietetic products, nutrition and allergies on a request from the commission related to the tolerable upper intake level of sodium (Request N° EFSA-Q-2003-018) Eur Food Saf J 209:1–26
European Food Safety Authority (2009) Scientific opinion of the panel on dietetic products, nutrition and allergies on a request from the commission related to the review of labelling reference intake values for selected nutritional elements (Question No EFSA-Q-2008-772) Eur Food Saf J 1008:1–14
European Parliament and Council (2006) Regulation (EU) No 1924/2006 of The European Parliament and of the Council of 20 December 2006 on nutrition and health claims made on foods. Off J Eur Union 18.1.2007: L12/3-L12/18
European Parliament and Council (2011) Regulation (EU) No 1169/2011 of the European Parliament and of the Council of 25 October 2011 on the provision of food information to consumers. Off J Eur Union 22.11.2011: L304/18-L304/63
FAO/WHO (2002) Human vitamin and mineral requirements. In: Report of a Joint FAO/WHO Expert Consultation Bangkok, Thailand. http://www.fao.org/docrep/004/y2809e/y2809e00.htm. Accessed 3 Jan 2015
Fennema OR (2000) Química de los alimentos, 2ª edn. Acribia, Zaragoza
FNB (Food and Nutrition Board) of the Institute of Medicine (2010) Dietary reference intakes for calcium and vitamin D. National Academy Press, Washington, DC
García-Herrera P (2014) Plantas silvestres de consumo tradicional. Caracterización de su valor nutricional y estimación de su actividad antifúngica. PhD Thesis. Universidad Complutense de Madrid, Spain

García-Herrera P, Sánchez-Mata MC, Cámara M, Fernández-Ruiz V, Díez-Marqués C, Molina M, Tardío J (2014) Nutrient composition of six wild edible Mediterranean Asteraceae plants of dietary interest. J Food Compos Anal 34:163–170

Gibson GR, Roberfroid MB (2008) Handbook of prebiotics. CRC Press. Boca Raton

Gropper SS, Smith JL, Groof JL (2008) Advanced nutrition and human metabolism, 5th edn. Cengage Learning, Belmont

Gjorgieva D, Kadifkova-Panovska T, Baeva K, Stafilov T (2011) Metalic trace elements in medicinal plants from macedonia. Middle-East J Sci Res 7(1): 109–114

Guil JL, Rodríguez-García I, Torija-Isasa ME (1997) Nutritional and toxic factors in selected wild edible plants. Plant Foods Hum Nutr 51(2):99–107

Guil-Guerrero JL (2001) Nutritional composition of *Plantago* species (*P. major* L., *P. lanceolata* L., and *P. media* L.). Ecol Food Nutr 40(5):481–495

Guil-Guerrero JL, Torija-Isasa ME (1997) Nutritional composition of leaves of *Chenopodium* species (*C. album* L, *C. murale* L and *C. opulifolium* Shraeder). Int J Food Sci Nutr 48(5):321–327

Guil-Guerrero JL, Torija-Isasa ME, Giménez-Martínez JJ (1996a) Nutritional composition of rock samphire (*Crithmum maritimum* L.). Alimentaria 272:65–72

Guil-Guerrero JL, Torija-Isasa ME, Giménez-Martínez JJ, Rodriguez-Garcia I, Giménez-Giménez A (1996b) Oxalic acid and calcium determination in wild edible plants. J Agric Food Chem 44(7):1821–1823

Guil-Guerrero JL, Giménez-Giménez A, Rodríguez-Garcia I, Torija-Isasa ME (1998a) Nutritional composition of *Sonchus* species (*S. asper* L, *S. oleraceus* L and *S. tenerrimus* L). J Sci Food Agric 76(4):628–632

Guil-Guerrero JL, Giménez-Martínez JJ, Torija-Isasa ME (1998b) Mineral nutrient composition of edible wild plants. J Food Compos Anal 11:322–328

Guil-Guerrero JL, Campra-Madrid P, Torija-Isasa ME (1999a) Mineral elements determination in wild edible plants. Ecol Food Nutr 38(3):209–222

Guil-Guerrero JL, Giménez-Martínez JJ, Torija-Isasa ME (1999b) Nutritional composition of wild edible crucifer species. J Food Biochem 23(3):283–294

Hacıseferoğulları H, Özcan MM, Arslan D, Unver A (2012) Biochemical compositional and technological characterizations of black and white myrtle (*Myrtus communis* L.) fruits. J Food Sci Technol 49(1):82–88

Hiçsönmez U, Ereeş FS, Özdemir C, Özdemir A, Çam S (2009) Determination of major and minor elements in the *Malva sylvestris* L. from Turkey using ICP-OES techniques. Biol Trace Elem Res 128:248–257

Judprasong K, Charoenkiatkul S, Sungpuag P, Vasanachitt K, Nakjamanong Y (2006) Total and soluble oxalate contents in Thai vegetables, cereal grains and legume seeds and their changes after cooking. J Food Compos Anal 19(4):340–347

Krstic-Pavlovic N, Dzamic R (1985) Zastupljenost adstrinogentnih i mineralnih materija u listu samonikle koprive (*Urtica dioica* L.) raznih prirodnih stanista. Agrohemija 3:191–198

Kwak H-S, Leeb W-J, Leeb M-R (2012) Revisiting lactose as an enhancer of calcium absorption. Int Dairy J 22(2):147–151

Mahan LK, Escott-Stump S, Raymond JL (2012) Krause Dietoterapia, 13th edn. Elsevier, Madrid

Marakoglu T, Arslan D, Özcan M, Hacıseferoğulları H (2005) Proximate composition and technological properties of fresh blackthorn (*Prunus spinosa* L. subsp *dasyphylla* (Schur.)) fruits. J Food Eng 68(2):137–142

Martin A (2001) Apports nutritionnels conseilles pour la population française 3rd edn. (Agence Française de Securité Sanitaire des Aliments (AFSSA)). Tec et Doc Lavoisier, París

Martínez-Para MC, Masoud TA, Torija-Isasa ME (1979) La ortiga en la alimentación. I) Composición centesimal y elementos minerales. An de Bromatol 31(3–4):327–331

Morales P (2011) Vegetales silvestres de uso alimentario: determinación de compuestos bioactivos y capacidad antioxidante. PhD Thesis. Universidad Complutense de Madrid, Spain

Morales P, Ferreira IC, Carvalho AM, Fernández-Ruiz V, Sanchez-Mata MC, Camara M, Morales R, Tardío J (2013) Wild edible fruits as a potential source of phytochemicals with capacity to inhibit lipid peroxidation. Eur J Lipid Sci 115(2):176–185

Morales P, Ferreira ICFR, Carvalho AM, Sánchez-Mata MC, Cámara M, Fernández-Ruiz V, Pardo-de-Santayana M, Tardío J (2014) Mediterranean non-cultivated vegetables as dietary sources of compounds with antioxidant and biological activity. LWT-Food Sci Technol 55:389–396

Moreiras O, Carbajal A, Cabrera L, Cuadrado C (2011) Tablas de composición de alimentos, 14th edn. Pirámide, Madrid

Özcan MM, Hacıseferoğulları H (2007) The strawberry (*Arbutus unedo* L.) fruits: chemical composition, physical properties and mineral contents. J Food Eng 78:1022–1028

Özcan MM, Unver A, Ucar T, Arslan D (2008) Mineral content of some herbs and herbal teas by infusion and decoction. Food Chem 106(3):1120–1127

Pereira C, Barros L, Carvalho AM, Ferreira IC (2013) Use of UFLC-PDA for the analysis of organic acids in thirty-five species of food and medicinal plants. Food Anal Methods 6(5):1337–1344

Poschenrieder C, Llugany M, Lombini A, Dinelli E, Bech J, Barceló J (2012) *Smilax aspera* L. an evergreen Mediterranean climber for phytoremediation. J Geochem Explor 123:41–44

Queralt I, Ovejero M, Carvalho ML, Marques AF, Llabres JM (2005) Quantitative determination of essential and trace element content of medicinal plants and their infusions by XRF and ICP techniques. X-Ray Spectrom 34(3):213–217

Ranfa A, Maurizi A, Romano B, Bodesmo M (2014) The importance of traditional uses and nutraceutical aspects of some edible wild plants in human nutrition: the case of Umbria (central Italy). Plant Biosyst 148(2):297–306

Romojaro A, Botella MA, Obon C, Pretel MT (2013) Nutritional and antioxidant properties of wild edible plants and their use as potential ingredients in the modern diet. Int J Food Sci Nutr 64(8):944–952

Ruiz-Rodríguez BM (2014) Frutos silvestres de uso tradicional en la alimentación. Evaluación de su valor nutricional, compuestos bioactivos y capacidad antioxidante. PhD Thesis. Universidad Complutense de Madrid, Spain

Ruiz-Rodríguez BM, Morales P, Fernández-Ruiz V, Sánchez-Mata MC, Cámara M, Díez-Marqués C, Pardo-de-Santayana M, Molina M, Tardío J (2011) Valorization of wild strawberry-tree fruits (*Arbutus unedo* L.) through nutritional assessment and natural production data. Food Res Int 44(5):1244–1253

Sacks FM, Svetkey LP, Vollmer WM, Appel LJ, Bray GA, Harsha D, Obarzanek E, Conlin PR, Miller ER, Simons-Morton DG, Karanja N, Lin PH, Aickin M, Most-Windhauser MM, Moore TJ, Proschan MA, Cutler JA (2001) Effects on blood pressure of reduced dietary sodium and the dietary approaches to stop hypertension (DASH) diet. N Engl J Med 344:3–10

Saklani S, Chandra S, Badoni PP, Dogra S (2012) Antimicrobial activity, nutritional profile and phytochemical screening of wild edible fruit of *Rubus ellipticus*. Int J Med Aromat Plants 2(2):269–274

Sánchez-Mata MC, Cabrera-Loera RD, Morales P, Fernández-Ruiz V, Cámara M, Díez- Marqués C, Pardo-de-Santayana M, Tardío J (2012) Wild vegetables of the Mediterranean area as valuable sources of bioactive compounds. Genet Resour Crop Evol 59(3):431–443

Schrager S (2005) Dietary calcium intake and obesity. J Am Board Fam Pract 18:205

Scientific Committee on Foods of the European Union (1992) Nutrient and energy intakes for the European community. Opinion adopted by the SCF on 11 December 1992. In: Reports of the SCF (ed), 31st series. European Commission, Luxembourg

Società Italianna di Nutrizione Umana (1998) Livelli di assunzione raccomandati di energia e nutrienti per la poplazione Italiani, revisione 1996. EDRA, Milano

Souci SW, Fachmann W, Kraut H (2008) Food composition and nutrition tables. Medpharm, Stutgart

Sun JB, Gao YG, Zang P, Yang H, Zhang LX (2013) Mineral elements in root of wild *Saposhnikovia divaricata* and its rhizosphere soil. Biol Trace Elem Res 153(1–3):363–370

Tardío J, Molina M, Aceituno-Mata L, Pardo-de-Santayana M, Morales R, Fernández-Ruiz V, Morales P, García P, Cámara M, Sánchez-Mata MC (2011) *Montia fontana* L. (Portulacaceae), an interesting wild vegetable traditionally consumed in the Iberian Peninsula. Genet Resour Crop Evol 58(7):1105–1118

Trichopoulou A, Vasilopoulou E, Hollman P, Chamalides Ch, Foufa E, Kaloudis T, Kromhout D, Miskaki Ph, Petrochilou I, Poulima E, Stafilakis K, Theophilou D (2000) Nutritional composition and flavonoid content of edible wild greens and green pies: a potential rich source of antioxidant nutrients in the Mediterranean diet. Food Chem 70:319–323

Trumbo P, Schlicker S, Yates AA, Poos M (2002) Dietary reference intakes for energy, carbohydrate, fiber, fat, fatty acids, cholesterol, protein and amino acids. J Am Diet Assoc 102:1621–1630

Villatoro-Pulido M, Moreno Rojas R, Munoz-Serrano A, Cardenosa V, Amaro Lopez MA, Font R, del Rio-Celestino M (2012) Characterization and prediction by near-infrared reflectance of mineral composition of rocket (*Eruca vesicaria* subsp *sativa* and *Eruca vesicaria* subsp *vesicaria*). J Sci Food Agric 92(7):1331–1340

Weaver CM, Heaney RP (1991) Isotopic exchange of ingested calcium between labeled sources: evidence that ingested calcium does not form a common absorptive pool. Calcif Tissue Int 49:244–247

Wetherilt H (1992) Evaluation of *Urtica* species as potential sources of important nutrients. Food Sci Hum Nutr 29:15–26

Wilman D, Riley JA (1993) Potential nutritive value of a wide range of grassland species. J Agric Sci 120:43–49

Yadav SK, Sehgal S (2003) Effect of domestic processing and cooking on selected antinutrient contents of some green leafy vegetables. Plant Foods Hum Nutr 58(3):1–11

Yildirim E, Dursuna A, Turan M (2001) Determination of the nutrition contents of the wild plants used as vegetables in Upper Çoruh Valley. Turk J Bot 25:367–371

Zeghichi S, Kallithraka S, Simopoulos AP, Kypriotakis Z (2005) Nutritional composition of selected wild plants in the diet of Crete. In: Simopoulos AP, Gopalan C (eds) Plants in human health and nutrition policy. Karger, Basel, pp 22–40

Chapter 8
Fatty Acid Profiles of Mediterranean Wild Edible Plants

José Luis Guil-Guerrero and María Esperanza Torija-Isasa

List of Abbreviations

AA	Arachidonic acid, 20:4*n-6*
ALA	α-linolenic acid, 18:3*n:3*
DHGLA	Dihomo-γ-linolenic, 20:3*n*-6
DHA	Docosahexaenoic acid, 22:6*n-3*
EFA	Essential fatty acid
EA	Erucic acid, 22:1*n*-9
EPA	Eicosapentaenoic acid, 20:5*n-3*
FA	Fatty acid
GLA	γ-linolenic, 18:3*n-6*
LA	Linoleic acid, 18:2*n-6*
MUFA	Monounsaturated fatty acid
OA	Oleic acid, 18:1*n*-9
PA	Palmitic acid, 16:0
PUFA	Polyunsaturated fatty acid
SA	Stearic acid, 18:0
SDA	Stearidonic acid, 18:4*n-3*
TAG	Triacylglycerol

J. L. Guil-Guerrero (✉)
Food Technology Division, Agronomy Department, University of Almería, Almería, Spain
e-mail: jlguil@ual.es

M. E. Torija-Isasa
Departamento de Nutrición y Bromatología II, Bromatología, Facultad de Farmacia, Universidad Complutense de Madrid, Madrid, Spain
e-mail: metorija@ucm.es

M. de C. Sánchez-Mata, J. Tardío (eds.), *Mediterranean Wild Edible Plants*,
DOI 10.1007/978-1-4939-3329-7_8

8.1 Sources and Role of Fatty Acids in Nutrition

The health benefits of omega-3 (*n*-3) and omega-6 (*n*-6) polyunsaturated fatty acids (PUFAs) are well established, as recent studies show (Wall et al. 2010; Asif 2011). Traditionally, two PUFAs have been considered as essential FAs (EFAs): linoleic acid (LA, 18:2*n*-6), which is the major PUFA component in our present diet, and α-linolenic acid (ALA, 18:3*n*-3), whose intake has declined in the modern societies (Das 2006; Kris-Etherton et al. 2000).

A basic overview of the pathways of FA biosynthesis is shown in Fig. 8.1. The enzymes for it are packaged together in a complex called FA synthase. The product of the enzyme is palmitic acid (16:0), and modifications of this primary FA lead to other longer or shorter FA, as well as to unsaturated ones. As shown, the FA molecule is synthesized 2 carbons at a time. For further step-wise 2-carbon extensions, acetyl CoA is first activated to malonyl CoA, a 3-carbon compound, by the addition of a CO_2 molecule.

From the two Δ6-desaturated FAs, LA and ALA, the body synthesizes *n*-3 and *n*-6 longer chain unsaturated homologues. However, in higher plants, γ-linolenic acid (GLA, 18:3*n*-6) and stearidonic acid (SDA, 18:4*n*-3) are the final compounds of the metabolic pathway (Fig. 8.1), although the route for successive elongation/desaturation continues in most algae and other cryptogams, to reach eicosapentaenoic (EPA, 20:5*n-3*) and docosahexaenoic (DHA, 22:6*n*-3) acids as *n*-3 metabolites, and dihomo-γ-linolenic (DHGLA, 20:3*n*-6) and arachidonic (AA, 20:4*n*-6) acids as *n*-6 (Guil-Guerrero 2007). The first step for both *n*-3 and *n*-6 pathways in which ALA is desaturated to SDA and LA to GLA by the enzyme Δ6-desaturase (Fig. 8.1) is rate-limiting in humans; therefore, both Δ6-desaturated PUFAs could be used to alleviate several deficiency-related diseases (Guil-Guerrero 2007; Horrobin 1992; Whelan 2009). In this regard, it has been observed that an adequate intake of FAs is crucial to maintain a good health status. Today, there is clear evidence that human beings had in the past a diet correctly balanced in relation to the *n*-6/*n*-3 FAs ratio, which was approximately 1, whereas in the present Western diets it ranges between 15/1 and 16.7/1 (Simopoulos 2002). Dietary intakes of ALA among Western adults are typically in the range of 0.5–2 g/d (Burdge and Calder 2006; Calder 2012), while LA is the main PUFA in most Western diets, which is typically consumed in 5- to 20-fold greater amounts than ALA (Burdge and Calder 2006; Calder 2012).

Current literature shows that a high ratio *n*-6/*n*-3 FA causes a progressive impairment of health. Furthermore, improving this ratio is essential for proper brain function and also to prevent cardiovascular diseases, arthritis, cancer, and inflammatory and autoimmune diseases (Simopoulos 2008).

However, commonly consumed oils are ineffective for the improvement of this ratio, taking into account that they usually contain LA as the more prominent FA. Thus, alternative oils based in an adequate content in the Δ6-desaturated FAs, GLA, and SDA, such as blackcurrant oil (*Ribes nigrum*), are now available to consumers (Guil-Guerrero 2007). In addition, some wild plants become cultivated, for example, Paterson's curse (*Echium plantagineum*), from which a seed oil rich in Δ6-desaturated PUFAs—GLA, SDA, and ALA—is obtained (Berti et al. 2007; Guil-Guerrero 2007).

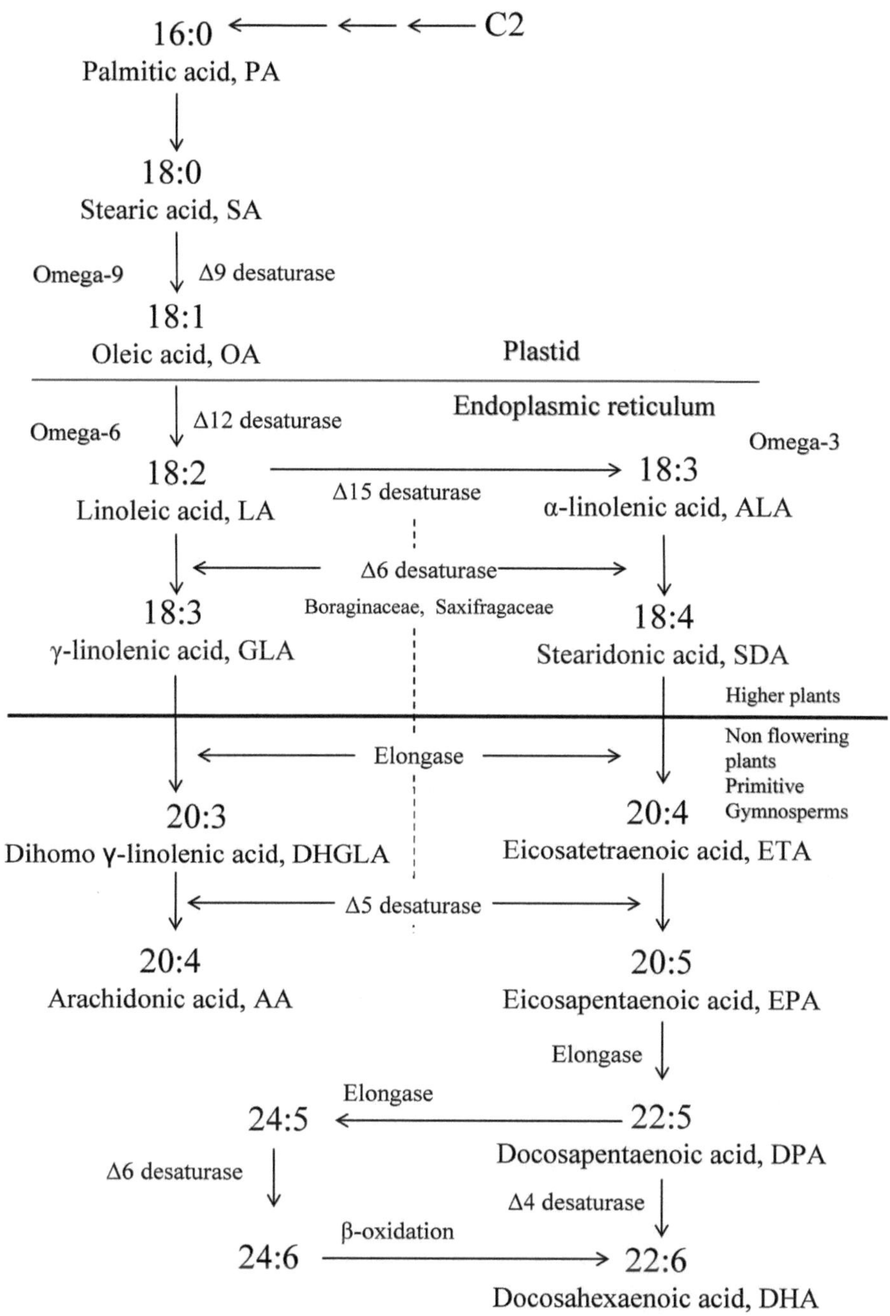

Fig. 8.1 Basic overview of the pathways for FA biosynthesis. Only the major steps are indicated for clarity; for detailed pathways, see Bates and Browse (2012) and Guil-Guerrero (2007). FA precursors are biosynthesized de novo in the chloroplast. Saturated FA synthesis proceeds 2 carbon units per cycle usually 8 or 9 times producing 16 or 18 carbon products. The relative abundance of each C18 PUFA in the tissues of higher plants is a consequence of the higher or lesser activity of the desaturase enzymes present in each given species, which is genetically regulated. Acyl-lipid desaturases introduce double bonds into FAs bonded to the glycerol moiety of polar glycerolipids,

Edible wild plants could help improve the above exposed ratio. According to each consumed organ, they provide different FAs. Leaves, shoots, and fruits constitute a rich source of ALA, while seeds typically contain LA and ALA as main FAs, although there are some exceptions. In any case, Mediterranean edible wild plants constitute a wide reservoir of new foods that can be consumed as vegetables, fruits, and spices. The following sections provide information about different organs (leaves, seeds, and fruits) of selected species of Mediterranean edible wild plants, which constitutes valuable sources of FAs.

8.2 Wild Green Leafy Vegetables as Source of Omega-3 FAs

The knowledge about the FA composition of Mediterranean edible wild plants started a few years ago, when Simopoulos (1986) promoted studies on the FA composition of purslane (*Portulaca oleracea*), an edible wild plant whose leaves and stems are usually consumed in Mediterranean countries. These studies were followed by others, such as the works of Guil et al. (1996), Guil-Guerrero and Rodríguez-García (1999), Pereira et al. (2011), and Morales et al. (2012).

A selection of Mediterranean wild species used as leafy vegetables arranged according to different botanical families is given in Table 8.1. Note that the pattern of FA composition denotes a similar trend in all exposed species, in which palmitic acid (PA) and stearic acid (SA) are the main saturated FAs, while OA is the predominant monounsaturated FA (MUFA), and LA and ALA are the common PUFAs. Their relative abundance is as follow: ALA > LA = PA > OA > SA, with some exceptions. However, there is no clear relationship between the amounts of each FA and the botanical family to which each species belongs. Therefore, the content of ALA

---

and they are widely distributed in plants. The desaturation of stearic acid yields oleic acid (*OA*, 18:1*n*-9), a step that is catalyzed by a Δ9 acyl desaturase in the stroma of plastids. The OA produced is transported to the thylakoid membrane for further desaturation in the lipid-bound form. This enzyme can desaturate both C16 and C18 FAs with the same efficiency (Los and Murata 1998). Then, two desaturase enzymes work sequentially to produce, first, linoleic acid (*LA*; Δ12 desaturase) and then α-linolenic acid (*ALA*; Δ15 desaturase). The greater or lesser degree of activity of these enzymes, which is characteristic of each taxonomic group, is responsible for the relative concentrations of the two PUFAs in the organs of each botanical species. However, few plants have Δ6 desaturase activity; among these are species of Boraginaceae, Saxifragaceae, Onagraceae, and some others. These can produce in their metabolic pathways, depending on the different affinities of their enzymes toward the substrates LA or ALA, γ-linolenic acid *(GLA)* and/or stearidonic acid *(SDA)*, respectively (Guil-Guerrero 2007). From the two last PUFAs stopped the metabolic pathways in higher plants, that is, lacking of the necessary elongase enzymes to obtain C20–22 PUFA, thus existing eicosapentaenoic acid *(EPA)*, docosahexaenoic acid *(DHA)*, and arachidonic acid *(AA)* in mosses and ferns as most evolved taxa. However, some species of gymnosperms barely evolved possess this capability, as is the case of *Agathis robusta*, an Australian Araucariaceae, which shows AA and EPA (Wolff et al. 1999). For higher plants, the ability to survive at low temperatures is correlated with the presence of PUFAs, which justifies that a great percentage of the leafy FAs are always PUFAs. Although the presence of elongase enzymes that could lead to C20 PUFAs or more appears to be restricted to higher plants, it is a recurring theme of biotechnology, and it has been obtained from transgenic soybean and other species displaying this capability (Alonso and Maroto 2000)

Table 8.1 Fat (%) and fatty acid composition (% of total FAs) of selected leaves of Mediterranean wild edible plants

Family	Species	Common name	Fat (%)	16:0	18:0	18:1*n*-9	18:2*n*-6	18:3*n*-3	Others	Source
Amaranthaceae	*Amaranthus viridis*	Amaranth	1.4[a]	21	3.0	9.0	20.0	24.0		Guil et al. 1996
Boraginaceae	*Anchusa azurea*	Alkanet	0.9[b]	10.5	1.7	2.2	12.2	64.7		Morales et al. 2012
	Borago officinalis	Borage	1.3[a]	12.0	2.3	2.1	9.5	12.3	10.8[c]	Pereira et al. 2011
Caryopllyllaceae	*Silene vulgaris*	Bladder campion	0.7[b]	13.5	0.5	2.4	22.4	54.5	3.3[d]	Alarcón et al. 2006
Chenopodiaceae	*Beta maritima*	Wild chard	2.2[a]	22.6	2.46	5.7	19.2	29.5		Guil et al. 1996
	Chenopodium album	Lamb's-quarters	2.3[a]	16.0	2.0	3.0	16.0	45.0		Guil et al. 1996
	Chenopodium murale	Nettleleaf goosefoot	2.4[a]	17.0	2.0	7.0	18.0	36.0		Guil et al. 1996
Compositae	*Chondrilla juncea*	Rush skeletonweed	1.5[b]	13.0	2.2	1.9	19.9	56.3		Morales et al. 2012
	Sonchus oleraceus	Sow thistle	5.0[a]	19	2.0	2.0	8.0	44.0		Guil-Guerrero et al. 1998
	Sonchus tenerrimus	Milk thistle	4.2[a]	19	1.0	4.0	9.0	30.0		Guil-Guerrero et al. 1998
	Taraxacum obovatum	Dandelion	1.2[b]	11.8	2.5	3.2	17.6	58.5		Morales et al. 2012
Cruciferae	*Alliaria petiolata*	Garlic mustard	2.1[a]	6.0	1.0	10.0	10	28.0	31[d]	Guil-Guerrero et al. 1999
	Capsella bursa-pastoris	Shepherd's purse	2.0[a]	15.1	2.6	1.4	11.1	50.7		Guil-Guerrero et al. 1999
	Sisymbrium irio	London rocket	2.7[a]	21.0	2.0	3.0	13.0	46.0		Guil-Guerrero et al. 1999
Malvaceae	*Malva sylvestris*	Common mallow	3.5[a]	2.1	2.1	1.7	10.4	42.2		Guil et al. 1996
Portulacaceae	*Montia fontana*	Water-blinks	3.1[a]	17.2	0.1	2.4	18.7	55.6		Pereira et al. 2011
	Portulaca oleracea	Purslane	3.7[a]	17.4	3.5	5.9	16.8	32.6		Guil et al. 1996
Umbelliferae	*Crithmum maritimum*	Rock samphire	3.7[a]	9.4	1.5	16.1	12.1	10.0		Guil-Guerrero and Rodríguez-García 1999
Urticaceae	*Urtica dioica*	Stinging nettle	2.1[a]	17.9	1.6	1.7	11.6	40.7		Guil-Guerrero et al. 2003

[a] Dry weight
[b] Fresh weight
[c] 18:3*n*-6
[d] 22:1*n*-9

seems to be more related to the degree of maturity of the harvested leafs or to climate parameters instead of taxonomic groups. Furthermore, the predominance of ALA over all FAs is necessary because ALA plays an important role in coordinating ion channel activities in cell membranes of plants (Gutla et al. 2012).

Considering that the amounts of ALA in green leafy vegetables are about double or triple those of LA, their intake is favorable for the improvement of the previously discussed *n*-6/*n*-3 ratio. Note that the amount of total FAs on dry matter in all species is quite low, and, although there exist some exceptions, a figure of 3 g/100 g could be considered as the more common value. By considering that the majority of the species have ALA at about 50 % of total FAs, a typical species stores approximately 1.5 g/100 g ALA on dry matter; and taken into account that moisture generally reaches about 90 % in leaves, the amount of ALA on fresh weight would achieve ~ 150 mg/100 g. Therefore, to obtain a quarter of the recommended daily intake of ~ 2 g ALA per day (European Food Safety Authority 2009), a consumption of about ~ 300 g of Mediterranean wild edible greens per day providing ~ 500 mg ALA, if bioavailable, seems to be necessary. However, such consumption is clearly excessive for most people in Western countries.

To the general scheme on FA occurrence in Mediterranean wild greens, it should be added that there are some other FAs found in variable quantities, which are characteristic of each botanical family, for example, in Boraginaceae species (such as *Borago officinalis*), GLA normally appears (Guil-Guerrero 2007); thus, regarding the general scheme of FA biosynthesis (Fig. 8.1), the occurrence of this FA and/or SDA in any future analyzed Boraginaceae species is expected. Erucic acid (EA, 22:1*n*-9), a toxic FA, is sometimes reported in Cruciferae, for example, in *Alliaria petiolata* (Guil et al. 1997), and is also found in smaller amounts in some Caryopllyllaceae species, such as in *Silene vulgaris* (Alarcón et al. 2006). This FA, together with eicosenoic (20:1*n*-9) and nervonic (24:1*n*-9) acids, has been cited as responsible for congestive heart failure; so their presence in the diet has been associated with cardiotoxicity in humans (Imamura et al. 2013). In any case, the low fat content of the leaves makes it unlikely that the occasional consumption of *A. petiolata* (31 % EA of total FAs) would cause heart problems. However, it seems reasonable to avoid regular consumption of leaves or seeds of this species.

8.3 Seeds from Mediterranean Wild Edible Plants: An Inexpensive Source of Some PUFAs

The seeds fatty acid composition of some of the main Mediterranean wild edible plants are presented in Table 8.2. Unlike wild green species, a reduced number of FA profiles of seeds from wild edible plants are reported. These agree, in general, with other profiles already known for their cultivated relatives. Few species may be considered as a source of *n*-3 FAs, which belong to Brassicaceae and Urticaceae families. With respect to the remaining botanical families, Anacardiaceae species constitutes a source of OA, taking into account the high percentage of oil content in their seeds. Poaceae species (*Avena fatua*) shows also a high percentage of OA on total FAs; however, the low amount of FAs on total seed reported makes it impossible to consider this species as a source of *n*-9 FAs. Boraginaceae is a botanical

Table 8.2 Fat (%) and fatty acid composition (% of total FAs) of selected seeds of Mediterranean wild edible plants

Family	Species	Common name	Fat (%)	16:0	18:0	18:1*n*-9	18:2*n*-6	18:3*n*-3	Others	Source
Amaranthaceae	*Amaranthus retroflexus*	Amaranth	7.2	9.7	2.0	23.3	61.5	1.1	–	Opute 1979
Anacardiaceae	*Pistacia atlantica*	Turpentine	15.3	13.1	2.3	50.7	29.8	0.6	0.3[a]	Givianrad et al. 2013
	Pistacia terebinthus	Turpentine	41.2	21.6	2.1	46.9	21.7	0.7	0.2[a]	Matthäus and Özcan 2006
Boraginaceae	*Borago officinalis*	Borage	28.9	11.7	4.4	19.8	36.8	–	19.5[b]	Guil-Guerrero et al. 2013
Brassicaceae	*Cardaria draba*	Hoary cress	7.0	17.9	9.2	12.3	18.9	38.9	–	Tonguç and Erbas 2012
	Diplotaxis tenuifolia	Lincoln weed	23.4	8.2	3.6	22.2	16.7	19.6	9.7[d]; 18.8[c]	Tonguç and Erbas 2012
	Eruca sativa	Salad rocket	24.2	5.3	2.1	10.3	7.0	8.3	10.1[d]; 56.6[c]	Tonguç and Erbas 2012
Chenopodiaceae	*Chenopodium album*	Lamb's-quarters	9.1	8.4	0.9	20.7	56.3	6.5	–	Daun and Tkachuk 1976
Compositae	*Carthamus dentatus*	Toothed thistle	15.4	9.8	3.9	19.9	66.2	–	–	Tonguç and Erbas 2012
	Centaurea depressa	Low cornflower	19.7	12.3	6.8	32.7	48.3	–	–	Tonguç and Erbas 2012
Pinaeae	*Pinus pinea*	Pine nut	44.9	6.5	3.5	38.6	47.6	0.7	–	Nergiz and Dönmez 2004
Poaceae	*Avena fatua*	Wild oat	1.4	23.4	3.3	40.9	29.5	–	–	Daun and Tkachuk 1976
Urticaceae	*Urtica dioica*	Stinging nettle	15.1	25.4	2.3	4.8	22.7	6.6	2.1[a]; 1.2[c]	Guil-Guerrero et al. 2003

[a] 20:1*n*-9
[b] 18:3*n*-6
[c] 22:1*n*-9
[d] 20:1*n*-9

family in which their seeds are widely exploited as a source of healthy oil, as occurs with *Borago officinalis*, a species consumed as vegetable, although its seeds are sporadically consumed. In this plant, the activity of the Δ6-desaturate enzyme leads to a GLA-enriched seed oil (Fig. 8.1). The high *n*-6/*n*-3 ratio of this plant should not be considered as disadvantageous by considering that GLA has multiple medicinal applications (Guil-Guerrero 2007), due to which this oil was being marketed several years ago.

The remaining reported families, that is, Amaranthaceae, Chenopodiaceae, Compositae, and Pinaeae, mainly contain LA, as well as variable amounts of ALA.

Considering both ALA percentage and seed oil content in species shown in Table 8.2, it is clear that some of them could contribute effectively to fulfill the daily needs of *n*-3 FAs; that is, the recommended daily intake of ~2 g ALA per day (European Food Safety Authority 2009) could be achieved by ingesting ~60 g of *Cardaria draba* seeds or ~200 g of *Urtica dioica* seeds, both amounts being reasonably affordable to consume in a regular diet in Western countries.

Overall, special caution should be observed regarding the consumption of Brassicaceae seeds, considering that they could contain potentially toxic FAs, as is the case of *Diplotaxis tenuifolia* and *Eruca sativa* seeds (Table 8.2), which contain high percentages of EA. Notwithstanding, some cruciferous wild species are reported as lacking of toxic FAs; however, sometimes not all FAs are correctly informed about their nature, so extreme caution regarding the consumption of seeds of this family is needed.

8.4 Fruits of Mediterranean Wild Edible Plants: A Balanced Contribution Between MFA and PUFA

There are very few species of Mediterranean edible wild plants acting as fruit providers being found only in the wild (Table 8.3). For example, strawberry tree is a wild species, but in certain areas it is farmed for fruit; therefore, it is difficult to categorize this as wild or not because its cultivation scarcely interferes with the natural biological cycle.

Among the reported families, Rosaceae has the largest number of species. The FAs of the three species here shown are closely similar, with a good balance among *n*-9, *n*-6, and *n*-3 FAs. Myrtaceae (*Myrtus communis*) is characterized by a high percentage of *n*-9 FAs, while *Ziziphus jujuba* (Rhamnaceae) shows medium-chain FAs as major fat constituents, which have been reported to improve health. This way, it appears that administration of triacylglycerols (TAGs), in which medium-chain FAs constitutes the main acylglycerol species, improves the long-term success of dietotherapy of obese patients (Hainer et al. 1994). Other results from longest controlled feeding of medium-chain FAs indicate that short-term feeding of diets enriched by medium-chain FAs increases total energy expenditure (White et al. 1999). This fact would have relevance in countries where jujube could have a key role in the diet, as occurs in some areas of Southeast Asia.

Table 8.3 Fat (%FW) and FA composition (% of total FAs) of selected fruits of Mediterranean wild edible plants

Family	Species	Common name	Fat (%)	12:0	14:0	16:0	18:0	18:1n-9	18:2n-6	18:3n-3	Others	Source
Ericaceae	*Arbutus unedo*	Strawberry tree	1.4	0.7	1.3	8.2	4.0	21.0	21.5	36.5	–	Barros et al. 2010
Myrtaceae	*Myrtus communis*	Myrtle	–	–	–	10.2	8.2	67.1	0.6	–	–	Serce et al. 2010
Rhamnaceae	*Ziziphus jujuba*	Jujube	1.1	15.1	2.5	6.2	0.3	1.4	1.7	0.4	6.7[a]; 46.8[b]; 10.7[c]	Guil-Guerrero et al. 2004
Rosaceae	*Prunus spinosa*	Blackthorn	2.0	0,1	0,1	6.5	2.5	57.6	23.6	2.8	–	Barros et al. 2010
	Rosa canina	Rose hips	0.7	0.6	0,1	0.4	2.4	14.4	39.5	26.3	–	Barros et al. 2010
	Rubus idaeus	Red raspberry	0.5	–	–	6	–	12.9	50.4	20.2	1.5[c]	Celik and Ercisli 2009

[a] 8:0
[b] 10:0
[c] 16:1*n*–7

Similarly to what happens in leafy greens, the content of FAs in fruit tissues is quite low in most species; and by considering *Arbutus unedo* as the more prominent species, which show 1.5 % FAs of fresh weight, by a daily intake of 200 g—an usual consumption for any fruit—and regarding the ALA percentage of 36 % of total FAs (Table 8.3), it is obtained approximately 100 mg ALA, that is a low figure, insufficient as to consider this fruit as a good source of ALA. On the other hand, other wild edible species (i.e. *Myrtus communis* and *Prunus spinosa*) provide a modest contribution of OA to the diet.

8.5 An Unclear Subject: Are Wild Edible Plants More Omega-3 FAs Enriched than Cultivated Ones?

Some authors argued that during the human evolution, our ancestors had a diet containing approximately an *n*-6/*n*-3 ratio close to 1, and due to modern agriculture, with its emphasis on production, the *n-3* FA content has decreased in many foods; so, in the Palaeolithic, the *n*-3 FAs were found more frequently in all foods: meat, wild plants, eggs, fish, nuts, and berries (Simopoulos 2002). Thus, it is believed that edible wild plants were used more frequently in the Palaeolithic than in the present (Diamond 2002), although today it is possible to find this feeding behavior in rural communities in several countries of the Mediterranean Basin (Leonti et al. 2006). In this sense, the knowledge about the healthy effect of the Mediterranean diet is widely documented; for example, its positive impact on lowering the rate of cardiovascular complications after myocardial infarction was established in the clinical trial "Lyon Diet Heart Study," in which it was concluded that ALA was responsible for this effect (De Lorgeril et al. 1994).

The benefits derived from the intake of *n*-3 FAs seem unquestionable. However, an unclear question is the one concerning the actual quantity of *n*-3 in wild plants versus cultivated ones. In this regard, Simopoulos and Salem (1986) and Simopoulos et al. (1992, 1995) reported an edible wild plant as an unmistakable source of *n*-3: purslane (*Portulaca oleracea*), which would contain important amounts of ALA. Based on this finding, it was argued that edible wild plants are the best sources of *n*-3 as compared to those farmed. However, although purslane is today widely reported as a good source of ALA, according to later studies (Liu et al. 2000; Palaniswamy et al. 2001), both the total FA percentage on fresh plant weight and ALA percentage of total FAs are in good agreement with other commonly consumed vegetables, either farmed or wild. Furthermore, EPA and DHA have not been found in subsequent analyses (Vaskovsky and Khotimchenko 1992).

To elucidate whether edible wild plants are better sources of *n*-3 than those farmed, it should be appropriate to check the FA content in the same organs of both plant types, edible and farmed, collected in the same area and belonging to the same or related species. However, few studies filled this premise. Among consulted literature, the work of Vardavas et al. (2006) is very useful in this regard.

The authors reported on the FA concentrations of 6 cultivated species and 48 wildly grown greens, all of them usually consumed in Crete. From their results, the authors highlighted that the highest percentage of *n*-3 was found in cultivated organic lettuce in which 60% of all FAs were ALA. Moreover, after selecting only the leafy greens among the reported species and discarding other organs that could lead to error, an *n*-6/*n*-3 ratio of 0.38 for cultivated species and 0.43 for wild ones was obtained. Thus, it seems that the traditional Cretan diet is composed mainly of vegetables, which explains its good nutritional qualities as well as the appreciable amount of ALA it provides, but which does not lead to the fact there is a decrease in the percentage of *n*-3 of crop plants.

Another particularly interesting work that could help resolve this question is the one conducted by Celik and Ercisli (2009) concerning the total lipids and FA composition of red raspberry fruits (*Rubus idaeus*). In that study, several wild grown raspberries were collected from their natural habitats, while other fruits were gathered from cultivated plants. The analyses showed that the yield of lipid was 0.40% as maximum for wild genotypes, while for the cultivated species it reached 0.63%. Moreover, ALA content was similar for the two kinds of raspberry, although all wild raspberry genotypes had higher amount of LA and PA than cultivated ones, thus slightly increasing the ratio *n*-6/*n*-3 for wild genotypes.

In this regard, a report about the FA composition of several jujube varieties from the southeast of Spain gave results similar to the previous studies (Guil-Guerrero et al. 2004) because wild jujubes showed lower percentages of ALA than five other cultivated varieties.

Therefore, although the previous data are very partial, the statement that edible wild plants are more enriched in *n*-3 than those farmed seems to be unfounded until now.

8.6 Conclusions

The intake of edible wild plants attracts nowadays the interest of many people who expect to find nutrients and phytochemicals in wild vegetables in higher amounts than in cultivated species. As exposed above, edible wild plants contribute to fulfill the daily need of *n*-3 FAs as well as constitute a good source of *n*-9 and *n*-6 ones. Various organs could be consumed with different contributions; that is, wild green leafy vegetables provide a significant portion of the daily dosing recommendations of *n*-3 FAs; seeds from Mediterranean wild edible plants constitute an inexpensive source of various PUFAs, Brassicaceae and Urticaceae include *n*-3-enriched oily species, Amaranthaceae and Compositae contain *n*-6 ones, while Anacardiaceae and Poaceae constitute a potential reservoir of *n*-9 FAs; finally, fruits of Mediterranean wild edible plants offer a balanced contribution between MUFA and PUFA, and most them constitute a good source of medium-chain FAs, whose intake reports unquestionable health benefits.

As shown, research on Mediterranean wild edible plants indicates good nutritional qualities in many of these. However, a greater emphasis should be given in research papers to elucidate potential toxicities, for example, by performing cytotoxic assays and/or animal experiments. Given that wild vegetables can contain a considerable diversity of active compounds whose analysis is not always done in research papers, such models could provide assurance about the safety of each targeted species.

References

Alarcón R, Ortiz LT, García P (2006) Nutrient and fatty acid composition of wild edible bladder campion populations [*Silene vulgaris* (Moench.) Garcke]. Int J Food Sci Technol 41:1239–1242

Alonso DL, Maroto FG (2000) Plants as 'chemical factories' for the production of polyunsaturated fatty acids. Biotechnol Adv 18:481–497

Asif M (2011) Health effects of omega-3, 6, 9 fatty acids: *Perilla frutescens* is a good example of plant oils. Orient Pharm Exp Med 11:51–59

Barros L, Carvalho AM, Morais JS, Ferreira ICFR (2010) Strawberry-tree, blackthorn and rose fruits: detailed characterization in nutrients and phytochemicals with antioxidant properties. Food Chem 120:247–254

Barros L, Dueñas M, Ferreira ICFR (2011) Use of HPLC–DAD–ESI/MS to profile phenolic compounds in edible wild greens from Portugal. Food Chem 127:169–173

Bates PD, Browse J (2012) The significance of different diacylgycerol synthesis pathways on plant oil composition and bioengineering. Front Plant Sci 3:147

Berti M, Johnson BL, Dash S, Fischer S, Wilckens R, Hevia F (2007) *Echium*: a source of stearidonic acid adapted to the Northern Great Plainsthe US. In: Janick J, Whipkey A (eds) Issues in new crops and new uses. ASHS Press, Alexandria, pp 120–125

Burdge GC, Calder PC (2006) Dietary α-linolenic acid and health-related outcomes: a metabolic perspective. Nutr Res Rev 19:26–52

Calder PC (2012) Mechanisms of action of *(n-3)* fatty acids. J Nutr 42:592–599

Celik F, Ercisli S (2009) Lipid and fatty acid composition of wild and cultivated red raspberry fruits (*Rubus idaeus* L.). J Med Plants Res 3:583–585

Das UN (2006) Essential fatty acids: a review. Curr Pharm Biotechnol 7:467–482

Daun JK, Tkachuk R (1976) Fatty acid composition of oils extracted from Canadian weed seeds. J Amer Oil Chem Soc 53:661–662

De Lorgeril MS, Renaud N, Mamelle E, Salen JL, Martin I, Monjaud J et al (1994) Mediterranean alpha-linolenic acid- rich diet in secondary prevention of coronary heart disease. Lancet 343:1454–1459

Diamond J (2002) Evolution, consequences and future of plant and animal domestication. Nature 418:700–707

European Food Safety Authority (2009) Scientific opinion: labelling reference intake values for *n-3* and *n-6* polyunsaturated fatty acids. EFSA J 1176:1–11

Givianrad MH, Saber-Tehrani M, Jafari Mohammadi SA (2013) Chemical composition of oils from wild almond *(Prunus scoparia)* and wild pistachio *(Pistacia atlantica)*. Grasas y aceites 64:77–84

Guil JL, Torija ME, Gimenez JJ, Rodríguez I (1996) Identification of fatty acids in edible wild plants by gas chromatography. J Chromat A 719:229–235

Guil JL, Rodríguez-García I, Torija E (1997) Nutritional and toxic factors in selected wild edible plants. Plant Foods Human Nutr 51:99–07

Guil-Guerrero JL (2007) Stearidonic acid (18:*4n-3*): metabolism, nutritional importance, medical uses and natural sources. Eur J Lipid Sci Technol 109:1226–1236

Guil-Guerrero JL, Rodríguez-García I (1999) Lipids classes, fatty acids and carotenes of the leaves of six edible wild plants. Eur Food Res Technol 209:313–316

Guil-Guerrero JL, Giménez-Giménez A, Rodríguez-García I, Torija-Isasa ME (1998) Nutritional composition of *Sonchus* species (*S. asper* L., *S. oleraceus* L. and *S. tenerrimus* L.). J Sci Food Agric 76:628–632

Guil-Guerrero JL, Giménez-Martínez JJ, Torija-Isasa ME (1999) Nutritional composition of wild edible crucifer species. J Food Biochem 23:283–294

Guil-Guerrero JL, Rebolloso-Fuentes MM, Torija-Isasa ME (2003) Fatty acids and carotenoids from Stinging Nettle (*Urtica dioica* L.). J Food Compos Anal 16:111–119

Guil-Guerrero JL, Díaz Delgado A, Matallana González MC, Torija Isasa ME (2004) Fatty acids and carotenes in some Ber (*Ziziphus jujuba* Mill) varieties. Plant Foods Hum Nutr 59:23–27

Guil-Guerrero JL, Rincón-Cervera MA, G-omez-Mercado F, Ramos-Bueno RP, Venegas-Venegas E (2013) New seed oils of Boraginaceae rich in stearidonic and gamma-linolenic acids from the Maghreb region. J Food Compos Anal 31:20–23

Gutla PVK, Boccaccio A, De Angeli A, Gambale F, Carpaneto A (2012) Modulation of plant TPC channels by polyunsaturated fatty acids. J Exp Bot 3:6187–6197

Hainer V, Kunesova M, Stich V, Zak A, Parizkova J (1994) The role of oils containing triacylglycerols and medium-chain fatty acids in the dietary treatment of obesity. The effect on resting energy expenditure and serum lipids. Cas Lek Cesk 133:373–375

Horrobin DF (1992) Nutritional and medical importance of gamma-linolenic acid. Prog Lipid Res 31:163–194

Imamura F, Lemaitre RN, King IB, Song X, Steffen LM, Folsom AR, Siscovick DS, Mozaffarian D (2013) Long-chain monounsaturated fatty acids and incidence of congestive heart failure in 2 prospective cohorts. Circulation 127:1512–1521

Kris-Etherton PM, Shaffer Taylor D, Yu-Poth S, Huth P, Moriarty K, Fishell V et al (2000) Polyunsaturated fatty acids in the food chain in the United States. Am J Clin Nutr 71:179–188

Leonti M, Nebel S, Rivera D, Heinrich M (2006) Wild gathered food plants in the European Mediterranean: a comparative analysis. Econ Bot 60:130–142

Liu L, Howe P, Zhou YF, Xu ZQ, Hocart C, Zhang R (2000) Fatty acids and β-carotene in Australian purslane *Portulaca oleracea* varieties. J Chromatogr A 893:207–213

Los DA, Murata N (1998) Structure and expression of fatty acid desaturases. Biochim Biophys Acta 1394:3–15

Matthäus B, Özcan MM (2006) Quantization of fatty acids, sterols, and tocopherols in turpentine (*Pistacia terebinthus* chia) growing wild in Turkey. J Agric Food Chem 54:7667–7671

Morales P, Ferreira IC, Carvalho AM, Sánchez-Mata MC, Cámara M, Tardío J (2012) Fatty acids profiles of some Spanish wild vegetables. Food Sci Technol Int 18:281–290

Nergiz C, Dönmez İ (2004) Chemical composition and nutritive value of *Pinus pinea* L. seeds. Food Chem 86:365–368

Opute FI (1979) Seed lipids of the grain amaranths. J Exp Bot 30:601–606

Palaniswamy UR, McAvoy RJ, Bible BB (2001) Stage of harvest and polyunsaturated essential fatty acid concentrations in purslane *(Portulaca oleraceae)* leaves. J Agric Food Chem 49:3490–3493

Pereira C, Barros L, Carvalho AM, Ferreira ICFR (2011) Nutritional composition and bioactive properties of commonly consumed wild greens: potential sources for new trends in modern diets. Food Res Int 44:2634–2640

Serce S, Ercisli S, Sengul M, Gunduz K, Orhan E (2010) Antioxidant activities and fatty acid composition of wild grown myrtle (*Myrtus communis* L.) fruits. Pharmacogn Mag 6:9

Simopoulos AP (2002) The importance of the ratio of omega-6/omega-3 essential fatty acids. Biomed Pharmacother 56:365–379

Simopoulos AP (2008) The importance of the omega-6/omega-3 fatty acid ratio in cardiovascular disease and other chronic diseases. Exp Biol Med 233:674–688

Simopoulos AP, Salem N Jr (1986) Purslane: a terrestrial source of omega-3 fatty acids. N Engl J Med 315:833

Simopoulos AP, Norman HA, Gillaspy JE, Duke JA (1992) Common purslane: a source of omega-3 fatty acids and antioxidants. J Am Coll Nutr 11:374–382
Simopoulos AP, Norman HA, Gillaspy JE (1995) Purslane in human nutrition and its potential for world agriculture. World Rev Nutr Diet 77:47
Tonguç M, Erbaş S (2012) Evaluation of fatty acid compositions and some seed characters of common wild plant species of Turkey. Turk J Agric For 36:673–679
Vardavas CI, Majchrzak D, Wagner KH, Elmadfa I, Kafatos A (2006) Lipid concentrations of wild edible greens in Crete. Food Chem 99:822–834
Vaskovsky V, Khotimchenko S (1992) Chemotaxonomic approach protects against mistakes in polyunsaturated fatty acid analyses in plants. J Amer Oil Chem Soc 69:1
Wall R, Ross RP, Fitzgerald GF, Stanton C (2010) Fatty acids from fish: the anti-inflammatory potential of long-chain omega-3 fatty acids. Nutr Rev 68:280–289
Whelan J (2009) Dietary stearidonic acid is a long chain (*n*-3) Polyunsaturated fatty acid with potential health benefits. J Nutr 39:5–10
White MD, Papamandjaris AA, Jones PJ (1999) Enhanced postprandial energy expenditure with medium-chain fatty acid feeding is attenuated after 14 d in premenopausal women. Am J Clin Nutr 69:883–889
Wolff RL, Christie WW, Pédrono F, Marpeau AM (1999) Arachidonic, eicosapentaenoic, and biosynthetically related fatty acids in the seed lipids from a primitive gymnosperm, *Agathis robusta*. Lipids 34:1083–1097

Chapter 9
Wild Edible Plants as Sources of Carotenoids, Fibre, Phenolics and Other Non-Nutrient Bioactive Compounds

Montaña Cámara, Virginia Fernández-Ruiz and Brígida María Ruiz-Rodríguez

9.1 Introduction

Plants have been part of the human diet since ancient times, and their selection is influenced by the attractive colours and flavours, in addition to nutrients that they provide. The nutritional quality of plant products depends on the amount and quality of the macronutrients (proteins, carbohydrates and lipids) and micronutrients (vitamins, minerals, fatty acids and essential amino acids).

Recent epidemiological and biochemical studies support the association between high intake of vegetables and better health status. This beneficial action of the vegetables is due to the presence of a wide variety of substances, which include macronutrients, micronutrients and other non-nutritive compounds (Tucker 2001). Although initially the protective role of these foods was exclusively attributed to its high nutritional density, making fruits and vegetables essential foods in a balanced diet, more recent research has focused attention on the presence therein of other bioactive compounds, called phytochemicals, that contribute to disease prevention and the improvement of the quality of life of the population. Currently, the term "bioactive compounds" has been coined to describe plant compounds with health beneficial properties, referring to those nutrients or non-nutrients capable of acting on the human body's physiological mechanisms, among which are vitamins, trace elements, dietary fibre and other active compounds. Such compounds may have different mechanisms, including modulation of detoxifying enzymes, stimulating the immune system, reduction of platelet aggregation, modulation of cholesterol synthesis and hormonal metabolism, reduced blood pressure, antioxidant, antibacterial and antiviral properties (FECYT 2005; Lampe 1999; Wargovich 2000).

V. Fernández-Ruiz (✉) · M. Cámara · B. M. Ruiz-Rodríguez
Departamento de Nutrición y Bromatología II, Bromatología, Facultad de Farmacia, Universidad Complutense de Madrid, Madrid, Spain
e-mail: vfernand@farm.ucm.es; mcamara@ucm.es; brigidamariaruiz@gmail.com

M. de C. Sánchez-Mata, J. Tardío (eds.), *Mediterranean Wild Edible Plants*,
DOI 10.1007/978-1-4939-3329-7_9

Bioactive compounds present in plant products include the following molecules (Ortega et al. 2007):

- Carotenoids, such as lycopene (abundant in tomatoes, watermelon and pink grapefruit varieties, involved in the prevention of prostate cancer) or lutein (found in vegetables and some fruits, involved in decreasing the risk of cataracts and other degenerative eye processes), among others (Rao 2006).
- Steroids, such as phytosterols (campesterol, sitosterol and stigmasterol) and stanols are present in appreciable amounts in some vegetables and fruits, with proven effect on inhibition of intestinal cholesterol absorption.
- Tocopherols and tocotrienols (in α, β, γ and δ forms), with antioxidant activity by inhibiting lipid peroxidation at the propagation phase.
- Phenolic compounds, with a diverse group structure, which includes flavonoids (flavonols, flavanols, flavanones, isoflavones and anthocyanins), phenolic acids, tannins, stilbenes, among others. This fraction also includes isoflavones, such as daidzein, genistein or coumestrol, with phytoestrogenic actions, present in soy, plums, figs, strawberries, melon or pears.
- Sulphur compounds: glucosinolates, and indole isothiocyanates (found in Cruciferae), allicin and allyl sulfides (abundant in *Allium* bulbs).

Many of these compounds (carotenoids, tocopherols, phenolic compounds and sulphur compounds) have antioxidant capacity, thus they are able to counteract the oxidative stress caused by the attack of high oxidizing molecules, such as free radicals, to the various tissues and biomolecules of organism (as genetic material or plasma lipoproteins and membrane), which triggers cellular aging processes, and the appearance of cardiovascular disease, cancer, cataracts or neurological disorders, among others (Wargovich 2000).

The beneficial effects of all these compounds are manifested by the consumption of a diet rich in fruits and vegetables; it cannot be attributed to a single compound or mixture of compounds, but to the synergistic effect of all of them. Numerous studies have shown that taking separate compounds in the form of food supplements do not produce the same effects as fruit and vegetables (Cervera Real 2008). Thus, it has been shown that the consumption of fruits and vegetables (for example garlic, grape juice, carrots, spinach, soy or cruciferous vegetables) is currently one of the most effective and safe strategies in preventing cardiovascular disease (Halliwel 2007; Lampe 1999; Suter 2000). In this connection, the so-called "cumulative antioxidant index" (CAI) has been defined as a parameter directly related to plasma concentrations of vitamin E, vitamin C, β-carotene and selenium, and inversely related to serum cholesterol, intake of fruits and vegetables, since its composition would help to increase the index (Bello 2006).

Several studies, such as SUppléments en VItamines et Minéraux AntioXydants (SUVIMAX) or European Prospective Investigation into Cancer and Nutrition (EPIC; Survey Research on Cancer and Nutrition), also show that the intake of a varied diet rich in fruits and vegetables is best to reduce the risk of several types of cancer (Agudo and González 2007; Hercberg et al. 2006). The World Foundation for Cancer Research has stated that of the 130,000 deaths per year caused by cancer,

30–40 % could have been prevented through a proper diet (Cotte 1999). Indeed, oxidative DNA damage is considered an important factor causing various types of cancer, and fruit and vegetables, for their high content of antioxidants, have been revealed as important chemopreventive agents (Clark et al. 1996). Furthermore, the presence of flavones, dithiothioles, thioethers, phenols, isothiocyanates and indoles suppress metabolic activity of carcinogens, and in this way, could reduce the risk of cancer. These beneficial components may be found in cruciferous vegetables (broccoli, brussels sprouts, cauliflower and cabbage), carrots and green leafy vegetables (Lampe 1999; Martínez et al. 2001; Williamson 1996).

9.2 Bioactive Compounds in Wild Edible Plants

Wild edible plants are important sources of bioactive compounds, such as carotenoids, fibre, phenolics and other non-nutrient compounds. Most of the compounds mentioned above have shown antioxidant activity, associated with health benefits due to the reduction in cardiovascular disease risk factors, the decrease of the incidence of many different types of cancers besides their protective role against a wide range of other chronic diseases. The concentration of these compounds in wild plants varies widely within botanical families, and it is also influenced by growing conditions, moisture and other factors. This chapter reviews the current knowledge about these bioactive compounds' content and contribution in wild edible plants.

9.2.1 Carotenoids

Carotenoids are one of the most important groups of pigments, widely distributed in nature, with important metabolic functions (Khachik et al. 2002; Van den Berg et al. 2000). Carotenoids, being C40 isoprenoids (tetraterpenes), play a key role in human diet due to their metabolism to vitamin A (retinol) besides other important activities due to their antioxidant properties.

Carotenoids can be divided into hydrocarbons and xanthophylls (oxygenated derivatives), and also into acyclic and alicyclic. Around 637 different carotenoids have been described, and possibly 70 of them might have an important role in human health. Most carotenoids occurring in nature meet all double bonds in *trans* configuration, although there are also small quantities of *cis*-isomers.

Carotenoids are considered of great interest due to the provitamin-A activity of some of them (α-carotene, β-carotene and cryptoxanthin, mainly) and also for other activities, such as the well documented antioxidant capacity (Stahl and Sies 2003, 2012) or other less documented features, based on the non-antioxidant effects of phytochemicals, like those related to the inhibition of inflammation (Elliott 2005). Their action as antioxidants is mainly by capturing free radicals and reducing oxidative stress in the body, which prevents mediated cell damage induced by radicals.

These properties are responsible for health-promoting effects, and most of the studies so far establish relationships between carotenoids and chronic diseases, such as cancer, atherogenesis, ocular degeneration and neuronal damage (Cantuti-Castelvetri et al. 2000; Krinsky and Johnson 2005; Maiani et al. 2009; Ramos et al. 1989). For that reason, the World Health Organization (WHO) recommends a β-carotene intake of 4–6 mg per person per day (Williamson 1996).

Among carotenoids, lycopene presently generates enormous interest for the important biological actions performed in the body, due to its specific chemical structure and properties. Recent epidemiological studies show its preventive role against cardiovascular disease and prostate, gastrointestinal tract, and epithelial cancers (Clinton 1998; Shi and Maguer 2000). This is due to not only its antioxidant activity, twice that of β-carotene, but also other hormonal or metabolic mechanisms, such as acting in intercellular communication (responsible for cell growth) and modulating immune functions (Rao and Agarwal 2000; Stahl and Sies 2012).

Total and individual carotenoids of cultivated plants or commercial species have been extensively studied (Deli et al. 2000; Granado et al. 1992; USDA 1998); however, there are very few studies on wild plants (Martins et al. 2011), and there are little previous data available on individual carotenoids in wild edible plants from the Mediterranean area.

Several authors studied the total carotenoid content in species traditionally consumed in the Mediterranean area (Fig. 9.1). From these wild edible species, a higher carotenoid content has been found in *Alliaria petiolata, Chenopodium opulifolium, C. album* and *C. murale*. As shown in Fig. 9.1, *A. petiolata* has the highest total carotenoid content (13.3 mg/100 g), while the lowest belong to *Bellis perennis* (2.5 mg/100 g).

García-Herrera et al. (2013), studied the edible young shoots of four species traditionally consumed as wild asparagus in Spain (*Asparagus acutifolius, Humulus lupulus, Bryonia dioica* and *Tamus communis*). The results are shown in Table 9.1. These wild edible young shoots present higher contents of xanthophylls (lutein, neoxanthin and violaxanthin) than β-carotene, and they are richer sources of carotenoids than many of the commercially available leafy vegetables often consumed in the Mediterranean area. Among the four species analysed, *B. dioica* has the highest carotenoid content (up to 11.63 mg/100 g).

Lutein is mainly present in green leafy vegetables, such as beet (4.2 mg/100 g) and spinach (1.5 mg/100 g), with content in a range similar to those found in most wild plants. The content of β-carotene in vegetables such as spinach (3.2 mg/100 g) and beet (1.1 mg/100 g) is also in the range similar to those reported in wild plants (Deli et al. 2000; Granado et al. 1992). Debussche et al. (1987) studied the variation in the composition of fleshy fruits of the Mediterranean region remarking the importance of the ripening season, life form, fruit type and geographical factors.

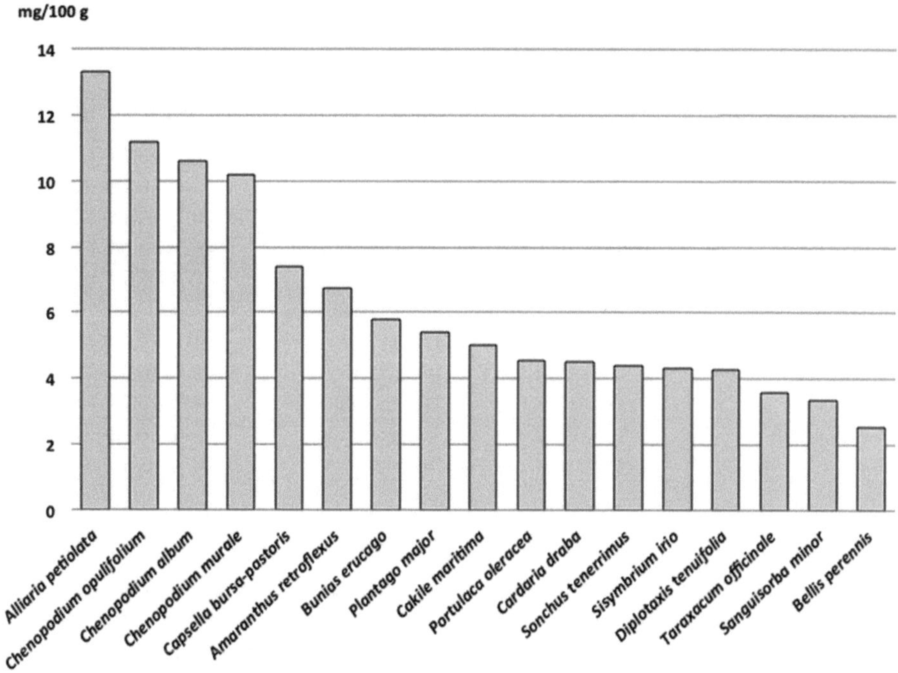

Fig. 9.1 Total carotenoid content in some wild Mediterranean vegetables (mg/100 g fresh edible material). (Data compiled from Aliotta and Pollio 1981; Guil-Guerrero et al. 1997; Guil-Guerrero and Rodríguez-García 1999; Ranfa et al. 2014)

9.2.2 *Fibre*

Fibre is not a simple and well defined chemical compound, but a combination of chemical substances on composition and structure, such as cellulose, hemicelluloses, and lignin, being defined as "edible parts of plants or analogous carbohydrates that are resistant to digestion and absorption in the human small intestine with complete or partial fermentation in the large intestine" (Mongeau and Brooks 2003). Thus, dietary fibre comprises food material, particularly plant material, that is not hydrolysed by enzymes secreted by the human digestive tract but that may be digested by microflora in the gut. Plant components that fall within this definition include non-starch polysaccharides (NSP) such as celluloses, some hemicelluloses, gums and pectins, as well as lignin, resistant dextrins and resistant starches (IFST 2007). According to the criteria followed by European Food Safety Authority (EFSA), the concept of dietary fibre refers to non-digestible carbohydrates plus lignin, including NSP: cellulose, hemicelluloses, pectins, hydrocolloids (i.e. gums, mucilages and glucans), resistant oligosaccharides (fructooligosaccharides, galacto-oligosaccharides and others), resistant starch (consisting of physically enclosed starch, some types of raw starch granules, retrograded amylose, chemically and/or physically modified starches), and lignin associated with the dietary fibre polysaccharides (EFSA 2010).

Table 9.1 Individual carotenoid concentrations (mg/100 g of fresh edible material) of some wild young shoots traditionally consumed in Mediterranean areas. (Data from García-Herrera et al. 2013, including variability from samples gathered in different years and sites of central Spain)

	Mean (range)		Mean (range)
Asparagus acutifolius		*Humulus lupulus*	
Lutein	0.59 (0.41–0.85)	Lutein	0.76 (0.19–1.72)
β-carotene	0.31 (0.16–0.42)	β-carotene	0.48 (0.16–1.26)
Neoxanthin	0.47 (0.04–0.98)	Neoxanthin	0.73 (0.13–1.57)
Violaxanthin	0.40 (0.26–0.57)	Violaxanthin	0.31 (0.04–1.01)
Total	*0.87–2.82*	Total	*0.52–5.56*
Bryonia dioica		*Tamus communis*	
Lutein	2.19 (0.68–3.70)	Lutein	1.14 (0.70–1.73)
β-carotene	0.81 (0.15–1.95)	β-carotene	0.44 (0.32–0.60)
Neoxanthin	1.96 (0.17–3.83)	Neoxanthin	1.19 (0.77–1.90)
Violaxanthin	1.14 (0.03–2.15)	Violaxanthin	0.62 (0.38–1.09)
Total	*1.03–11.63*	Total	*2.13–5.32*

Fibre in human nutrition gained great interest from Burkitt's work (Burkitt et al. 1974), which studied the relationship between inadequate fibre intake and the progressive increase in degenerative diseases in developed societies. Nowadays, research shows that the ingestion of suitable quantities of food fibre produces many beneficial effects on the digestive tract, such as the regulation of the intestinal function, improvement of the tolerance to glucose in diabetics, or prevention of chronic diseases such as colon cancer (Mongeau and Brooks 2003; Pérez Jiménez et al. 2008). In addition, soluble fibre (SF, mainly pectins) influences fat level and arteriosclerosis in humans and animals (Yamada 1996). Diverse in vivo studies have demonstrated that an ingestion of insoluble fibre (IF) from fruits and vegetables can produce a significant decrease in the plasmatic concentration of cholesterol, which implies a decrease in the risk of suffering cardiovascular disease, colon cancer, diabetes and obesity. Also, a regulatory activity of the immune system has been attributed to dietary fibre (Brett and Waldron 1996). For all these reasons, the current fibre intake recommended by the Food and Nutrition Board (Trumbo et al. 2002) in adults is 21–38 g/day, depending on lifestage groups. The current Europe consumption of fibre is estimated to be 20 g/person/day, so an increase in fibre consumption is needed, and a way to achieve that goal is by supplementing the diet with commercial fibre-rich products.

One of the most important nutritional aspects of fruits and vegetables is the fact that, together with cereals and legumes, they are one of the main sources of dietary fibre in the diet, with contents ranging between 1 and 11 %, which is insoluble fibre distributed between cellulose, hemicellulose, lignin (predominant in vegetables), and soluble fibre (pectin, gums and mucilages) more important in fruits (Cámara et al. 2003). Within the soluble fibre fraction, prebiotic compounds such as short-chain fructans (fructooligosaccharides) or long-chain fructans (inulin) are also included.

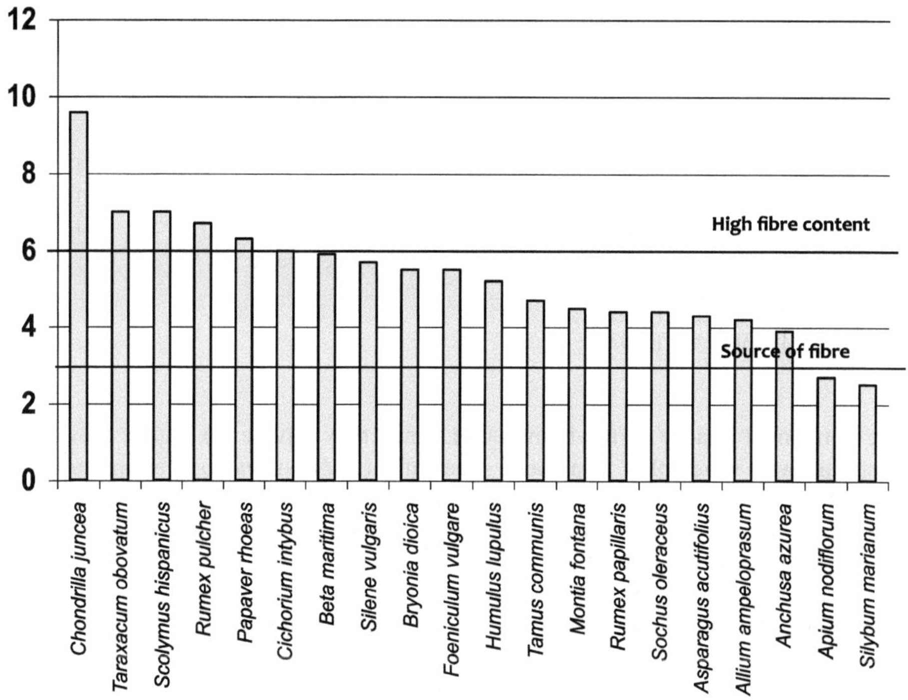

Fig. 9.2 Total fibre content (g/100 g of fresh edible material) in different Mediterranean wild vegetable species. (Data compiled from García-Herrera 2014)

Most cultivated food plants generally have a fibre content between 1.2 and 4 g/100 g, except for special cases such as artichokes which reach values of 10.8 g/100 g (Souci et al. 2008). Fibre, from the nutritional standpoint, is the most important component of wild plants. As shown in Fig. 9.2, some wild species studied by García Herrera et al. (2014) presented mean values which ranged from 3.9 to 9.6 g/100 g, except for *Silybum marianum*, with a lower average fibre contribution (2.5 g/100 g) due to its high water content. The plant with the highest fibre content was *Chondrilla juncea*, 100 g portion covering about 30 % of daily recommendations of fibre for adults, given by Trumbo et al. (2002) as adequate intake (AI). It can be seen that the species studied exceed usual fibre contents in most vegetables. The Regulation 1924/2006 requires a minimum of 3 g/100 g of fibre to declare that a food is "source of fibre" and 6 g/100 g to claim it as "high fibre". Under this regulation, all edible wild plants shown in Fig. 9.2, except *Apium nodiflorum* and *S. marianum*, could be considered "sources of fibre", standing out for its high content some Asteraceae species (*C. juncea, Taraxacum obovatum* and *Scolymus hispanicus*) as well as *Rumex pulcher* and *Papaver rhoeas*.

Ruiz-Rodríguez (2014) studied the soluble and insoluble fibre content of four wild species whose fruits are traditionally consumed in Spain (Table 9.2), to evaluate their potential contribution to fibre daily recommendations, or adequate intakes. Results showed that these wild fruits are interesting sources of fibre for improving

Table 9.2 Soluble and insoluble fibre in some wild fruits. Data from Ruiz Rodriguez (2014)

	Arbutus unedo	*Crataegus monogyna*	*Rubus ulmifolius*	*Prunus spinosa*
Insoluble fibre (g/100 g)	13.20	10.11	9.87	9.54
Soluble fibre (g/100 g)	2.98	4.37	2.01	2.19
% Adequate intake EFSA[a]	64.72	57.92	47.52	46.92
% Adequate intake FNB males[b]	42.58	38.10	31.26	30.87
% Adequate intake FNB females[c]	64.72	57.92	47.52	46.92

ESFA European Food Safety Authority, *FNB* Food and Nutrition Board
[a] 25 g/day, EFSA (2010)
[b] 38 g/day , Trumbo et al. (2002)
[c] 25 g/day , Trumbo et al. (2002)

the nutritional quality of the human diet, since they have even higher fibre content than wild vegetables. Strawberry-tree fruits (*Arbutus unedo*) were reported as one of the best contributors to fibre recommendations among wild edible fruits (16.28 g/100 g), especially insoluble fibre, contributing more than 42 % of fibre AI in a 100 g portion (Ruíz-Rodríguez 2014).

As in some cases already mentioned, all the studied fruits provide interesting fibre amounts, since 100 g could cover between 30 and 65 % of the AI recommended for adults. In addition, and following the guidelines of Regulation (EC) No 1924/2006 of the European Parliament and of the Council of 20 December 2006 on nutrition and health claims made on foods (European Parliament and Council 2006), it could be stated that all of them, even fresh, are "sources of fibre" and also could be defined as "high fibre" (over 3 and 6 g/100 g, respectively, for each of the statements).

Within the fibre fraction, we find that in conventional fruits, a relation FS/FI is approximately equal parts in strawberry and cherry (*Prunus cerasus*), but in most other fruits the ratio is more favourable to the FI, for example grapes, pineapple, banana, apple, being even more favourable to the FI, and up to 90 % in black currant and watermelon (Souci et al. 2008). In the case of the wild fruits, we also observed this higher amount of FI, with values ranging from 30/70 in *Crataegus monogyna* to 20/80 in *Arbutus unedo, Prunus spinosa* and *Rubus ulmifolius*.

9.2.3 Phenolic Compounds

Phenolic compounds, as secondary plant metabolites, include a wide spectrum of molecules that contain an aromatic group and one or more hydroxyl groups on the aromatic ring. This family includes simple phenolic acids, which are derivatives of benzoic acid or cinnamic acid, and extends to the complex family of flavonoids and tannins. Due to their antioxidant properties, total phenols and flavonoids have an important role in antioxidant defence mechanisms of biological systems (Nile and Park 2014; Proestos et al. 2006).

Polyphenolic compounds are gaining greater prominence as bioactive agents. They are a complex group of substances including flavonols, catechins and anthocyanins, and are found in plants isolated or attached to sugars (glycosides; Cheynier 2012). Flavonoids are a group of polyphenolic compounds widely distributed throughout the plant kingdom that inhibit the oxidation chain initiation and prevent chain propagation by acting as free radical scavengers (Bravo 1998; Mertz et al. 2007; Scalbert et al. 2002). They are usually found as their glycoside derivatives in plants, with various monosaccharides and disaccharides as parts of their structure.

Flavonols are the most ubiquitous flavonoids in fruits, and quercetin and kaempferol are the main representatives being the richest sources: onions (up to 1.2 g/kg edible portion), curly kale, leeks, broccoli and blueberries. Flavanols exist in both the monomer form (catechins) and the polymer form (proanthocyanidins). Catechins are found in many types of fruit (apricots, which contain 250 mg/kg, are one of the richest sources). They are also present in red wine (up to 300 mg/L), but green tea and chocolate are by far the richest sources. Proanthocyanidins, which are also known as condensed tannins, are dimers, oligomers and polymers of catechins that are bound together by links between C4 and C8 (or C6) and are responsible for the astringent character of fruit.

Anthocyanins are pigments dissolved in the vacuolar sap of the epidermal tissues of flowers and fruit, to which they impart a pink, red, blue or purple colour. Cyanidin is the most common anthocyanidin in foods. Food contents are generally proportional to colour intensity and reach values up to 2–4 g/kg of edible portion of blackcurrants or blackberries; these values increase as the fruit ripens (Harborne and Williams 2000; Manach et al. 2004). The most significant are the anthocyanins found in red grapes, strawberries, pomegranates, blackberries and blueberries, quercetin in fruits and onions, and resveratrol present in grapes (Bravo 1998).

There are a great number of in vitro and in vivo studies, using cell cultures of animal and human cells, suggesting that these bioactive compounds of wild plants positively show an effect on human health. The presented characteristics of various berry fruits point to vast differences in the type of their bioactive compounds. Such differences are observed with regard to both the content and the qualitative composition of those compounds. The most significant health benefits described for phenolic compounds are shown in Fig. 9.3, which result mostly from different biological activities such as antioxidant, antitumoral, antimutagenic, antimicrobial, anti-inflammatory and neuroprotective properties (Nile and Park 2014; Vasco et al. 2009).

Some of the most important beneficial effects of phenolic compounds are the prevention of malignancies by inhibiting the formation of nitrosamines or even decreasing their capacity for action, when they are formed; they have antioxidant properties to be effective in preventing oxidation of the fraction low-density lipoprotein (LDL) cholesterol, thereby preventing atherosclerosis and other cardiovascular diseases; and they are also capable of blocking the allergic response of the body to inhibit histamine and exhibit anti-inflammatory and diuretic properties (Williamson 1996). However, it should be noted that the antioxidant capacity of

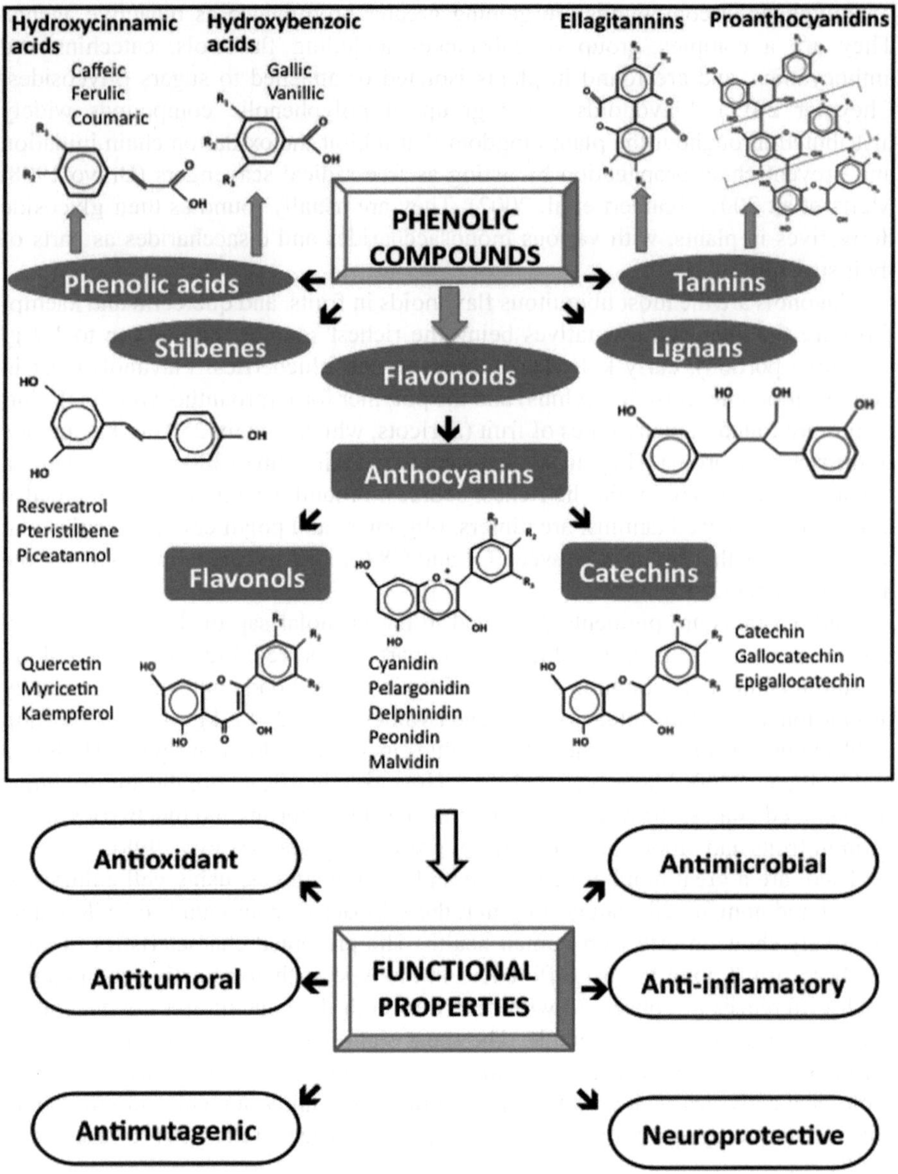

Fig. 9.3 The most significant health benefits of phenolic compounds. (Data from Nile and Park 2014)

fruits and vegetables cannot be attributed to a particular group of flavonoids but to all of them (see Chap. 10), and still there is a need for further research to assess the synergies or antagonisms between the different compounds (Heim et al. 2002; Rice-Evans et al. 1997).

Phenolic compounds are present in vegetables (brussels sprouts, cauliflower, leeks, onion, parsley, tomato and celery) and fruits (berries, cherries, blueberries, plums, raspberries, strawberries, grapes, grapefruit, orange, peach and apple; Cheynier 2012; Manach et al. 2004). We must also make special mention of soy isoflavones, which promote bone mineralization and are preventive factors of atherosclerosis and some cancers (Murkies et al. 2000; Raschke et al. 2006; Wangen et al. 2000).

There is great interest in the study of phenolic compounds in 'lesser-known' wild plants, since several studies revealed the important role they played in human health. However, it is difficult to compare results obtained by different authors due to the great differences in the units and way of expressing the results obtained. Furthermore, it is also important to take into account the natural variability of wild plants and a wide variability even within the same species, which substantiate the importance of analysing several batches of them.

Several authors studied the total phenolic content in wild species traditionally consumed in the Mediterranean area (Conforti et al. 2008; Conforti et al. 2011; García-Herrera 2014; Martins et al. 2011; Morales 2011; Morales et al. 2012; Morales et al. 2014; Pereira et al. 2011; Salvatore et al. 2005; Vardavas et al. 2006; Wojdyło et al. 2007; Zeghichi et al. 2005). Figure 9.4 shows the averages of data available from these studies for different wild vegetables expressed as gallic acid equivalents (GAE)

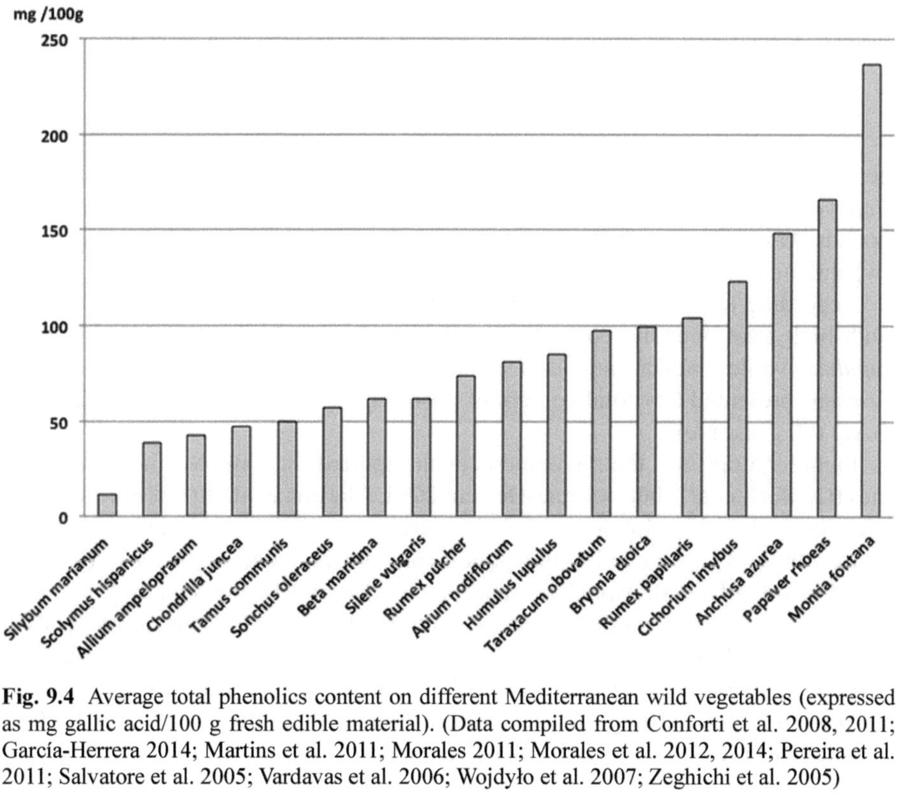

Fig. 9.4 Average total phenolics content on different Mediterranean wild vegetables (expressed as mg gallic acid/100 g fresh edible material). (Data compiled from Conforti et al. 2008, 2011; García-Herrera 2014; Martins et al. 2011; Morales 2011; Morales et al. 2012, 2014; Pereira et al. 2011; Salvatore et al. 2005; Vardavas et al. 2006; Wojdyło et al. 2007; Zeghichi et al. 2005)

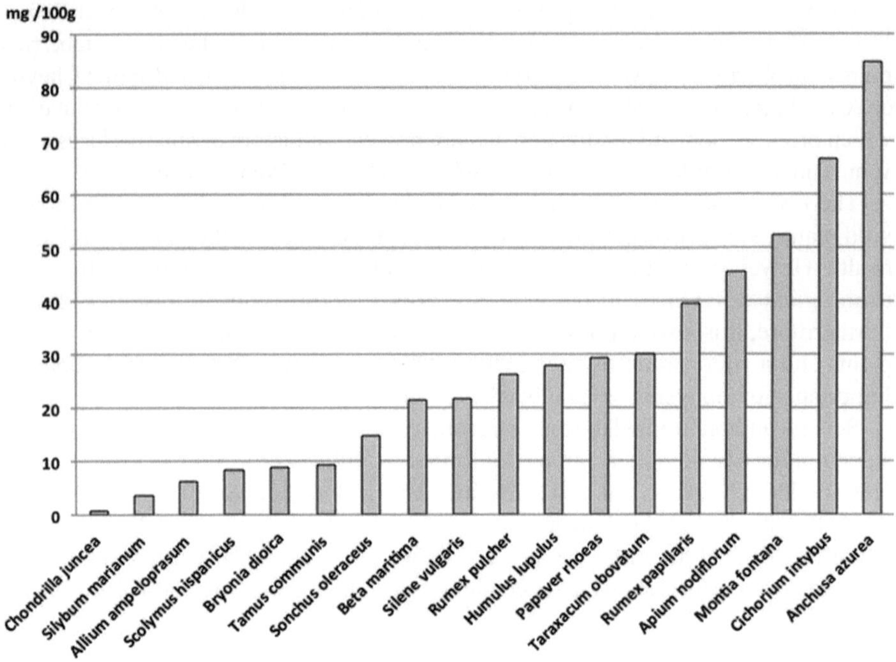

Fig. 9.5 Average total flavonoid content in different Mediterranean wild vegetables (expressed as mg catechin/100 g). (Data compiled from Conforti et al. 2008; García-Herrera 2014; Martins et al. 2011; Morales 2011; Morales et al. 2012, 2014; Pereira et al. 2011; and Salvatore et al. 2005)

per 100 g of edible fresh product. The highest content was found in *Montia fontana, Papaver rhoeas* and *Anchusa azurea*; they are therefore rich sources of these compounds. *Montia fontana* had the highest total phenolics content (237 mg GAE/100 g fw) and the lowest corresponded to *Silybum marianum* (11.6 mg GAE/100 g fw).

According to Morales (2011), besides water-blinks (*M. fontana*), other species with high total phenolic content were fennel (*Foeniculum vulgare*) and hop (*H. lupulus*), while bladder campion (*Silene vulgaris*) and wild leek (*Allium ampeloprasum*) presented low total content of phenolic compounds (62.38 and 42.24 mg GAE/100 g sample).

Figure 9.5 presents the flavonoid content reported for some Mediterranean species (Conforti et al. 2008; García-Herrera 2014; Martins et al. 2011; Morales 2011; Morales et al. 2012; Morales et al. 2014; Pereira et al. 2011; Salvatore et al. 2005). Among these wild vegetables, the highest content was found in *A. azurea* (84.8 mg catechin/100 g fw), followed by, *Cichorium intybus* and *M. fontana*; they are therefore rich sources of flavonoids. On the contrary, the lowest content corresponds to *C. juncea* (0.78 mg catechin/100 g fw). In the study by Morales (2011), the highest level of flavonoids, expressed as catechin equivalents (CE), was found in *A. nodiflorum* (143 mg CE/100 g sample), whereas the lower content (6.29 mg CE/100 g sample) was presented by wild leek (*A. ampeloprasum*).

Morales (2011) compared the composition of phenolics and total flavonoids in samples of *F. vulgare* from Spain with those from other origins, such as Portugal, Finland, Italy and Iran (Barros et al. 2009; Conforti et al. 2009; Hinneburg et al. 2006; Motamed and Naghibi 2010). In comparison, the Spanish fennel showed a higher content of total phenols (42.16 mg GAE/g extract), much greater than fennel from Portuguese, Finnish and Iranian origin (8.61, 7.74 and 30.3 mg GAE/g extract, respectively) but lower than its Italian relative (80 mg GAE/g). However, although the Iranian plants showed lower content of total phenols than others, they presented a higher concentration of total flavonoids (16.49 mg CE/g) than the samples collected in Spain.

In relation to young shoots, some Spanish samples of *H. lupulus* had a high content of phenolic compounds (55.83 mg GAE/g extract), significantly higher than that reported for the same species from Poland (7.14 mg GAE/g extract) studied by Wojdyło et al. (2007). Comparing the results obtained for wild asparagus (*A. acutifolius*), white and black bryony (*B. dioica* and *T. communis*) from Portugal and Spain, it was observed that the Portuguese samples had a higher content of phenolic compounds (624, 258 and 759 mg GAE/g of extract, respectively) and flavonoids (57.8, 18.1 and 150 mg CE/g extract, respectively) than samples from Spain (Martins et al. 2011; Morales et al. 2012).

In the case of wild leek (*A. ampeloprasum*), total phenols are around 5.77 mg GAE/g extract (García-Herrera 2014), while other species of the same family have shown a higher level of total phenols, around 9–15.87 mg GAE/g extract. Gorinstein et al. (2009) found total flavonoid level of 1.31 mg CE/g extract for *Allium cepa*, 0.86 mg CE/g extract for its wild relatives, and 0.56 mg CE/g for *Allium sativum*. Furthermore, Santas et al. (2008) reported values of 2.58 mg GAE/g extract for *calçot* (Spanish variety of *A. cepa* typically consumed in Catalonia), slightly below the wild leek levels.

With regard to wild fruits, blueberries and blackberries are rich sources of phenolics with antioxidant capacity: phenolic acids (gallic acid, p-hydroxybenzoic acid, caffeic acid, ferulic acid and ellagic acid) were found to range from 0.19 to 259 mg/100 g (Sellappan et al. 2002). Flavonoids (catechin, epicatechin, myricetin, quercetin and kaempferol) ranged from 2.50 to 387 mg/100 g. Total polyphenols ranged from 261 to 929 mg/100 g, and total anthocyanins ranged from 12.70 to 197 mg/100 g, all values expressed on fresh basis (Lim 2012). Some recent studies reported values of phenolic content in Mediterranean wild fruits (*A. unedo, R. ulmifolius, P. spinosa* and *C. monogyna*). The analysis of the samples, collected from two different sites of Spain during three different seasons, showed higher values than those found in many cultivated fruits (Ruiz-Rodríguez 2014). The specific phenolic profile of these wild fruits is presented in Fig. 9.6. The fruits of *P. spinosa* showed total phenolic compounds ranging from 1851 to 3825 mg/100 g, characterised by a high content of anthocyanins and phenolic acids. The fruits of *C. monogyna* presented 449–1438 mg total phenolic compounds/100 g, characterised by a high content of phenolic acids and flavonols (Ruiz-Rodríguez et al. 2014a). *A. unedo* fruits showed higher phenolic content than those of *R. ulmifolius* (773–1621 mg total phenolic compounds/100 g *vs.* 376–1326 mg total phenolic

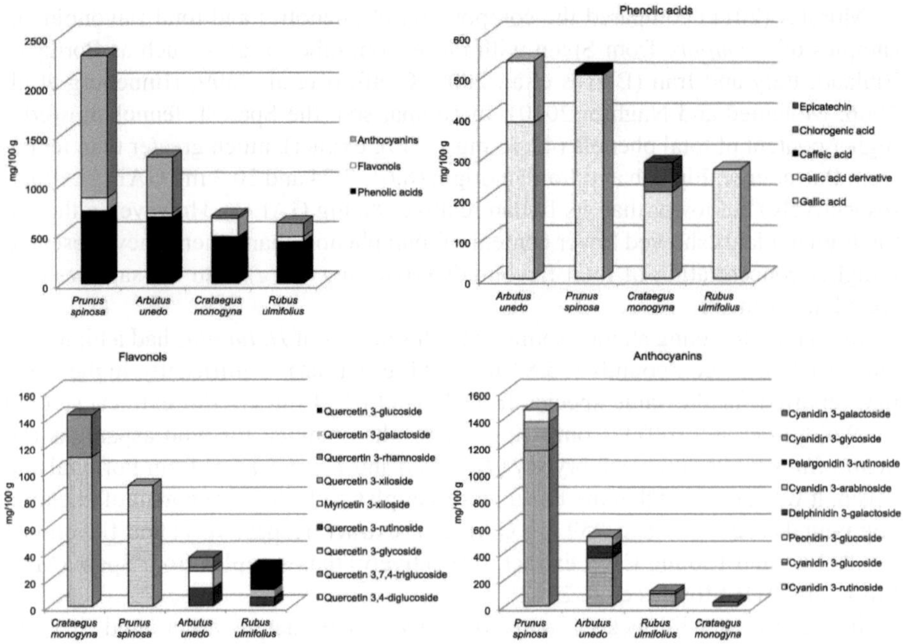

Fig. 9.6 Phenolic profile of wild *Arbutus unedo, Rubus ulmifolius, Prunus spinosa* and *Crataegus monogyna* fruits. (Data from Ruiz-Rodríguez 2014)

compounds/100 g, respectively). Phenolic acids and anthocyanins are the major groups of phenolic compounds found in both species, with gallic acid and cyanidin 3-glucoside as the main compounds (Ruiz-Rodríguez et al. 2014b). Those values are often in the highest range or even above those reported for the majority of the berries (192–929 mg/100 g), according to several studies (Česonienė et al. 2009; Phenol-Explorer Database 2013; Sellappan et al. 2002). These previous studies have reported lower values in many cultivated and wild berries, total anthocyanins ranging from 12.70 to 262 mg cyanidin 3-GE/100 g. Also, lower values of flavonols and anthocyanins (7.60 mg RE/100 g and 100.56 mg cyanidin 3-GE/100 g fw, respectively) in *R. ulmifolius* wild fruits have been reported by Ganhao et al. (2010). Therefore, these wild fruits, and especially *P. spinosa* and *C. monogyna*, should be reconsidered as new valuable sources of safe and inexpensive antioxidants.

9.2.4 *Other Compounds*

Wild edible plants are also sources of other important components considered as non-nutrient compounds. These compounds, such as sulphur and allyl disulfide, found in plants of *Allium* genus, or benzyl isothiocyanate, and indole-3-carbinol phenethyl isothiocyanate in *Brassica* species, have the ability to enhance the body's own antioxidant defences (Williamson 1996). Recent studies have shown significant interest due to their reported health benefits, including anticancer properties (El-Bayoumy et al. 2006; Nian et al. 2009).

Other important compounds are steroids such as phytosterols (campesterol, sitosterol and stigmasterol) and phytostanols, in some cases present in appreciable amounts in vegetables and fruits, have demonstrated effects in the inhibition of intestinal absorption of cholesterol (Moghadasian and Jiri Frohlich 1999).

References

Agudo A, González CA (2007) Fruit and vegetable intakes, dietary antioxidant nutrients, and total mortality in Spanish adults: findings from the Spanish cohort of the European Prospective Investigation into Cancer and Nutrition (EPIC-Spain). Am J Clin Nutr 85:1634–1642

Aliotta G, Pollio A (1981) Vitamin A and C content in some edible wild plants in Italy. Riv Ital EPPOS 63:47–48

Barros L, Heleno SA, Carvalho AM, Ferreira ICFR (2009) Systematic evaluation of the antioxidant potential of different parts of *Foeniculum vulgare* Mill. from Portugal. Food Chem Toxicol 47:2458–2464

Bello J (2006) La ciencia de los alimentos saludables: una panorámica actual. Ars Pharmaceutica 47(2):137–171

Bravo L (1998) Polyphenols: chemistry, dietary sources, metabolism, and nutritional significance. Nutr Rev 56(11):317–333

Brett CT, Waldron KW (1996) Physiology and biochemistry of plant cell walls. In: Black M, Charlwood B (eds) Topics in plant functional biology. Chapman and Hall, London

Burkitt DP, Walker AR, Painter NS (1974) Dietary fibre and disease. JAMA 229(8):1068–1074

Cámara M, Sánchez-Mata MC, Isasa ME (2003) Frutas y verduras, fuente de salud. Servicio de promoción de la salud, Instituto de Salud Pública. Consejería de Sanidad y Consumo, Comunidad de Madrid

Cantuti-Castelvetri I, Shukkit-Hale B, Joseph JA (2000) Neurobehavioral aspects of antioxidants in aging. Int J Dev Neurosci 18:367–381

Cervera Real P (2008) Fomento del consumo de frutas y hortalizas: vehículo de promoción de la salud y prevención de trastornos y patologías prevalentes en la sociedad actual. Alimentación, Nutrición y Salud 15(2):39–48

Česonienė L, Jasutienė I, Šarkinas A (2009) Phenolics and anthocyanins in berries of European cranberry and their antimicrobial activity. Medicina (Kaunas) 45:992–999

Cheynier V (2012) Phenolic compounds: from plants to foods. Phytochem Rev 11:153–177

Clark LC, Combs GF, Turnbull BW (1996) Effects of selenium supplementation for cancer prevention in patients with carcinoma of the skin. JAMA 276:1957–1963

Clinton SK (1998) Lycopene: chemistry, biology, and implications for human health and disease. Nutr Rev 56(2):35–51

Conforti F, Sosa S, Marrelli M, Menichini F, Statti GA, Uzunov D, Tubaro A, Menichini F, Loggia RD (2008) In vivo anti-inflammatory and in vitro antioxidant activities of Mediterranean dietary plants. J Ethnopharmacol 116(1):144–151

Conforti F, Sosa S, Marrelli M, Menichini F, Statti GA, Uzunov D, Tubaro A, Menichini F (2009) The protective ability of Mediterranean dietary plants against the oxidative damage: the role of radical oxygen species in inflammation and the polyphenol, flavonoid and sterol contents. Food Chem 112:587–594

Conforti F, Marrelli M, Carmela C, Menichini F, Valentina P, Uzunov D, Statti GA, Duez P, Menichini F (2011) Bioactive phytonutrients (omega fatty acids, tocopherols, polyphenols), in vitro inhibition of nitric oxide production and free radical scavenging activity of non-cultivated Mediterranean vegetables. Food Chem 129(4):1413–1419

Cotte P (1999) Tackling diet-related disease by promoting fruit and vegetables. Nutr Food Sci 4:173–177

Debussche M, Cortez J, Rimbault I (1987) Variation in fleshy fruit composition in the Mediterranean region. The importance of ripening season, life-form, fruit type and geographical distribution. Oikos 49:244–252

Deli J, Matus Z, Toth G (2000) Carotenoid composition in the fruits of *Asparagus* officinalis. J Agric Food Chem 48:2793–2796

EFSA Panel on Dietetic Products, Nutrition, and Allergies (NDA) (2010) Scientific opinion on dietary reference values for carbohydrates and dietary fibre. EFSA J 8(3):1462–1538

El-Bayoumy K, Sinha R, Pinto JT, Rivlin RS (2006) Cancer chemoprevention by garlic and garlic-containing sulfur and selenium compounds. J Nutr 136(3):864–869

Elliott R (2005) Mechanisms of genomic and non-genomic actions of carotenoids. Biochim Biophys Acta 1740:147–154

European Parliament and Council (2006) Regulation (EU) No 1924/2006 of the European Parliament and of the Council of 20 December 2006 on nutrition and health claims made on foods. Off J Eur Union, 18.1.2007: L12/3–L12/18

FECYT (2005) Alimentos Funcionales. Fundación Española para la Ciencia y la Tecnología, Madrid

Ganhao R, Morcuende D, Estevez M (2010) Protein oxidation in emulsified cooked burger patties with added fruit extracts: influence on colour and texture deterioration during chill storage. Meat Sci 85:402–409

García-Herrera P, Sánchez-Mata MC, Cámara M, Tardío J, Olmedilla-Alonso B (2013) Carotenoid content of wild edible young shoots traditionally consumed in Spain (*Asparagus acutifolius* L., *Humulus lupulus* L., *Bryonia dioica* Jacq. and *Tamus communis* L.). J Sci Food Agric 93(7):1692i Foo. (Erratum published. J Sci Food Agric (2014)) 94:1914–16

García-Herrera P (2014) Plantas silvestres de consumo tradicional. Caracterización de su valor nutricional y estimación de su actividad antifúngica. PhD Thesis. Universidad Complutense de Madrid, Facultad de Farmacia, Departamento de Nutrición y Bromatología II

Gorinstein S, Park YS, Heo BG, Namiesnik J, Leontowicz H, Leontowicz M (2009) A comparative study of phenolic compounds and antioxidant and antiproliferative activities in frequently consumed raw vegetables. Eur Food Res Technol 228:903–911

Granado F, Olmedilla B, Blanco I, Rojas-Hidalgo E (1992) Carotenoid composition in raw and cooked Spanish vegetables. J Agric Food Chem 40:2135–2140

Guil-Guerrero JL, Rodríguez-García I (1999) Lipids classes, fatty acids and carotenes of the leaves of six edible wild plants. Eur Food Res Technol 209(5):313–316

Guil-Guerrero JL, Rodríguez-García I, Torija-Isasa ME (1997) Nutritional and toxic factors in selected wild edible plants. Plant Foods Hum Nutr 51(2):99–107

Halliwel B (2007) Dietary polyphenols: good, bad, or indifferent for your health? Cardiovasc Res 73(2):341–347

Harborne JB, Williams CA (2000) Advances in flavonoid research since 1992. Biofactors 6:481–504

Heim KE, Tagliaferro AR, Bobilya DJ (2002) Flavonoid antioxidants: chemistry, metabolism and structure-activity relationship. J Nutr Biochem 13(10):572–584

Hercberg S, Czernichow S, Galan P (2006) Antioxidant vitamins and minerals in prevention of cancers: lessons from the SUVIMAX study. Br J Nutr 96(1):S28–S30

Hinneburg I, Dorman HJD, Hiltunen R (2006) Antioxidant activities of extracts from selected culinary herbs and spices. Food Chem 97:122–129

IFST (2007) The Institute of Food Science & Technology, Public Affairs and Technical & Legislative Committees. http://www.ifst.org. Accesed Feb 2014

Khachik F, Carvalho L, Bernstein PS, Muir GJ, Zhao DY, Katz NB (2002) Chemistry, distribution, and metabolism of tomato carotenoids and their impact on human health. Exp Biol Med 227(10):845–851

Krinsky NI, Johnson EJ (2005) Carotenoid actions and their relation to health and disease. Mol Asp Med 26:459–516

Lampe JW (1999) Health efffects of vegetables and fruits: assesing the mechanisms of action in human experiments studies. Am J Clin Nutr 70:475S–90S

Lim TK (2012) Edible medicinal and non-medicinal plants: volume 2. Fruits. Springer, London

Maiani G, Periago M, Catasta G, Toti E, Goñi I, Bysted A, Granado-Lorencio F, Olmedilla-Alonso B, Knuthsen P, Valoti M, Volker B, Mayer-Miebach E, Behsnilian D, Schlemmer U (2009) Carotenoids: actual knowledge on food sources, intakes, stability and bioavailability and their protective role in humans. Mol Nutr Food Res 53:S194–S218

Manach C, Scalbert. A, Morand Ch, Rémésy Ch, Juménez L (2004) Polyphenols: food resources and bioavailability. Am J Clinic Nutr 74:727–747

Martínez A, Haza AI, Morales P (2001) Frutas y verduras como agentes preventivos de la dieta. II. Actividad antimutagénica y anticancerígena. Alimentaria 319:33–40

Martins D, Barros L, Carvalho AM, Ferreira ICFR (2011) Nutritional and in vitro antioxidant properties of edible wild greens in Iberian Peninsula traditional diet. Food Chem 125(2):488–494

Mertz C, Cheynier V, Günata Z, Brat P (2007) Analysis of phenolic compounds in two blackberry species (*Rubus glaucus* and *Rubus adenotrichus*) by high-performance liquid chromatography with diode array detection and electrospray ion trap mass spectrometry. J Agric Food Chem 55(21):8616–8624

Moghadasian MH, Jiri Frohlich J (1999) Effects of dietary phytosterols on cholesterol metabolism and atherosclerosis: clinical and experimental evidence. Am J Med 107(6):588–594

Mongeau R, Brooks PJ (2003) Properties, sources and determination. In: Caballero B, Turgo LZ, Finglas PM (eds) Encyclopedia of food science and nutrition, vol 3, 2nd edn. Elsevier Science BV, Amsterdam, pp. 1813 scie

Morales P (2011) Vegetales silvestres de uso alimentario: determinación de compuestos bioactivos y capacidad antioxidante. PhD Thesis. Universidad Complutense de Madrid, Facultad de Farmacia, Departamento de Nutrición y Bromatología II

Morales P, Carvalho AM, Sánchez-Mata MC, Cámara M, Molina M, Ferreira ICFR (2012) Tocopherol composition and antioxidant activity of Spanish wild vegetables. Genet Resour Crop Evol 59:851–863

Morales P, Ferreira ICFR, Carvalho AM, Sánchez-Mata MC, Cámara M, Fernández-Ruiz V, Pardo-de-Santayana M, Tardío J (2014) Mediterranean non-cultivated vegetables as dietary sources of compounds with antioxidant and biological activity. LWT-Food Sci Technol 55:389–396

Motamed SM, Naghibi F (2010) Antioxidant activity of some edible plants of the Turkmen Sahra region in northern Iran. Food Chem 119:1637–1642

Murkies A, Dalais FS, Briganti EM, Burger HG, Healy DL, Wahlqvist ML, Davis SR (2000) Phytoestrogens and breast cancer in postmenopausal women: a case control study. Menopause 7:289–296

Nian H, Delage B, Ho E, Dashwood RH (2009) Modulation of histone deacetylase activity by dietary isothiocyanates and allyl sulfides: studies with sulforaphane and garlic organosulfur compounds. Environ Mol Mutagen 50(3):213–221

Nile SH, Park SW (2014) Edible berries: review on bioactive components and their effect on human health. Nutrition 30(2):134–144

Ortega Anta RM, Requejo Marcos AM, Martínez García RM (2007) Nutrición y alimentación en promoción de la salud. Consejería de Sanidad, Comunidad de Castilla La-Mancha, Cuenca

Pereira C, Barros L, Carvalho AM, Ferreira IC (2011) Nutritional composition and bioactive properties of commonly consumed wild greens: potential sources for new trends in modern diets. Food Res Int 44(9):2634–2640

Pérez Jiménez J, Serrano J, Tabernero M, Arranz S, Díaz-Rubio E, García-Diz L, Goñi I, Saura-Calixto F (2008) Effects of grape antioxidant dietary fibre in cardiovascular disease risk factors. Nutrition 24(7–8):646–653

Phenol-Explorer (2013) Data base on polyphenol content in Foods. http://www.phenol-explorer.eu/contents/food/70. Accessed Oct 2013

Proestos C, Boziaris IS, Nychas G-JE, Komaitis M (2006) Analysis of flavonoids and phenolic acids in Greek aromatic plants: investigation of their antioxidant capacity and antimicrobial activity. Food Chem 95:664–671

Ramos M, González J, Reuvers T Ruiz E (1989) Contenido de vitaminas A, C y E en diferentes tipos de alimentos de mercado español. Alimentaria 26:51ime

Ranfa A, Maurizi A, Romano B, Bodesmo M (2014) The importance of traditional uses and nutraceutical aspects of some edible wild plants in human nutrition: the case of Umbria (central Italy). Plant Biosyst 148(2):297–306

Rao AV (2006) Tomatoes, lycopene and human health. Caledonian Science Press, UK

Rao AV, Agarwal S (2000) Role of antioxidant lycopene in cáncer and heart disease. J Am Coll Nutr 19(5):563–569

Raschke M, Wahala K, Beatrice L (2006) Reduced isoflavone metabolites formed by the human gut microflora suppress growth but do not affect DNA integrity of human prostate cancer cells. Br J Nutr 96(3):426–434

Rice-Evan CA, Miller NJ, Paganga G (1997) Antioxidant properties of phenolics compounds. Trends in Plant Sci 2(4):152–159

Ruiz-Rodríguez BM (2014) Frutos silvestres de uso tradicional en la alimentación. Evaluación de su valor nutricional, compuestos bioactivos y capacidad antioxidante. PhD Thesis. Universidad Complutense de Madrid, Facultad de Farmacia, Departamento de Nutrición y Bromatología II

Ruiz-Rodríguez BM, De Ancos B, Sánchez-Moreno C, Fernández-Ruiz V, Sanchez-Mata MC, Camara M, Tardío J (2014a) Wild blackthorn (*Prunus spinosa* L.) and hawthorn (*Crataegus monogyna* Jacq.) fruits as valuable sources of antioxidants. Fruits 69(1):61–73

Ruiz-Rodríguez BM, Sánchez-Moreno C, De Ancos B, Sánchez-Mata MC, Fernández-Ruiz V, Camara M, Tardío J (2014b) Wild *Arbutus unedo* L. and *Rubus ulmifolius* Schott fruits are underutilized sources of valuable bioactive compounds with antioxidant capacity. Fruits 69(6):435–448

Salvatore S, Pellegrini N, Brenna OV, Del Río D, Frasca G, Brighenti F, Tumino R (2005) Antioxidant characterization of some Sicilian edible wild greens. J Agric Food Chem 53(24):9465–9471

Santas J, Carbó R, Gordon MH, Almajano MP (2008) Comparison of the antioxidant activity of two Spanish onion varieties. Food Chem 107:1210–1216

Scalbert A, Morand C, Manach C, Rémésy C (2002) Absorption and metabolism of polyphenols in the gut and impact on health. Biomed Pharmacother 56:276–282

Sellappan S, Akoh CC, Krewer GP (2002) Phenolic compounds and antioxidant capacity of Georgia-Grown blueberries and blackberries. J Agric Food Chem 50:2432–2438

Shi J, Maguer M (2000) Lycopene in tomatoes: chemical and physical properties affected by food processing. Crit Rev Biotechnol 20:293–334

Souci SW, Fachmann W, Kraut H (2008) Food composition and nutrition tables. Medpharm, Stuttgart

Stahl W, Sies H (2003) Antioxidant activity of carotenoids. Mol Asp Med 24:345–351

Stahl W, Sies H (2012) Photoprotection by dietary carotenoids: concept, mechanisms, evidence and future development. Mol Nutr Food Res 56(2):287–295

Suter PM (2000) Effect of vitamin E, vitamin C, and β-carotene on stroke risk. Nutr Rev 58(6):184–187

Trumbo P, Schlicker S, Yates AA, Poos M (2002) Dietary reference intakes for energy, carbohydrate, fibre, fat, fatty acids, cholesterol, protein and amino acids. J Am Diet Assoc 102(11):1621–1630

Tucker KL (2001) Eat a variety of healthful foods: old advice with new support. Nutr Rev 59(5):156–158

USDA-NCC Carotenoid Database for U.S. Foods (1998) http://www.nal.usda.gov/fnic/foodcomp/Data/car98/car98.html. Accessed 1 July 2012

Van den Berg H, Faulks R, Fernando-Granado H, Hirschberg J, Olmedilla B, Sandmann G, Southon S, Stahl W (2000) The potential for the improvement of carotenoid levels in foods and the likely systemic effects. J Sci Food Agri 80:880–912

Vardavas CI, Majchrzak D, Wagner KH, Elmadfa I, Katafos A (2006) The antioxidant and phylloquinone content of wildly grown greens in Crete. Food Chem 99(4):813–821
Vasco C, Riihinen K, Ruales J, Kamal-Eldin A (2009) Phenolic compounds in Rosaceae fruits from Ecuador. J Agri Food Chem 57(4):1204–1212
Wangen TE, Duncan AM, Merz-Demlow BE, Xu X, Marcus R, Phipps WR, Kurzer MS (2000) Effects of soy isoflavones on markers of bone turnover in premenopausal and postmenopausal women. J Clin Endocrinol Metab 85:3043–3048
Wargovich JM (2000) Anticancer properties of fruits and vegetables. Hortsci 35(4):573–575
Williamson G (1996) Protective effects of fruits and vegetables in the diet. Nutr Food Sci 1:6–10
Wojdyło A, Oszmianski J, Czemerys R (2007) Antioxidant activity and phenolic compounds in 32 selected herbs. Food Chem 105(3):940–949
Yamada H (1996) Contributions of pectins on health care. Pectins and pectinases. Prog Biotechnol 14:173–190
Zeghichi S, Kallithraka S, Simopoulos AP, Kipriotakis Z (2005) Nutritional composition of selected wild plants in the diet of Crete. In: Simopoulos AP, Gopalan C (eds) Plants in human health and nutrition policy. World review of nutrition and dietetics, vol 91. Karger, Basel, pp 22–40

Part III
Biological Activities

Chapter 10
Antioxidant Potential of Wild Plant Foods

Lillian Barros, Patricia Morales, Ana Maria Carvalho and Isabel C. F. R. Ferreira

10.1 Oxidative Stress and Antioxidant Defenses

10.1.1 Reactive Species and the Condition of Oxidative Stress

A free radical is defined as any species containing one or more unpaired electrons (electrons singly occupying an atomic or molecular orbital), whereas reactive species is the collective term for radicals and some other non-radical derivatives of oxygen, nitrogen, or sulfur that can easily generate free radicals and/or cause oxidative damage (Halliwell 2012).

As shown in Table 10.1, reactive oxygen species (ROS) include free radicals such as hydroperoxyl radical ($HO_2^{\bullet}$), superoxide anion radical ($O_2^{-\bullet}$), hydroxyl radical ($HO^{\bullet}$) and peroxyl radical ($ROO^{\bullet}$; e.g., lipid derived), and other species such as hydrogen peroxide (H_2O_2), singlet oxygen ($^1O_2$), hypochlorous acid (HOCl), and peroxide (ROOR; Lü et al. 2010; Carocho and Ferreira 2013).

The "primary" ROS $O_2^{-\bullet}$ is formed by the addition of one electron to molecular oxygen; this addition occurs in or outside mitochondria and involves different endogenous enzymatic systems such as NADPH oxidases or xanthine oxidases (Ferreira et al. 2009). At pH 7, $HO_2^{\bullet}$, also formed from molecular oxygen, dissociates to $O_2^{-\bullet}$. This radical is not very active, but it can interact with other molecules generating "secondary" ROS, such as H_2O_2 (by superoxide dismutase, SOD in Haber–Weiss reaction) and then $HO^{\bullet}$ (by Fenton reaction—electron transfer from

I. C. F. R. Ferreira (✉) · L. Barros · A. M. Carvalho
Mountain Research Centre, Polytechnical Institute of Bragança, BragançaPortugal
e-mail: iferreira@ipb.pt; lillian@ipb.pt; anacarv@ipb.pt

P. Morales
Departamento de Nutrición y Bromatología II, Bromatología, Facultad de Farmacia, Universidad Complutense de Madrid, MadridSpain
e-mail: patricia.morales@farm.ucm.es

M. de C. Sánchez-Mata, J. Tardío (eds.), *Mediterranean Wild Edible Plants*,
DOI 10.1007/978-1-4939-3329-7_10

Table 10.1 Radical and non-radical reactive oxygen, nitrogen, and sulfur species involved in oxidative stress

Reactive species		Formation
Reactive oxygen species (ROS)		
Radicals	Superoxide anion ($O_2^{-\bullet}$)	$O_2 + e^- \rightarrow O_2^{-\bullet}$ $HO_2^\bullet \rightarrow H^+ + O_2^{-\bullet}$ (pH 7.4) Mitochondria electron transport chain; NADPH oxidases; Xanthine oxidase
	Hydroperoxyl radical ($HO_2^\bullet$)	$O_2 + e^- + H^+ \rightarrow HO_2^\bullet$
	Hydroxyl radical ($HO^\bullet$)	$H_2O_2 + Fe^{2+} \rightarrow HO^- + HO^\bullet + Fe^{3+}$ (Fenton reaction)
	Peroxyl radical ($ROO^\bullet$)	$RH + O_2^{-\bullet} \rightarrow R^\bullet$ $R^\bullet + O_2 \rightarrow ROO^\bullet$
Non-radicals	Hydrogen peroxide (H_2O_2)	$2O_2^{-\bullet} + 2H^+ \rightarrow H_2O_2 + O_2$ (SOD)
	Singlet oxygen ($^1O_2$)	$OCl^- + H_2O_2 \rightarrow Cl^- + H_2O + {}^1O_2$
	Hypochlorous acid (HOCl)	$Cl^- + H_2O_2 \rightarrow OCl^- + H_2O$ (myeloperoxidase) $OCl^- + H^+ \rightarrow HOCl$
	Hydroperoxide (ROOH)	$ROO^\bullet \rightarrow ROOH$
Reactive nitrogen species (RNS)		
Radicals	Nitric oxide ($NO^\bullet$)	$Arginine + NADPH + H^+ \rightarrow NO^\bullet + Citrulline + NADP^+$ Nitric oxide synthases
Non-radicals	Peroxynitrite ($ONOO^-$)	$NO^\bullet + O_2^{-\bullet} \rightarrow ONOO^-$
Reactive sulfur species (RSS)		
Radicals	Thiyl radical ($RS^\bullet$)	$RSH \rightarrow RS^\bullet + e^- + H^+$ (ROS; RNS)
	Sulfoxyl radical ($RSOO^\bullet$)	$RS^\bullet + O_2 \rightarrow RSOO^\bullet$
	Sulfinyl radical ($RSO^\bullet$)	$RSOO^\bullet + RSH \rightarrow RSO^\bullet + RSOH$
	Sulfonyl peroxyl radical ($RSO_3O^\bullet$)	$RSOO^\bullet + O_2 \rightarrow RSO_3O^\bullet$
Non-radicals	Thiol (RSH; e.g. cysteine)	$RSH + NO^\bullet \rightarrow RSNO$ $RSNO + GSH \rightarrow RSH + GSNO$
	Disulfide (RSSR)	RSH + (RSOH or ROS) → RSSR
	Sulfenic acid (RSOH)	RSH + (ROS) → RSOH $RSH \rightarrow RS^- + H^+$; $RS^- + H_2O_2 + H^+ \rightarrow RSOH + H_2O$
	Thiosulfinate (disulfide-S-monoxide) (RS(O)SR)	RSOH + (ROS) → RSOOH (Sulfinic acid) RSOOH + (ROS) → RSO_2OH (Sulfonic acid) $2RSOH \rightarrow RS(O)SR + H_2O$ $2RS^- + H_2O_2 + H^+ \rightarrow RS(O)SR$ RSSR + (ROS) → RS(O)SR
	Thiosulfonate (disulfide-S-dioxide) ($RS(O)_2SR$)	RS(O)SR + (ROS) → $RS(O)_2SR$

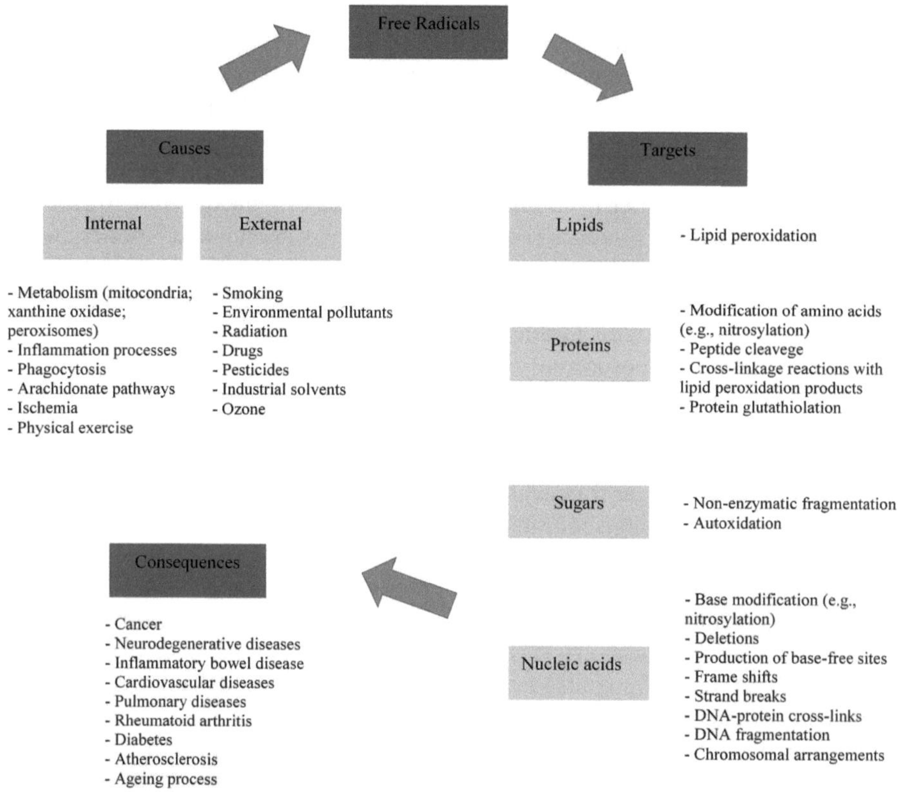

Fig. 10.1 Internal and external causes for overproduction of free radicals, main cellular targets, and related consequences

transition metals to H_2O_2), with the latter considered the most toxic ROS (Valko et al. 2007; Ferreira et al. 2009; Flora 2009; Carocho and Ferreira 2013). In the presence of H_2O_2 and the chloride ion (Cl^-), the enzyme myeloperoxidase produces OCl^- and then HOCl; the latter species can generate $HO^•$ by reacting with $O_2^{-•}$ or Fe^{2+}. OCl^- can also react with H_2O_2 to make singlet O_2 ($^1O_2$) (Halliwell 2006). Lipid peroxidation promotes the production of different types of ROS such as $R^•$ that can react with O_2 to form $ROO^•$; if not neutralized, these radicals react with other adjacent lipids producing hydroperoxide lipids (ROOH) that can easily be decomposed to form new $R^•$, initiating a process that is known as chain propagation reaction (Ferreira et al. 2009).

Table 10.1 also lists the reactive nitrogen species (RNS), including nitric oxide radical ($NO^•$), generated in biological tissues by specific nitric oxide synthases (NOS) that transform arginine to citrulline. $NO^•$ reacts with $O_2^{•-}$ to yield peroxynitrite ($ONOO^-$), a non-radical RNS (Ghafourifar and Cadenas 2005; Ferreira et al. 2009).

As shown in Fig. 10.1, reactive species are produced in mitochondria or peroxisomes within metabolic processes or by xanthine oxidase activity, inflammation processes, phagocytosis, arachidonate pathway, ischemia, and physical exercise. Smoking, environmental pollutants, radiation, drugs, pesticides, industrial solvents, and ozone are examples of external factors that promote the production of free radicals (Halliwell 2011; Carocho and Ferreira 2013).

The main targets of reactive species are lipids, proteins, sugars, and nucleic acids (Lü et al. 2010). Lipid peroxidation (attack on membrane lipids) occurs mainly as a result of the action of $HO^{\bullet}$ or $^{1}O_2$, but also of $ONOO^{-}$. Proteins can be oxidatively modified in specific amino acids (e.g., nitrosylation with $NO^{\bullet}$) by free radical-mediated peptide cleavage or by formation of protein cross-linkages due to reaction with lipid peroxidation products (Ferreira et al. 2009; Carocho and Ferreira 2013). In particular, *S*-nitrosation of glutathione (GSH) produces *S*-nitrosoglutathione (GSNO), which itself is capable of *S*-nitrosating cysteine residues in proteins to make cysteine-*S*-nitrosothiol (Table 10.1). Protein glutathionylation is a prominent consequence of RSS exposure and consists in the redox reaction of protein cysteinyl residues with the tripeptide glutathione, resulting in a protein–glutathione mixed disulfide (Giles et al. 2001; Gruhlke and Slusarenko 2012). The formation of ROS could also contribute to glycoxidative damage; during the initial stages of nonenzymatic glycosylation, sugar fragmentation produces short-chain species such as glycolaldehyde, whose chain is too short to cyclize and is therefore prone to autoxidation (Benov and Beema 2003; Carocho and Ferreira 2013). The damage in nucleic acids induced by reactive species includes production of base-free sites, deletions, modification of all bases, frame shifts, strand breaks, DNA–protein cross-links, and chromosomal arrangements. $HO^{\bullet}$ is known to react with all the components of the DNA molecule, intervening also in DNA oxidation, whereas $ONOO^{-}$ is related to DNA fragmentation (Ferreira et al. 2009; Carocho and Ferreira 2013).

The damage to cells and tissues caused by reactive species, mostly ROS, is called oxidative damage and is a consequence of oxidative stress, a serious imbalance between the generation of ROS and antioxidant protection in favor of the former (Halliwell 2012). Recently, the mentioned author answered the question "does the oxidative stress that is likely to occur as a result of the tissue damage play any role at all in the disease pathology?", with a firm yes for cancer and neurodegenerative diseases, with a probably yes for inflammatory bowel disease, rheumatic arthritis, chronic granulomatous disease, and with a maybe for atherosclerosis and diabetes.

10.1.2 Endogenous Antioxidant Defenses

As shown in Fig. 10.2, humans produce many endogenous antioxidant systems (enzymes such as SOD (superoxide dismutases), CAT (catalases), Prx (peroxiredoxins), GPx (glutathione peroxidase), GRed (glutathione reductase) and GST (glutathione-S-transferases), or nonenzymatic antioxidant defenses, namely GSH (reduced glutathione), Q10, and uric acid) but also obtain some other antioxidants from the diet, such as vitamin E, vitamin C, polyphenols, and carotenoids (Halliwell 2011; Carocho and Ferreira 2013).

SOD converts $O_2^{\bullet -}$ into H_2O_2 through a dismutation reaction, which is then detoxified to water either by CAT in the peroxisomes or by GPx in the mitochon-

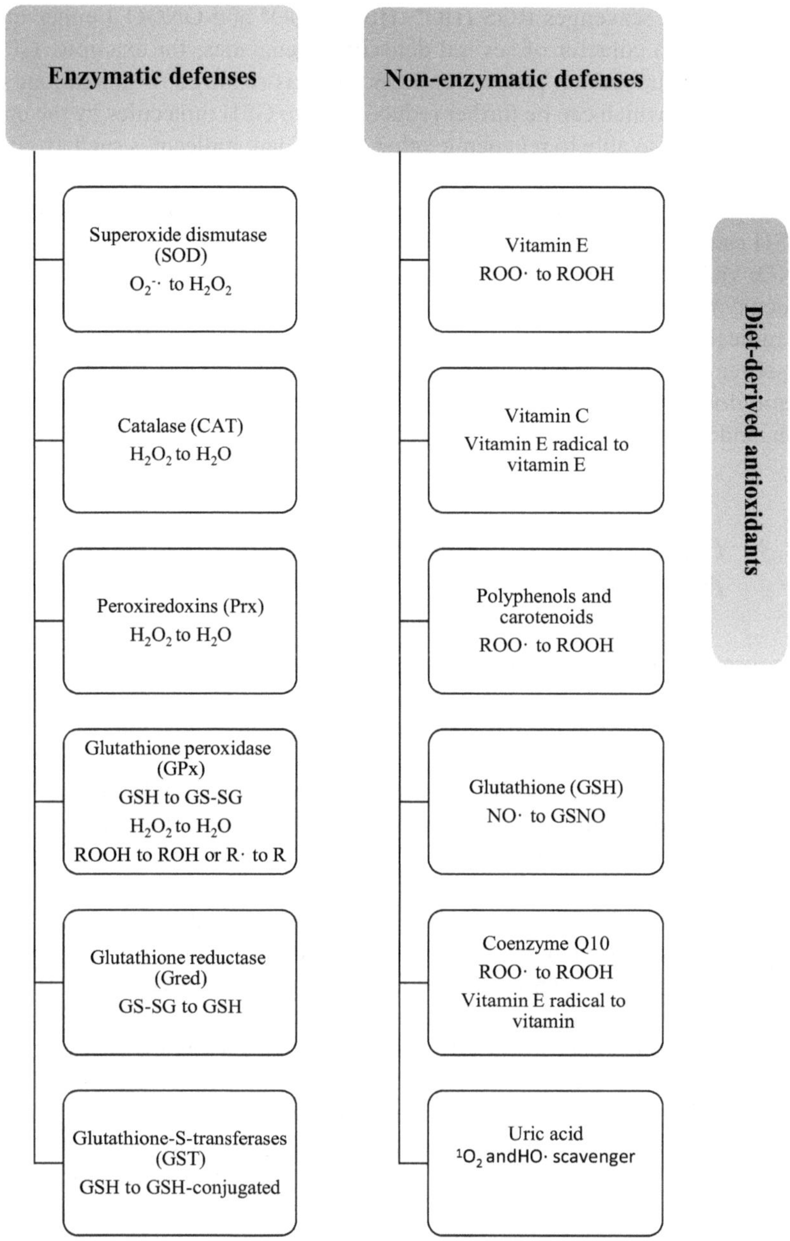

Fig. 10.2 Major endogenous and exogenous enzymatic and nonenzymatic antioxidant defenses. *GS-SG* glutathione disulphide, *GSNO* S-nitrosoglutathione, *R* (e.g., lipid)

dria, cytosol, or nucleus. Other enzymes that reduce H_2O_2 are peroxiredoxins (with cysteine in the active site). GRed regenerates GSH that is used as a hydrogen donor by GPx; the latter can also transform hydroperoxide lipids into alcohols (ROH).

GSH effectively scavenges ROS ($HO^\bullet$, H_2O_2, $LOO^\bullet$ and $ONOO^-$) either directly or indirectly as a cofactor of several detoxifying enzymes, for example, GPx and GST. In the neutralization process of ROS, GSH is oxidized to glutathione disulphide (GS-SG), which can be further reduced to two GSH molecules by the enzyme GRed. GSH is also able to regenerate other antioxidant molecules such as vitamins C and E and react with a variety of electrophilic xenobiotics in reactions catalyzed by GST, generating products with higher solubility (thus easier to eliminate). Finally, GSH can also neutralize $NO^\bullet$, resulting in the formation of S-nitrosoglutathione (GSNO; Valko et al. 2007; Ferreira et al. 2009).

Coenzyme Q10 acts by preventing the formation of or by neutralizing lipid peroxyl radicals, but its ability to regenerate vitamin E has also been reported. Uric acid prevents the overproduction of oxo-heme oxidants that result from the reaction of hemoglobin with peroxides. It also prevents the lysis of erythrocytes by peroxidation, and it is a potent scavenger of $HO^\bullet$ or $^1O_2$ (Carocho and Ferreira 2013).

10.1.3 Contribution of Plants as Exogenous Antioxidant Defenses

Plants, used since ancient times due to their medicinal properties, may be considered as a source of bioactive compounds with antioxidant potential. These properties have been studied in the past few years, proving their potential to act as functional foods (Krishnaiah et al. 2011).

As shown in Fig. 10.2, different compounds, such as vitamin E, vitamin C, polyphenols, and carotenoids, have been reported to help the endogenous antioxidant defense system as exogenous sources. Vitamin E, a liposoluble vitamin present in the membranes, plays an important role in the prevention of lipid peroxidation. ROS (e.g., $HO^\bullet$ and $LOO^\bullet$) react with vitamin E, generating vitamin $E^\bullet$. Then, vitamin C reacts with vitamin $E^\bullet$ (Fig. 10.3) producing vitamin $C^\bullet$ (that could be eliminated by semidehydroascorbate reductase), regenerating vitamin E. Both radicals (vitamin $E^\bullet$ and vitamin $C^\bullet$) are poorly reactive species (Ferreira et al. 2009).

The antioxidant properties of polyphenols, mostly flavonoids and phenolic acids, are conferred by the phenolic hydroxyl groups attached to the ring structures. They can act as reducing agents; hydrogen donators; singlet oxygen quenchers; peroxynitrites, superoxide, hydroxyl, and peroxyl radical scavengers; and even as metal chelators. They also activate antioxidant enzymes, reduce vitamin E radicals,

$$LOO^\bullet + \text{Vit. E} \rightarrow LOOH + \text{Vit. E}^\bullet$$
$$\text{Vit. E}^\bullet + \text{Vit. C} \rightarrow \text{Vit. E} + \text{Vit. C}^\bullet$$

$LOO^\bullet$= Lipid peroxyl radical; LOOH= Lipid hydroperoxide.

Fig. 10.3 Vitamin E regeneration mediated by vitamin C

inhibit oxidases, mitigate nitrosative stress, and increase levels of uric acid and low-molecular-weight molecules (Procházková et al. 2011; Carocho and Ferreira 2013).

The main antioxidant potential of carotenoids is due to singlet oxygen quenching. The only free radicals that completely destroy these pigments are peroxyl radicals. Carotenoids are relatively unreactive but may also decay and form nonradical compounds that may terminate free radical attacks by binding to these radicals (Paiva and Russel 1999; Carocho and Ferreira 2013).

According to Halliwell (2012), our endogenous antioxidant defenses are inadequate to prevent oxidative damage completely; hence, dietary sources of antioxidants are especially important to avoid diseases related to oxidative stress. Nevertheless, the contribution of some of them (e.g., polyphenols and carotenoids) to the beneficial dietary effect of plants is uncertain, as suggested by the limited and confusing literature on their in vivo effects, except possibly in the stomach, small intestine, and colon (Halliwell 2011, 2012).

The extracts obtained from plant materials (whole herb or roots, young stems, leaves, basal leaves, shoots, aerial parts, flower buds, flowers, inflorescences, fruits, seeds, and wood) might be used as antioxidants due to the chemical diversity of their phytochemicals and synergistic effects. In fact, the beneficial effects of diet-derived antioxidants may be maximally exerted when they are consumed at currently recommended dietary intakes, rather than in large amounts. Plants are a rich source of antioxidants, nonetheless the protective effect may not be the same by pulling up one or two individual antioxidant molecules into a high-dose pill (Halliwell 2012).

In this perspective, the antioxidant potential of several plants from Portugal and Spain has been extensively reviewed (Barros et al. 2009, 2010a, b, 2011a, b, c; Martins et al. 2011; Morales et al. 2012, 2013a, b; Pereira et al. 2011), revealing very promising results.

10.1.4 Measurement of Antioxidant Activity in Plants

Regarding the study of antioxidant activity in plants, there is not one method that can provide unequivocal and defining results, which necessitates the use of various methods instead of a one-dimension approach; each method has its specific target within the matrix and its advantages and disadvantages. Some of these procedures use synthetic antioxidants or free radicals; some are specific for lipid peroxidation and require animal or plant cells (Carocho and Ferreira 2013).

One of the methods most frequently used to evaluate the antioxidant properties of different wild edible plants is the Folin–Ciocalteu assay. This method has been often used to evaluate total phenolic content in natural products. However, as it is based on the measurement of the reducing capacity of a sample, nowadays it is being used for antioxidant capacity determination (Huang et al. 2005). It follows the reaction below:

$$\mathrm{Mo(VI)(yellow)} + \mathrm{e-} \rightarrow \mathrm{Mo(V)(blue)}. \tag{10.1}$$

Molybdotungstate (Mo) reagent oxidizes phenols and yields a colored product with an absorption maximum at 745–750 nm. The reagent contains heteropolyphosphotungstates-molybdates that, under basic conditions, react with phenolic compounds to form a phenolate anion, possibly (phenol-$MoW_{11}O_{40})^{4-}$, by dissociation of a phenolic proton. This sequence of reversible one- or two-electron reduction reactions leads to blue-colored products (Huang et al. 2005; Prior et al. 2005).

The screening of antioxidant properties can also be measured using chemical assays (i and ii) or assays related to lipid peroxidation (iii and iv):

i. DPPH-scavenging activity:

$$X^{\bullet} + AH \rightarrow XH + A^{\bullet} \quad (10.2)$$

where $X^{\bullet}$ represents a DPPH radical and AH represents antioxidants present in the sample (plant tissue). Antioxidants donate a hydrogen atom to the DPPH radical, decreasing its absorbance at 517 nm (Antolovich et al. 2002).

ii. Reducing power:

$$Fe(CN)_6^{\ 3-} + AH \rightarrow Fe(CN)_6^{\ 4-} + AH^{+} \quad (10.3)$$

$$Fe(CN)_6^{\ 4-} + Fe^{3+} \rightarrow Fe\left[Fe(CN)_6\right]^{-} \quad (10.4)$$

where $Fe(CN)_6^{3-}$ is the compound with the ferric form, and $Fe(CN)_6^{4-}$ is the compound with the ferrous form. Antioxidants present in the wild plants transfer an electron to ferricyanide complex, reducing Fe^{3+} to Fe^{2+}. The second reaction allows the measurement of the absorbance at 700 nm; higher absorbance corresponds to higher reducing power (Huang et al. 2005; Prior et al. 2005).

iii. β-carotene bleaching inhibition:

$$\beta - carotene - H(orange) + LOO^{\bullet} \rightarrow \beta - carotene^{\bullet}(bleached) + LOOH \quad (10.5)$$

$$\beta - carotene - H(orange) + LOO^{\bullet} + AH \rightarrow \beta - carotene - H(orange) + LOOH + A^{\bullet} \quad (10.6)$$

where $LOO^{\bullet}$ represents the linoleate free radical. Antioxidants present in the plants donate a hydrogen atom neutralizing the linoleate free radical formed in the system avoiding its attack on the highly unsaturated β-carotene and therefore inhibiting β-carotene bleaching (Prior et al. 2005).

iv. TBARS formation inhibition:

$$MDA + TBA \rightarrow MDA - TBA_2 \quad (10.7)$$

$$MDA + TBA + A \rightarrow MDA + TBA_2 \quad (10.8)$$

where MDA represents malondialdehyde and TBA, thiobarbituric acid. The antioxidants present in the sample (plant tissue) will inhibit the formation of the MDA-TBA_2 complex. The TBARS assay measures the MDA formed as the split product of an endoperoxide of unsaturated fatty acids resulting from oxidation of a lipid substrate. The MDA reacts with TBA to form a pink pigment that is measured spectrophotometrically at 532 nm (Fernández et al. 1997).

As different methods measure the antioxidant activity through different mechanisms, the combined analysis of samples through different assays is often needed to have an overall idea of the properties of a given sample. Therefore, the interpretation of the data of antioxidant activity of foods is complex, because of the variety of assays available, and the different units used for expression of the results make comparison of the data difficult.

In this chapter, data about antioxidant properties of 32 different wild species widespread in the Mediterranean area have been reviewed and compared and information is gathered in Tables 10.2, 10.3, 10.4, and it will be carefully reviewed in the followed sections. For that purpose, they have been grouped according to the different plant parts traditionally used.

10.2 Overview of the Wild Food Plants with Antioxidant Potential

Since prehistoric times, when our ancestors used hunting and gathering to provide sustenance, wild food plants have played a central role in human diet and nutrition.

All over the world, wild edibles have been used as dietary supplements and are particularly important in times of famine and food shortage. Other plants were used to preserve food and for seasoning regional and traditional recipes. Natural flavors of some species have lent flavor to very poor, insufficient, and monotonous daily meals for decades. Therefore, these plants and the knowledge and practices associated with them are part of an interesting biocultural heritage (Carvalho 2010; Carvalho and Morales 2010; Barros et al. 2010a, 2011a, d).

Demand for natural products and ingredients of high quality has drawn people's attention to wild edibles, and in many different regions, their use and consumption are becoming more widespread (Łuczaj et al. 2012).

In recent years, experimental research based largely on ethnobotanical surveys and empirical traditional knowledge has shown that wild plants are interesting sources of nutrients and phytochemicals (see also Chap. 9), with significant antioxidant properties (e.g. Barros et al. and Morales et al. cited works) that prevent various illnesses, especially age-related diseases (The Local Food-Nutraceuticals Consortium 2005; Guarrera and Savo 2013).

Table 10.2 Antioxidant activity of several wild vegetables: wild leafy vegetables and wild asparaguses

Species	Origin	Folin–Ciocalteu (mg GAE/g dry extract)	EC_{50} values (mg dry extract/mL methanol)				References
			DPPH scavenging activity	Reducing power	β-carotene bleaching inhibition	TBARS inhibition	
Anchusa azurea Mill.	Spain	148.62±2.00	0.02±0.00	0.01±0.00	0.02±0.00	0.03±0.00	1
Apium nodiflorum (L.) Lag.	Spain	80.47±4.41	0.07±0.00	0.02±0.00	0.02±0.00	0.04±0.00	2
Beta maritima L.	Spain	61.91±7.51	1.35±0.03	0.47±0.00	0.38±0.00	0.05±0.00	1
Borago officinalis L.	Italy	97±1.03[a]	0.06±0.00[a]	–	0.004[a,b,c]	0.04±0.00[a]	3
	Portugal	113.58±0.92	0.07±0.00	0.23±0.01	0.13±0.02	0.14±0.00	4
Chondrilla juncea L.	Spain	37.66±2.40	1.64±0.15	0.34±0.01	0.38±0.02	0.12±0.00	1
Cichorium intybus L.	Italy	190.00±2.03	0.026[c]	–	0.10[b,c]	0.074[c]	5
	Spain	73.68±0.66	1.11±0.05	0.57±0.01	0.45±0.01	0.02±0.00	1
Foeniculum vulgare Mill. leaves	Italy	80±0.95	0.148[c]	–	0.046[b,c]	0.244[c]	5
	Portugal	39.49±0.62	6.88±0.70	1.17±0.07	1.14±0.03	0.22±0.02	6
F. vulgare young stems with leaves	Portugal	65.85±0.74	1.34±0.07	0.48±0.02	0.49±0.03	0.13±0.03	6
	Spain	42.16±0.98	2.75±0.06	1.10±0.02	0.47±0.00	0.02±0.00	2
F. vulgare stems	Portugal	8.61±0.09	12.16±0.94	2.82±0.04	2.38±0.12	0.27±0.01	6
Glechoma hederacea L.	Portugal	196.61±6.09	0.39±0.02	0.22±0.00	0.87±0.10	0.11±0.01	7
Montia fontana L.	Portugal	47.47±1.62	0.22±0.01	0.84±0.02	0.46±0.04	0.25±0.01	4
	Spain	75.53±7.05	1.49±0.07	0.36±0.01	0.48±0.01	0.02±0.00	2
Papaver rhoeas L.	Italy	72±0.76	0.049[c]	–	0.007[b,c]	0.283[c]	5
	Spain	25.86±3.52	1.28±0.03	0.40±0.00	0.56±0.11	0.02±0.00	1
Rorippa nasturtium-aquaticum (L.) Hayek	Portugal	50.42±2.77	0.13±0.03	0.74±0.02	0.85±0.16	0.38±0.06	4
	Turkey	74.18±1.72[d]	0.287[c]	0.20[c]	–	–	8
Rumex acetosella L.	Portugal	141.58±3.67	0.03±0.00	0.16±0.01	0.12±0.01	0.11±0.02	4
Rumex induratus Boiss. & Reut.	Portugal	117.08±2.54	0.03±0.00	0.22±0.01	0.19±0.03	0.10±0.01	4

Table 10.2 (Continued)

Species	Origin	Folin–Ciocalteu (mg GAE/g dry extract)	EC_{50} values (mg dry extract/mL methanol)				References
			DPPH scavenging activity	Reducing power	β-carotene bleaching inhibition	TBARS inhibition	
Rumex papillaris Boiss. & Reut.	Spain	104.18±4.17	2.45±0.09	0.60±0.01	0.30±0.01	0.03±0.00	1
Rumex pulcher L.	Spain	73.44±5.32	3.31±0.10	0.84±0.01	0.34±0.00	0.02±0.00	1
Scolymus hispanicus L.	Spain	21.51±1.51	4.97±0.08	5.97±0.04	0.65±0.01	0.04±0.00	1
Silene vulgaris (Moench) Garcke	Spain	26.72±1.63	3.31±0.07	0.84±0.01	0.62±0.08	0.02±0.00	2
Silybum marianum (L.) Gaertn.	Spain	3.72±0.36	13.09±0.04	1.82±0.01	0.44±0.03	0.02±0.00	1
Sonchus oleraceus L.	Italy	61±0.65	0.164[c]	–	0.065[b,c]	0.435[c]	5
	Spain	51.33±1.75	1.36±0.02	0.89±0.05	0.03±0.00	0.05±0.00	1
Taraxacum obovatum (Willd.) DC.	Spain	58.26±0.90	0.79±0.10	0.48±0.01	0.37±0.00	0.07±0.00	1
Asparagus acutifolius L.	Portugal	623.73±27.68	0.42±0.02	0.19±0.01	0.17±0.01	0.10±0.00	9
	Spain	17.60±0.29	4.87±0.38	1.62±0.00	0.47±0.04	0.07±0.02	2
Bryonia dioica Jacq.	Portugal	258.24±21.95	0.64±0.05	0.20±0.01	0.37±0.01	0.20±0.01	9
	Spain	35.10±2.43	4.43±1.29	1.44±0.01	0.47±0.03	0.08±0.01	2
Humulus lupulus L.	Spain	55.83±1.34	1.36±0.02	0.80±0.01	0.48±0.02	0.03±0.00	2
Tamus communis L.	Portugal	758.99±28.96	0.20±0.03	0.07±0.00	0.07±0.01	0.09±0.01	9
	Spain	49.51±4.07	3.59±0.93	1.32±0.01	0.49±0.15	0.05±0.01	2

GAE gallic acid equivalents; [a] ethanolic extract; [b] data obtained after 60 min of incubation; [c] data was showed as graphical values; [d] μg pyrocatechol/1000 mg extract. Data expressed as mean ± standard deviation. Plant names according to Flora Iberica (http://www.floraiberica.es), except for Asteraceae, which is according to The Plant List (2010) (http://www.theplantlist.org). References: 1:Morales et al. 2013a; 2:Morales et al. 2012; 3:Conforti et al. 2008; 4:Pereira et al. 2011; 5:Conforti et al. 2009; 6:Barros et al. 2009; 7:Barros et al. 2010a; 8:Özen 2009; 9:Martins et al. 2011

Table 10.3 Antioxidant activity of several wild plants used for seasoning and flavoring

Species	Origin	Folin-Ciocalteu (mg GAE/g dry extract)	EC_{50} values (mg dry extract/mL methanol)				References
			DPPH scavenging activity	Reducing power	β-carotene bleaching inhibition	TBARS inhibition	
Foeniculum vulgare Mill.	Portugal	34.68±0.74	7.72±0.87	1.02±0.02	1.29±0.03	0.25±0.01	Barros et al. 2009
Mentha pulegium L.	Portugal	331.69±19.63	0.56±0.05	0.12±0.01	0.01±0.00	0.08±0.00	Fernandes et al. 2010
	Portugal	71.7±2.1[a,b]	0.025±0.00	–	0.165±0.00[b]	–	Mata et al. 2007
	Portugal	6[c]	0.1[c]	<0.140[c]	–	–	Teixeira et al. 2012
	Spain	–	0.037±0.00	–	–	–	Lopéz et al. 2007
Origanum vulgare L.	Portugal	368.58±18.18	0.16±0.03	0.18±0.00	0.45±0.05	0.01±0.00	Barros et al. 2010a
	Portugal	13.5±0.3[b]	0.233±0.06[b]	–	–	–	Teixeira et al. 2013
	Serbia	135±1.08[b]	0.035±0.00[b]	–	–	–	Ličina et al. 2013
	Spain	–	0.186±0.00	–	–	–	Lopéz et al. 2007
	Spain	120	0.014±0.00	–	–	–	Rodríguez-Meizoso et al. 2006
	Turkey	220	0.01±0.00	–	3.125[c]	–	Şahin et al. 2004
Rosa canina L.	Portugal	270.28±35.54	0.22±0.01	0.24±0.03	0.12±0.03	0.03±0.00	Barros et al. 2011a
Sambucus nigra L.	Portugal	92.73±4.66	0.57±0.03	0.27±0.01	0.16±0.01	0.12±0.01	Barros et al. 2011b
Thymus mastichina L.	Portugal	165.29±1.11	0.69±0.04	0.23±0.00	0.90±0.09	0.43±0.02	Barros et al. 2010a

GAE gallic acid equivalents

[a] mg pyrogallol per g of sample

[b] ethanol extract

[c] Data were showed as graphical values. Data expressed as mean ± standard deviation. Plant names according to The Plant List (2010) (http://www.theplantlist.org/)

Table 10.4 Antioxidant activity of edible fruits of several wild edible plants

Species	Origin	Folin-Ciocalteu (mg GAE/g dry extract)	EC_{50} values (mg dry extract/mL methanol)				References
			DPPH scavenging activity	Reducing power	β-carotene bleaching inhibition	TBARS inhibition	
Arbutus unedo L.	Algeria	104.98±4.59[a]	0.006±0.00[a]	0.001±0.00[a]	–	NA	Boulanouar et al. 2013
	Italy	922±38[b]	4.5±1.1[c]	–	–	–	Tuberoso et al. 2013
	Portugal	126.83±6.66	0.45±0.00	0.41±0.00	0.77±0.00	0.09±0.00	Barros et al. 2010b
	Portugal	14.6±0.9	–	–	–	–	Silva et al. 2001
	Portugal	48.26±4.49	0.37±0.02	1.09±0.05	–	–	Oliveira et al. 2011
	Portugal	16.7±0.4	0.790±0.016	2.894±0.049	0.732±0.452	–	Mendes et al. 2011
	Spain	–	–	–	0.10±0.00	0.03±0.00	Morales et al. 2013b
	Spain	16.53	–	–	–	–	Ruiz-Rodriguez et al. 2011
	Spain	586±53[d]	0.59±0.18	–	–	–	Ganhão et al. 2010
Crataegus monogyna Jacq.	France	1226.3±33.7[e]	5.40±0.40[e]	–	–	–	Froehlicher et al. 2009
	Lithuania	182	0.2[f]	–	–	–	Bernatonienė et al. 2008
	Portugal	247.03±9.32	0.13±0.01	0.08±0.00	0.10±0.01	0.05±0.00	Barros et al. 2011c
	Serbia	35.4±2.48[g]	1.470±0.00[g]	–	–	–	Tadić et al. 2008
	Spain	–	–	–	0.02±0.00	0.02±0.00	Morales et al. 2013b
	Spain	820.55[d]	1.54[h]			–	Ruiz-Rodriguez et al. 2014
	Spain	600±105[d]	0.70±0.16	–	–	–	Ganhão et al. 2010

Table 10.4 (Continued)

Species	Origin	Folin-Ciocalteu (mg GAE/g dry extract)	EC_{50} values (mg dry extract/mL methanol)				References
			DPPH scavenging activity	Reducing power	β-carotene bleaching inhibition	TBARS inhibition	
Prunus spinosa L.	Portugal	83.40 ± 2.75	0.60 ± 0.00	0.61 ± 0.00	0.99 ± 0.00	0.15 ± 0.00	Barros et al. 2010b
	Spain	–	–	–	0.09 ± 0.00	0.02 ± 0.00	Morales et al. 2013b
	Spain	326 ± 29[d]	1.98 ± 0.32	–	–	–	Ganhão et al. 2010
	Spain	2255.57[d]	1.14[h]	–	–	–	Ruiz-Rodriguez et al. 2014
Rosa canina L.	Portugal	143.17 ± 5.25	0.43 ± 0.00	0.17 ± 0.00	0.40 ± 0.00	0.09 ± 0.00	Barros et al. 2010b
	Spain	2377 ± 492[d]	0.18 ± 0.05	–	–	–	Ganhão et al. 2010
Rubus ulmifolius Schott	Spain	–	–	–	0.03 ± 0.00	0.02 ± 0.00	Morales et al. 2013b
	Spain	871 ± 80[d]	0.41 ± 0.05	–	–	–	Ganhão et al. 2010

NA not active. Data expressed as mean ± standard deviation. Plant names according to The Plant List (2010) (http://www.theplantlist.org/)

GAE gallic acid equivalents

[a] Hydro-alcoholic extract (70 % ethanol)

[b] mg GAE/L

[c] DPPH (mmol TEAC/L) results are expressed as TEAC millimolar concentration, obtained from a Trolox solution having an antiradical capacity equivalent

[d] mg of GAE/100 g of fruit fresh matter

[e] Results expressed in dry weight basis

[f] Around 0.2 g/mL of ethanolic extract provide around 50 % of inhibition effect; the measure was made after 5 min of incubation

[g] Ethanolic extract

[h] mmol of Trolox Eq/100 g fw

10.2.1 Wild Species Providing Vegetables

The Mediterranean flora is a combination of taxa of various biogeographical origins and evolutionary histories effected over time by climatic events and anthropogenic actions with increasing impact (Cowling et al. 1996). Besides floristic diversity, the richness of habitats, different cultures and mores, and landscape management and historic development around the Mediterranean Basin resulted in the common use of many different plants that met basic dietary needs and are interesting food resources, well adapted to local diets and folk traditions (Rivera et al. 2005; Hadjichambiset al. 2008).

Although it is reported that some tree organs (leaves, young shoots, and flowers) are also used (e.g. Tardío et al. 2006), vegetables are mainly the edible product of herbaceous plants and can be roots or underground stems (tubers, bulbs, or rhizomes), whole immature plants (sprouts), stems, whorled basal leaves, expanded leaves, leaf sheaths, midribs and veins, flower heads, and unripe fruits and seeds. Green, vibrant colored or yellowish white, vivid flavored or having very little taste, these edibles are traditionally used raw or cooked for preparing soups, broths, stews, stir-fries, accompaniments, salads, and sometimes desserts.

Several botanical families provide leafy vegetables that play a significant role in different local Mediterranean cuisines and have interesting phytochemical profiles and promising bioactive properties (Guarrera and Savo 2013). Therefore, species of the Asteraceae, Polygonaceae, Brassicaceae, and Amaryllidaceae families are some of the most gathered and consumed edible greens at least in the three large southern European peninsulas (e.g. Pieroni et al. 2005; Tardío et al. 2006; Hadjichambis et al. 2008; Łuczaj et al. 2012). Some examples are basal leaves, midribs, and soft leafy stems of genera such as *Chondrilla* L., *Cichorium* L., *Hypochaeris* L., *Scolymus* L., *Scorzonera* L. and *Sonchus* L. and species from Carduoideae and Asteroideae subfamilies; leaves of docks and sorrels (genus *Rumex* L.); tender leaves, stems and small flowers from annuals such as *Eruca vesicaria* (L.) Cav., *Raphanus raphanistrum* L., *Capsella bursa-pastoris* (L.) Medik., mustards and wild cabbage; the bulbs, bulbils and linear, channelled or flat leaf blades of *Allium* species (see Chaps. 4 and 13).

Sweet immature pods and seeds from wild species of the Fabaceae family (e.g. genera *Vicia* L., *Lathyrus* L., *Astragalus* L.) are frequently considered as organoleptically interesting wild food (Carvalho and Telo 2012; Tardío et al 2006) with perceived health benefits.

Other noteworthy leafy vegetables are the vernal leaves of some species of Amaranthaceae (genera *Atriplex* L. and *Chenopodium* L.), Boraginaccac (c.g. *Borago officinalis* L. and *Anchusa* sp. pl.), and Apiaceae (such as *Foeniculum vulgare* Mill. and *Apium nodiflorum* (L.) Lag.).

Small amounts of many of these edibles can safely be eaten raw. Different cooking processes such as soaking or boiling are, sometimes, able to remove most traces of different toxins and alkaloids, for instance. However, some plants have both edible and toxic parts (e.g. bryonies); other botanical families such as Apiaceae have species that can be very poisonous. Many leafy greens have high oxalic acid content (e.g. sorrels), and some accumulate chemicals from several contaminants. Thus, consumers must avoid potential hazards, act with extreme caution, and be informed about the risks and learn how to distinguish between edible and poisonous species, organs, or parts of plants.

Data of antioxidant activity in basal leaves as well as leaves accompanied by other aerial parts of 17 different edible species are shown in Table 10.2.

Anchusa azurea Mill. (bugloss) is a Boraginaceae, commonly consumed after cooking as well as for medicinal purposes against gastralgia, cold, kidney stones, pain, and skin problems (Benítez et al. 2008; Carvalho and Morales 2010; Carvalho 2010; Tardío 2010), and samples of this species gathered from Spain presented the lowest EC_{50} values (highest antioxidant activity) for DPPH scavenging activity and reducing power (0.02 and 0.01 mg/mL, respectively) (Morales et al. 2013) when compared with all the plant species shown in Table 10.2. Leaves of *Borago officinalis* L. (borage, also a consumed wild Boraginaceae) from Italy registered the best results in β-carotene bleaching inhibition assays (0.004 mg/mL), according to Conforti et al. (2008).

For the TBARS assay, the leaves of *Cichorium intybus* L. (chicory), *Papaver rhoeas* L. (poppy), *Rumex pulcher* L. (fiddle dock), and *Silybum marianum* (L.) Gaertn. (milk thistle), also gathered from Spain, presented the lowest EC_{50} values (0.02 mg/mL; Morales et al. 2013). Traditionally, most of them are eaten cooked; however, they are sometimes boiled and fried in olive oil with garlic, as the fleshy midribs of *Silybum marianum. Cichorium intybus* and *Papaver rhoeas* are eaten raw in salads as well. Chicory and poppy are also used in traditional medicine due to their effectiveness in digestive disorders, nervousness, insomnia, respiratory disorders, among others (Benítez et al. 2008; Carvalho and Morales 2010; Carvalho 2010; Tardío 2010). Furthermore, *Glechoma hederacea* L. from Portugal revealed the highest phenolic content (196.61 mg GAE/g).

Borago officinalis samples obtained from Italy had lower phenolic content (97 mg chlorogenic acid equivalents/g extract) but slightly higher DPPH scavenging activity and β-carotene bleaching inhibition capacity (Conforti et al. 2008) when compared with the samples of the same species obtained from Portugal (Pereira et al. 2011).

Cichorium intybus samples obtained from Italy presented higher phenolic content, expressed in gallic acid equivalents (GAE), DPPH scavenging activity, and β-carotene bleaching inhibition (190 mg GAE/g; 0.026 and 0.10 mg/mL, respectively) (Conforti et al. 2009) than those gathered from Spain and Greece. Moreover, chicory obtained from Spain had the highest reducing power and TBARS formation inhibition (0.57 and 0.02 mg/mL, respectively). *Foeniculum vulgare* Mill. (fennel), *Papaver rhoeas* L. and *Sonchus oleraceus* L. (smooth sow thistle) from Italy presented the highest DPPH scavenging activity (0.148, 0.049 and 0.164 mg/mL, respectively) and β-carotene bleaching inhibition capacity (0.046, 0.007 and 0.065 mg/mL, in that order) when compared with the same species from other countries (Table 10.2). Fennel and poppy leaves also presented the highest activity for TBARS assay (0.24 and 0.283 mg/mL, respectively), according to Conforti et al.

(2009), and *Papaver rhoeas* from Italy presented the highest levels of phenolics (72 mg GAE/g). *Rorippa nasturtium-aquaticum* (L.) Hayek (watercress) from Portugal showed higher phenolics (50.42 mg GAE/g extract) and DPPH scavenging activity (0.13 mg/mL) but lower reducing power (0.20 mg/mL) than the sample from Turkey (Özen 2009).

It is remarkable that the edible leaves of *Anchusa azurea, Borago officinalis, Cichorium intybus, Papaver rhoeas, Rumex pulcher*, and *Silybum marianum* exhibit higher antioxidant activity measured by one or more different assays than the other species. Special attention should be paid to *Anchusa azurea* leaves gathered in Spain (Morales et al. 2013), which revealed the highest antioxidant activity in all the assays reviewed.

10.2.2 Wild Species Providing Wild Asparaguses

Sprouts and young shoots (asparaguses) of different plants are also considered edible in southern countries. *Asparagus acutifolius* L. (Asparagaceae), *Bryonia dioica* Jacq (Cucurbitaceae), *Humulus lupulus* L. (Cannabaceae), *Tamus communis* L. (Dioscoreaceae), and *Rubus* species (Rosaceae) are some examples of these type of wild edibles usually gathered in early spring while tender and still lacking flower buds (Tardío et al. 2006; Carvalho 2010).

Data on antioxidant activity in young shoots of four wild Mediterranean species are also shown in Table 10.2. The highest phenolic content was found in *Tamus communis* L. (black bryony) from Portugal (758.99 mg GAE/g).

Asparagus acutifolius L. (wild asparagus), *Bryonia dioica* Jacq. (white bryony), and *Tamus communis* from Portugal (Martins et al. 2011) had higher antioxidant activity, with the exception of the TBARS assay, when compared with samples from Spain (Morales et al. 2012).

Edible parts of *Tamus communis* should be remarked for the coincident results found through the different antioxidant assays performed, revealing a high antioxidant potential.

10.2.3 Plants Used for Seasoning and Flavoring

Seasoning and preserving food are still common procedures that have an influence on the traditional cuisine and are fundamental to many regional recipes, particularly in rural areas. Many species with natural flavors are used as additives for enhancing the taste and smell of food but their bioactive properties are also important as food preserves (Dias and Dias 2006; Pardo-de-Santayana et al. 2007).

In former times, such species were related to nutritional needs, especially during those famine periods when wild edible plants were the main source of nourishment for rural families. Moreover, different sauces and pastes were prepared with herbs and stored in glass bottles to use all through the year (Póvoa et al. 2009; Carvalho and Morales 2010). An interesting example is the traditional *piso* from southern Portugal, a paste made of *Mentha pulegium* L. or *M. cervina* L (Lamiaceae), crushed with garlic and salt and then covered with olive oil. Experimental assays proved that after a 6-month storage period, food sauces retained their physicochemical properties and could be used for seasoning (Póvoa et al. 2009).

At least in the Iberian Peninsula, Lamiaceae and Apiaceae are undoubtedly the botanical families providing a great number of species used as seasoning and flavoring agents although many others are also used, such as some Fabaceae like *Pterospartum tridentatum* (L.) Willk. and *Cytisus* sp. pl., *Alliaria petiolata* (M. Bieb.) Cavara & Grande (Brassicaceae) and *Allium ampeloprasum* L., *A. schoenoprasum* L. and *A. triquetum* L. (Amaryllidaceae) (Pardo-de-Santayana et al. 2007; Carvalho 2010). Many of these species are also included in the preparation of different fresh beverages, liqueurs, and herbal teas drunk daily or after meals (Carvalho 2010; Sõukand et al 2013)

Data regarding flowers and inflorescences of *Foeniculum vulgare, Mentha pulegium* L. (pennyroyal), *Origanum vulgare* L. (oregano), *Rosa canina* L. (dog rose), *Sambucus nigra* L. (elder), and *Thymus mastichina* L. (mastic thyme), widespread Mediterranean perennial herbs traditionally used for medicinal purposes and seasoning, are also listed in Table 10.3.

Of all the species mentioned in Table 10.3, *Mentha pulegium* and *Origanum vulgare* are two of the most studied species. *Mentha pulegium* from Portugal presented the highest reducing power and β-carotene bleaching inhibition capacity (0.12 and 0.01 mg/mL, respectively, Fernandes et al. 2010); *Origanum vulgare* from Portugal had the highest amount of total phenolics (368 mg GAE/g) and the lowest EC_{50} values (highest antioxidant activity) for the TBARS assay (0.01 mg/mL, Barros et al. 2010a), whereas the same species from Turkey presented the highest DPPH scavenging activity (0.01 mg/mL, Şahin et al. 2004). The flowers of these two species should be highlighted for their high antioxidant capacity measured by different assays, with coincidences among studies performed with samples from different origins, showing their higher antioxidant potential compared with flowers of other species.

10.2.4 Wild Edible Fruits

Regarding wild Mediterranean fruits, five wild species were reviewed: *Arbutus unedo* L. (strawberry-tree), *Crataegus monogyna* Jacq. (common hawthorn), *Prunus spinosa* L. (blackthorn), *Rosa canina* L., and *Rubus ulmifolius* Schott (blackberry), and the results are summarized in Table 10.4.

Arbutus unedo from Algeria presented the highest antioxidant potential for radical scavenging activity and reducing power (0.006 and 0.001 mg/mL, Boulanouar et al. 2013), but *Crataegus monogyna* from Spain showed the lowest EC_{50} values for lipid peroxidation assays, such as β-carotene bleaching inhibition and TBARS assay (0.02 mg/mL, Morales et al. 2013a). The sample from Portugal revealed the highest phenolic content, expressed in mg per g of extract (247 mg GAE/g, Barros et al 2011c). The different units used for expression of the results of antioxidant activity of plant material make the comparison of data difficult (Table 10.4). However, *Arbutus unedo* and *Crataegus monogyna* are two of the fruits which reported the highest values in the Folin–Ciocalteu assay.

10.2.5 Underutilized and Underexploited Species

Many plant resources, growing as wild plants or that have been naturalized growing on their own, are well adapted to different ecological situations and have great potential to be exploited. This is the case of many leafy vegetables from the Fabaceae, Brassicaceae, and Amaryllidaceae families that have a surprising number of edible species and varieties. A number of them are naturalized from old crops and long-time introduced specimens. Lentils, peas and wormseed (*Chenopodium ambrosioides* L.) are some examples. Leeks and wild garlic, for instance, also have great potential that sometimes is forgotten and not used (Carvalho 2010).

Wild fruits and aromatic species used to be commonly preserved and stored for consumption during long and hard winters. These are species, such as *Rosa* sp. pl. and many others from woods, scrubland, riversides, and natural prairies or meadows, that have become underutilized because various staple products from the retail market or cultivated for daily meals are now offered or more accessible (Carvalho and Morales 2010). However, experimental research shows that some of them have great antioxidant potential (Barros et al. 2010b, 2011a, b).

In rural areas from the Iberian Peninsula, people have brought some of the most popular wild plants used as food additives and beverages from the wild to grow in their home gardens, in order to make them easily available (Carvalho and Morales 2010). This behavior shows that some species can be easily adapted to cultivation providing sustainable use without endangering wild populations (see Chap. 5).

10.3 Concluding Remarks

A widespread traditional use of many wild botanicals (e.g. leafy vegetables, flowers, fruits and seeds) as food was the starting point of experimental research on the antioxidant potential of several plants in Mediterranean regions. The diversity of phytochemicals, present in different edible parts of selected species, provides antioxidant properties with potential health benefits. Whenever possible, data obtained were systematically compared with other studies already published. The antioxidant activity results, using the same methodology, can be expressed differently (e.g. EC_{50}

or as trolox equivalents); therefore, in some cases it is very difficult to compare results.

Overall, leaves of *Anchusa azurea, Apium nodiflorum, Borago officinalis, Cichorium intybus, Papaver rhoeas, Rumex pulcher*, and *Silybum marianum*; young shoots of *Tamus communis*; flowers and inflorescences of *Mentha pulegium* and *Origanum vulgare*; and fruits of *Arbutus unedo* and *Crataegus monogyna* stand out among others for their antioxidant potential. Moreover, it can be highlighted that, in general, fruits were the most active plant part (EC_{50} values of all the assays ranged between 0.001 and 5.4 mg/mL), whereas in most cases, leaves reported lower antioxidant activity (EC_{50} values between 0.01 and 13 mg/mL).

There are also numerous publications of the antioxidant potential and bioactive compounds present in the same parts of the reviewed 32 species from other countries that do not belong to the Mediterranean. *Asparagus acutifolius* from Brazil (Tiveron et al. 2012), China (Shou et al. 2007), USA (Sun et al. 2007a and b), and Poland (Vinson et al. 1998); *Borago officinalis* from Lithuania (Bandoniene and Murkovic 2002; Bandoniene et al. 2005); *Crataegus monogyna* fruits from France (Froehlicher et al. 2009); *Foeniculum vulgare* from Iran (Motamed and Naghibi 2010); *Rorippa nasturtium-aquaticum* from Iran (Bahramikia and Yazdanparast 2010), Denmark (Justesen and Knuthsen 2001), Brazil (Hassimotto et al. 2009), and Australia (Lako et al. 2007); *Rosa canina* from Austria (Wenzig et al. 2008) and Denmark (Kirkeskov et al. 2011); *Sambucus nigra* from Austria (Rieger et al. 2008) and USA (Wu et al. 2004) are some examples of wild species studied out of the Mediterranean area.

References

Antolovich M, Prenzler PD, Patsalides E, McDonald S, Robards K (2002) Methods for testing antioxidant activity. Analyst 127:183–198

Bahramikia S, Yazdanparast R (2010) Antioxidant efficacy of Nasturtium officinale extracts using various in vitro assay systems. J Acupunct Meridian Stud 3:283–290

Bandoniene D, Murkovic M (2002) The detection of radical scavenging compounds in crude extract of borage (*Borago officinalis* L.) by using an on-line HPLC-DPPH method. J Biochem Biophys Methods 53:45–49

Bandoniene D, Murkovic M, Venskutonis PR (2005) Determination of Rosmarinic acid in sage and borage leaves by high-performance liquid chromatography with different detection methods. J Chromatogr Sci 43:372–376

Barros L, Heleno SA, Carvalho AM, Ferreira ICFR (2009) Systematic evaluation of the antioxidant potential of different parts of *Foeniculum vulgare* mill from Portugal. Food Chem Toxicol 47:2458–2464

Barros L, Heleno SA, Carvalho AM, Ferreira ICFR (2010a) Lamiaceae often used in Portuguese folk medicine as a source of powerful antioxidants: vitamins and phenolics. LWT—Food Sci Tech 43:544–550

Barros L, Carvalho AM, Morais JS, Ferreira ICFR (2010b) Strawberry-tree, blackthorn and rose fruits: detailed characterization in nutrients and phytochemicals with antioxidant properties. Food Chem 120:247–254

Barros L, Carvalho AM, Ferreira ICFR (2011a) Exotic fruits as a source of important phytochemicals: improving the traditional use of *Rosa canina* fruits in Portugal. Food Res Inter 44:2233–2236

Barros L, Cabrita L, Vilas-Boas M, Carvalho AM, Ferreira ICFR (2011b) Chemical, biochemical and electrochemical assays to evaluate phytochemicals and antioxidant activity of wild plants. Food Chem 127:1600–1608

Barros L, Carvalho AM, Ferreira ICFR (2011c) Comparing the composition and bioactivity of *Crataegus Monogyna* flowers and fruits used in folk medicine. Phytochem Anal 22:181–188

Barros L, Carvalho AM, Ferreira ICFR (2011d) From famine plants to tasty and fragrant spices: three Lamiaceae of general dietary relevance in traditional cuisine of Trás-os-Montes (Portugal). LWT—Food Sci Technol 44(2):543–548

Benítez G, González-Tejero MR, Molero-Mesa J (2008) Pharmaceutical ethnobotany in the western part of Granada province (southern Spain): ethnopharmacological synthesis. J Ethnopharm 129:87–105

Benov L, Beema AF (2003) Superoxide-dependence of the short chain sugars-induced mutagenesis. Free Radic Bio Med 34:429–433

Bernatonienė J, Masteikova R, Majienė D, Savickas A, Kėvelaitis E, Bernatonienė R, Dvoráčková K, Civinskienė G, Lekas R, Vitkevičius K, Pečiūra R (2008) Free radical-scavenging activities of *Crataegus monogyna* extracts. Medicina 44:706–771

Boulanouar B, Abdelaziz G, Aazza S, Gago C, Miguel MC (2013) Antioxidant activities of eight Algerian plant extracts and two essential oils. Ind Crops Prod 46:85–96

Carocho M, Ferreira ICFR (2013) A review on antioxidants, pro-oxidants and related controversy: natural and synthetic compounds, screening and analysis methodologies and future perspectives. Food Chem Toxicol 51:15–25

Carvalho AM (2010) Plantas y sabiduría popular del Parque Natural de Montesinho. Un estudio etnobotánico en Portugal. Biblioteca de Ciencias 35. Consejo Superior de investigaciones Científicas, Madrid

Carvalho AM, Morales R (2010) Persistence of wild food and wild medicinal plant knowledge in a north-eastern region of Portugal. In: Pardo-de-Santayana M, Pieroni A, Puri R (eds) Ethnobotany in the New Europe: people, health and wild plant resources. Berghahn Books, Oxford pp 147–171

Carvalho AM, Telo M (2012) Etnoflora da Terra de Miranda. Cultibos, Yerbas i Saberes. FRAUGA-Istituto Politécnico de Bragança, Bragança

Conforti F, Sosa S, Marrelli M, Menichini F, Statti GA, Uzunov D, Tubaro A, Menichini F, Loggia RD (2008) In vivo anti-inflammatory and in vitro antioxidant activities of Mediterranean dietary plants. J Ethnopharm 116:144–151

Conforti F, Sosa S, Marrelli M, Menichini F, Statti GA, Uzunov D, Tubaro A, Menichini F (2009) The protective ability of Mediterranean dietary plants against the oxidative damage: the role of radical oxygen species in inflammation and the polyphenol, flavonoid and sterol content. Food Chem 112:587–594

Cowling RM, Rundel PW, Lamont BB, Arroyo MK, Arianoutsou M (1996) Plant diversity in Mediterranean-climate regions. Trends Ecol Evol 11:362–366

Dias LS, Dias AS (2006) Herbs and spices in traditional recipes of Alentejo (Portugal). In: Ertug ZF (ed) Proceedings of the IVth International Congress of Ethnobotany (ICEB 2005) pp 69–72

Fernandes ASF, Barros L, Carvalho AM, Ferreira ICFR (2010) Lipophilic and hydrophilic antioxidants, lipid peroxidation inhibition and radical scavenging activity of two Lamiaceae food plants. Eur J Lipid Sci Technol 112:1115–1129

Fernández J, Perez-Alvarez JA, Fernández-Lopez JA (1997) Thiobarbituric acid test for monitoring lipid oxidation in meat. Food Chem 59:345–353

Ferreira ICFR, Barros L, Abreu RMV (2009) Antioxidants in wild mushrooms. Curr Med Chem 16:1543–1560

Flora SJS (2009) Structural, chemical and biological aspects of antioxidants for strategies against metal and mettaloid exposure. Oxid Med Cell Longev 2:191–206

Froehlicher T, Hennebelle T, Martin-Nizard F, Cleenewerck P, Hilbert J-L, Trotin F, Grec S (2009) Phenolic profiles and antioxidative effects of hawthorn cell suspensions, fresh fruits, and medicinal dried parts. Food Chem 115:897–903

Ganhão R, Estévez M, Kylli P, Heinonen M, Morcuende D (2010) Characterization of selected wild mediterranean fruits and comparative efficacy as inhibitors of oxidative reactions in emulsified raw pork burger patties. J Agric Food Chem 58:8854–8861
Ghafourifar P, Cadenas E (2005) Mitochondrial nitric oxide synthase. Trends Pharmacol Sci 26:190–195
Giles GI, Tasker KM, Jacob AC (2001) Hypothesis: the role of reactive sulfur species in oxidative stress. Free Rad Biol Med 31:1279–1283
Gruhlke MCH, Slusarenko AJ (2012) The biology of reactive sulfur species (RSS). Plant Physiol Biochem 59:98–107
Guarrera PM, Savo V (2013) Perceived health properties of wild and cultivated food plants in local and popular traditions of Italy: a review. J Ethnopharm 146:659–680
Hadjichambis A, Paraskeva-Hadjichambi D, Della A et al (2008) Wild and semi-domesticated food plant consumption in seven circum-Mediterranean areas. Int J Food Sci Nutr 59(5):383–341
Halliwell B (2006) Phagocyte-derived reactive species: salvation or suicide? Trend Biochem Sci 31:509–515
Halliwell B (2011) Free radicals and antioxidants—quo vadis? Trend Pharmacol Sci 32:125–129
Halliwell B (2012) Free radicals and antioxidants: updating a personal review. Nutr Rev 70:257–265
Hassimotto NMA, Genovese MI, Lajolo FM (2009) Antioxidant capacity of Brazilian fruit, vegetables and commercially-frozen fruit pulps. J Food Comp Anal 22:394–396
Huang D, Ou B, Prior RL (2005) The chemistry behind antioxidant capacity assays. J Agric Food Chem 53:1841–1856
Justesen U, Knuthsen P (2001) Composition of flavonoids in fresh herbs and calculation of flavonoid intake by use of herbs in traditional Danish dishes. Food Chem 73:245–250
Kirkeskov B, Christensen R, Bügel S, Bliddal H, Danneskiold-Samsoe, Christensen LP, Andersen JR (2011) The effects of rose hip *(Rosa canina)* on plasma antioxidative activity and C-reactive protein in patients with rheumatoid arthritis and normal controls: a prospective cohort study. Phytomedicine 18:953–958
Krishnaiah D, Sarbatly R, Nithyanandam R (2011) A review of the antioxidant potential of medicinal plant species. Food Bioprod Proc 89:217–233
Lako J, Trenerry VC, Wahlqvist M, Wattanapenpaiboon N, Sotheeswaran S, Premier R (2007) Phytochemical flavonols, carotenoids and the antioxidant properties of a wide selection of Fijian fruit, vegetables and other readily available foods. Food Chem 101:1727–1741
Ličina BZ, Stefanović OD, Vasić SM, Radojevic ID, Dekić MS, Comić LR (2013) Biological activities of the extracts from wild growing *Origanum vulgare* L. Food Control 33:498–504
Lopéz V, Akerreta S, Casanova E, Garcia-Mina JM, Cavero RY, Calvo MI (2007) In vitro antioxidant and anti-rhizopus activities of Lamiaceae herbal extracts. Plant Foods Hum Nutri 62:151–155
Lü J, Lin PH, Yao Q, Chen C (2010) Chemical and molecular mechanisms of antioxidants: experimental approaches and model systems. J Cell Mod Med 14:840–860
Łuczaj L, Pieroni A, Tardío J, Pardo-de-Santayana M, Sõukand R, Svanberg I, Kalle R (2012) Wild food plant use in 21st century Europe: the disappearance of old traditions and the search for new cuisines involving wild edible. Acta Soc Bot Pol 81(4):359–370
Martins D, Barros L, Carvalho AM, Ferreira ICFR (2011) Nutritional and in vitro antioxidant properties of edible wild greens in Iberian Peninsula traditional diet. Food Chem 125:488–494
Mata AT, Proença C, Ferreira AR, Serralheiro MLM, Nogueira JMF, Araújo MEM (2007) Antioxidant and antiacethylcholinesterase activities of five plants used as Portuguese food spices. Food Chem 103:778–786
Mendes L, Freitas V, Baptista P, Carvalho M (2011) Comparative antihemolytic and radical scavenging activities of strawberry tree (*Arbutus unedo* L.) leaf and fruit. Food Chem Toxicol 49:2285–2291
Morales P, Carvalho AM, Sánchez-Mata, MC Cámara M, Molina M, Ferreira ICFR (2012) Tocopherol composition and antioxidant activity of spanish wild vegetables. J Gen Resour Crop Evol 59:851–863

Morales P, Ferreira ICFR, Carvalho AM, Fernández-Ruiz V, Sánchez-Mata MC, Cámara M, Morales R, Tardío J (2013) Wild edible fruits as a sources of phytochemicals with capacity to inhibit lipid peroxidation. Eur J Lipid Sci Technol 115:176–185

Morales P, Ferreira ICFR, Carvalho AM, Sánchez-Mata MC, Cámara M, Fernández-Ruiz V, Pardo-de-Santayana M, Tardío J (2014) Mediterranean non-cultivated vegetables as dietary sources of compounds with antioxidant and biological activity. LWT—Food Sci Technol 55:389–396

Motamed SM, Naghibi F (2010) Antioxidant activity of some edible plants of the Turkmen Sahra region in northern Iran. Food Chem 119:1637–1642

Oliveira I, Baptista P, Malheiro R, Casal S, Bento A, Pereira JA (2011) Influence of strawberry tree (*Arbutus unedo* L.) fruit ripening stage on chemical composition and antioxidant activity. Food Res Int 44:1401–1407

Özen T (2009) Investigation of antioxidant properties of *Nasturtium officinale* (watercress) leaf extracts. Acta Pol Pharm 66:187–193

Paiva SAR, Russell RM (1999) β-Carotene and other carotenoids as antioxidants. J Am Coll Nutr 18:426–433

Pardo-de-Santayana M, Tardío J, Blanco E, Carvalho AM, Lastra JJ, San Miguel E, Morales R (2007) Traditional knowledge on wild edible plants in the northwest of the Iberian Peninsula (Spain and Portugal): a comparative study. J Ethnobiol Ethnomed 3:27

Pereira C, Barros L, Carvalho AM, Ferreira ICFR (2011) Nutritional composition and bioactive properties of commonly consumed wild greens: potential sources for new trends in modern diets. Food Res Inter 44:2634–2640

Pieroni A, Nebel S, Santoro RF, Heinrich M (2005) Food for two seasons: culinaryuses of non-cultivated local vegetables and mushrooms in a south Italian village. Int J Food Sci Nutr 56(4):245–272

Póvoa O, Ribeiro G, Rodrigues L, Lobato P, Monteiro P, Monteiro A, Moldão-Martins M (2009) Effect of storage time on physico-chemical characteristics of *Mentha pulegium* L. and *Mentha cervina* L. Acta Hortic 826:193–200

Prior RL, Wu XL, Schaich K (2005) Standardized methods for the determination of antioxidant capacity and phenolics in foods and dietary supplements. J Agric Food Chem 53:4290–4302

Procházková D, Boušová I, Wilhelmová N (2011) Antioxidant and prooxidant properties of flavonoids. Fitoterapia 82:513–523

Rieger G, Muller M, Guttenberger H, Bucar F (2008) Influence of altitudinal variation on the content of Phenolic compounds in wild populations of *Calluna vulgaris, Sambucus nigra*, and *Vaccinium myrtillus*. J Agric Food Chem 56:9080–9086

Rivera D, Obón C, Inocencio C, Heinrich M, Verde A, Fajardo J, Llorach R (2005) The ethnobotanical study of local Mediterranean food plants as medicinal resources in Southern Spain. J Physiol Pharmacol 56(1):97–114

Rodríguez-Meizoso I, Marin FR, Herrero M, Señorans FJ, Reglero G, Cifuentes A, Ibáñez E (2006) Subcritical water extraction of nutraceuticals with antioxidant activity from oregano. Chemical and functional characterization. J Pharm Biomed Anal 41:1560–1565

Ruiz-Rodriguez BM, Morales P, Fernández-Ruiz V, Sánchez-Mata MC, Cámara M, Díez-Marqués C, Pardo-de-Santayana M, Molina M, Tardío J (2011) Valorization of wild strawberry-fruit (*Arbutus unedo* L.) though nutritional assessment and natural production data. Food Res Inter 44:1244–1253

Ruiz-Rodriguez BM, De Ancos B, Sánchez-Moreno C, Férnandez-Ruiz MC, Cámara M, Tardio J (2014) Wild Blackthirn (*Prunus spinosa* L.) and Hawthorn (*Crataegus monogyna* Jacq.) fruits as valuable sources of antioxidants. Fruits 96:61–73

Şahin F, Güllüce M, Daferera D, Sökmen A, Sökmen M, Polissiou M, Agar G, Özer H (2004) Biological activities of the essential oils and methanol extract of *Origanum vulgare* ssp. vulgare in the Eastern Anatolia region of Turkey. Food Control 15:549–557

Shou S, Lu G, Huang X (2007) Seasonal variations in nutritional components of green asparagus using the mother fern cultivation. Sci Hortic 112:251–257

Silva AEMLCMM, Leitao AEB, Azinheira HG, Leitao MCA (2001) The arbutus berry: studies on its color and chemical characteristics at two mature stages. J Food Comp Anal 14:27–35

Sõukand R, Quave CL, Pieroni A et al (2013) Plants used for making recreational tea in Europe: a review based on specific research sites. J Ethnobiol Ethnomed 9(1):58

Sun T, Tang J, Powers JR (2007a) Antioxidant activity and quality of asparagus affected by microwave-circulated water combination and conventional sterilization. Food Chem 100:813–819

Sun T, Powers JR, Tang J (2007b) Evaluation of the antioxidant activity of asparagus, broccoli and their juices. Food Chem 105:101–106

Tadić VM, Dobrić S, Marković GM, Dordević SM, Arsić IA, Menković NR, Stević T (2008) Anti-inflammatory, gastroprotective, free-radical-scavenging, and antimicrobial activities of hawthorn berries ethanol extract. J Agric Food Chem 56:7700–7709

Tardío J (2010) Spring is coming. The gathering and consumption of wild vegetables in Spain. In: Pardo-de-Santayana M, Pieroni A, Puri R (eds) Ethnobotany in the New Europe. Berghahn Books, Oxford, pp 211–237

Tardío J, Pardo-de-Santayana M, Morales R (2006) Ethnobotanical review of wild edible plants in Spain. Bot J Linn Soc 152(1):27–72

Teixeira B, Marques A, Ramos C, Batista I, Serrano C, Matos O, Nenge NR, Nogueira JMF, Saraiva JA, Nunes ML (2012) European pennyroyal *(Mentha pulegium)* from Portugal: chemical composition of essential oil and antioxidant and antimicrobial properties of extracts and essential oil. Ind Crops Prod 36:81–87

Teixeira B, Marques A, Ramos C, Serrano C, Matos O, Neng NR, Nogueira JMF, Saraiva JA, Nunes ML (2013) Chemical composition and bioactivity of different oregano *(Origanum vulgare)* extracts and essential oil. J Sci Food Agric 93:2707–2714

The Local Food-Nutraceuticals Consortium, Heinrich M (coord) (2005) Understanding local Mediterranean diets: a multidisciplinary pharmacological and ethnobotanical approach. Pharmacol Res 52:353–366

The Plant List (2010) Version 1. http://www.theplantlist.org. Accessed Aug 2013

Tiveron AP, Melo PS, Bergamaschi KB, Vieira TMFS, Regitano-d'Arce MAB, Alencar SM (2012) Antioxidant activity of Brazilian vegetables and its relation with phenolic composition. Int J Mol Sci 13:8943–8957

Tuberoso CIG, Boban M, Bifulco E, Budimir D, Pirisi FM (2013) Antioxidant capacity and vasodilatory properties of Mediterranean food: the case of Cannonau wine, myrtle berries liqueur and strawberry-tree honey. Food Chem 140:686–691

Valko M, Leibfritz D, Moncol J, Cronin MT, Mazur M, Telser J (2007) Free radicals and antioxidants in normal physiological functions and human disease. Int J Biochem Cell Bio 39:44–84

Vinson JA, Hao Y, Su X, Zubik L (1998) Phenol antioxidant quantity and quality in foods: vegetables. J Agric Food Chem 46:3630–3634

Wenzig EM, Widowitza U, Kunert O, Chrubasik S, Bucar F, Knauder E, Bauer R (2008) Phytochemical composition and in vitro pharmacological activity of two rose hip (*Rosa canina* L.) preparations. Phytomedicine 15:826–835

Wu X, Gu L, Prior RL, McKay S (2004) Characterization of anthocyanins and proanthocyanidins in some cultivars of Ribes, Aronia, and Sambucus and their antioxidant capacity. J Agric Food Chem 52:7846–7856

Chapter 11
Antimicrobial Potential of Wild Edible Herbaceous Species

Donato Di Venere, Maria Antonia Gatto, Antonio Ippolito and Vito V. Bianco

11.1 Wild Edible Herbaceous Species in the Mediterranean Countries

All countries overlooking the Mediterranean basin possess a varied germplasm of wild edible herbaceous species, with Turkey, Spain, and Italy having the richest floras (Aedo et al. 2013). According to unpublished data gathered by one of us (VV Bianco), Italy probably has one of the highest numbers of wild herbs utilizable as food and condiment. Studies on botanical, agronomical, gastronomic, and nutritional aspects of wild edible herbaceous plants existing in Italy have been undertaken by Bianco since 1964. In the first published work (Bianco 1969), 113 species were reported. In the successive study (Bianco 1989), 360 *taxa* belonging to 66 families and 230 genera were listed. Afterwards, the results released by Bianco and Machakova (2002) showed that the number of *taxa* has reached 808, belonging to 91 families and 403 genera. Nowadays, according to the latest data of Bianco, the *taxa* utilizable as food and condiment herbs may reach 1078, corresponding to 14% of the total number of *taxa* surveyed for the Italian vascular flora (Conti et al. 2005). They belong to 96 families, the most important being the following: Asteraceae (18.7%), Brassicaceae (7.7%), Lamiaceae (7.1%), Apiaceae (6.6%), Fabaceae (5.5%), Polygonaceae (3.8%), Chenopodiaceae (3.2%), Boraginaceae (2.5%), Alliaceae (2.4%), and Malvaceae (2.0%), and 451 genera, where the most represented are the following: *Allium* (26), *Rumex* (21), *Lathyrus* (15), *Chenopodium* (14), *Crepis* (13), *Vicia* (12), *Amaranthus, Malva*, and *Plantago* (9). Leaves are the most used organ in culinary preparation (about 70% of the *taxa*), followed by tender stems (45%), flowers (18%), and roots (16%); fruits, rhizomes, bulbs,

D. Di Venere (✉) · M. A. Gatto · V. V. Bianco
CNR—Institute of Sciences of Food Production (ISPA), Bari, Italy
e-mail: donato.divenere@ispa.cnr.it; mariantonia.gatto@ispa.cnr.it; vvbian36@gmail.com

A. Ippolito
Department of Soil, Plant, and Food Sciences, University of Bari Aldo Moro, Bari, Italy
e-mail: antonio.ippolito@uniba.it

M. de C. Sánchez-Mata, J. Tardío (eds.), *Mediterranean Wild Edible Plants*,
DOI 10.1007/978-1-4939-3329-7_11

inflorescences, tender pods, petioles, and midribs show values less than 3 %. Of the total number of *taxa*, about 42 % are used to prepare salads, while 23 % are used for making soups, 21 % are boiled, 18 % are eaten raw, 17 % are used as condiments, 10 % are prepared as omelets, 5 % are pickled in vinegar or olive oil, 3 % are roasted, and 3 % are used for filling pies. With reference to the period in which they were reported for the first time in Italy, about 13 % were mentioned by authors of the Roman period, 9 % from the Middle Ages, and 30 % were reported for the first time in 1500–1599. The remaining were reported after 1600. With the aim to protect the endangered species and to meet the growing demands of consumers, the cultivation of about 20 % of species has been recently attempted.

Although the number of *taxa* of wild edible herbs present in each one of the 20 countries bordering the Mediterranean Sea amounts to several hundreds, only a relatively small number of them are used in those countries. A survey carried out to verify the presence of wild edible species common in all 20 Mediterranean countries showed that 20 species were the most representative, being present and utilized for several purposes in almost all Mediterranean countries: *Amaranthus retroflexus* L., *Beta vulgaris* L. subsp. *maritima* (L.) Arcang., *Borago officinalis* L., *Capsella bursa-pastoris* (L.) Medik., *Cichorium intybus* L., *Eruca vesicaria* (L.) Cav., *Foeniculum vulgare* Mill., *Malva sylvestris* L., *Nasturtium officinale* R. Br., *Papaver rhoeas* L., *Portulaca oleracea* L., *Raphanus raphanistrum* L., *Rumex crispus* L., *Scolymus hispanicus* L., *Silybum marianum* (L.) Gaertn., *Sinapis alba* L., *Sonchus oleraceus* L., *Taraxacum officinale* Weber ex F.H. Wigg., *Urospermum picroides* (L.) Scop. ex F.W. Schmidt, and *Urtica dioica* L. To our knowledge, all the abovementioned species are present in all 20 countries of the Mediterranean basin, except for *U. dioica* at Malta and Cyprus, *F. vulgare* at Cyprus, and *A. retroflexus* in Tunisia, where the occurrence of these species is doubtful. Most of the 20 species belong to Asteraceae (6) and Brassicaceae (5) families, which are the most important families within the vascular flora of the 20 considered countries; the other represented families are Amaranthaceae (2), Apiaceae (1), Boraginaceae (1), Malvaceae (1), Papaveraceae (1), Polygonaceae (1), Portulacaceae (1), and Urticaceae (1).

The abovementioned species are used to prepare traditional dishes and show good nutritional properties. They are also utilized as condiments, as wildflowers (for ornamental purposes, found at such places as urban traffic islands, edge of roads, and highways), or in traditional medicine to treat various ailments. Some of them (*Capsella bursa-pastoris, Cichorium intybus, M. sylvestris, Papaver rhoeas, Portulaca oleracea, Scolymus hispanicus, Silybum marianum, Sonchus oleraceus, T. officinale*, and *U. dioica*) are utilized as food for animals; some of them (*B. vulgaris, C. intybus, F. vulgare, M. sylvestris, S. alba*, and *T. officinale*) are visited by bees to produce honey; most of them show antioxidant and antimicrobial activities. All the organs of the plant are used, including sprouted seeds and immature pods. Twelve of these 20 species (*Beta vulgaris, Borago officinalis, C. intybus, E. vesicaria, F. vulgare, M. sylvestris, N. officinale, R. crispus, Scolymus hispanicus, Sonchus oleraceus, Urospermum picroides*, and *Urtica dioica*) were already known and used in traditional popular medicine in the Greek and Roman periods.

All the considered species are rich in nutritionally important vitamins, some of which (i.e., vitamin C) also provide antioxidant activity (see Chaps. 6 and 10). Some studies report the vitamin C content of *P. rhoeas, A. retroflexus*, and *U. dioica* (more than 150 mg/100 g) in the range of sweet pepper, one of the richest sources of this vitamin among vegetables, and also *B. vulgaris, E. vesicaria, M. sylvestris, C. bursa-pastoris* between 80 and 100 mg/100 g (Bianco and Machacova 2002; Bianco et al. 2007). They are also reported to have considerable pro-vitamin A content, in the range 119–508 µg RAE/100 g (*E. vesicaria* and *T. officinale*, respectively), compared to the content of 835 µg RAE/100 g reported for carrot (USDA, National Nutrient Database 2011). As for α-tocopherol, the strongest antioxidant form among all tocopherols, the highest content (3.1 mg/100 g) was recorded in *E. vesicaria* by Vardavas et al. (2006); it was found in the range 0.29–1.70 mg/100 g in *B. vulgaris, C. intybus, F. vulgare, P. rhoeas, S. oleraceus*, and *U. picroides* (Simopoulos 2004; Vardavas et al. 2006; Morales et al. 2014).

Besides the abovementioned compounds, the considered species are also very rich in polyphenols, which represent another important class of bioactive molecules with antioxidant (see Chap. 10) and antimicrobial activities. Due to the presence of considerable amounts of polyphenols, the studied species are interesting both from a nutraceutical point of view, being used as food or natural drugs, as well as for agro-industrial applications, as a source of antimicrobial substances potentially exploitable as natural biopesticides or food preservatives (Gatto et al. 2011).

11.2 Phenolic Compounds: Natural Antimicrobials Widely Spread in the Plant Kingdom

Phenolic compounds are important secondary metabolites occurring in plants (see Chap. 9) with a wide variety of polyphenolic structures (Bravo 1998). Polyphenols are very widely distributed in plant foods, including fruits, vegetables, nuts, seeds, flowers, and barks. In regards to the beneficial properties for human health, many phenolic compounds have been reported to possess potent antioxidant activity and to have anticarcinogenic/antimutagenic, antiatherosclerotic, antibacterial, antiviral, and anti-inflammatory activities (Veeriah et al. 2006; Baidez et al. 2007; Han et al. 2007; Miccadei et al. 2008; Huang et al. 2010; Mileo et al. 2012). Moreover, in the agricultural field, phenolic compounds represent a rich source of biocides and preservatives that have been explored for a long time as postharvest alternative control means (Ippolito and Nigro 2003; Lattanzio 2003). In particular, many studies have pointed out the antimicrobial efficacy of certain classes of phenolic compounds, such as hydroxybenzoic acid derivatives (Lattanzio et al. 1996; Amborabé et al. 2002; Veloz-García et al. 2010), coumaric and caffeic acid derivatives (Zhu et al. 2004; Widmer and Laurent 2006; Korukluoglu et al. 2008), flavonoids and coumarins (Ojala et al. 2000; Ortuño et al. 2006; Sanzani et al. 2009a; Sanzani et al. 2014), catechin, epicatechin, proanthocyanidins, and tannins (Di Venere et al. 1998; Terry et al. 2004; Engels et al. 2009; Parashar et al. 2009; Yoshida et al. 2009). Moreover, some authors have studied the relationship between molecular structure and antimicrobial activity of some phenolic compounds (Lattanzio et al. 1994; Amborabé et al. 2002; Bisogno et al. 2007).

Plants contain different groups of phenolic compounds including simple phenols, phenolic acids (e.g., rosmarinic, carnosic acid), anthocyanins (delphinidin), hydroxybenzoic acids (vanillic acid), hydroxycinnamic acid (ferulic and chlorogenic acids), tannins (procyanidin, tannic acid), lignans (sesaminol), stilbenes (resveratrol), coumarins (α-coumarin), essential oil components (limonene, carvacrol—also called cymophenol, and eugenol), flavonoids (apigenin, quercetin, catechin, rutin) (Charles 2013). The synthesis of mono- and polyphenolic compounds is from carbohydrates by way of shikimic acid, phenylpropanoid, and flavonoid biosynthetic pathways (Lattanzio 2003; Fresco et al. 2006).

Catechol and pyrogallol are hydroxylated phenols with two and three hydroxyl (–OH) groups, respectively; both of them were shown to be toxic to microorganisms. The site(s) and number of hydroxyl groups on the aromatic ring are probably responsible for their relative toxicity to microorganisms, with evidence that increased hydroxylation results in increased toxicity (Cowan 1999). In addition, a positive relationship between the presence of oxidized phenols and the inhibitory activity has been reported (Scalbert 1991).

Phenolic acids are a major class of phenolic compounds, widely occurring in the plant kingdom, and include hydroxybenzoic acids (e.g., gallic acid, *p*-hydroxybenzoic acid, protocatechuic acid, vanillic acid, and syringic acid) and hydroxycinnamic acids (e.g., ferulic acid, caffeic acid, *p*-coumaric acid, chlorogenic acid, and sinapic acid). Natural phenolic acids, either occurring in the free or conjugated forms, usually appear as esters or amides.

Phenolic compounds possessing a C3 side chain at a lower level of oxidation and containing no oxygen are classified as essential oils and often cited as antimicrobial as well. Eugenol is a well-characterized representative considered bacteriostatic against both fungi and bacteria (Cowan 1999).

Several other polyphenols are considered as phenolic acid analogues or derivatives, such as capsaicin, rosmarinic acid, gingerol, gossypol, tyrosol, hydroxytyrosol, ellagic acid, chlorogenic acid, dicaffeoylquinic acids, chicoric acid, and salvianolic acid B (Cai et al. 2004, 2006; Fresco et al. 2006; Han et al. 2007). Red fruits (blueberry, blackberry, strawberry, black currant, red currant, etc.) are rich in hydroxycinnamic acids (caffeic, ferulic, *p* -coumaric acid) and *p*–hydroxybenzoic and ellagic acid, and several vegetables (artichoke, chicory, lettuce, etc.) are rich in caffeic acid derivatives. Rosmarinic acid is found in many dietary spices belonging to Lamiaceae family such as mint, sweet basil, oregano, rosemary, sage, and thyme (Shan et al. 2005), and in some herbaceous species belonging to Boraginaceae family used for food or medicinal purposes, according to unpublished data collected by one of us (D Di Venere) and other authors (Petersen and Simmonds 2003; Mehrabani et al. 2005; Garcia-Herrera 2014).

Flavonoids have been recognized as one of the largest and most widespread groups of plant secondary metabolites, with marked antioxidant and antimicrobial properties. They are found in leaf epidermis and fruit skins in high concentrations and have important functions in plants as secondary metabolites involved in processes such as pigmentation, protection against UV radiation, and disease resistance. They are characterized by a C6–C3–C6 configuration consisting of two aromatic rings (A and B) and can readily participate in hydrogen-donating, radical-scavenging, and metal-chelating mechanisms (Lattanzio 2003).

The major subclasses of flavonoids are the flavonols, flavones, flavanols, chalcones, flavanones, flavanonols, anthocyanins, and isoflavonoids. Flavonols are the most widespread flavonoids and numerous flavonol conjugates exist with more than 200 different sugar conjugates of kaempferol. Anthocyanins, including anthocyanidins (cyanidin, delphinidin, malvidin, peonidin, and pelargonidin) and their glycosides, are widely distributed in fruits, vegetables, and cereals, but they are scarcely represented in herbs. On the contrary, chalcones (butein, phloretin sappanchalcone, carthamin, etc.) are detected in herbs (Charles 2013).

The presence of phenolic compounds in wild edible herbs has been widely reported in the literature (Di Venere et al. 2004; Katalinic et al. 2006; Surveswaran et al. 2007; Wojdyło et al. 2007; Antal 2010). Moreover, many papers showed significant correlations between phenolic content and antioxidant activity of herbs and vegetables (Moreno et al. 2006; Borneo et al. 2009; Chao et al. 2014; Lin et al. 2014). Many authors evaluated the total phenolic content of numerous species and reported results not always in agreement with each other. This was caused by different reasons, such as the use of different analytical methods and ways to express the quantitative results but principally by the different factors (environmental, pedoclimatic, seasonal, etc.) influencing the phenolic content of herbs (Schena et al. 2007). On the other hand, the qualitative phenolic composition of wild edible herbs has been less investigated and little information is available on this topic. Phenolic composition is scarcely influenced by the abovementioned factors; so, High Performance Liquid Chromatography (HPLC) phenolic profile is generally a characteristic of a certain species and could be a useful chemotaxonomic means for species identification (Di Venere et al. 2009).

As for the abovementioned 20 edible species representative of all Mediterranean countries, many authors have reported data regarding their total phenolic content. Simopoulos et al. (2004) found *Foeniculum vulgare* richer in total phenols (83 mg/100 g) than *Sonchus oleraceus, Urospermum picroides*, and *Papaver rhoeas*. The phenolic content of *F. vulgare* and *S. oleraceus* (about 300 mg/100 g) was also evaluated by Vanzani et al. (2011), which also reported a content of about 1050 mg/100 g for *Rumex crispus*. The phenolic content of *F. vulgare*, expressed as dry weight (DW), was also reported by Barros et al. (2009) (66 mg/g DW), Morales et al. (2012) (42 mg GAE/g extract, GAE = gallic acid equivalent), and Hinneburg et al. (2006) (30 mg GAE/g extract).

Di Venere et al. (2004) evaluated by HPLC the total phenolic content of *Sonchus oleraceus* and *Cichorium intybus* (420 and 868 mg/100 g, respectively); the contribution of the sole caffeic acid derivatives to the total phenolic content was estimated as 272 and 588 mg/100 g, respectively. Comparable values for phenolic content of *S. oleraceus* and *C. intybus* (75–157 and 48–107 mg/g DW, respectively) were reported by Schaffer et al. (2005), which also reported a phenolic content of 34–286 mg/g DW for *Papaver rhoeas*.

The phenolic content of *Eruca vesicaria* was reported by Lee et al. (2004) (136 mg/100 g) and Arbos et al. (2010) (109 mg GAE/100 g). Proestos et al. (2005) found 4.5 mg GAE/g DW of total phenolics in *Malva sylvestris*, whereas Wojdyło et al. (2007) 4.8 mg GAE/g DW in *Silybum marianum*; Aberoumand and Deokule (2008) evaluated in 586 mg/100 g DW the total phenolic content of *Portulaca oleracea*.

The phenolic content of *Taraxacum officinale* was evaluated by many authors with different results; Proestos et al. (2005) and Wojdyło et al. (2007) reported a phenolic content of 5.4 mg GAE/g DW and 12.6 mg GAE/100 g DW, respectively. On the other hand, Gatto et al. (2011) evaluated by HPLC the total phenolic content of *T. officinale, Sonchus oleraceus, Borago officinalis*, and other species, finding values of about 2720, 2200, and 580 mg/100 g DW, respectively. For *B. officinalis* they found that about 40 % of total phenolic content is represented by caffeic acid derivatives, principally by one of them, probably rosmarinic acid, as already identified (Wettasinghe et al. 2001); the remaining phenolics are represented by quercetin and kaempferol derivatives in about equal proportion. Moreover, they found that about 90 and 60 % of the total phenolic content for *T. officinale* and *S. oleraceus*, respectively, can be ascribed to chicoric acid, and the remaining 10 % for *T. officinale* and about 20 % for *S. oleraceus* to other caffeic acid derivatives (i.e., chlorogenic and caftaric acids).

Regarding the flavonoids, Gatto et al. (2011) reported the presence of luteolin-7-glucoside in *T. officinale* (40 mg/100 g DW) and of luteolin-7-glucoside and apigenin-7-glucoside in *S. oleraceus* (360 and 170 mg/100 g DW, respectively). Yang et al. (2008) reported the total flavonoid content of *Eruca vesicaria, Nasturtium officinale, Amaranthus retroflexus*, and *T. officinale* (49.1—37.5—9.2—6.2 mg/100 g, respectively). Moreover, they specified the distribution in the different types of flavonoids; in *Eruca vesicaria* they found quercetin and kaempferol derivatives (8.7 and 36.5 mg/100 g, respectively), in *N. officinale* the presence of only kaempferol derivatives (35.1 mg/100 g), whereas in *A. retroflexus* they found only quercetin derivatives (9.2 mg/100 g), and in *T. officinale* only luteolin derivatives (6.2 mg/100 g), this latter result being in accordance with Gatto et al. (2011). Lee et al. (2004) reported, for *E. vesicaria*, a total flavonoid content of 46 mg/100 g, in accordance with Yang et al. (2008). Glycosides of quercetin, kaempferol, and isorhamnetin in *E. vesicaria* were also reported by Di Venere et al. (2000) and Cartea et al. (2011), the latter authors also mentioning the presence of seven quercetin and ten kaempferol glycosides in *N. officinale*. Trichopoulou and Vasilopoulou (2000) reported the presence of good amounts of quercetin, myricetin, isorhamnetin, and kaempferol in *Foeniculum vulgare* (47-20-9 and 7 mg/100 g, respectively). Morales et al. (2012) evaluated the total flavonoid content of *F. vulgare* (10 mg catechin equivalent/g extract), whereas Wojdyło et al. (2007) that of *Silybum marianum* (239 mg/100 g DW). In *Cichorium intybus*, quercetin and kaempferol derivatives (222 and 58 mg/100 g, respectively) were found by Di Venere et al. (2004). In *Cap-*

sella bursa-pastoris, Grosso et al. (2011) found kaempferol-3-*0*-rutinoside, quercetin 3-*0*-glucoside, and quercetin 6-*C*-glucoside (2180–1241 and 564 mg/kg DW).

The presence of phenolic compounds in wild edible species, as highlighted by several studies, may have a close relationship with their biological activity against many harmful microorganisms, as it will be described below.

11.3 Antimicrobial Activity of Phenol-Rich Extracts from Wild Edible Herbs

Natural products extracted from plants, either as pure compounds or as standardized extracts, provide unlimited opportunities to control microbial growth owing to their chemical diversity. Plant extracts have shown a considerable promise in a range of applications in the food industry and several plant extracts enjoy generally recognized as safe (GRAS) status (Negi 2012). Many herbs and spice extracts possess antimicrobial activity against a range of bacteria, yeast, and moulds (Raybaudi-Massilia et al. 2009; Tajkarimi et al. 2010; Gatto et al. 2011; 2013). The mechanism of action for the antimicrobial activity of natural preservatives is not fully understood; however, membrane disruption by terpenoids and phenolics, metal chelation by phenols and flavonoids, and effect on genetic material by coumarin and alkaloids are thought to inhibit the growth of microorganisms (Cowan 1999).

Evidence from literature shows the powerful antimicrobial action exerted by phenolics against several pathogens of human as well as plant interest. However, when total plant extracts were used as antimicrobial agents, their complex composition and the possible contribution of all natural components to the activity have to be considered (Negi 2012).

Phenolic compounds change membrane functioning, influencing protein-to-lipid ratios in the membrane (Keweloh et al. 1990) and inducing efflux of potassium ions (Heipieper et al. 1991). Catechins have been shown to disrupt membrane integrity, as they cause leakage from liposomes (Ikigai et al. 1993). Catechins and epigallocatechin gallate interact in the outer polar zone of lipid bilayers in liposomes and cause membrane disruption (Terao et al. 1994; Hashimoto et al. 1999). Vanillin showed antimicrobial effect by affecting membrane functions (Vaara 1992) and through the inhibition of respiration in several bacteria (Fitzgerald et al. 2004).

The antimicrobial activity against numerous pathogens of selected flavonoids and the antimicrobial mechanisms of action have been reviewed by Cushnie and Lamb (2005). They reported many discrepancies present in the literature on the flavonoid antimicrobial activity as well as many references about structure-activity relationship. As the mechanism of action of the different flavonoids, they reported the inhibition of some important cellular functions (i.e., nucleic acid synthesis, cytoplasmic membrane functionality, and energy metabolism) as the most common cause of their antimicrobial activity (Cushnie and Lamb 2005).

Testing the effectiveness of seven phenolic compounds (esculetin, ferulic acid, quercetin, resveratrol, scopoletin, scoparone, and umbelliferone) in controlling

Penicillium expansum growth and patulin accumulation, two of them (quercetin and umbelliferone) proved to be effective in reducing patulin production, particularly when applied in combination, without consistently affecting mycelial growth (Sanzani et al. 2009a). The same substances were tested also in vivo on Granny Smith and Golden Delicious apples, with quercetin providing a better control of both incidence of decay and disease severity as compared to umbelliferone. Studies on the mode of action of quercetin and umbelliferone demonstrated that a downregulation of genes likely involved in patulin biosynthesis, such as isoepoxydon dehydrogenase (IDH), 6-methylsalicylic acid synthase (msas), and an ATP-binding cassette transporter (peab1) was responsible for the observed activity (Sanzani et al. 2009b). Moreover, by suppression subtractive hybridization (SSH), it was demonstrated that quercetin was also able to induce resistance in apples against blue mould, by acting on the transcription level of various genes involved in several distinct metabolic processes coding pathogenesis-related proteins (RNase-like PR10 and PR8), or proteins expressed under stress conditions (Sanzani et al. 2010).

11.3.1 Antibacterial Activity

Since antiquity, plant extracts and, in particular, essential oils have been used to treat common diseases as infectious pathologies in the respiratory system, urinary tract, and gastrointestinal and biliary systems, as well as on the skin. Scientific literature is full of papers reporting the antimicrobial properties of every kind of plant extracts and essential oils.

One of the first studies on this topic, dealing with the antibacterial activity of aqueous extracts from *Crepis vesicaria* against *Staphylococcus aureus*, dates back to the 1940s (Heatley 1944).

Recently, Borchardt et al. (2008) reported data regarding the antimicrobial activity of extracts obtained from leaves of several species common in the USA against important bacteria, that is, *Staphylococcus aureus, Escherichia coli*, and *Pseudomonas aeruginosa*.

Kokoskova et al. (2011) reported the effectiveness of essential oils from *Melissa officinalis, Mentha arvensis, Nepeta cataria, Origanum vulgare*, and *Thymus vulgaris* against *Erwinia amylovora, Pseudomonas syringae, Pseudomonas fluorescens, Pantoea dispersa*, and *Pantoea agglomerans*.

The efficacy of extracts from *Bidens pilosa, Eclipta prostrata*, and *Plantago major* was tested against some important infectious bacteria (i.e., *Staphylococcus aureus, Bacillus anthracis, Salmonella typhi, E. coli*, and *Streptococcus fecalis*); among them, *Bidens pilosa* extract showed the best antibacterial activity against all the tested microorganisms (Wong-Leung 1988). Adelapo et al. (2011) reported the antibacterial activity of acetone, methanol, and water extracts obtained from leaves of *Bidens pilosa* and *Chenopodium album* against several Gram-positive (*Bacillus cereus, Staphylococcus epidermis, Staphylococcus aureus, Micrococcus kristinae,* and *Streptococus pyrogens*) and Gram-negative (*E. coli, Salmonella pooni, Serratia marcescens, Pseudomonas aeruginosa*, and *Klebsiella pneumoniae*) bacteria.

Proestos et al. (2005) studied the antimicrobial activity of many extracts from different plants (i.e., *Salvia officinalis, Thymus vulgaris, Origanum majorana*, etc.) against *E. coli, Salmonella enteridis, Staphylococcus aureus, Listeria monocitogenes*, and *B. cereus*.

Furthermore, the extract obtained from *Teucrium chamaedrys* was shown as the most active against *Listeria monocitogenes* (Proestos et al. 2005).

The extract of *Thalictrum minus* was found highly active against *Staphylococcus aureus, S. epidermis*, and *Micrococcus luteus* (Lotfipour et al. 2008).

Leaf extracts from some *Amaranthus* spp. (i.e., *A. hybridus, A. spinosus*, and *A. caudatus*) showed antimicrobial activity against *E. coli, Salmonella typhi, Pseudomonas aeruginosa, Proteus mirabilis*, and *Klebsiella pneumoniae* (Maiyo et al. 2010).

Among the 20 previously mentioned Mediterranean species, most of them are reported to have antimicrobial properties. The main findings regarding their antibacterial activity can be summed up as follows:

Beta vulgaris Its extracts were found to be active against *Staphylococcus aureus, B. cereus, Citrobacter freundii*, and *Salmonella typhimurium* (Velićanski et al. 2011), and against *Klebsiella pneumoniae, E. coli, Salmonella typhi, Staphylococcus aureus*, and *Bacillus subtilis* (Hussain et al. 2011).

Borago officinalis The antibacterial activity of the aqueous extracts obtained from leaves was recently assayed against several bacteria (42 strains of *Listeria monocitogenes*, 35 strains of *Staphylococcus aureus*, 38 strains of *Enterobacter* spp., and 18 strains of *Salmonella enterica*) commonly associated with foodborne diseases (Miceli et al. 2014). Aqueous and acetone extracts from flowers were also found active against *L. monocitogenes* and *Bacillus subtilis*, respectively (Aliakbarlu and Tajik 2012).

Capsella bursa-pastoris The activity of its extract against several bacteria and yeast was shown by El-Abyad et al. (1990); they suggested such effectiveness could be caused by some alkaloids (i.e., yohimbine and ergocristine) and some flavonoids (i.e., diosmin) found in the extract. Ethanolic extracts were also found active against *Streptococcus mutans, Streptococcus sanguis, Actinomyces viscosus, Enterococcus faecalis, Staphylococcus aureus, E. coli* (Soleimanpour et al. 2013), and against *Pseudomonas aeruginosa* and *Klebsiella pneumoniae* (Hasan et al. 2013).

Cichorium intybus Its aqueous extracts were found to inhibit *Agrobacterium tumefaciens, Erwinia carotovora, Pseudomonas fluorescens*, and *P. aeruginosa* (Petrovic et al. 2004). Ethanolic extracts (both from basal leaves and roots) were found active also against a methicillin-resistant *Staphylococcus aureus* (MRSA) strain (Quave et al. 2008), whereas methanolic extracts from roots and leaves showed a strong antibacterial activity against *E. coli* and *P. aeruginosa* (Verma et al. 2013). A slight antimicrobial activity of extracts from aerial parts has been also reported by Kokoska et al. (2002). Furthermore, antimalarial activity of lactucin and lactucopicrin, two sesquiterpene lactones isolated from *C. intybus* roots, was reported by Bishoff et al. (2004). Ethyl acetate extracts of this plant were found highly active against *P. aeruginosa*, followed by *Bacillus subtilis, Enterobacter cloacae, S. aureus, Proteus mirabilis*, and *E. coli* (Stefanovic et al. 2012), while Akroum et al. (2009) showed the efficacy of methanolic extracts against *B. subtilis, B. cereus*, and *S. aureus*.

Eruca vesicaria Its essential oil was found very active against *Staphylococcus aureus, Pseudomonas aeruginosa, E. coli* (Khoobchandani et al. 2010; Gulfraz et al. 2011), *B. subtilis* and *Shighella flexneri* (Khoobchandani et al. 2010), and *Salmonella typhi* (Gulfraz et al. 2011). Moreover, methanolic extracts of aerial and root parts showed slight activity against *E. coli, P. aeruginosa*, and *Shigella flexneri* (aerial) and *Staphylococcus aureus* and *B. subtilis* (root) (Khoobchandani et al. 2010).

Foeniculum vulgare The essential oil was proved to have antibacterial activity against *Xanthomonas campestris* and *Pseudomonas syringae* (Lo Cantore et al. 2004) and against *E. coli, Listeria monocitogenes, Salmonella typhimurium*, and *Staphylococcus aureus* (Dadalioğlu and Evrendilek 2004). Moreover, methylene chloride/methanol 1:1 extracts were found active against *E. coli, Proteus vulgaris*, and *Staphylococcus saprophyticus* (Zellagui et al. 2011).

Malva sylvestris Proestos et al. (2005) found a slight antimicrobial activity of its extracts against *E. coli, Listeria monocitogenes*, and *B. cereus*, whereas Dulger and Gonuz (2004) showed a good activity, besides *B. cereus*, against *Staphylococcus aureus, Klebsiella pneumoniae, Pseudomonas aeruginosa*, and *Mycobacterium smegmatis*.

Nasturtium officinale The antibacterial activity of its extracts against *S. aureus, B. subtilis*, and *P. aeruginosa* was reported (Penecilla and Magno 2011), as well as their effectiveness against some foodborne bacteria (*Klebsiella pneumoniae, Shigella* spp., *Listeria monocitogenes*, and *Xanthomonas vesicatoria*) and some plant-born bacteria (*Pseudomonas tomato* and *Erwinia carotovora*) (Iseri et al. 2014).

Papaver rhoeas Different extracts were found active against *Staphylococcus aureus* (Ünsal et al. 2009) and against *S. aureus, E. coli*, and *Pseudomonas aeruginosa* (Kostic et al. 2010).

Portulaca oleracea Elkhayat et al. (2008) reported the antibacterial efficacy of methanol extracts against some Gram-positive (*B. subtilis, Staphylococcus aureus*, and *Streptococcus faecalis*—namely *Enterococcus faecalis*) and Gram-negative (*E. coli, Pseudomonas aeruginosa*, and *Neisseria gonorrhea*) strains.

Raphanus raphanistrum The extracts were found active against *Campylobater jejuni*, one of the most common causes of acute enteritis in humans (Kurekci et al. 2012).

Rumex crispus Its extracts were found to be active against *Staphylococcus aureus* with an efficacy of about 50 % compared to chloramphenicol (Borchardt et al. 2008);

it was found active also against a methicillin-resistant *S. aureus* (MRSA) strain by Quave et al. (2008) and against *S. aureus* and *B. subtilis* (Yildirim et al. 2001).

Silybum marianum The ethanolic extracts of this plant, known to contain silymarin (a mixture of the three flavolignans silybin, silycristin, and silydianin), were found to be active against *B. subtilis, Staphylococcus aureus*, and *Proteus vulgaris* (Mukarram Shah et al. 2011).

Sinapis alba Methanolic extracts from seeds showed a good antibacterial activity against *E. coli* (Ono et al. 1998).

Sonchus spp. Six species of this genus were investigated for their antibacterial activity against several bacteria; among them, *Sonchus oleraceus* showed the highest activity and was found to be effective in inhibiting Gram-positive *Staphylococcus aureus* and Gram-negative *E. coli, Salmonella enterica*, and *Vibrio parahaemoliticus* (Xia et al. 2011). Jimoh et al. (2011) also compared the antibacterial activity of acetone, methanol, and water extracts of *Sonchus oleraceus* and *S. asper* against Gram-positive *B. cereus, Staphylococcus epidermis, S. aureus, Micrococcus kristinae*, and *Streptococus pyrogens* and Gram-negative *E. coli, Serratia marcescens*, and *Pseudomonas aeruginosa*. The water extract obtained from *S. oleraceus* did not show activity against all the tested bacteria; unlike *Sonchus asper* water extract showed to be active against *Staphylococcus aureus, M. kristinae, Streptococcus pyrogens*, and *S. marcescens*. All bacteria showed to be susceptible to different extents to the other extracts, except *Streptococus pyrogens* for *Sonchus oleraceus* and *Staphylococcus epidermis* for *S. asper* (Jimoh et al. 2011).

Taraxacum officinale Extracts obtained from this plant showed their effectiveness against *E. coli* and *Salmonella abony* (Ionescu et al. 2013) and against *Staphylococcus aureus*, a multiresistant *S. aureus* (MRSA) strain, and *B. cereus* (Kenny et al. 2014).

Urtica dioica Its water extract was tested for antimicrobial activity against several bacteria, showing a good effectiveness in inhibiting *Micrococcus luteus, Staphylococcus epidermis, E. coli, Proteus mirabilis, Citrobacter koseri, Staphylococcus aureus, Streptococcus pneumoniae*, and *Enterobacter aerogenes* (Gülçin et al. 2004). A study on the antimicrobial activity of hexane, methanol, and chloroform extracts showed the activity of the different extracts against *B. cereus*, MRSA, and *Vibrio parahaemoliticus* (Modarresi-Chahardehi et al. 2012).

11.3.2 Antifungal Activity

Despite the large number of papers present in literature about the antibacterial activity of plant extracts and essential oils against many dangerous bacteria causing food spoilage as well as serious diseases in humans, not many studies are available on the antifungal properties of such extracts and oils. Some of their main findings are reported below.

The antifungal effects of wild fennel (*Foeniculum vulgare*) essential oil against the mycelial growth of *Alternaria alternata, Fusarium oxysporum*, and *Rhizoctonia solani* were reported by Özcan et al. 2006.

Belardi et al. (1986) reported a strong activity of *Rosmarinus officinalis* and *Mentha arvensis* extracts against *Candida albicans*, while Diaz Dellavalle et al. (2011) reported the efficacy of *R. officinalis, Salvia officinalis*, and *S. sclarea* extracts in inhibiting *Alternaria* spp.

Iqbal et al. (2012) showed the antifungal activity of *Amaranthus viridis* extracts against *Rhizopus oligosporus*.

The fungistatic activity of aqueous extracts of *Malva sylvestris, Mentha x piperita*, and *Chamaemelum nobile* on the growth of *Aspergillus candidus, A. niger, Fusarium culmorum*, and *Penicillium* spp. was assessed by Magro et al. (2006).

Kostic et al. (2010) reported the activity of *Papaver rhoeas* ethanol extract against the yeast *C. albicans*.

Hexane and aqueous extracts of *Portulaca oleracea* showed antifungal activity against *Fusarium* spp., while ethanol and chloroform extracts of the same herb inhibited the growth of *Rhizopus artocarpi* (Banerjee and Mukherjee 2002). The antifungal activity of *P. oleracea* extracts against *Aspergillus niger, A. fumigatus, Candida albicans*, and *Trichophyton mentagrophytes* (Oh et al. 2000) and against *C. albicans* (Elkhayat et al. 2008) was also reported.

Extracts from *Raphanus raphanistrum* roots were found active against *Cladosporium cucumerinus* (Schreiner and Koide 1993).

Ethanol extracts of *Rumex crispus* showed to be active against *Candida albicans* and *Trichophyton menthagrophytes* (Kosikowska et al. 2011).

The effectiveness of postharvest treatments with *Urtica dioica* extracts to control storage decay caused on strawberry by *Botrytis cinerea, Rhyzopus stolonifer*, and *Penicillium* spp. and on sweet cherry by *B. cinerea, R. stolonifer*, and *Monilinia laxa* were recently reported by Romanazzi et al. (2013) and by Feliziani et al. (2013), respectively. Moreover, *U. dioica* water extract was found to be active in inhibiting *Candida albicans* growth (Gülçin et al. 2004).

The in vitro and in vivo activity of extracts obtained from nine wild edible herbaceous species (*Borago officinalis, Orobanche crenata, Plantago coronopus, P. lanceolata, Sanguisorba minor, Silene vulgaris, Sonchus asper, S. oleraceus*, and *Taraxacum officinale*) against some important postharvest fungi (i.e., *Botrytis cinerea, Monilinia laxa, Penicillium digitatum, P. expansum, P. italicum, Aspergillus carbonarius*, and *A. niger*) were investigated by Gatto et al. (2011). In this study, extracts from *Sanguisorba minor* and *Orobanche crenata* showed the highest efficacy, with *S. minor* completely inhibiting conidial germination of *M. laxa, P. digitatum, P. italicum*, and *A. niger* and strongly reducing *B. cinerea*, and *O. crenata* extract showing a lower but significant reduction in conidial germination on all the tested fungi. In trials performed on wounded fruit, *S. minor* extract completely inhibited brown rot on apricots and nectarines; *O. crenata* extract strongly reduced grey mould on table grapes, brown rot on apricots and nectarines, and green mould on oranges (Gatto et al. 2011). Further studies on these two extracts provided the purification of two pure active phenolic compounds (verbascoside and isoverbascoside) from *O. crenata* extracts, showing strong in vitro antifungal activity against the abovementioned postharvest pathogens (Gatto et al. 2013).

The antifungal activity of *Anchusa azurea* and *Beta vulgaris* phenolic extracts against some postharvest fungi were recently studied by Garcia-Herrera (2014); in this study, preliminary results were reported about a good response of both extracts against *Penicillium digitatum* and *P. italicum* in terms of inhibition of conidial growth and germ tube elongation.

11.4 Future Perspectives

There is much evidence that different antimicrobials of plant origin can effectively reduce or inhibit pathogenic and spoilage microorganisms, and thus have the potential to become a good alternative to synthetic antimicrobials. Several underutilized plants (i.e., wild edible plants) have thousands of years of history and their non-toxicity, at least at an oral level, is proven. This safety feature is very important in formulations of such products for commercial purposes because it has an impact on the cost of development and registration of a new pesticide product (Tripathi and Dubey 2004). The knowledge of a plant extract antimicrobial activity represents first preliminary information that should be followed by the identification of active principle(s) by means of a bio-guided assay (Ríos and Recio 2005). The development of cost-effective isolation and purification procedures avoiding loss of functional properties of active compounds will aid in wider use and acceptance of plant extracts as natural preservatives (Negi 2012).

The marker compounds in extracts are affected by plant variety, geographical origin, used plant parts, plant age and growth condition, methods of extraction or drying, preparation, packaging, and storage. The starting material must be accurately identified in order to ensure that the plant materials for food use are consistent with respect to quality and quantity of active ingredient, and the method of preparation must meet good manufacturing practices. Risk assessment of natural products may require adequate specification of identity and composition as it may be the whole plant, extracts thereof, or purified components. However, the variability among plant source and the process used to obtain the constituents will be a limiting factor in adopting a generic approach to their risk assessment (Negi 2012). Generally, for herbs or complex extracts, it is not possible to make a risk assessment on the basis of a single active component as more than one component may be of toxicological significance, and food matrix may affect their bioavailability. A decision tree has been suggested as an aid to make the safety evaluation process for plant material intended for food use (Walker 2004), and the general framework for safety assessment of botanicals has been described (Speijers et al. 2010; van den Berg et al. 2011). It is hoped for the future setting-up of new and more specific protocols to clearly and univocally assess composition, biological activity, and safety of plant extracts used for applications that can potentially affect human health.

References

Aberoumand A, Deokule SS (2008) Comparison of phenolic compounds of some edible plants of Iran and India. Pak J Nutr 7(4):582–585

Adelapo A, Jimoh F, Afolayan A (2011) Comparison of the nutritive value and biological activities of the acetone, methanol, and water extracts of the leaves of *Bidens pilosa* and *Chenopodium album*. Acta Pol Pharm 68(1):83–92

Aedo C, Medina L, Fernández-Albert M (2013) Species richness and endemicity in the Spanish vascular flora. Nord J Bot 31(4):478–488

Akroum S, Satta D, Lalaoui K (2009) Antimicrobial, antioxidant, cytotoxic activities and phytochemical screening of some Algerian plants. Eur J Sci Res 31(2):289–295

Aliakbarlu J, Tajik H (2012) Antioxidant and antibacterial activities of various extracts of *Borago officinalis* flowers. J Food Process Pres 36:539–544

Amborabé BE, Fleurat-Lessard P, Chollet J-F, Roblin G (2002) Antifungal effects of salicylic acid and other benzoic acid derivatives towards *Eutypa lata*: structure–activity relationship. Plant Physiol Biochem 40:1051–1060

Antal DS (2010) Medicinal plants with antioxidant properties from Banat region (Romania): a rich pool for the discovery of multi-target phytochemicals active in free-radical related disorders. Analele Universităţii din Oradea—Fascicula Biologie, Tom. XVII/1, 14–22. http://www.bioresearch.ro/revistaen.html

Arbos KA, de Freitas RJS, Stertz SC, Dornas MF (2010) Antioxidant activity and phenolic content in organic and conventional vegetables. Food Sci. Technol (Campinas) 30(2):501–506

Baidez AG, Gomez P, Del Rio JA, Ortuño A (2007) Dysfunctionality of the xylem in *Olea europaea* L. plants associated with the infection process by *Verticillium dahliae* Kleb. Role of phenolic compounds in plant defense mechanism. J Agric Food Chem 55:3373–3377

Banerjee C, Mukherjee A (2002) Biological activity of a common weed—*Portulaca oleracea* L. II. Antifungal activity. Acta Bot Hung 44(3–4):205–208

Barros L, Heleno SA, Carvalho AM, Ferreira ICFR (2009) Systematic evaluation of the antioxidant potential of different parts of *Foeniculum vulgare* Mill. from Portugal. Food Chem Toxicol 47(10):2458–2464

Belardi M, Marini D, Finesi R (1986) Caratteristiche chimiche ad attività antibatteriche di alcuni oli essenziali. Erboristeria Domani 4:74–77

Bianco VV (1969) Vegetable crops in Mediterranean countries. Lecture delivered to the Course 150, Vegetable as World Food Crops, 13 p, Dept. Veg. Crops Univ. California, Davis, CA (USA)

Bianco VV (1989) Wild plants utilizable as vegetables and condiment herbs in Italy. Proc Int Symp Hort Germplasm cultivated and wild, 1989 Beijing (China), Int. Acad. Publishers, Beijing (China) 2, 55–64

Bianco VV, Machacova M (2002) Specie spontanee della flora italiana utilizzabili come ortaggi e piante da condimento. VI Giornate Scientifiche SOI, 2002 Spoleto, PG (Italy) 2, 435–436

Bianco VV, Lopedota O, Pace B (2007) Acido ascorbico in specie spontanee commestibili. Atti 2° Convegno Nazionale "Valorizzazione delle risorse e sviluppo sostenibile", 2004 Agrigento, 565–573

Bishoff TA, Kelley CJ, Karchesy Y, Laurantos M, Nguyen-Dinh P, Arefi AG (2004) Antimalarial activity of lactucin and lactucopicrin: sesquiterpene lactones isolated from *Cichorium intybus* L. J Ethnopharmacol 95:455–457

Bisogno F, Mascoti L, Sanchez C, Garibotto F, Giannini F, Kurina-Sanz M, Enriz R (2007) Structure-antifungal activity relationship of cinnamic acid derivatives. J Agric Food Chem 55:10635–10640

Borchardt JR, Wyse DL, Sheaffer CC, Kauppi KL, Fulcher RG, Ehlke NJ, Biesboer DD, Bey RF (2008) Antimicrobial activity of native and naturalized plants of Minnesota and Wisconsin. J Med Plants Res 2(5):98–110

Borneo R, Léon AE, Aguirre A, Ribotta P, Cantero JJ (2009) Antioxidant capacity of medicinal plants from the Province of Córdoba (Argentina) and their in vitro testing in a model food system. Food Chem 112:664–670

Bravo L (1998) Polyphenols: chemistry, dietary sources, metabolism and nutritional significance. Nutr Rev 56:317–333

Cai YZ, Luo Q, Sun M, Corke H (2004) Antioxidant activity and phenolic compounds of 112 traditional Chinese medicinal plants associated with anticancer. Life Sci 74:2157–2184

Cai YZ, Sun M, Xing J, Luo Q, Corke H (2006) Structure-radical scavenging activity relationships of phenolic compounds from traditional Chinese medicinal plants. Life Sci 78:2872–2888

Cartea ME, Francisco M, Soengas P, Velasco P (2011) Phenolic compounds in *Brassica* vegetables. Molecules 16:251–280

Chao PY, Lin SY, Lin KH, Liu YF, HsuJI, Yang CM, Lai JY (2014) Antioxidant activity in extracts of 27 indigenous Taiwanese vegetables. Nutrients 6:2115–2130

Charles DJ (2013) Natural antioxidants (Chap. 3, pp 39–64) In: Charles DJ Antioxidant properties of spices, herbs and other sources. Springer, New York, p 610

Conti F, Abbate G, Alesandrini A, Blasi C (2005) An annotated checklist of the Italian vascular flora. Palombi Editori, Rome, p 420

Cowan MM (1999) Plant products as antimicrobial agents. Clin Microbiol Rev 12:64–582

Cushnie TP, Lamb AJ (2005) Antimicrobial activity of flavonoids. Int J Antimicrob Agents 26(5):343–356

Dadalioğlu I, Evrendilek GA (2004) Chemical compositions and antibacterial effects of essential oils of Turkish oregano (*Origanum minutiflorum*), bay laurel (*Laurus nobilis*), Spanish lavender (*Lavandula stoechas* L.), and fennel (*Foeniculum vulgare*) on common foodborne pathogens. J Agric Food Chem 52:8255–8260

Di Venere D, Linsalata V, Ippolito A, Nigro F, Arcuti P, Lattanzio V (1998) Endogenous phenolics, ripening and susceptibility of strawberry fruits *(Fragaria x ananassa* Duch.) to post-harvest diseases. In: Charbonnier F, Delacotte JM, Rolando C (eds) Polyphenols Communications 98. Groupe Polyphenols, Bordeaux, pp 459–460

Di Venere D, Calabrese N, Linsalata V, Cardinali A, Bianco VV (2000) Influence of sowing time on phenolic composition of rocket. Acta Hortic 533:343–349

Di Venere D, Linsalata V, Sergio L, Cardinali A, Pieralice M, Bianco VV (2004) Composizione fenolica ed attività antiossidante di alcune specie spontanee eduli della Macchia Mediterranea. Italus Hortus 11(4):128–131

Di Venere D, Sergio L, Linsalata V, Pieralice M, Cardinali A, Cascarano N, Bianco VV (2009) Antioxidant properties of wild edible species. It J Agron 4(4):635–640

Díaz Dellavalle P, Cabrera A, Alem D, Larrañaga P, Ferreira F, Dalla Rizza M (2011) Antifungal activity of medicinal plant extracts against phytopatogenic fungus *Alternaria* spp. Chilean J Agric Res 71(2):231–239

Dulger B, Gonuz A (2004) Antimicrobial activity of certain plants used in Turkish traditional medicine. Asian J Plant Sci 3(1):104–107

El-Abyad MS, Morsi NM, Zaki DA, Shaaban MT (1990) Preliminary screening of some Egyptian weeds for antimicrobial activity. Microbios 62(250):47–57

Elkhayat ES, Ibrahim SRM, Aziz MA (2008) Portulene, a new diterpene from *Portulaca oleracea* L. J Asian Nat Prod Res 10(11):1039–1043

Engels C, Knödler M, Zhao YY, Carle R, Gänzle MG, Schieber A (2009) Antimicrobial activity of gallotannins isolated from mango (*Mangifera indica* L.) kernels. J Agric Food Chem 57:7712–7718

Feliziani E, Santini M, Landi L, Romanazzi G (2013) Pre- and postharvest treatments with alternative to synthetic fungicides to control postharvest decay of sweet cherry. Postharv Biol Technol 78:133–138

Fitzgerald DJ, Stratford M, Gasson MJ, Ueckert J, Bos A, Narbad A (2004) Mode of antimicrobial action of vanillin against *Escherichia coli, Lactobacillus plantarum* and *Listeria innocua*. J Appl Microbiol 97:104–113

Fresco P, Borges F, Diniz C, Marques MPM (2006) New insights on the anticancer properties of dietary polyphenols. Med Res Rev 26:747–766

Garcia-Herrera P (2014) Plantas silvestres de consumo tradicional en España. caracterización de su valor nutricional y estimación de su actividad antifúngica. Doctoral Thesis, Universidad Complutense de Madrid, Facultad de Farmacia, Madrid 2014, p 220

Gatto MA, Ippolito A, Linsalata V, Cascarano NA, Nigro F, Vanadia S, Di Venere D (2011) Activity of extracts from wild edible herbs against postharvest fungal diseases of fruit and vegetables. Postharv Biol Technol 61(1):72–82

Gatto MA, Sanzani SM, Tardia P, Linsalata V, Pieralice M, Sergio L, Di Venere D (2013) Antifungal activity of total and fractionated phenolic extracts from two wild edible herbs. Nat Sci 5(8):895–902

Grosso C, Vinholes J, Silva LR, Guedes de Pinho P, Gonçalves RF, Valentão P, Jäger AK, Andrade PB (2011) Chemical composition and biological screening of *Capsella bursa-pastoris*. Rev Bras Farmacogn 21(4):635–644

Gülçin I, ÖI, Oktay M, Büyükokuroğlu ME (2004) Antioxidant, antimicrobial, antiulcer and analgesic activities of nettle (*Urtica dioica* L.). J Ethnopharmacol 90:205–215

Gulfraz M, Sadiq A, Tariq H, Imran M, Qureshi R, Zeenat A (2011) Phytochemical analysis and antibacterial activity of *Eruca sativa* seed. Pak J Bot 43(2):1351–1359

Han XZ, Shen T, Lou HX (2007) Dietary polyphenols and their biological significance. Int J Mol Sci 8:950–988

Hasan RN, Ali MR, Shakier SM, Khudhair AM, Hussin MS, Kadum YA, Mohammed AI, Abbas AA (2013) Antibacterial activity of aqueous and alcoholic extracts of *Capsella Bursa* against selected pathogenic bacteria. Am J BioSci 1(1):6–10

Hashimoto T, Kumazawa S, Nanjo F, Hara Y, Nakayama T (1999) Interaction of tea catechins with lipid bilayers investigated with liposome systems. Biosci Biotechnol Biochem 63:2252–2255

Heatley NG (1944) An antibiotic from *Crepis taraxacifolia* (Thuill.). Brit J Exp Path 25:208–211

Heipieper HJ, Keweloh H, Rehm HJ (1991) Influence of phenols on growth and membrane permeability of free and immobilized *Escherichia coli*. Appl Env Microbiol 57:1213–1217

Hinneburg I, Dorman HD, Hiltunen R (2006) Antioxidant activities of extracts from selected culinary herbs and spices. Food Chem 97:122–129

Huang WY, Cai YZ, Zhang Y (2010) Natural phenolic compounds from medicinal herbs and dietary plants: potential use for cancer prevention. Nutr Cancer 62(1):1–20

Hussain Z, Mohammad P, Sadozai SK, Khan KM, Nawaz Y, Perveen S (2011) Extraction of antipneumonia fractions from the leaves of sugar beets *Beta vulgaris*. J Pharm Res 4(12):4783

Ikigai H, Nakae T, Hara Y, Shimamura T (1993) Bactericidal catechins damage the lipid bilayer. Biochem Biophys Acta 1147:132–136

Ionescu D, Predan G, Rizea GD, Mihele D, Dune A, Ivopol G, Ionită C (2013) Antimicrobial activity of some hydroalcoholic extracts of artichoke *(Cynara scolymus)*, burdock *(Arctium lappa)* and dandelion *(Taraxacum officinale)*. Bull Transilvania Univ Braşov, Ser II, 6:113–120

Ippolito A, Nigro F (2003) Natural antimicrobials in postharvest storage of fresh fruits and vegetables. In: Roller S (ed) Natural antimicrobials for the minimal processing of foods. Woodhead, Cambridge, pp 201–234, 306

Iqbal MJ, Hanif S, Mahmood Z, Anwar F, Jamil A (2012) Antioxidant and antimicrobial activities of Chowlai (*Amaranthus viridis* L.) leaf and seed extracts. J Med Plants Res 6(27):4450–4455

Iseri ÖD, Körpe DA, Sahin FI, HM (2014) Screening of *Nasturtium officinale* extracts for biological activities: implications for plant pathogens. J Biol Act Prod Nat 4(1):18–28

Jimoh FO, Adelapo AA, Afolayan AJ (2011) Comparison of the nutritive value, antioxidant and antibacterial activities of *Sonchus asper* and *Sonchus oleraceus*. Rec Nat Prod 5(1):29–42

Katalinic V, Milos M, Kulisic T, Jukic M (2006) Screening of 70 medicinal plant extracts for antioxidant capacity and total phenols. Food Chem 94(4):550–557

Kenny O, Smyth TJ, Walsh D, Kelleher CT, Hewage CM, Brunton NP (2014) Investigating the potential of under-utilised plants from the *Asteraceae* family as a source of natural antimicrobial and antioxidant extracts. Food Chem 161:79–86

Keweloh H, Weyrauch G, Rehm HN (1990) Phenol-induced membrane changes in free and immobilized *Escherichia coli*. Appl Microbiol Biotechnol 33:66–71

Khoobchandani M, Ojeswi BK, Ganesh N, Srivastava MM, Gabbanini S, Matera R, Iori R, Valgimigli L (2010) Antimicrobial properties and analytical profile of traditional *Eruca sativa* seed oil: comparison with various aerial and root plant extracts. Food Chem 120:217–224

Kokoska L, Polesny Z, Rada V, Nepovim A, Vanek T (2002) Screening of some Siberian medicinal plants for antimicrobial activity. J Ethnopharmacol 82:51–53

Kokoskova B, Pouvova D, Pavela R (2011) Effectiveness of plant essential oils against *Erwinia amylovora, Pseudomonas syringae* pv. *syringae* and associated saprophytic bacteria on/in host plants. J Plant Pathol 93(1):133–139

Korukluoglu M, Sahan Y, Yigit A (2008) Antifungal properties of olive leaf extracts and their phenolic compounds. J Food Saf 28:76–87

Kosikowska U, Wegiera M, Smolarz HD, Malm A (2011) Screening of antifungal activity of *Rumex* L. species. Annales Universitatis Mariae Curie-Slodowska Lublin-Polonia Vol XXIV, 4(10):99–104

Kostic DA, Mitic SS, Mitic MN, Zarubica A R, Velickovic JM, Dordevic AS Randelovic SS (2010) Phenolic contents, antioxidant and antimicrobial activity of *Papaver rhoeas* L. extracts from Southeast Serbia. J Med Plants Res 4(17):1727–1732

Kurekci C, Bishop-Hurley SL, Vercoe PE, Durmic Z, Al Jassim RAM, McSweeney CS (2012) Screening of Australian plants for antimicrobial activity against *Campylobacter jejuni*. Phytother Res 26:186–190

Lattanzio V (2003) Bioactive polyphenols: their role in quality and storability of fruit and vegetables. J Appl Bot 77:128–146

Lattanzio V, De Cicco V, Di Venere D, Lima G, Salerno M (1994) Antifungal activity of phenolics against fungi commonly encountered during storage. Ital J Food Sci 6:23–30

Lattanzio V, Di Venere D, Linsalata V, Lima G, Ippolito A, Salerno M (1996) Antifungal activity of 2,5-dimethoxybenzoic acid on postharvest pathogens of strawberry fruits. Postharv Biol Technol 3:325–334

Lee J, Koo N, Min DB (2004) Reactive oxygen species aging and antioxidative nutraceuticals. Compr Rev Food Sci Food Saf 3:1–33

Lin KH, Yeh H, Lin SY, Yang CM, Tsai HJ, Tsai JJ, Chao PY (2014) Antioxidant activities of methanol extracts from selected Taiwanese herbaceous plants. J Food Nutr Res 2(8):435–442

Lo Cantore P, Iacobellis NS, De Marco A, Capasso F, Senatore F (2004) Antibacterial activity of *Coriandrum sativum* L. and *Foeniculum vulgare* Miller var. *vulgare* (Miller) essential oils. J Agric Food Chem 52:7862–7866

Lotfipour F, Nazemiyeh H, Fathi-Azad F, Garaei N, Arami S, Talat S, Sadegpour F, Hasanpour R (2008) Evaluation of antibacterial activities of some medicinal plants from North-West Iran. Iranian J Basic Med Sci 11(2):80–85

Magro A, Carolino M, Bastos M, Mexia A (2006) Efficacy of plant extracts against stored products fungi. Rev Iberoam Micol 23(3):176–178

Maiyo ZC, Ngure RM, Matasyoh JC, Chepkorir R (2010) Phytochemical constituents and antimicrobial activity of leaf extracts of three *Amaranthus* plant species. Afr J Biotechnol 9(21):3178–3182

Mehrabani M, Shams-Ardakani M, Ghannadi A, Dehkordi NG, Jazi SES (2005) Production of rosmarinic acid in *Echium amoenum* Fisch. and C.A. Mey. cell cultures. Iranian J Pharm Res 2:111–115

Miccadei S, Di Venere D, Cardinali A, Romano F, Durazzo A, Foddai MS, Fraioli R, Mobarhan S, Maiani G (2008) Antioxidative and apoptotic properties of polyphenolic extracts from edible part of artichoke (*Cynara scolymus* L.) on cultured rat hepatocytes and on human hepatoma cells. Nutr Cancer 60(2):276–283

Miceli A, Aleo A, Corona O, Sardina MT, Mammina C, Settanni L (2014) Antibacterial activity of *Borago officinalis* and *Brassica juncea* aqueous extracts evaluated *in vitro* and *in situ* using different food model systems. Food Control 40:157–164

Mileo AM, Di Venere D, Linsalata V, Fraioli R, Miccadei S (2012) Artichoke polyphenols induce apoptosis and decrease the invasive potential of the human breast cancer line MDA-MB231. J Cell Physiol 227:3301–3309

Modarresi-Chahardehi A, Ibrahim D, Fariza-Sulaiman S, Mousavi L (2012) Screening antimicrobial activity of various extracts of *Urtica dioica*. Rev Biol Trop 60(4):1567–1576

Morales P, Carvalho AM, Sanchez-Mata MC, Camara M, Molina M, Ferreira ICFR (2012) Tocopherol composition and antioxidant activity of Spanish wild vegetables. Gen Res Crop Evol 59:851–863

Morales P, Ferreira ICFR, Carvalho AM, Sánchez-Mata MC, Cámara M, Fernández-Ruiz V, Pardo-de-Santayana M, Tardío J (2014) Mediterranean non-cultivated vegetables as dietary sources of compounds with antioxidant and biological activity. LWT-Food Sci Technol 55:389–396

Moreno S, Scheyer T, Romano CS, Vojnov AA (2006) Antioxidant and antimicrobial activities of rosemary extracts linked to their polyphenol composition. Free Rad Res 40(2):223–231

Mukarram Shah SM, Khan FA, Hassan Shah SM, Chishti KA, Saifur Shah Pirzada SM, Asif Khan M, Farid A (2011) Evaluation of phytochemicals and antimicrobial activity of white and blue capitulum and whole plant of *Silybum marianum*. World Appl Sci J 12(8):139–1144

Negi PS (2012) Plant extracts for the control of bacterial growth: efficacy, stability and safety issues for food application. Int J Food Microbiol 156:7–17

Oh KB, Chang IM, Hwang KJ, Mar W (2000) Detection of antifungal activity in *Portulaca oleracea* by a single-cell bioassay system. Phytother Res 14:329–332

Ojala T, Remes S, Haansuu P, Vuorela H, Hiltunen R, Haahtela K, Vuorela P (2000) Antimicrobial activity of some coumarin containing herbal plants growing in Finland. J Ethnopharmacol 73:299–305

Ono H, Tesaki S, Tanabe S, Watanabe M (1998) 6-Methylsulfinylhexil isothiocyanate and its homologues as food-originated compounds with antibacterial activity against *Escherichia coli* and *Staphylococcus aureus*. Biosci Biotechnol Biochem 62(2):363–365

Ortuño A, Báidez A, Gómez P, Arcas MC, Porras I, García-Lidón A, Del Río JA (2006) *Citrus paradisi* and *Citrus sinensis* flavonoids: their influence in the defence mechanism against *Penicillium digitatum*. Food Chem 98:351–358

Özcan MM, Chalchat JC, Arslan D, Ateş A, Ünver A (2006) Comparative essential oil composition and antifungal effect of bitter fennel (*Foeniculum vulgare* ssp. *piperitum*) fruit oils obtained during different vegetation. J Med Food 9(4):552–561

Parashar A, Gupta C, Gupta SK, Kumar A (2009) Antimicrobial ellagitannin from pomegranate (*Punica granatum*) fruits. Int J Fruit Sci 9:226–231

Penecilla GL, Magno CP (2011) Antibacterial activity of extracts of twelve common medicinal plants from the Philippines. J Med Plants Res 5(16):3975–3981

Petersen M, Simmonds MSJ (2003) Rosmarinic acid. Phytochemistry 62:121–125

Petrovic J, Stanojcovic A, Comic Lj, Curcic S (2004) Antibacterial activity of *Cichorium intybus*. Fitoterapia 75:737–739

Proestos C, Chorianopoulos N, Nychas GJE, Komaitis M (2005) RP-HPLC analysis of the phenolic compounds of plant extracts. Investigation of their antioxidant capacity and antimicrobial activity. J Agric Food Chem 53:1190–1195

Quave CL, Plano LRW, Pantuso T, Bennett BC (2008) Effects of extracts from Italian medicinal plants on planktonic growth, biofilm formation and adherence of methicillin-resistant *Staphylococcus aureus*. J Ethnopharmacol 118:418–428

Raybaudi-Massilia RM, Mosqueda-Melgar J, Soliva-Fortuny R, Martin-Belloso O (2009) Control of pathogenic and spoilage microorganisms in fresh cut fruits and fruit juices by traditional and alternative natural antimicrobials. Compr Rev Food Sci Food Saf 8:157–180

Ríos JL, Recio MC (2005) Medicinal plants and antimicrobial activity. J Ethnopharmacol 100:80–84

Romanazzi G, Feliziani E, Santini M, Landi L (2013) Effectiveness of postharvest treatment with chitosan and other resistance inducers in the control of storage decay of strawberry. Postharv Biol Technol 75:24–27

Sanzani SM, De Girolamo A, Schena L, Solfrizzo M, Ippolito A, Visconti A (2009a) Control of *Penicillium expansum* and patulin accumulation on apples by quercetin and umbelliferone. Eur Food Res Technol 228:81–389

Sanzani S M, Schena L, Nigro F, De Girolamo A, Ippolito A (2009b) Effect of quercetin and umbelliferone on the transcript level of genes involved in patulin biosynthesis. Eur J Plant Pathol 125:223–233

Sanzani SM, Schena L, De Girolamo A, Ippolito A, González-Candelas L (2010) Characterization of genes associated to induced resistance against Penicillium expansum in apple fruit treated with quercetin. Postharv Biol Technol 56:1–11
Sanzani SM, Schena L, Ippolito A (2014) Effectiveness of phenolic compounds against citrus green mould. Molecules 19:12500–12508
Scalbert A (1991) Antimicrobial properties of tannins. Phytochemistry 30:3875–3883
Schaffer S, Schmitt-Schillig S, Müller WE, Eckert GP (2005) Antioxidant properties of Mediterranean food plant extracts: geographical differences. J Phys Pharmacol 56(1):115–124
Schena L, Nigro F, Ippolito A (2007) Natural antimicrobials to improve storage and shelf-life of fresh fruit, vegetables and cut flowers. In: Ray RC, Ward OP (eds) Microbial biotechnology in horticulture, vol 2. Science Publisher, Enfield, pp 259–303
Schreiner RP, Koide RT (1993) Antifungal compounds from the roots of mycotrophic non-mycotrophic plant species. New Phytol 123:99–105
Shan B, Cai YZ, Sun M, Corke H (2005) Antioxidant capacity of 26 spice extracts and characterization of their phenolic constituents. J Agric Food Chem 53:7749–7759
Simopoulos AP (2004) Omega-3 fatty acids and antioxidants in edible wild plants. Biol Res 37:263–277
Soleimanpour S, Sedighinia FS, Afshar AS, Zarif R, Asili J, Ghazvini K (2013) Synergistic antibacterial activity od *Capsella bursa-pastoris* and *Glycirrhiza glabr*a against oral pathogens. Jundishapur J Microbiol 6(8):e7262
Speijers G, Bottex B, Dusemund B, Lugasi A, Toth J, Amberg-Muller J, Galli CL, Silano V, Rietjens IMCM (2010) Safety assessment of botanicals and botanical preparations used as ingredients in food supplements: testing an European Food Safety Authority-tiered Approach. Molec Nutr Food Res 54:175–185
Stefanović O, Radojević I, Vasić S, Čomić L (2012) Antibacterial activity of naturally occurring compounds from selected plants. In: Bobbarala V (ed) Biochemistry, Genetics and Molecular Biology—"Antimicrobial Agents" (Chap. 1). Rijeka, InTech Publ—www.intechopen.com, pp 1–24
Surveswaran S, Cai Y-Z, Corke H, Sun M (2007) Systematic evaluation of natural phenolic antioxidants from 133 Indian medicinal plants. Food Chem 102(3):938–953
Tajkarimi MM, Ibrahim SA, Cliver DO (2010) Antimicrobial herb and spice compounds in food. Food Control 21:1199–1218
Terao J, Piskula M, Yao Q (1994) Protective effect of epicatechin, epicatechin gallate, and quercetin on lipid peroxidation in phospholipid bilayers. Arch Biochem Biophys 308:78–284
Terry LA, Joyce DC, Adikaram NKB, Khabay BPS (2004) Preformed antifungal compounds in strawberry fruit and flower tissues. Postharv Biol Technol 31:201–212
Trichopoulou A, Vasilopoulou E (2000) Mediterranean diet and longevity. Br J Nutr 84(2):S205–S209
Tripathi P, Dubey NK (2004) Exploitation of natural products as an alternative strategy to control postharvest fungal rotting of fruit and vegetables. Postharv Biol Technol 32:235–245
Ünsal Ç, Özbek B, Saniar G, Mat A (2009) Antimicrobial activity of four annual *Papaver* species growing in Turkey. Pharm Biol 47(1):4–6
USDA (2011) National Nutrient Database for Standard Reference, Release 27, Vegetables and Vegetable Products
Vaara M (1992) Agents that increase the permeability of the outer membrane. Microbiol Rev 56:395–411
van den Berg SJPL, Serra-Majem L, Coppens P, Rietjens IMCM (2011) Safety assessment of plant food supplements (PFS). Food Funct 2:760–768
Vanzani P, Rossetto M, De Marco V, Sacchetti LE, Paoletti MG, Rigo A (2011) Wild Mediterranean plants as traditional food: a valuable source of antioxidants. J Food Sci 76(1):C46–C51
Vardavas CI, Majchrizak D, Wagner KH, Elmadfa I, Kafatos A (2006) The antioxidant and phylloquinone content of wildly grown greens in Crete. Food Chem 99:813–821

Veeriah S, Kautenburger T, Habermann N, Sauer J, Dietrich H (2006) Apple flavonoids inhibit growth of HT29 human colon cancer cells and modulate expression of genes involved in the biotransformation of xenobiotics. Mol Carcinog 45:164–174

Velićanski AS, Cvetković DD, Markov SL, Vulić J, Djilas SM (2011) Antibacterial activity of *Beta vulgaris* L. pomace extract. APTEFF 42:263–269

Veloz-García R, Marín-Martínez R, Veloz-Rodríguez R, Rodríguez-Guerra R, Torres-Pacheco I, González-Chavira MM, Anaya-López JL, Guevara-Olvera L, Feregrino-Pérez AA, Loarca-Pĩna G, Guevara-González RG (2010) Antimicrobial activities of cascalote (*Caesalpinia cacalaco*) phenolics-containing extract against fungus *Colletotrichum lindemuthianum*. Ind Crops Prod 31:134–138

Verma R, Rawat A, Ganie SA, Agnihotri RK, Sharma R, Mahajan S, Gupta A (2013) *In vitro* antibacterial activity of *Cichorium intybus* against some pathogenic bacteria. Brit J Pharm Res 3(4):767–775

Walker R (2004) Criteria for risk assessment of botanical food supplements. Toxicol Lett 149: 187–195

Wettasinghe M, Shahidi F, Amarowicz R, Abou-Zaid MM (2001) Phenolic acids in defatted seeds of borage (*Borago officinalis* L.). Food Chem 75:49–56

Widmer TL, Laurent N (2006) Plant extracts containing caffeic acid and rosmarinic acid inhibit zoospore germination of *Phytophthora* spp. pathogenic to *Theobroma cacao*. Eur J Plant Pathol 115:377–388

Wojdyło A, Oszmianski J, Czemerys R (2007) Antioxidant activity and phenolic compounds in 32 selected herbs. Food Chem 105(3):940–949

Wong-Leung YL (1988) Antibacterial activity of some Hong-Kong plants used in Chinese medicine. Fitoterapia 59:11–16

Xia DZ, Yu XF, Zhu ZY, Zou ZD (2011) Antioxidant and antibacterial activity of six edible wild plants (*Sonchus* spp.) in China. Nat Prod Res 25(20):1893–1901

Yang RY, Lin S, Kuo G (2008) Content and distribution of flavonoids among 91 edible plants species. Asia Pac J Clin Nutr 17:275–279

Yildirim A, Mavi A, Kara AA (2001) Determination of antioxidant and antimicrobial activities of *Rumex crispus* L. extracts. J Agric Food Chem 49(8):4083–4089

Yoshida T, Hatano T, Ito H, Okuda T (2009) Structural diversity and antimicrobial activities of ellagitannins. In: Quideau S (ed) Chemistry and biology of ellagitannins: an underestimated class of bioactive polyphenols. World Scientific Publishing Co. Pte Ltd, Hackensack, pp 55–93

Zellagui A, Gherraf N, Elkhateeb A, Hegazy MEF, Mohamed TA, Touil A, Shaat AA, Rhouati S (2011) Chemical constituent from Algerian *Foeniculim vulgare* aerial parts and evaluation of antimicrobial activity. J Chil Chem Soc 56(3):759–763

Zhu X, Zhang H, Lo R (2004) Phenolic compounds from the leaf extract of artichoke (*Cynara scolymus* L.) and their antimicrobial activities. J Agric Food Chem 52:7272–7278

Chapter 12
Recent Advances in Research on Wild Food Plants and Their Biological–Pharmacological Activity

Michael Heinrich, Sarah Kerrouche and Kawaldeep Singh Bharij

12.1 Introduction

It is again becoming a commonplace that food and medicine (at least if they are derived from plant and fungal sources) are linked intrinsically. 'Let food be thy medicine and medicine be thy food', attributed to Hippocrates, 431 BC, is the first in a long row of statements which highlight these links. To cite just another one, in 1991, Etkin and Ross stated, 'Consideration of the dietary contexts of local "medicines" is central to this wider perspective.' However, in this discussion, one easily blurs some important lines dividing—certainly in a post-Renaissance scientific context—the prevention and treatment of diseases. Food is generally consumed to nourish and it may (or may not) have additional health benefits, which help prevent or limit (mostly chronic) diseases. Medicines are used specifically to treat and overcome certain health problems, and are used with a very specific goal in mind: to get rid of a health problem (Heinrich and Prieto-Garcia 2008; Heinrich et al. 2012).

In the context of this chapter, the focus is on foods with a certain health beneficial function, that is, not as foods that serve simply the sustenance or an exclusive source of pleasure and enjoyment (like sweets). What constitutes a functional food can be defined from a wide range of perspectives, for example, regulatory, a social science or a food science perspective (cf. Heinrich and Prieto-Garcia 2008), and any definition remains controversial.

In the European Union (EU), a working definition of functional food has been proposed (European Commission Concerted Action on Functional Food Science in Europe (FUFOSE)): '…a food that beneficially affects one or more target functions

M. Heinrich (✉) · S. Kerrouche · K. S. Bharij
Centre for Pharmacognosy and Phytotherapy/Research Cluster 'Biodiversity and Medicines', UCL School of Pharmacy, University of London, London, UK
e-mail: m.heinrich@ucl.ac.uk; k.sarah19@hotmail.co.uk; ksb.mpharm@gmail.com

M. de C. Sánchez-Mata, J. Tardío (eds.), *Mediterranean Wild Edible Plants*,
DOI 10.1007/978-1-4939-3329-7_12

in the body beyond adequate nutritional effects in a way that is relevant to either an improved state of health and well-being and/or reduction of risk of disease. It is consumed as part of a normal food pattern. It is not a pill, a capsule or any form of dietary supplement' (EU 2010).

Consequently, in this discussion about recent advances in research on wild food plants and their biological–pharmacological activity, the focus is on health benefits for certain target conditions, for example, in preventing cardiovascular diseases, metabolic syndrome/obesity/diabetes, or maintaining brain health, and not on generic claims about 'healthy' food. The latter would in essence require an epidemiological approach and not a plant-species-specific discussion.

Key to this discussion are local and traditional food plants, generally gathered, but in a few cases also grown locally and used as part of a regular diet. The use may be for food purposes, but depending on the context, a certain species may also have a specific use to treat certain conditions.

Local food plants are species that are generally derived from local or regional traditions. Some are used widely, for example, throughout the Mediterranean, but their usage is generally based not on a commercial exchange but on the local production and processing of these resources. For example, Leonti et al. (2006) identified a core group of 18 culinary used wild gathered plant species. Their use is embedded in a complex historical tradition and highlights that there is both a commonality in the traditions and there are species which are used widely throughout the Mediterranean, but also a continuous exchange of ideas. As such, 'local' should not be considered as implying that these species must be restricted to a specific area or that they are part of one tradition. As in many other aspects of cultures, these usages have been formed as part of historical development and the continuous exchange of knowledge, plants and people in Europe and adjacent regions.

Estimates about the total number of higher plant species vary, but 250,000 is a reasonable approximation (Greuter 1991). If we assume that only 5–10% of these may have some sort of a use as a food or in drinks, we would arrive at about 12,500–25,000 species used locally for nutritional purposes. As stated by Tardío and Pardo-de-Santayana in this book (see Chap. 4), only in mainland Spain and the Balearic Islands, there are a bit more than 500 'wild' species used for food purposes.

Covering all commonly used species in a single and relatively short overview would be impossible. Therefore, a selective, monograph-based approach was used in order to assess the current state of the art on local food plants consumed mostly in the western Mediterranean.

Local food uses are broad and generally vary, but both in terms of the quantities used and their cultural importance, some core groups stand out:

- As a fruit
- As a vegetable (including salads)
- As a part of beverages
- As an occasional snack

Using nine examples of commonly used plants, we discuss the existing pharmacological evidence and how this is linked to the species' composition in order to assess the evidence base for these species. Since most of these local foods are normally consumed in relatively small quantities, this review does not focus on the species' food properties and thus on their nutritional value (including their contents in minerals and vitamins); rather, it focuses on secondary and primary metabolites directly relevant for health.

12.2 Biological–Pharmacological Evidence on Some Common Local Food Plants

12.2.1 Species Used Because of Their Fruit

Among the species used throughout the Iberian Peninsula, three have attained a particular relevance: the strawberry tree, the common beech and the holly oak. All are managed species, actively encouraged and occasionally planted in order to secure a good supply of the fruit.

The strawberry tree (*Arbutus unedo* L.), Ericaceae, is one of the iconic plants of the Mediterranean, well known as part of the coat of arms of the city of Madrid (El oso y el madroño), and with a long track record of local and traditional uses. It is an important local food plant (Rivera et al. 2006a, b) and the fruits are edible, eaten either raw or prepared as jams. In some regions, a liqueur is prepared using the mashed fruits in alcohol. Medicinal uses include, for example, as a urinary tract antiseptic, for cystitis, as an astringent, for treating a range of gastrointestinal (GI) disorders including diarrhoea and dysentery, for cardiovascular problems and diabetes, and as an anti-inflammatory agent (Chevallier 2001; van Wyk and Wink 2004; Kivcak et al. 2001; Morales et al. 2013; Oliveira et al. 2009; Legssyer et al. 2004).

A wide range of studies attest to some relevant biological and pharmacological activities (Box 1), and, interestingly, most research has been done on extracts from the leaves and, to a lesser degree, on the fruits that are used as a food. Many studies assess antioxidant effects, a range of activities which may have some health-protective and disease-preventing effects, but which generally cannot be linked to treating specific diseases. Key to the species' activity is arbutin, a hydroquinone often found in the Ericaceae and known to have a wide range of activities but best known for antibacterial effects (especially in the urinary tract). An extract showed cardioprotective effects by interfering with platelet aggregation, hence decreasing the risk of blocked blood vessels, and indirectly lowering blood pressure (Mekhfi et al. 2006).

Another particularly relevant effect is the species' activity against glycaemia in diabetic rats at a dose of 0.4 g/L (0.4 mg/mL), where *Arbutus unedo* has shown far superior antihyperglycaemic activity by decreasing plasma glucose levels by 31.6 %, compared to 13 % in the case of *Urtica dioica* (Bnouham et al. 2010).

Box 1: Recorded Pharmacological Activities of *Arbutus unedo* L.

Known Bioactive Compounds and Other Metabolites:

Arbutin, methylarbutin, hydroquinones, α-Tocopherol, α-Terpineol, tannins, malic acid, ascorbic acid, oxalic acid, kaempferol, quercetin, catechin gallate (Chevallier 2001; Wyk and Wink 2004; Kivcak et al. 2001; Morales et al. 2013; Oliveira et al. 2009; Legssyer et al. 2004)

Relevant Pharmacological and Microbiological Studies

Antihyperglycaemic Activity

The aqueous extract of the roots showed in vivo antihyperglycaemic activity in a rat model. The administration of 500 mg/kg of the extract 30 min prior to glucose loading decreased glycaemia for the oral glucose tolerance test. In the intravenous glucose tolerance test, 500 mg/kg of the extract showed a marked decrease in the jejuna absorption of glucose by 31.6 % (Bnouham et al. 2007). In chronic treatment of neonatal streptozotocin-induced diabetic rats, an aqueous extract lowered plasma glucose levels by 31.6 % at a concentration of 0.4 g/L (drinking water). At 1 mg/mL, the extract also increased glucose consumption in combination with insulin (Bnouham et al. 2010).

Cardioprotective Activity

Acetone/H_2O/Diethyl ether/EtOAc extracts: These leaf extracts showed in vitro anti-aggregant effects on human platelets by antagonizing thrombin, hence potential cardiovascular benefits. This was achieved by decreasing the production of reactive oxygen species, preventing the phosphorylation of tyrosine, and decreasing the mobilization of Ca^{2+} ions (Haouari et al. 2007).

Another study showed effects of aqueous extracts of leaves and roots in vivo in a rat model which resulted in a reduction in increased systolic blood pressure development, and the root extract also reduced ventricular hypertrophy (Afkir et al. 2008; Ziyyat and Boussairi 1998).

A MeOH extract (leaves) showed in vitro vasorelaxant activity at 87 ± 4 % in rat aorta. This is primarily due to the tannin content; as without the tannins, there was only a 42 ± 8 % vasorelaxant effect (Legssyer et al. 2004). Tannins from a MeOH extract of the leaves also showed a strong anti-aggregant effect on human platelets with an IC_{50} of 0.7 ± 0.08 g/L and a 75.3 ± 1.4 % inhibition (Mekhfi et al. 2006).

Spasmolytic Activity

An EtOH (70 %) extract of the leaves showed in vitro spasmolytic activity by decreasing the basal tone of a rat ileum. This was achieved through reducing the contractile response by antagonizing the action of acetylcholine and the inhibition of calcium channels (Pavlovic et al. 2011).

Antimicrobial Activity

An aqueous extract of the leaves showed inhibitory activity against *Staphylococcus aureus* and *Aspergillus parasiticus* (Orak et al. 2011) and an EtOH extract of the leaves showed antibacterial action by growth inhibition against *Mycobacterium aurum, M. smegmatis* and *M. bovis* (Ouarti et al. 2012).

Cytotoxic Activity

An aqueous extract of the leaves showed in vitro inhibitory action on interferon-elicited STAT1 activation in human fibroblasts and MDA-MB-231 human breast cancer cells, via downregulation of the STAT1 tyrosine protein phosphorylation. SHP2 tyrosine phosphatase phosphorylation may also contribute to this inhibitory effect. This shows a strong anti-inflammatory action of *Arbutus unedo* and its use in inflammatory diseases (Mariotto et al. 2008).

Antioxidant Activity

An aqueous extract of the leaves showed a 2,2-diphenyl-1-picrylhydrazyl (DPPH•) radical scavenging activity at an EC_{50} of 0.4872 mg/mL, β-carotene bleaching inhibition at an EC_{50} of 0.67 mg/mL, iron chelation capacity at 14.46 ± 0.17 % and an H_2O_2 scavenging activity at 26.02 ± 1.91 % (Orak et al. 2011). An EtOH extract showed a DPPH• radical scavenging activity at an EC_{50} of 0.6554 mg/mL and β-carotene bleaching inhibition at an of 0.6660 mg/mL, iron chelation capacity at 6.06 ± 0.13 % and an H_2O_2 scavenging activity at 21.23 ± 0.20 % (Orak et al. 2011); in the Trolox equivalent antioxidant capacity (TEAC) assay (2,2'-azino-bis(3-ethylbenzothiazoline-6-sulphonic acid), ABTS radical scavenging activity), the extract showed an antioxidant capacity of 2.25 ± 0.047 mM. A leaf extract had a reducing power at an EC_{50} of 0.318 ± 0.007 mg/mL, antihaemolytic activity at an IC_{50} of 0.062 ± 0.002 mg/mL and a lipid peroxidation inhibition at an IC_{50} of 0.075 ± 0.014 mg/mL. The fruit had a reducing power at an EC_{50} of 2.85 ± 0.049 mg/mL, antihaemolytic activity at an IC_{50} of 0.430 ± 0.091 mg/mL and a lipid peroxidation inhibition at an IC_{50} of 0.732 ± 0.452 mg/mL (Mendes et al. 2011).

An extract of the leaves showed lipid peroxidation inhibition (74.31 %) at 125 μg/mL, a radical scavenging activity at an IC_{50} of 7.17 ± 0.46 μg/mL and a fluorescence recovery after photobleaching (FRAP) value of 5.11 ± 0.09 μmol Fe^{2+}/g (Pavlovic et al. 2009). The methanolic extract of the leaves showed a DPPH• radical scavenging activity at an EC_{50} concentration of 0.42 mg/mL, β-carotene bleaching inhibition at an EC_{50} concentration of 0.7173 mg/mL, iron chelation capacity at 14.05 ± 1.18 % and H_2O_2 scavenging activity at 28.22 ± 1.18 % (Orak et al. 2011); in the TEAC assay (ABTS radical scavenging activity), the extract showed an antioxidant capacity of 1.71 ± 0.013 mM (Pabuccuoglu et al. 2003).

In a large-scale screening for health beneficial properties, an extract of the fruit only showed moderate activity (Heinrich et al. 2005).

Another example shows the other extreme in terms of our knowledge about commonly used food species important in Western Europe—*Fagus sylvatica* L., Fagaceae, is more widely distributed in Temperate Oceanic and sub-Mediterranean regions and there it is commonly used as a food (e.g. Tardío et al. 2005). Medical uses include as an antiseptic, for chronic bronchitis and skin diseases, for example, psoriasis (Weiss 2001; Grieve 1995; van Wyk and Wink 2004). The raw or toasted fruits are used as a snack and after grinding used in a mixture, for example, as flour. The seeds are rich in polysaccharides, which, however, seem to be little studied. Interestingly, no pharmacological studies have been conducted, which could shed light on potential benefits of its use.

The third example is one of the Mediterranean species of oak: *Quercus ilex* L., holly oak, most importantly *Quercus ilex* L. subsp. *ballota* (Desf.) Samp., Fagaceae. The acorns (*bellotas* in Spanish) again are a resource with a broad range of uses: crude and peeled they are consumed as a (often somewhat bitter) snack; toasted fruits are used in a similar way; in combination with honey and other sweet materials, they make excellent sweets (Rivera et al. 2006a, b); and the ground toasted fruits are used in a way similar to coffee. They can also be prepared as part of soups and stews (Rivera et al. 2005). Some of the medical uses are linked to its astringent properties (in essence found in all plant parts) and include as an antimicrobial and for diarrhoea. It is also known as a galactagogue.

Box 2: Recorded Pharmacological Activities of *Quercus ilex* L.

Known Bioactive Compounds and Other Metabolites

α-Tocopherol, tannins, polyphenol triloside (*trans-, cis-*), kaempferol glucosides, quercilicoside A, astragalin (Rivera et al. 2006b; Chevolleau et al. 1993; Loggia et al. 1989)

Relevant Pharmacological and Microbiological Studies

Anti-Inflammatory Activity

Selected constituents were tested in vivo in a mouse model for their anti-inflammatory activity (IC_{50} values: kaempferol 0.266 μM, astragalin 1.123 μM, tiliroside 0.154 μM and ADC 0.036 μM, compared to indomethacin and hydrocortisone at 0.122 and 0.009 μM, respectively; Loggia et al. 1989).

Antimicrobial Activity

A MeOH extract of the leaves showed in vitro antimicrobial activity against 15 bacteria and 1 yeast but none stood out as being particularly strong (Gulluce et al. 2004).

In vitro antibacterial activity of three extracts on 11 bacterial organisms again only showed moderate effects, with the best effect shown with the EtOH extract (*Staphylococcus aureus* and *Staphylococcus epidermidis*—128 μg/mL; Berahou et al. 2007).

Gastroprotective Activity

An aqueous extract (bark) showed an in vivo gastroprotective effect in a rat model against ethanol-induced gastric damage at 50 mL/kg. When administered with 50 % ethanol, the aqueous extract provided 18.9 % gastroprotection compared to the control (water). Given 60 min before the administration of ethanol, the aqueous extract gave 50.8 % protection. When given with ethanol, the aqueous extract showed an increase in stomach acid content, but a decrease in acid content when given 60 min prior to ethanol administration (Gharzouli et al. 1999).

In a large-scale screening for health beneficial properties, an extract of the fruit only showed moderate activity (Heinrich et al. 2005).

12.2.2 Species Mostly Used as Cooked Vegetables

Asparagus acutifolius L. and some other closely related species are among the most widely appreciated local vegetables in many Mediterranean regions, and they are commonly consumed as a vegetable, for their health benefits including diuretic effects, and used to treat bile duct stones, rheumatism, arteriosclerosis and as a laxative. They are still collected from the wild and are highly appreciated as elements of a wide range of dishes including omelettes and often mixed with other vegetables. Very limited pharmacological information has become available for this species, aside from a range of studies on its (often *in silico,* i.e., chemical) antioxidant effects.

Box 3: Recorded Pharmacological Activities of *Asparagus acutifolius* L.

Known Bioactive Compounds and Other Metabolites

Lutein, β-carotene, quercetin, isorhamnetin, kaempferol, α-Tocopherol, β-Tocopherol, γ-Tocopherol, δ-Tocopherol, myricetin-3-glucoside, myricetin, glycosylated quercetin, steroidal saponins (Salvatore et al. 2005; Martins et al. 2011; Sautour et al. 2007; Rivera et al. 2006b)

Relevant Pharmacological and Microbiological Studies

Antioxidant Activity

An aqueous extract of the spears showed *in silico* antioxidant DPPH• radical scavenging activity at an IC_{50} of 72.4 μg/mL, H_2O_2 radical scavenging activity at an IC_{50} of > 2000 μg/mL, NO radical scavenging activity at an IC_{50} of 56.7 μg/mL, compared to α-Tocopherol with IC_{50} values of 15.1, 13.1 and 11.1 μg/mL, respectively (Ferrara et al. 2011).

An aqueous preparation of the aerial part (1:10, grams of sample in millilitres of tap water boiled for 15 min) showed in vitro antioxidant activity in the FRAP assay at 11.23 ± 1.90 mmol Fe^{2+}/kg, TRAP assay at 5.29 ± 1.63 mmol Trolox/kg and TEAC assay (ABTS radical scavenging activity) at 3.01 ± 0.46 mmol Trolox/kg (Salvatore et al. 2005).

An MeOH extract (1) of the shoots in vitro showed antioxidant DPPH• radical scavenging activity at an EC_{50} of 4.87±0.38 mg/mL, reducing power at an EC_{50} of 1.62±0.00 mg/mL, β-carotene bleaching inhibition (lipid peroxidation inhibition) at an EC_{50} of 0.47±0.04 mg/mL, and thiobarbituric acid reactive substance (TBARS) inhibition (lipid peroxidation inhibition) at an EC_{50} of 0.07±0.02 mg/mL (Morales et al. 2012a).

Another MeOH extract (2) of the shoots in vitro showed antioxidant DPPH• radical scavenging activity at an EC_{50} of 423±24 μg/mL, reducing power at an EC_{50} of 191±12 μg/mL, β-carotene bleaching inhibition (lipid peroxidation inhibition) at an EC_{50} of 166±7 μg/mL and TBARS inhibition (lipid peroxidation inhibition) at an EC_{50} of 105±4 μg/mL (Martins et al. 2011).

Cardioprotective Activity

An EtOH (70%) extract of the stems in vitro showed pancreatic lipase inhibition at an IC_{50} of >10 mg/mL (Conforti et al. 2012).

In a large-scale screening for health beneficial properties, an extract of young shoots only showed weak activity (Heinrich et al. 2005).

Reichardia picroides (L.) Roth (French Scorzonera, Asteraceae) is used both as a cooked vegetable and a salad commonly eaten together with other plants. Overall, this is a commonly used species that still has received very limited attention in the scientific literature.

This species is rich in a wide range of luteolin—and apigenin-glycosides including many O-glycosides—as well as in caffeic acid, chlorogenic acid, isochlorogenic acid and 3,4-dicaffeoylquinic acid (Recio 1992). The aqueous extract of the leaves improved platelet sensitivity in vivo in patients with metabolic syndrome, which implies a reduced risk of thrombosis and related conditions in this high-risk group EC_{50} (Fragopoulou et al. 2012).

The leaf stalks of *Scolymus hispanicus* L. (golden thistle, Asteraceae) are a highly appreciated vegetable known to contain only very low amounts of ascorbic acid, dehydroascorbic acid, as well as oxalic acid, malic acid, citric acid and fumaric acid (Sanchez-Mata et al. 2012; Ozkol 2013). A methyl extract of the aerial parts showed in vivo a significant decrease of the elevated fasting blood glucose levels in streptozotocin-induced diabetes in Sprague–Dawley male rats, which showed that *S. hispanicus* may have a role through elevating insulin secretion of intact pancreatic beta cells. Therefore, this species possesses hypoglycaemic effect (Ozkol 2013).

12.2.3 Species Mostly Used in Salads

There can be no doubt that there is a tremendous overlap between this and the previous group. *Rorippa nasturtium-aquaticum* (L.) Hayek (*syn.: Nasturtium officinale* R. Br.), watercress, Brassicaceae, is typically used in salads and its unique spicy taste is highly appreciated. It may be used alone or in combination with other 'greens', and again medical uses have been reported including diuretic, antibiotic, tonic effects and to relieve indigestion, chronic bronchitis, to stimulate the appetite, for detoxifying, as a febrifuge and bechic (cough relieving).

It is another so far little-studied local (medicinal) food. Extracts of *Rorippa nasturtium-aquaticum* (and *Urtica dioica*) have been shown to have anti-inflammatory effects by downregulating iNOS and COX-2 (Rose et al. 2005) and inhibiting pro-inflammatory leukotrienes (ESCOP 2003), respectively. An EtOH extract has shown in vivo cardio-protective effects by decreasing LDL cholesterol (52.9 %), triglycerides (30.1 %) and total serum cholesterol (34.2 %) at a comparatively high dose of 500 mg/kg/day in (Bahramikia and Yazdanparast 2008).

Box 4: Recorded Pharmacological Activities of *Rorippa nasturtium-aquaticum* (L.) Hayek

Known Bioactive Compounds and Other Metabolites

Phenylethyl glucosinolate, allyl isothiocyanate, myristicin, limonene, α-terpinolene, caryophyllene oxide, p-cymene-8-ol, β-Phenylethyl isothiocyanate (PEITC), 8-methylsulphinyloctyl isothiocyanate (Chevallier 2001; Moriyama et al. 2003; Amiri 2012; Yazdanparast et al. 2008; Bahramikia and Yazdanparast 2008; Chen et al. 1996; Rivera et al. 2005; Rose et al. 2005)

Relevant Pharmacological and Microbiological Studies

Cardioprotective Activity

The EtOH (50 %) extract of the whole plant showed in vitro activity on platelet aggregation; cardioprotective properties were shown in vivo in a hypercholesterolaemic rat model at 500 mg/kg/day for 10 days for an EtOH/H_2O (70:30) extract. The extract showed a significant decrease in total serum cholesterol by 34.2 %, triglycerides 30.1 % and low-density lipoprotein cholesterol 52.9 %. After 10 days of treatment, there was a raised serum high-density lipoprotein cholesterol level by 27.0 % (Bahramikia and Yazdanparast 2008).

Chemoprotective Activity

PEITC and 8-methylsulphinyloctyl isothiocyanate (MSO) showed chemopreventative activity by decreasing the expression of iNOS and COX-2 proteins via inactivation of NFκB and the stabilization of IκBα. This was shown by the decrease in nitrite and PGE-2 production in lipopolysaccharide-stimulated Raw-264.7 macrophages. This led to a downregulation of pro-inflammatory mediators (Rose et al. 2005). PEITC also showed chemoprotective activity in another study by mediating apoptosis in HepG2 human hepatoma cells (Neo et al. 2005).

Antimicrobial Activity

Antibacterial activity of *Rorippa nasturtium-aquaticum* at a concentration of 1000 μg/disc against four bacterial organisms was expressed by mm of inhibition (Penecilla and Magno 2011). With such high values, the relevance of the data is minimal.

Cytotoxic Activity

An MeOH extract of the whole plant showed cytotoxicity against human colon carcinoma cells in vitro at an IC_{50} of 54.2 μg/mL (Seoud et al. 2003).

Antioxidant Activity

Two EtOH (80 and 70%) extracts of the aerial part showed in vitro anti-lipid peroxidation activity by reducing TBARS formation at an EC_{50} of 273.5 μg/mL, iron chelating activity at an EC_{50} of 538.6 μg/mL, antioxidant DPPH• radical scavenging activity at an EC_{50} of 114.7 μg/mL, ABTS•+ radical scavenging activity at an EC_{50} of 60.8 μg/mL, NO scavenging activity at an EC_{50} of 395.2 μg/mL and H_2O_2 scavenging activity at an EC_{50} of 312.4 μg/mL (Bahramikia and Yazdanparast 2010).

An MeOH/(5 mL, 1:1 v/v) extract of freeze-dried leaves in vitro showed antioxidant activity in the ABTS assay at 260 mg/100 g fw, FRAP assay at 209 mg/100 g fw and DPPH assay at 244 mg/100 g fw (Martinez-Sanchez et al. 2008).

Cichorium intybus L. (Chicory, Asteraceae) may well be one of the most widely used local food plants in Europe. It is, of course, best known for its use as a fresh salad using varieties with variegated red or red and green leaves (radicchio) or the 'Belgian endives' or chicory, cultivars grown under special conditions in the dark or the root chicory (var. *sativum*) important as a coffee substitute. In the context of this discussion, the focus is on noncultivated forms. These are generally relatively bitter, which is linked to the presence of sesquiterpene lactones, especially lactucin and lactucopicrin. Commonly, the young leaves are collected and used raw in salads or as a cooked vegetable. Limited research is available, and the existing data are mostly based on the cultivated varieties. As in other cases, antioxidant and antimicrobial effects have been at the centre of attention, and the lack of research on the noncultivated varieties makes it difficult to draw any specific conclusions.

Box 5: Recorded Pharmacological Activities of *Cichorium intybus* L.

Known Bioactive Compounds and Other Metabolites

The roots are particularly rich in carbohydrates. Alkaloids, tannins, glycosides, amines, tridecanoic acid, myristoleic acid, caprylic acid, capric acid, undecanoic acid, lauric acid, myristic acid, pentadecanoic acid, palmitic acid, palmitoleic acid, heptadecanoic acid, stearic acid, oleic acid, linoleic acid, γ-linolenic acid, α-linolenic acid, arachidic acid, eicosenoic acid, heneicosanoic acid, tricosanoic acid, nervonic acid, lactucin, lactucopicrin—two sesquiterpene lactose responsible for the bitter taste (Van Wyk and Wink 2004)

Relevant Pharmacological and Microbiological Studies

Cardioprotective Activity

An alcoholic extract of the roots showed in vivo a decrease in the amplitude and rate of the toad heart at the dose of 0.4 mL/L of extract (obtained from 10 g of air-dried material extracted with 70 % alcohol with the total volume adjusted to 100 ml). (Balbaa et al. 1973)

Antioxidant Activity

An ethanol extract (1) of the roots showed in vitro antioxidant effects (DPPH assay) at a final concentration of 2.5 mg/mL, radical scavenging activity at IC_{50} of 281.00±983.33 µg/mL (Liu et al. 2013). Another EtOH extract (2) of the whorls showed in vitro antioxidant DPPH at a final concentration of 2.5 mg/mL, radical scavenging activity at EC_{50} of 22 mg/ml (Pieroni et al. 2002b).

Chicory-supplemented diet: The dried plants were added to the standard diet of male rats, which showed at the end of the experiment that chicory is a natural substance for ameliorating the oxidative stress and hepatic injury induced by nitrosamine compounds (Hassan 2010).

Antimicrobial Activity

The EtOH extract (1) of the roots showed in vitro an antimicrobial activity at concentration of 10 mg/disc against gram-positive and gram-negative bacteria. The minimun inhibitory concentration (MIC) values were 0.1625, 1.25, and 2.5 mg/mL of *Salmonella typhi, Escherichia coli* and *Staphylococcus aureus,* respectively (Liu et al. 2013).

Anthelmintic Activity

Sesquiterpene lactones extracted from the leaves showed in vitro anthelmintic activity against a predominantly *Haemonchus contortus* egg population at a concentration of 10 mg/mL (Foster et al. 2011).

Using, for example, cv. Grasslands Puna of *Cichorium intybus,* toxic effects on metazoic parasites (*H. contortus*) are well known and used in the management of livestock by adding it to the fodder (Heckendorn et al. 2007). These effects may well be linked to the presence of sesquiterpene lactones, too.

In a large-scale screening for health beneficial properties, three extracts of the aerial parts obtained from Spain, Italy and Greece, respectively, showed moderate activity (Heinrich et al. 2005).

12.2.4 Species Mostly Used as Infusions

Malva sylestris L. (common mallow, Malvaceae) is a common weedy plant in many European regions, but also an important medicinal plant used products for treating respiratory conditions and externally as a vulnerary (Rivera et al. 2006b). In many European regions, it is a common element of herbal (medical) teas (Edwards et al. 2015). This species serves as an example for a species which, in fact, has a very wide use and which is well known for phytotherapeutic practices (Gasparetto et al. 2012).

It acts as a demulcent and pectoral. An infusion is used for colds and coughs, and the mucilage from the leaves is anti-inflammatory with anticomplement activity. Its medical importance is evident from its inclusion in a range of pharmacopoeias, including the European Pharmacopoeia. However, aside from these more widely recognized uses, a wide range of other uses are known. For example, in soups, it is used to enhance uterine contractions during birth (Pieroni et al. 2002). The immature fruits are eaten by children as a snack (Rivera et al. 2006b; Tardío et al. 2006) and throughout Europe a wide range of therapeutic benefits are associated with the use, most importantly of the leaves—as antiphlogistic, astringent, demulcent, diuretic, emollient, expectorant and laxative, and salve. Mucilage, sulphated flavonol glycosides such as gossypin-3-sulphate, hypolaetinglucoside-3'-sulphate and others, and anthocyanins (malvin, the diglucoside of malvidin and delphinidin) are key constituents relevant for the observed effects. Even though there are considerable gaps in terms of the species' biological–pharmacological profile, it is a relatively well-studied food plant species from the Mediterranean (Gasparetto et al. 2012).

Box 6: Recent Reports on Pharmacological Activities of *Malva sylvestris* L.

Known Bioactive Compounds and Other Metabolites

Mucilage polysaccharides (6 to >10%) composed of neutral and acidic monosaccharide residues including rhamnose, galactose, arabinose, galacturonic acid and glucuronic acid. Anthocyanins (6–7%), mainly malvidin 3,5-diglucoside, malvidin 3-glucoside, malvidin 3-(6"-malonylglucoside)-5-glucoside and delphinidin 3-glucoside with traces of petunidin and cyanidin glycosides, scopoletin, ursolic acid and phytosterols (ESCOP 2003)

Relevant Pharmacological and Microbiological Studies

A review by Gasparetto et al. (2012) covers most of the relevant data up to 2011.

Antioxidant Activity

An aqueous extract of the leaves showed an antioxidative potential activity in vivo because of its richness in phenolic compounds. The extract was given at a dose of 0.2 g dry mallow/kg body weight of the rats during 90 days (Marouane 2011).

An EtOH extract of the leaves showed an antioxidant activity at a concentration of 150 mg/L (Samavati and Zadeh 2013).

Antibacterial Activity

An MeOH extract both of flowers and leaves showed antibacterial effects against the plant pathogen *Erwinia carotovora* with MIC of 128 and 256 μg/mL, respectively (Razavi et al. 2011).

The MeOH extract of the flowers showed antibacterial effects against human pathogen bacteria strains such as *Staphylococcus aureus, Streptococcus agalactiae* and *Enterococcus faecalis,* with MIC of 192, 200 and 256 μg/ml, respectively (Razavi et al. 2011).

12.3 Conclusions

These nine examples highlight the overall situation with respect to our current understanding of food plants used locally in the Mediterranean. Even for species that are widely used and common throughout the region, only limited data exist. Only species that have found entry into the more recognized medical or naturopathic practice, such as *Malva sylvestris* (or *Foeniculum vulgare*, not discussed in this short chapter), have a relatively robust body of evidence associated with it. Clinical data or information from intervention studies are generally lacking. Also, there is clearly a need for more high-quality and mechanistic studies and for intervention studies focusing on disease-preventive effects.

Their ambivalent status both as a popular local food and a medicine and their limited economic importance seems to have resulted in many of the species not having been studied bioscientifically in detail. However, the sheer number of species that are of importance poses a challenge and a systematic study of only the most important species will require a coordinated and systematic research effort backed by adequate funding.

There also are challenges in terms of what activities will be of particular relevance. Antioxidant and antimicrobial effects are a good and general first level of evidence, but they cannot be used for understanding specific diseases' preventive or therapeutic effects. More systematic studies will be key for assessing the relevance of these local products, most notably on the prevention or amelioration of aging-related disorders. The benefits will, however, most likely be linked to their use as an element of a 'Mediterranean diet' and not as a single 'superfood' (Heinrich et al. 2011).

In conclusion, such species collectively contribute to health and well-being, but we need to study the species' specific effects much more systematically.

Acknowledgements We would like to thank Drs. Manuel Pardo-de-Santayana, Marco Leonti, Michelle Viteri and Rocio Alarcon for many fruitful and interesting discussions on this and related topics. The project received no external funding and is based on two MPharm dissertations at the Centre for Pharmacognosy and Phytotherapy, UCL School of Pharmacy, University of London.

References

Afkir S, Nguelefack TB, Aziz M, Zoheir J, Cuisinaud G, Bnouham M, Mekhfi H, Legssyer A, Lahlou S, Ziyyat A (2008) *Arbutus unedo* prevents cardiovascular and morphological alterations in L-NAME-induced hypertensive rats—Part 1: cardiovascular and renal hemodynamic effects of *Arbutus unedo* in L-NAME-induced hypertensive rats. J Ethnopharmacol 116:288–295

Amiri H (2012) Volatile constituents and antioxidant activity of flowers, stems and leaves of *Nasturtium officinale* R. Br. Nat Prod Res 26:109–115

Bahramikia S, Yazdanparast R (2008) Effect of hydroalcoholic extracts of *Nasturtium officinale* leaves on lipid profile in high-fat diet rats. J Ethnopharmacol 115:116–121

Bahramikia S, Yazdanparast R (2010) Antioxidant efficacy of *Nasturtium officinale* extracts using various *in vitro* assay systems. J Acupunct Meridian Stud 3:283–290

Balbaa SI, Zaki AY, Abdel-Wahab SM, El-Denshary ES, Motazz-Bellah M (1973) Preliminary phytochemical and pharmacological investigations of the roots of different varieties of *Cichorium intybus*. Planta Medica 24:133–144

Berahou A, Auhmani A, Fdil N, Benharref A, Jana M, Gadhi CA (2007) Antibacterial activity of *Quercus ilex* bark's extracts. J Ethnopharmacol 112:426–429

Bnouham M, Merhfour FZ, Legssyer A, Mekhfi H, Maallem S, Ziyyat A (2007) Anti hyperglycemic activity of *Arbutus unedo*, *Ammoides pusilla* and *Thymelaea hirsuta*. Pharmazie 62:630–632

Bnouham M, Merhfour FZ, Ziyyat A, Aziz M, Legssyer A, Mekhfi H (2010) Antidiabetic effect of some medicinal plants of Oriental Morocco in neonatal non-insulin-dependent diabetes mellitus rats. Hum Exp Toxicol 29:865–871

Chen LS, Mohr SN, Yang CS (1996) Decrease of plasma and urinary oxidative metabolites of acetaminophen after consumption of watercress by human volunteers. Clin Pharmacol Therapeut 60:651–660

Chevallier A (2001) Encyclopedia of medicinal plants. Dorling Kindersley Limited, Great Britain

Chevolleau S, Mallet JF, Debal A, Ucciani E (1993) Antioxidant activity of Mediterranean plant-leaves—occurrence and antioxidant importance of alpha-tocopherol. J Am Oil Chem Soc 70:807–809

Conforti F, Perri V, Menichini F, Marrelli M, Uzunov D, Statti GA, Menichini F (2012) Wild Mediterranean dietary plants as inhibitors of pancreatic lipase. Phytother Res 26:600–604

Edwards S, Rocha I, Williamson EM, Heinrich M (2015) Herbal medicines—an evidence-based guide to phytopharmacy. Wiley, Chichester. (In press)

ESCOP (2003) ESCOP monographs. The Scientific Foundation for Herbal Medicinal Products. Thieme, Stuttgart

Etkin N, Ross PJ (1991) Should we set a place for diet in ethnopharmacology. J Ethnopharmacol 32:25–36

EU (2010) Functional food. Publications Office of the European Union, Luxembourg. ISBN 978-92-79-14239-0

Ferrara L, Dosi R, Di Maro A, Guida V, Cefarelli G, Pacifico S, Mastellone C, Fiorentino A, Rosati A, Parente A (2011) Nutritional values, metabolic profile and radical scavenging capacities of wild asparagus (*A. acutifolius* L.). J Food Compos Anal 24:326–333

Foster JG, Cassida KA, Turner KE (2011) In vitro analysis of the anthelmintic activity of forage chicory (*Cichorium intybus* L.) sesquiterpene lactones against a predominantly haemonchus contortus egg population. Vet Parasitol 180:298–306

Fragopoulou E, Detopoulou P, Nomikos T, Pliakis E, Panagiotakos DB, Antonopoulou S (2012) Mediterranean wild plants reduce postprandial platelet aggregation in patients with metabolic syndrome. Metab Clin Exp 61(3):325–334

Gasparetto JC, Martins CA, Hayashi SS, Otuky MF, Pontarolo R (2012) Ethnobotanical and scientific aspects of *Malva sylvestris* L.: a millennial herbal medicine. J Pharm Pharmacol 64(2):172–189. doi: 10.1111/j.2042-7158

Gharzouli K, Khennouf S, Amira S, Gharzouli A (1999) Effects of aqueous extracts from *Quercus ilex* L. root bark, *Punica granatum* L. fruit peel and *Artemisia herba-alba* Asso leaves on ethanol-induced gastric damage in rats. Phytother Res 13:42–45

Greuter W (1991) Botanical diversity, endemism, rarity, and extinction in the Mediterranean area: an analysis based on the published volumes of med-checklist. Bot Chron 10:63–79

Grieve M (1995) A modern herbal. Dover Publications, New York. (Online)

Gulluce M, Adiguzel A, Ogutcu H, Sengul M, Karaman I, Sahin F (2004) Antimicrobial effects of *Quercus ilex* L. extract. Phytother Res 18:208–211

Haouari ME, Lopez JJ, Mekhfi H, Rosado JA, Salido GM (2007) Antiaggregant effects of *Arbutus unedo* extracts in human platelets. J Ethnopharmacol 113:325–331

Hassan HA, Hanaa A, Yousef MI (2010) Ameliorating effect of chicory (*Cichorium intybus* L.)-supplemented diet against nitrosamine precursors-induced liver injury and oxidative stress in male rats. Food Chem Toxicol 48:2163–2169

Heckendorn F, Häring DA, Maurer V, Senn M, Hertzberg H (2007) Individual administration of three tanniferous forage plants to lambs artificially infected with *Haemonchus contortus* and *Cooperia curticei*. Vet Parasitol 146(1–2):123–134

Heinrich M, Prieto-Garcia J (2008) Diet and healthy ageing 2100: will we globalise local knowledge systems? Ageing Res Rev 7:249–274

Heinrich M, Leonti M, Nebel S, Peschel W, Pieroni A, Smith F, Rivera D, Obon C, Inocencio C, Verde A, Fajardo J, Llorach R, Muller WE, Eckert GP, Schaffer S, Schmitt-Schillig S, Antonopoulou S, Kypriotakis Z, Manios Y, Nomikos T, Kaliora A, Sidossis L, Galli C, Visioli F, Grande S, Bogani P, De Saizieu A, Fluhmann B, D'Orazio D, Fowler A, Koj A, Bereta J, Dulak J, Guzdek A, Kapiszewska M, (The Local Food-Nutraceutical Consortium) (2005) Understanding local Mediterranean diets: a multidisciplinary pharmacological and ethnobotanical approach. Pharmacol Res 52:353–366

Heinrich M, Dhanji T, Casselman I (2011) Açai (*Euterpe oleracea* Mart.)—a phytochemical and pharmacological assessment of the species' health claims. Phytochem Lett 4:10–21. doi:10.1016/j.phytol.2010.11.005

Heinrich M, Barnes J, Gibbons S, Williamson EM (2012) 2004 Fundamentals of pharmacognosy and phytotherapy. Churchill Livingston, Edinburgh (Elsevier)

Kivcak B, Mert T, Demirci B, Baser KHC (2001) Composition of the essential oil of *Arbutus unedo*. Chem Nat Compd 37:445–446

Legssyer A, Ziyyat A, Mekhfi H, Bnouham M, Herrenknecht C, Roumy V, Fourneau C, Laurens A, Hoerter J, Fischmeister R (2004) Tannins and catechin gallate mediate the vasorelaxant effect of *Arbutus unedo* on the rat isolated aorta. Phytother Res 18:889–894

Leonti M, Nebel S, Rivera D, Heinrich M (2006) Wild gathered food in the European Mediterranean: a comparison. Econ Bot 60(2):130–142

Liu H, Wang Q, Liu Y, Chen G, Cui J (2013) Antimicrobial and antioxidant activities of *Cichorium intybus* root extract using orthogonal matrix design. J Food Sci 78:M258–M263

Loggia RD, Negro PD, Bianchi P, Romussi G, Tubaro A (1989) Topical anti-inflammatory activity of some flavonoids from *Quercus ilex* leaves. Planta Med 55:109–110

Mariotto S, Ciampa AR, De Prati AC, Darra E, Vincenzi S, Sega M, Cavalieri E, Shoji K, Suzuki H (2008) Aqueous extract of *Arbutus unedo* inhibits STAT1 activation in human breast cancer cell line MDA-MB-231 and human fibroblasts through SHP2 activation. Med Chem 4:219–228

Marouane W, Soussi A, Murat J-C, Bezzine S, El Feki A (2011) The protective effect of *Malva sylvestris* on rat kidney damaged by vanadium. Lipids Health Dis 10:65

Martinez-Sanchez A, Gil-Izquierdo A, Gil MI, Ferreres F (2008) A comparative study of flavonoid compounds, vitamin C, and antioxidant properties of baby leaf Brassicaceae species. J Agric Food Chem 56:2330–2340

Martins D, Barros L, Carvalho AM, Ferreira ICFR (2011) Nutritional and in vitro antioxidant properties of edible wild greens in Iberian Peninsula traditional diet. Food Chem 125:488–494

Mekhfi H, Haouari ME, Bnouham M, Aziz M, Ziyyat A, Legssyer A (2006) Effects of extracts and Tannins from *Arbutus unedo* leaves on rat platelet aggregation. Phytother Res 20:135–139

Mendes L, De Freitas V, Baptista P, Carvalho M (2011) Comparative antihemolytic and radical scavenging activities of strawberry tree (*Arbutus unedo* L.) leaf and fruit. Food Chem Toxicol 49:2285–2291

Morales P, Carvalho AM, Sanchez-Mata MC, Camara M, Molina M, Ferreira ICFR (2012a) Tocopherol composition and antioxidant activity of Spanish wild vegetables. Genet Resour Crop Evol 59:851 863

Morales P, Ferreira ICFR, Carvalho AM, Fernandez-Ruiz V, Sanchez-Mata MC, Camara M, Morales R, Tardío J (2013) Wild edible fruits as a potential source of phytochemicals with capacity to inhibit lipid peroxidation. Eur J Lipid Sci Technol 115:176–185

Moriyama H, Iizuka T, Nagai M, Hoshi K, Murata Y, Taniguchi A (2003) Platelet aggregatory effects of *Nasturtium officinale* and *Solanum torvum* extracts. Nat Med 57:133–138

Neo JCH, Rose P, Ong CN, Chung MCM (2005) Beta-phenylethyl isothiocyanate mediated apoptosis: a proteomic investigation of early apoptotic protein changes. Proteomics 5:1075–1082

Oliveira I, Coelho V, Baltasar R, Pereira JA, Baptista P (2009) Scavenging capacity of strawberry tree (*Arbutus unedo* L.) leaves on free radicals. Food Chem Toxicol 47:1507–1511

Orak HH, Yagar H, Isbilir SS, Demirci AS, Gumus T, Ekinci N (2011) Evaluation of antioxidant and antimicrobial potential of strawberry tree (*Arbutus unedo* L.) leaf. Food Sci Biotechnol 20:1249–1256

Ouarti AE, Haouat AC, Sqalli H, Haggoud A, Ennabili A, Ibnsouda S, Iachgar M, Iraqui M (2012) Extra- and intracellular antimycobacterial activity of *Arbutus unedo* L. Afr J Microbiol Res 6:1283–1290

Ozkol H, Tuluce Y, Dilsiz N, Koyuncu İ (2013) Therapeutic potential of some plant extracts used in Turkish traditional medicine on streptozocin-induced type 1 diabetes mellitus in rats. J Membr Biol 246, 47–55

Pabuccuoglu A, Kivcak B, Bas M, Mert T (2003) Antioxidant activity of *Arbutus unedo* leaves. Fitoterapia 74:597–599

Pavlovic DR, Lakusic B, Doslov-Kokorus Z, Kovacevic N (2009) Arbutin content and antioxidant activity of some Ericaceae species. Pharmazie 64:656–659

Pavlovic DR, Brankovic S, Kovacevic N, Kitic D, Veljkovic S (2011) Comparative study of spasmolytic properties, antioxidant activity and phenolic content of *Arbutus unedo* from Montenegro and Greece. Phytother Res 25:749–754

Penecilla GL, Magno CP (2011) Antibacterial activity of extracts of twelve common medicinal plants from the Philippines. J Med Plants Res 5:3975–3981

Pieroni A, Nebel S, Quave C, Heinrich M (2002a) Ethnopharmacology of *Liakra*: traditional, weedy vegetables of the Arbëreshë of the vulture area in southern Italy. J Ethnopharmacol 81:165–185

Pieroni A, Janiak V, Dürr CM, Lüdeke S, Trachsel E, Heinrich M (2002b) In vitro antioxidant activity of non-cultivated vegetables of ethnic Albanians in southern Italy. Phytother Res 16:467–473

Razavi SM, Zarrini G, Molavi G, Ghasemi G (2011) Bioactivity of *Malva sylvestris* L., a medicinal plant from Iran. Iran J Basic Med Sci 14:574–579

Recio M, Giner RM, Hermenegildo M, Peris JB, Mañez S, Rios J-L (1992) Phenolics of *Reichardia* and their taxonomic implications. Biochem Syst Ecol 20(5):449–452

Rivera D, Obon C, Inocencio C, Heinrich M, Verde A, Fajardo J, Llorach R (2005) The ethnobotanical study of local Mediterranean food plants as medicinal resources in Southern Spain. J Physiol Pharmacol 56(S.11):97–114

Rivera D, Obón C, Heinrich M, Inocencio C, Verde A, Fajardo J (2006a). Gathered Mediterranean food plants—ethnobotanical investigations and historical development. In: Heinrich M, Müller WE, Galli C (eds) Local Mediterranean food plants and nutraceuticals. Kargerm, Basel

Rivera D, Verde A, Fajardo J, Inocencio C, Obón C, Heinrich M (eds) (2006b) Guía etnobotánica de los alimentos locales recolectados en la provincia de Albacete. Instituto de Estudios Albacetenses 'Don Juan Manuel' de la Excma. Diputación de Albacete. Alabacete, Spain. Serie Estudios No. 167. Alabacete ISBN: 84-95394-92-8

Rose P, Won YK, Ong CN, Whiteman M (2005) Beta-phenylethyl and 8-methylsulphinyloctyl isothiocyanates, constituents of watercress, suppress LPS induced production of nitric oxide and prostaglandin E2 in RAW 264.7 macrophages. Nitric Oxide-Biol Chem 12:237–243

Salvatore S, Pellegrini N, Brenna OV, Rio DD, Frasca G, Brighenti F, Tumino R (2005) Antioxidant characterization of some Sicilian edible wild greens. J Agric Food Chem 53:9465–9471

Samavati V, Zadeh AM (2013) Polysaccharide extraction from *Malva sylvestris* and its anti-oxidant activity. Int J Biol Macromol 60:427–436. (PubMed ID 23612362)

Sánchez-Mata MC, Cabrera-Loera RD, Morales P, Fernández-Ruiz V, Cámara M, Díez-Marqués C, Pardo-de-Santayana M, Tardío J (2012) Wild vegetables of the Mediterranean area as valuable sources of bioactive compounds. Genet Resour Crop Evol 59(3):431–443

Sautour M, Miyamoto T, Lacaille-Dubois MA (2007) Steroidal saponins from *Asparagus acutifolius*. Phytochemistry 68:2554–2562

Seoud KAEHAE, Bibby MC, Shoeib N, Wright CW (2003) Evaluation of some Egyptian plant species for in vitro antimycobacterial and cytotoxic activities. Pharm Biol 41:463–465

Tardío J, Pascual H, Morales R (2005) Wild food plants traditionally used in the province of Madrid, Central Spain. Econ Bot 59:122–136

Tardío J, Pardo-de-Santayana M, Morales R (2006) Ethnobotanical review of wild edible plants in Spain. Bot J Linn Soc 152(1):27–72

Van Wyk B-E, Wink M (2004) Medicinal plants of the world: an illustrated scientific guide to important medicinal plants and their uses. Timber, Cambridge, p 8

Weiss RF (2001) Weiss's herbal medicine. Thieme, Stuttgart

Yazdanparast R, Bahramikia S, Ardestani A (2008) *Nasturtium officinale* reduces oxidative stress and enhances antioxidant capacity in hypercholesterolaemic rats. Chem Biol Interact 172:176–184

Ziyyat A, Boussairi EH (1998) Cardiovascular effects of *Arbutus unedo* L. in spontaneously hypertensive rats. Phytother Res 12:110–113

Part IV
Descriptive Dossier: Botany, Uses and Food Composition Tables

Chapter 13
Ethnobotanical and Food Composition Monographs of Selected Mediterranean Wild Edible Plants

Javier Tardío, María de Cortes Sánchez-Mata, Ramón Morales, María Molina, Patricia García-Herrera, Patricia Morales, Carmen Díez-Marqués, Virginia Fernández-Ruiz, Montaña Cámara, Manuel Pardo-de-Santayana, María Cruz Matallana-González, Brígida María Ruiz-Rodríguez, Daniel Sánchez-Mata, María Esperanza Torija-Isasa, Jose Luis Guil-Guerrero and Noureddine Boussalah

J. Tardío (✉)
Departamento de Investigación Agroalimentaria, Instituto Madrileño de Investigación y Desarrollo Rural, Agrario y Alimentario (IMIDRA), Alcalá de Henares, Madrid, Spain
e-mail: javier.tardio@madrid.org

M. d. C. Sánchez-Mata · P. García-Herrera · P. Morales · C. Díez-Marqués · V. Fernández-Ruiz · M. Cámara · M. C. Matallana-González · B. M. Ruiz-Rodríguez · M. E. Torija-Isasa
Departamento de Nutrición y Bromatología II, Bromatología, Facultad de Farmacia, Universidad Complutense de Madrid, Madrid, Spain
e-mails: cortesm@ucm.es; patrigarcia31@hotmail.com; patricia.morales@farm.ucm.es; cardimar@ucm.es; vfernand@farm.ucm.es; mcamara@ucm.es; mcmatall@ucm.es; brigidamariaruiz@gmail.com; metorija@ucm.es

R. Morales
Departamento de Biodiversidad y Conservación, Real Jardín Botánico, CSIC, Madrid, Spain
e-mail: morales@rjb.csic.es

M. Molina · M. Pardo-de-Santayana
Departamento de Biología (Botánica), Facultad de Ciencias, Universidad Autónoma de Madrid (UAM), Madrid, Spain
e-mails: m.molina.simon@gmail.com; manuel.pardo@uam.es

D. Sánchez-Mata
Departmento de Biología Vegetal II, Facultad de Farmacia, Universidad Complutense de Madrid, Madrid, Spain
e-mail: dsmata@ucm.es

J. L. Guil-Guerrero
Food Technology Division, Agronomy Department, Universidad de Almería, Almería, Spain
e-mail: jlguil@ual.es

N. Boussalah
Laboratoire de Biomathematique, Biophysique, Biochimie et Scientométrie (L3BS), Université de Bejaia, Bejaia, Algeria
e-mail: noureddine.boussalah@univ-bejaia.dz

M. de C. Sánchez-Mata, J. Tardío (eds.), *Mediterranean Wild Edible Plants*,
DOI 10.1007/978-1-4939-3329-7_13

13.1 Introduction

13.1.1 Wild Plants in the Mediterranean Diet

The high consumption of fruits and vegetables, characteristic of the Mediterranean diet, has traditionally included wild fruits and especially wild vegetables, as shown by many ethnobotanical studies (Leonti et al. 2006; Rivera et al. 2006; The Local Food-Nutraceutical Consortium 2005). These wild edible plants have historically played an important role in complementing staple agricultural foods (e.g. Pieroni et al. 2002; Tardío et al. 2005), having also considered as a worthy part of the diet of this vast region (e.g. Vanzani et al. 2011; Zeghichi et al. 2003).

According to Rivera et al. (2006), ethnobotanical research has identified about 2300 different plant and fungi taxa, which are gathered and consumed in the Mediterranean region. However, the floristic and cultural richness of this wide area makes that the consumption of a great number of these species (about 1000) is strictly local. Many wild species are shared by several countries, but only a few taxa (30) are eaten in most of the Mediterranean region (Rivera et al. 2006). Most of these widespread species have been traditionally consumed since ancient times, often with a dual purpose: nutrition and medicine (e.g. Leonti 2012; Pieroni and Price 2006). In this chapter, we present a dossier about the majority of these plants, joining some of their available ethnobotanical, historical, and nutritional composition data.

13.1.2 Food Composition Databases

The first food composition studies were carried out to determine the chemical nature of the food components that affect human health (McCollum 1957). In this context, the first Food Composition Tables were based on analyses carried out during the nineteenth century, in the laboratories of researchers such as C. Von Voit in Germany, W. O. Atwater in USA, and R.H.A. Plimmer in the UK. The first European Food Composition Table was published in Germany in 1878 (Konig 1878); later, the trend in the elaboration of Food Composition Tables moved towards compiling tables from scrutinized data produced by a number of laboratories. At the present moment, it is recognized that Food Composition Databases (FCDs) are basic tools for quantitative nutrition research, dietary evaluation, and development of food and nutrition policies (Greenfield and Southgate 2003). As it has been recognized by Widdowson and McCance in 1943, "there are two schools about food tables. One tends to regard the figures in term as having the accuracy of atomic weight determinations; the other dismisses them on the ground that a foodstuff may be so modified by the soil, the season or its rate of growth that no figure can be a reliable guide to its composition. The truth, of course, lies somewhere between these two points of view".

FCDs are used as a reference for food monitoring and serve the food industry and food scientists as a tool for product development and nutritional labelling. They provide reference values for nutrient calculations used in dietetics and epidemiology and are used to evaluate the results of food consumption surveys and to promote or encourage the consumption of certain types of foods attending to their nutritional interest (Church 2006; Somogyi 1974; Souci et al. 2008). Uncultivated foods as well as cooked foods are two of the main missing groups of foods in conventional FCD. Both groups constitute the main focus of this chapter.

13.2 Methodology

13.2.1 Selection of Species

As formerly pointed out, the selection of the species to be included in the present chapter has been mainly based on the cultural importance of these species in the Mediterranean region and on the existence of food composition data.

To check the widespread edible use of the wild plants in the study area, besides our own database that included information from nearly 90 Spanish ethnobotanical sources, 44 additional ethnobotanical works from other countries have been consulted, and a database has been created covering information about 22 Mediterranean countries with both databases. Data about 41 plant species has been included in the 38 monographs presented in the chapter (Fig. 13.1). As shown in the figure, the majority of species have been extensively consumed in the Mediterranean region, being mentioned from 4 to 16 different countries (most of them, 8 or more). We have included, however, a few species more locally used, such as *Montia fontana* and *Rumex papillaris*. The current use of the first species has been only registered in Spain and Portugal, although there are references that report its consumption in the past in other European countries. The second one (*R. papillaris*) is endemic to the Iberian Peninsula, although very close to other more widely distributed sorrels, such as *R. acetosa*, with multiple and similar references of consumption in the Mediterranean region.

Regarding nutritional data, most of the wild species (24) have been subject matter for our research group, gathering and analyzing several raw and cooked samples of these plants (García-Herrera 2014; Morales 2011; Ruiz-Rodríguez 2014), in some cases, with the collaboration of other laboratories. To facilitate the search through the chapter, the monographs have been ordered alphabetically by the scientific name of the selected plants, and they have been listed at the begining of Section 13.3.

13.2.2 Botanical and Ethnobotanical Data in the Monographs

The first part of the monographs includes the botanical and ethnobotanical data of the selected species. The nomenclature of *Flora iberica* (Castroviejo et al. 1986–2015) was followed for the species included therein, and *Flora Europaea* (Tutin et al. 1964–1980) for the remaining ones. Nevertheless, the family names adopted were those of the APGIII system (Angiosperm Phylogeny Group III system).

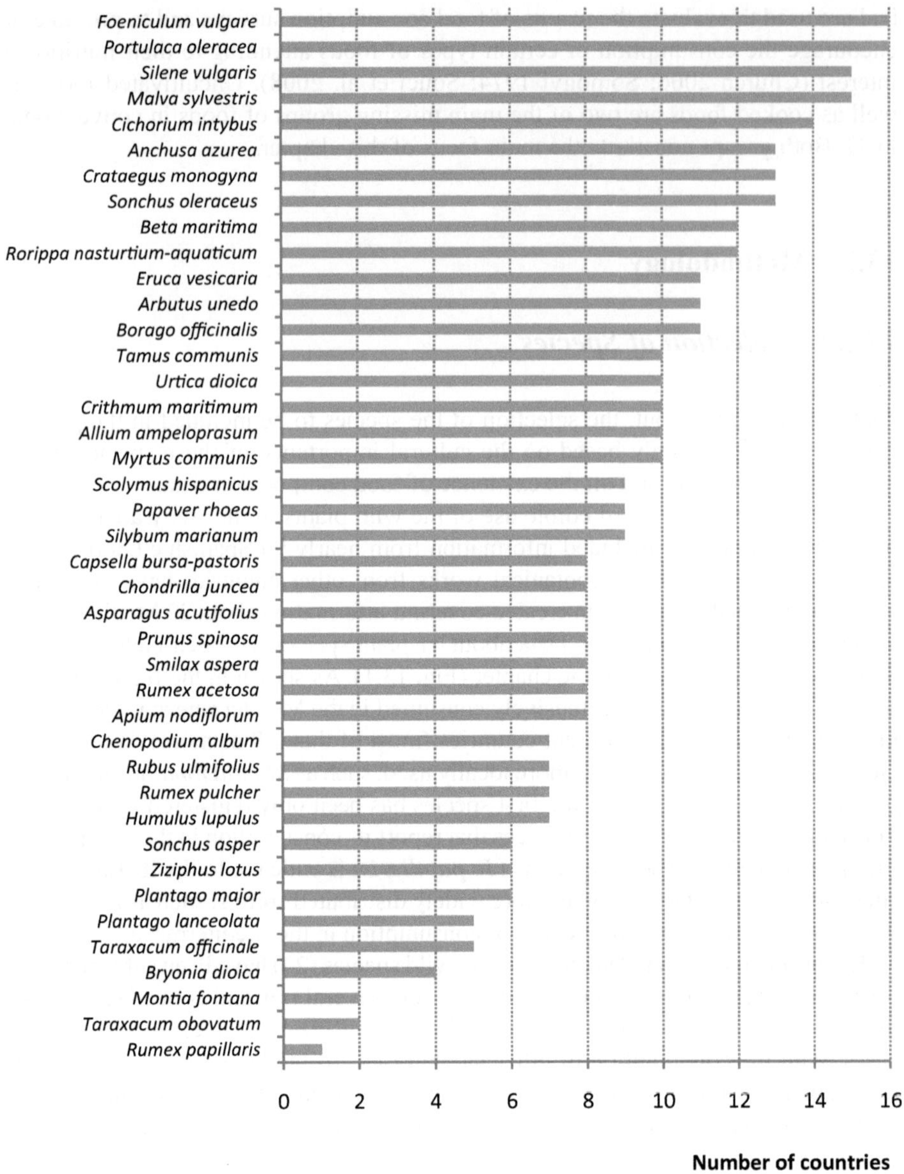

Fig. 13.1 Cultural importance of the 41 wild edible plants selected for the monographs, measured by the number of Mediterranean countries where its use have been reported

A list of selected common names used for the plant in different languages are presented, firstly its English name, followed by a name in some of the main languages of each of the Mediterranean countries, sorted in clockwise order, starting in Portugal and finishing in Morocco. These languages are indicated in brackets by a code of two characters (ISO code 639-1): Arabic (ar), Albanian (sq), Bosnian (bs), Croatian (hr), French (fr), Greek (el), Hebrew (he), Italian (it), Portuguese (pt), Spanish (es), and Turkish (tr). In the search of the common names, besides the ethnobotanical studies used for each species, other general works have been consulted, such as *Flora Agrícola* (Sánchez-Monge 1991), *Mansfeld's Encyclopedia of Agricultural and Horticultural Crops* (Hanelt and Institute of Plant Genetics and Crop Plant Research 2001), *Plants and Humans in the Near East and the Caucasus* (Rivera et al. 2012), and *A Guide to Medicinal Plants in North Africa* (IUCN 2005).

As an aid in the identification of each species, some photographs and a short description of the plant are presented, together with a paragraph of ecology and distribution, where the places that the plant can be found throughout the world are described.

The section Food Uses summarizes different aspects of the traditional human consumption of the plant, such as where the plant is eaten in the Mediterranean region (countries sorted in clockwise order), what part of the plant is used, and how it is consumed (mode of preparation, traditional recipes, etc.).

In the section Other Uses, a summary of other interesting uses (medicinal, technological, cosmetic, etc.) are related. Finally the section of Historical References tries to present (when possible) a brief outline on the use of the species in the past (especially its edible use) that enable us to better understand its present use.

As usual in ethnobotanical compilations, some taxonomical problems have been found in the elaboration of the monographs, related with the allocation of the botanical species (scientific name) and the ethnospecies (the popular species). Some genera (e.g. *Rubus, Taraxacum*) are taxonomically very complicated. In the case of *Rubus*, the most frequent species are *R. ulmifolius* grex in the West Mediterranean area and *R. fruticosus* grex in the East, and due to its similarity they have been sometimes confused. Moreover, both taxa have nearly the same uses, and it is likely that people use them indistinctly. In this monograph we have considered the species *R. ulmifolius*, but in some cases the cited uses could have been referred to other closely related species. Similar comments could be done in the case of the genus *Taraxacum*. We have accepted and compiled the ethnobotanical data published under the quite ambiguous name *T. officinale*, though this botanical name is not presently accepted by the specialized botanists.

Other problems encountered are related to the different botanical taxonomies followed in the diverse ethnobotanical works. This is the case of the wild chard that can be considered an independent species (*Beta maritima* L.), feral individuals of the cultivated *B. vulgaris* L., or a subspecies, variety or even a form of *B. vulgaris*. Therefore, if we find *B. vulgaris* for a wild chard in an ethnobotanical work, it is not possible to know if it refers to *B. maritima*. So, we only have included information named as *B. maritima* or *B. vulgaris* subsp. *maritima*.

13.2.3 Elaboration of Food Composition Tables

For their elaboration, most of the criteria recommended by Greenfield and Southgate (2003) for production, management and use of FCD have been approached, especially those concerning to nomenclature, consistency of units (International

System), significant figures, factors and procedures used in rounding values. These tables are therefore not a simple compilation of data. Their elaboration requires experience in food analysis and in FCD management in order to correctly scrutinize literature data to assess their suitability for inclusion in these databases.

The methodology for the elaboration of the present Food Composition Tables follows a combination of direct and indirect methods, containing original analytic values together with other values taken from the literature and a few ones from other databases (Greenfield and Southgate 2003).

FCD should include all the nutrients or other bioactive food components that are known to be important in human nutrition. As this ideal can rarely be achieved, decisions must be made on priorities, based on the state of knowledge in nutritional and toxicological sciences, the availability of existing data, the existence of analytical methods, or the feasibility of analytical work (Greenfield and Southgate 2003). In this way, data of a great variety of macronutrients, micronutrients, and bioactive compounds have been compiled, as is described below.

Foods, being biological materials, exhibit variations in composition. Thus, as nutrient levels may vary greatly because of genetic or environmental factors, the variability in food composition must be covered in FCD, especially in the most affected parameters, for example, carbohydrates and mineral contents. For that reason, some of the analytical data recorded were obtained by the authors from samples gathered in at least two different sites and two different years for each species. In these analyses, good laboratory practices, in agreement with the criteria described by Greenfield and Southgate (2003), have been followed in the collection, handling, and analysis of the samples, since the primary objective of FCD is to provide their users accurate compositional information. Many of the analyses were performed in the laboratories of Departamento de Nutrición y Bromatología II-Bromatología (Facultad de Farmacia, Universidad Complutense de Madrid, Spain), with the collaboration of several other laboratories (Centro de Investigação de Montanha-Instituto Politécnico de Bragança, Portugal; Instituto de Ciencia y Tecnología de Alimentos y Nutrición-Consejo Superior de Investigaciones Científicas, Madrid, Spain). Some of these results have been presented as internal laboratory reports or doctoral thesis.

Other data have been taken from different scientific literature sources, either to complete the directly obtained analytical data or to have information on other non-analyzed species. These data were compiled by searching and evaluating a variety of scientific literature sources, including a thorough revision of scientific databases, taking special care in designing the search strategies and keywords. Different types of data sources were found: primary publications (mainly journals on Food Science and Nutrition) and secondary publications (reviews and other published compilations of compositional data, including Food Composition Tables, computerized databases, and material published in books). All these bibliographic materials have been scrutinized and evaluated following quality criteria, especially in that related to identification and origin of the plants, edible parts analyzed, sampling, analytical procedures, and general reliability of the data presented. As a result, some have been rejected and other incorporated into the present tables, recording all the relevant information. Several sets of compatible values have been obtained for each wild plant studied, which have been re-evaluated to detect "outlier" values (those who lie outside a reasonable variability of the values). In this case, a decision was taken, judging the values with a higher level of confidence, and selecting a reasonable range of variation based on scientific criteria or analytical experience.

Thus, in agreement with Greenfield and Southgate (2003), data used in the elaboration of FCD may be:

- Original analytical values. Many data herein presented are original analytical values obtained by a combination of direct (analytically) or indirect (compiled) methods. They have been assimilated into the database unmodified or as a selection or average of analytical values. Some original calculated values may enter this category (e.g. protein values calculated by multiplying nitrogen by the suitable factor).
- Imputed values. Estimated data derived from analytical values obtained from incomplete or partial analyses of a food (e.g. values expressed on fresh basis obtained from dry basis values).
- Calculated values. For example, average of different analytical values coming from different sources, or data about fatty acids categories obtained through the sum of individual compounds.
- Borrowed data, taken from other tables or databases. These data are rare, since the presence of wild plant foods in conventional FCD is scarce, and they do not always indicate the origin of the samples.

In every case, original analytical data (from either directly made analysis or rigorously scrutinized sources) are the most frequent in the tables presented; borrowed data have been included only when original analytical data were not available or were considered of insufficient quality. Other types of values that can be found in the tables presented are, for example, trace values (traces), which mean that the constituent is present but at a level that cannot be measured adequately; non-detected (nd) values, which are used when analytical work has shown that a constituent is not presumably present in the food (below detection limit of the analytical technique applied).

For each species presented, the composition of the edible parts are described and expressed per 100 g of edible portion (fresh weight basis). In some cases, due to differences in the expression of the results in literature, transformations have been done with the given moisture content to obtain the final data. Sometimes an aproximation has been done, estimating fresh basis values from the average moisture content of the vegetable (when the original source presents dry basis data, without a moisture value); when the only data appearing in the table is obtained this way, it is presented as aproximate by the symbol "≈". The criteria of Greenfield and Southgate (2003), in terms of units and number of significant digits, have been generally followed.

A wide variability is often found in literature for each specific parameter and species, depending on the natural variation of plants, the analytical methodology applied, and other possible sources of variation. In order to take into account such variability, the tables include an average value (arithmetic mean), obtained from all the analytical reports compiled, and a variation range from individual data (as an interval between the lowest and the highest values reported) in those cases where enough data were available. In some cases, a wide variation range is recorded, when data from different populations of the studied species are compiled. On the contrary, when a small variation range is recorded, it may be due to a high stability in a given parameter or due to the fact that only one result was found in the literature. In the last case, the variation range depends only on the analytical technique and does not include natural variability. When no range is given means that, just one value with no deviation is found in the scientific literature.

As many wild plants are eaten cooked, and cooking process may lead to composition changes in the foods (mainly nutrient losses by either solubilisation on the surrounding liquid or thermal degradation), the composition of some cooked plants is also given in separate composition tables when available (as directly obtained original analytical data). In those cases, a standard cooking process (boiling in water) has been taken as a reference to estimate the nutrient composition of the final boiled product (it should be taken into account that variations in the traditional cooking preparations may occur and thus this would lead to differences in the final composition of the food).

13.2.4 Description of Food Composition Tables in the Monographs

All the parameters recorded for most of the selected species have been usually organized in three tables: the first for main constituents, the second for vitamins and other constituents (minor compounds, bioactive substances), and the third for fatty acids profile. At the end, there is a box of remarks. When available, two additional tables that presents the nutritional data of cooked material and other box of remarks are presented.

Table of Main Constituents

This table records data on proximal composition of the edible material: water, available carbohydrates, dietary fibre, proteins, lipids and minerals. Although the proximate system for routine analysis, devised in the middle nineteenth century (Henneberg and Stohmann 1860), has evolved to improve some of the methodologies used, the concept of proximal composition still remains nowadays. The term "proximate" may be useful to represent the gross components that make up foods, being the basis for the analysis of food for several purposes, including food legislation and labelling (Greenfield and Southgate 2003), and thus it is included for the species recorded in these tables. In many cases, subcategories of main constituents have been described, for example, individual sugars or mineral elements. The analytical methodology applied for the quantification of these components in plant foods may vary within the different bibliographic sources, but only scientifically accepted methodologies have been considered.

These tables start with the energy value, calculated from the parameters of proximal composition (lipids, available carbohydrates, proteins, fibre, and available organic acids such as citric, lactic, and malic acids), using the energy equivalents recommended by the Regulation (EU) No 1169/2011 of the European Union (European Parliament and Council 2011) on the provision of food information to consumers (9 kcal/g for lipids; 4 kcal/g for proteins and available carbohydrates; 2 kcal/g for dietary fibre; and 3 kcal/g for available organic acids being malic, citric, and lactic acids). Alcohol and polyols have been excluded because of the absence of reliable data of these nonrelevant compounds in the wild greens and fruits included in these tables.

It is essential to give values for water (moisture) content in published food composition data at all levels of data management, as variations in water content are important determinants of the levels of other components, and water content values make possible to compare nutrient values on a similar moisture basis (Greenfield and Southgate, 2003). Moisture analysis by desiccation methods (oven, vacuum oven, freeze-drying) has been considered as suitable for inclusion in the present tables.

The earlier practice of including carbohydrates "by difference" in FCD was frequent according to the state of knowledge of carbohydrate chemistry at the time of Weende proximate system of food analysis. Nowadays, carbohydrates contents may be identified much more in detail, and for that reason this practice has proven to be scientifically unsound, and according to Food and Agriculture Organization of the United Nations/World Health Organization (FAO/WHO 1998) should be phased out as soon as possible. For this reason, in the present tables, at least two values are recorded for most cases:

- Available carbohydrates, that is, the sum of all carbohydrates (monosaccharides, oligosaccharides, and starch) which are physiologically utilized. This content is usually measured by hydrolysis of high molecular weight carbohydrates and quantifying soluble sugars by colorimetry or other techniques. Some individual carbohydrates, such as mono and disaccharides, have been indicated when available, coming from more specific techniques, such as high performance liquid chromatography, with slight variations between the different studies. These contents may be more relevant for fruits and may be quite variable depending on the maturity stage.
- Dietary fibre, understood as all high molecular substances, which cannot be split by the enzymes of the human digestive system. Basically they include certain non-digestible oligosaccharides and polysaccharides as well as some other compounds (e.g. some phenolic compounds) that may be associated with them. It must be distinguished between water soluble (e.g. pectins, gums) and insoluble (e.g. cellulose, hemicellulose, lignin) dietary fibre. The different analytical methods generally used for the determination of the dietary fibre may lead to different results. The data recorded in these tables, come mostly from enzimatic–gravimetric methods.

Crude protein content usually includes low-molecular nitrogen compounds such as free amino acids and peptides, as well as other nitrogen compounds in low amounts. This value is obtained from the total N content determined by the Kjeldahl method (acid-catalyzed digestion, distillation, and titration), multiplied by 6.25, which is the most widely used factor applied to fruits and vegetables analysis and recommended for nutrition labelling by Regulation (EU) No 1169/2011 (European Parliament and Council 2011), on the provision of food information to consumers.

Crude lipid contents are included, usually coming from continuous or discontinuous extraction, using an organic solvent, and determined gravimetrically after evaporation.

Regarding mineral content, data may be given as ash (obtained by dry mineralization at temperatures of 450 °C or higher) or as a more detailed mineral profile including main macro and microelements, usually obtained by a more specific technique such as atomic absorption spectroscopy.

Table of Vitamins and Other Constituents

As previously explained in Chap. 6, "vitamin" is a physiological rather than a chemical term, expressing a certain physiological activity, due to a group of chemical compounds, usually related structurally to one another (vitamers). The different vitamers are often separately measured using many different techniques, and sometimes a value is calculated for a total vitamin activity. The available vitamin data for the selected species have been recorded, including:

- Hydrosoluble vitamins (including different vitamers, when available), such as vitamins C (ascorbic and dehydroascorbic acids) and B_9 (folic acid and folates).
- Liposoluble vitamins (including different vitamers, when available), such as vitamin E (tocopherols) and provitamin A (carotenoids). When possible, vitamin A activity has been calculated as retinol activity equivalents (RAE), rather than retinol equivalents, in agreement to Mahan et al. (2012), calculated as [1].

$$[1]\ \text{RAE} = (\beta\text{-carotene})/12 + (\alpha\text{-carotene} + \beta\text{-cryptoxhantine})/24$$

Other food constituents included in the present tables are:

- Vitamin-related compounds, such as carotenoids, either with or without provitamin A activity (when available). They are relevant due to their well-known antioxidant properties (Isler 1971). In some cases (specified), carotenoids are given as the sum of the whole carotenoids, expressed as β-carotene.
- Other compounds of interest, such as organic acids, considered as relevant for fruits and vegetables FCD (with specification of the individual compounds and special interest in oxalic acid as an antinutrient), phenolics (as total amount and/or individual compounds, depending on the information available), and any other food constituent of interest from a point of view of the food use of the described species.

Phenolic data are complex, due to the high variety of compounds involved, the limitations in methodology, and the form of expressing the results (as an individual compound, or as a sum of different identified compounds; fresh weight, dry weight, or amount per g of dry extract in a given solvent), which make data comparisons very difficult. Total phenolic contents are included in the tables, sometimes expressed in different units, and in some cases, information about different families of compounds (e.g. phenolic acids or anthocyanins) are given.

Table of Fatty Acids Profile

The values given represent the amounts of the individual fatty acids, expressed as percentage (%) of total fatty acids content. These data usually come from analysis made by gas chromatography of the methyl esters of the fatty acids prepared by transmethylation of the lipid extracts from foods, with slight variations among the different studies. Identification difficulties derived from methodology have been taken into account when compiling fatty acids proportions. Although other minor compounds may be present, only the main components of fatty acids fraction have been included in tables:

C12:0	Lauric acid
C14:0	Myristic acid
C16:0	Palmitic acid
C16:1	Palmitoleic acid
C18:0	Stearic acid
C18:1*n*−7	11-Octadecenoic acid
C18:1*n*−9	Oleic acid
C18:2*n*−6	Linoleic acid
C18:3*n*−3	α-Linolenic acid
C18:3*n*−6	γ-Linolenic acid
C18:4*n*−3	Stearidonic acid
C20:0	Arachidic acid
C20:1*n*−9	Eicosenoic acid
C22:0	Behenic acid
C22:1*n*−9	Erucic acid
C24:0	Lignoceric acid
C24:1*n*−9	Nervonic acid

These tables also include a separated section of categories with some useful calculated values. Firstly, the distribution of fatty acids in saturated fatty acids (SFA), monounsaturated fatty acids (MUFA), and polyunsaturated fatty acids (PUFA). Secondly, the distribution of unsaturated fatty acids in three categories related to the position of the double bonds: the omega-3 fatty acids (*n*−3), the omega-6 fatty acids (*n*−6), and the omega-9 fatty acids (*n*−9), which has nutritional importance due to their different properties in the human body.

References

Each table presents the citations of the consulted references used to calculate the average and range values for each compound. The complete references can be found in the list of references at the end of the chapter.

When available, the origin of the plants studied and the number of samples analysed (if more than one batch has been sampled) is also included. This may have interest in terms of representativeness of the results.

In all the cases, special care has been taken to compile only studies that indicate the wild origin of the samples. Usually, only data of plants gathered from the wild in the Mediterranean region and surrounding areas are recorded in the tables. However, other data have been exceptionally recorded when there is no information about a given parameter from Mediterranean plants, being in these cases indicated as a footnote.

Remarks
This part concisely records in a box, the most interesting features of the plants described, from a nutritional point of view, in the light of the current Nutrient Reference Values (NRV) or other claims established by European Regulations about nutrition and health claims and food information to consumers (European Parliament and Council 2006, 2011). In this sense, a given plant food has been considered as a source of a vitamin or mineral if 100 g of fresh material provides 15 % of NRV for that vitamin or mineral (taking into account the range of variability in the literature data). However, if the average value but not the whole range of variability reaches 15 % NRV, or if there is no range of variability given, it is mentioned that the plant may be sometimes a source of that nutrient.

Some specific remarks require some calculations:

- For plants being vitamin A sources, the calculated values of RAE according to [1].
- The contribution of tocopherols to vitamin E activity has been calculated according to FAO/WHO (2001) from D-α-tocopherol (100 % vitamin E activity), β-tocopherol (50 %), γ-tocopherol (10 %), and δ-tocopherol (3 %) contents.
- Mentions about *n*–3 fatty acids contents come from estimations made with the relative percentage of fatty acids, the amount of lipids in the sample and a conversion factor of 0.8 for vegetables and fruits given by Greenfield and Southgate (2003) to calculate the total fatty acids amount in the fresh plant.

This part also contains some information related to cautions in the consumption of some species that have been reported to be potentially dangerous for groups of population, due to the presence of some hazardous compounds, for example, oxalic acid that should be avoided by people suffering of some kidney diseases, or the presence of high levels of nitrates, whose intake should be limited in the diet of infants (Concon 1988).

13.3 Ethnobotanical and Nutritional Dossier of Selected Mediterranean Wild Edible Plants

The following pages present 38 monographs about 41 selected wild edible plants traditionally consumed in different countries of the Mediterranean region. Each one has been elaborated by different authors, according to the following list (referred as ethnobotanical part; Food Composition Tables responsibles). The coordination and edition of all of them have been responsibility of Javier Tardío and María de Cortes Sánchez-Mata as coordinators of this chapter.

1 *Allium ampeloprasum*, J Tardío, M Molina; P García-Herrera, MC Sánchez-Mata
2 *Anchusa azurea*, J Tardío; P Morales, M Cámara
3 *Apium nodiflorum*, M Molina, J Tardío; P Morales, M Cámara
4 *Arbutus unedo*, J Tardío, M Molina; V Fernández-Ruiz, BM Ruiz-Rodríguez

5 *Asparagus acutifolius*, J Tardío, R Morales; P García-Herrera, M Cámara
6 *Beta maritima*, M Molina, J Tardío; P Morales, V Fernández-Ruiz
7 *Borago officinalis*, J Tardío; E Torija, JL Guil-Guerrero
8 *Bryonia dioica*, M Molina, J Tardío; C Díez-Marqués, MC Sánchez-Mata
9 *Capsella bursa-pastoris*, M Molina, J Tardío; C Díez-Marqués, MC Sánchez-Mata
10 *Chenopodium album*, M Molina, J Tardío; D Sánchez-Mata, MC Sánchez-Mata
11 *Chondrilla, juncea*, J Tardío; P García-Herrera, M Cámara
12 *Cichorium intybus*, M Molina, J Tardío; P García-Herrera, V Fernández-Ruiz
13 *Crataegus monogyna*, J Tardío, R Morales, M Pardo-de-Santayana; V Fernández-Ruiz, BM Ruiz-Rodríguez
14 *Crithmum maritimum*, R Morales, J Tardío; E Torija, MC Sánchez-Mata
15 *Eruca vesicaria*, J Tardío, M Molina; D Sánchez-Mata, C Díez-Marqués
16 *Foeniculum vulgare*, M Molina, J Tardío; P Morales, C Díez-Marqués
17 *Humulus lupulus*, M Molina, J Tardío; P García-Herrera, M Cámara
18 *Malva sylvestris*, J Tardío, R Morales; MC Matallana, MC Sánchez-Mata
19 *Montia fontana*, J Tardío, M Pardo-de-Santayana; P Morales, MC Sánchez-Mata
20 *Myrtus communis*, J Tardío; D Sánchez-Mata, N Boussalah
21 *Papaver rhoeas*, J Tardío, R Morales; P Morales, C Díez-Marqués
22 *Plantago lanceolata* and *P. major*, J Tardío, M Molina, M Pardo-de-Santayana; D Sánchez-Mata, C Díez-Marqués, E Torija, JL Guil-Guerrero
23 *Portulaca oleracea*, J Tardío, R Morales; E Torija, JL Guil-Guerrero
24 *Prunus spinosa*, M Pardo-de-Santayana, J Tardío, R Morales; V Fernández-Ruiz, BM Ruiz-Rodríguez
25 *Rorippa nasturtium-aquaticum*, J Tardío, M Molina; MC Matallana, MC Sánchez-Mata
26 *Rubus ulmifolius*, M Pardo-de-Santayana, J Tardío, R Morales; V Fernández-Ruiz, BM Ruiz-Rodríguez
27 *Rumex acetosa* and *R. papillaris*, J Tardío, R Morales; P Morales, M Cámara
28 *Rumex pulcher*, J Tardío, R Morales; P Morales, M Cámara
29 *Scolymus hispanicus*, J Tardío, M Pardo-de-Santayana; P García-Herrera, V Fernández-Ruiz
30 *Silene vulgaris*, J Tardío; P Morales, MC Sánchez-Mata
31 *Silybum marianum*, J Tardío; P García-Herrera, MC Sánchez-Mata
32 *Smilax aspera*, R Morales, J Tardío; MC Matallana, MC Sánchez-Mata
33 *Sonchus asper*, M Molina, J Tardío; C Díez-Marqués, MC Sánchez-Mata
34 *Sonchus oleraceus*, J Tardío, R Morales; P García-Herrera, MC Sánchez-Mata
35 *Tamus communis*, J Tardío; P García-Herrera, MC Sánchez-Mata
36 *Taraxacum officinale* and *T. obovatum*, J Tardío, M Molina, R Morales; MC Matallana, C Díez-Marqués, P García-Herrera, P Morales
37 *Urtica dioica*, R Morales, J Tardío; C Díez-Marqués, MC Sánchez-Mata
38 *Ziziphus lotus*, J Tardío, R Morales; N Boussalah, MC Sánchez-Mata

Regarding the authors of photographs, all of them has been taken by Javier Tardío, except for 3.2, 3.3 (*Apium nodiflorum*), 17.1 and 17.4 (*Humulus lupulus*) by María Molina, and 22.3 (*Plantago major*), 30.2 (*Silene vulgaris*), and 32.2 (*Smilax aspera*) by Ramón Morales.

13.3.1 Allium ampeloprasum L. (Amaryllidaceae)

Common Names Wild leek, *ajoporro* (es), *poireau d'été* (fr), *porro selvatico* (it), *agriopraso* (el), *pırasa* (tr), *thum barri* (ar).

Description Perennial and bulbous plant, which can reach more than 1 m tall when in bloom. The bulb is surrounded by numerous bulblets nearly spherical. It has a thick cylindrical stem with leaves that reach half its height. They are lanceolate, wide, flat, glabrous, and with a strong smell of garlic. The inflorescence is a characteristic umbel, almost spherical, about 8 cm in diameter, with numerous tiny white to purple flowers. Its height and broad leaves folded along the midrib clearly distinguishes this species from other wild and also edible garlics.

Ecology and Distribution It grows in almost all types of soils with preference for well drained and rich in organic matter ones, in clearings, thickets, meadows, rocky outcrops, dunes, and also roadsides and path verges, gardens, and waste ground. Native to the Mediterranean region, it has been introduced in other regions of the world, such as North and South America and Australia.

Food Uses The bulb and the pseudostem formed by the overlapping basal leaves are traditionally gathered before blooming and consumed both as a vegetable and as a condiment in the Mediterranean region. Its use has been registered at least in Spain (e.g. Cobo and Tijera 2011; Perera López 2005; Tardío et al. 2002; Tejerina 2010), Italy (e.g. Guarrera 2006; Lentini and Venza 2007; Leonti et al. 2006; Picchi and Pieroni 2005), Slovenia (Cerne 1992), Croatia (Łuczaj et al. 2013b), Greece (Hadjichambis et al. 2008; Leonti et al. 2006), Cyprus (Della et al. 2006), Turkey (Dogan et al. 2004; Ertuğ 2014), Jordan (Al-Qura'n 2010), Palestine (Ali-Shtayeh et al. 2008), and Tunisia (Le Floc'h 1983). As a vegetable, it is consumed raw, directly from the field or used in salads, and much more frequently cooked. The edible part is usually boiled and then consumed in different ways, such as seasoned with olive oil and vinegar, fried in the pan, sometimes with scrambled eggs or in omelettes (e.g. Couplan 2009a; Dogan 2012; Hadjichambis et al. 2008; Tardío et al. 2006). Occasionally, they are preserved in vinegar (Barandiarán and Manterola 1990) or in oil, after being boiled for some minutes in water and vinegar (Picchi and Pieroni 2005). As a condiment, it has been used as a garlic substitute being included in different recipes, such as soups, sauces, chicken broth, potato stews, or as a seasoning for olives (e.g. Cobo and Tijera 2011; Lentini and Venza 2007; Molina 2001; Tardío et al. 2002; Tejerina 2010).

Other Uses Though not widely extended, it also has medicinal uses (Benítez 2009; Guarrera 2006; Guenaoui et al. 2013; Triano et al. 1998). It has been mentioned for its antihelminthic and diuretic properties and its use against hypertension (Guarrera 2006). Its consumption has been also popularly considered healthy for its digestive properties (Benítez 2009) or for lowering fat content (Triano et al. 1998).

Historical References This plant is known since antiquity. Although Theophrastus (third century BC) does not include a clear description of the species when he treats the cultivated leek, onion, and garlic, it is likely that he referred to *A. ampeloprasum* when he mentioned other wild similar species (Teofrasto 1988). Nevertheless, Dioscorides (first century AD) clearly specified the medicinal properties of the wild leek, a plant known as *ampeloprason* by the Greeks. He said that it is worse for the stomach than leeks but warmer and more diuretic, expelling the menstrual blood, and also being good for those bitten by poisonous beasts (Laguna 1555; Osbaldeston 2000). Andres Laguna, the Spanish translator of Dioscorides, explained in his own comments (Laguna 1555) that this wild species typically grows inside the vineyards, hence called *ampeloprason* (vineyard leek). According to the Spanish botanist Bernardo Cienfuegos (1627), this species was consumed in the seventeenth century in Spain and was described as a plant, with intermediate characteristics between garlic and leek.

Food Composition Tables for raw bulbs and pseudostem of *Allium ampeloprasum* (Tables 13.1, 13.2 and 13.3).

Table 13.1 Main constituents, per 100 g of fresh bulbs and pseudostems of *A. ampeloprasum*

	Units	Average	Range	References
Energy (calculated value)	kcal	85	65–103	
Moisture	g	78.3	76.3–80.3	1, 2
Available carbohydrates	g	16.6	12.8–19.7	2
Dietary fibre	g	4.23	3.72–4.74	2
Proteins	g	1.67	1.31–2.03	2
Lipids	g	0.34	0.13–0.61	2, 3
Ash	g	0.79	0.59–0.99	2
K	mg	455	145–600	2, 4
Na	mg	32.7	10.8–67.3	2, 4
Ca	mg	75.6	30.2–88.7	2, 4
Mg	mg	17.1	7.80–21.4	2, 4
Fe	mg	0.54	0.28–0.92	2, 4
Cu	μg	110	40–250	2
Mn	μg	110	70–152	2
Zn	μg	752	30–1672	2, 4

1 Sánchez-Mata et al. (2012), Spain (2 samples), *2* García-Herrera et al. (2014a), Spain (4 samples), *3* Morales et al. (2012b), Spain (4 samples), *4* Romojaro et al. (2013), Spain

Table 13.2 Vitamins and other constituents, per 100 g of fresh bulbs and pseudostems of *A. ampeloprasum*

	Units	Average	Range	References
Vitamin B_9 (total folates)	μg	145	100–190	1
Vitamin C	mg	6.69	2.71–12.67	2, 3
Ascorbic acid	mg	3.44	1.66–6.94	2, 3
Dehydroascorbic acid	mg	2.14	0.53–3.83	2, 3
Vitamin E				
α-tocopherol	mg	0.03	0.02–0.04	3
β-tocopherol	mg	nd	–	3
γ-tocopherol	mg	nd	–	3
δ-tocopherol	mg	0.02	0.02–0.02	3
Organic acids				
Oxalic acid	mg	50.3	13.32–239	2, 3
Glutamic acid	mg	51.7	5.90–160	3
Malic acid	mg	70.9	7.57–275	2, 3
Citric acid	mg	24.4	9.44–58.3	2, 3
Fumaric acid	mg	0.85	0.42–1.29	2
Succinic acid	mg	2.14	Traces–3.23	3
Phenolics (total)	mg	42.2[a]	37.6–46.8	3
Flavonoids	mg	6.30[b]	5.93–6.66	3

1 Morales et al. (2015), Spain (4 samples), *2* Sánchez-Mata et al. (2012), Spain (2 samples), *3* García-Herrera et al. (2014a), Spain (4 samples); *nd* non-detected values

[a] Expressed as gallic acid

[b] Expressed as catechin

Table 13.3 Fatty acids profile (% of total fatty acids) of fresh bulbs and pseudostems of *A. ampeloprasum*

Individual compounds	Average	Range	References
12:0	0.18	0.15–0.21	1, 2
14:0	0.64	0.61–0.67	1, 2
16:0	26.4	26.1–26.7	1, 2
16:1	0.22	0.20–0.24	1, 2
18:0	3.30	2.95–3.65	1, 2
18:1$n-9$	7.39	6.97–7.81	1, 2
18:2$n-6$	53.4	53.2–53.7	1, 2
20:0	0.80	0.58–1.02	1, 2
22:0	2.75	2.70–2.80	1, 2
24:0	1.73	1.25–2.21	1, 2
Categories (calculated values)			
SFA	37.0	36.3–37.6	–
MUFA	7.9	7.6–8.1	–
PUFA	55.2	54.2–56.2	–
$n-3$	nd	–	–
$n-6$	55.2	54.2–56.2	–
$n-9$	7.9	7.6–8.1	–

1 Morales et al. (2012b) Spain (4 samples); *2* García-Herrera et al. (2014a) Spain (4 samples)*SFA* saturated fatty acids, *MUFA* monounsaturated fatty acids, *PUFA* polyunsaturated fatty acids; *nd* non-detected values

Remarks (bulbs and pseudostems of *A. ampeloprasum*)

- High available carbohydrate content, compared to other vegetables.
- High proportion of $n-6$ and low of $n-3$ fatty acids, compared to other vegetables.
- Source of dietary fibre (>3 g/100 g).
- Low Na content (<120 mg/100 g). It can be considered as a source of vitamin B_9 (folates); sometimes also a source of K and Cu.

13.3.2 *Anchusa azurea* Mill. (Boraginaceae)

Common Names Bugloss, *lenguaza* (es), *buglosse azurée* (fr), *buglossa* (it), *agoglossos* (el), *siğirdili* (tr), *mussais* (ar).

Description Perennial, herbaceous, and bristly plant, with a deep taproot that produces a rosette of basal leaves in springtime. It has acute, ovate-lanceolate leaves, and one or several stems that grow up to 1–1.5 m high. The lower leaves are petiolate, 30 cm long, while the apical ones are sessile and smaller. It has bright blue or violet flowers, with a white centre, forming loose racemes (cymes). They are 5-lobed, with densely hairy protuberances in the insertion of the calyx. The fruits are small and rough textured.

Ecology and Distribution It grows in waste grounds, roadsides, and arable lands, occasionally in scrublands. It is a drought- and frost-resistant species, preferring open sunny positions. Flowering occurs during late spring and early summer. Its range covers central, southern, and western Europe, northern Africa and western Asia, although it also grows naturalized in northern Europe.

Food Uses This species have been traditionally consumed as vegetable in some Mediterranean countries, such as Spain (e.g. Benítez 2009; Tardío et al. 2006), Italy (Guarrera 2006), Cyprus (Della et al. 2006; Hadjichambis et al. 2008), Turkey (Ertuğ 2014; Dogan 2012), Jordan (Al-Qura'n 2010), Tunisia (Le Floc'h 1983), and Morocco (Nassif and Tanji 2013). Though the plant is completely covered by bristles, they disappear after boiling. The majority of the ethnobotanical references mention the use of the basal leaves (sometimes only the midribs), but there are also references of the consumption of the young flowering shoots and even the roots. The whole leaves are usually cooked, either alone (Dogan 2012; Guzmán 1997; Hadjichambis et al. 2008; Tardío et al. 2002) or included in mixed vegetable recipes with up to 20 wild vegetables, as recorded in Spain (Verde et al. 1998) and Morocco (Nassif and Tanji 2013). In the south of Spain, the midribs of the leaves, the young shoots and even the roots are consumed as fritters, battered and fried in olive oil (Benítez 2009; Blanco and Cuadrado 2000; Fernández Ocaña 2000; Triano et al. 1998). It is said that its flavour is similar to fish fritters, so they are sometimes called *boquerones del campo* (field anchovies).

The sucking of nectar found in the basal part of the flowers is other minor but widely extended food use of this species, especially by children, at least in Spain (Benítez 2009; Fajardo 2008; Tardío et al. 2006), Italy (Couplan 2009a), Albania (Hadjichambis et al. 2008), and Jordan (Al-Qura'n 2010). This is the reason of one of its Spanish names, *chupamiel* (honey suck).

The consumption of other species of the same genus have been mentioned in the Mediterranean basin, such as *A. officinalis* L. in Italy (Picchi and Pieroni 2005) and south of France (Couplan 2009a), *A. strigosa* Banks & Sol. in Palestine (Ali-Shtayeh et al. 2008) and Cyprus (Della et al. 2006), and *A. arvensis* (L.) M. Bieb. in Croatia (Łuczaj et al. 2013b).

Other Uses The use of the crushed root against bleeding and for healing wounds seems to have been extended in the past, at least in the south of Spain (Benítez et al. 2010; Fernández Ocaña 2000; Mesa 1996). It has also been used for its anti-inflammatory properties, against snake bites, and, less frequently mentioned, against animal and human respiratory complaints (Benítez et al. 2010; Fajardo et al. 2007; Guarrera 2006). Some people consider that the consumption of the cooked leaves or roots is helpful against kidney stones (Benítez et al. 2010), an interesting reference of the popular use of this species as a medicinal food.

The aerial part cooked and mixed with flour or bread waste was widely used to feed animals in southern Spain, especially turkeys (Benítez 2009; Fernández Ocaña 2000; Sánchez-Romero 2003). It is also a well-known honeybee plant (Fernández Ocaña 2000; Tardío et al. 2002).

Historical References Although the leaves of this species were surely used as a vegetable since ancient times, the oldest reference we have found is from a Spanish botanist of the seventeenth century (Cienfuegos 1627). He mentioned that this species, called *lengua de buey* (ox tongue) or *lenguaza*, together with other wild vegetables, was consumed like spinach especially in times of scarcity.

However, the species of this genus were well known in the antiquity for their red roots, which were used as cheek colouring for women (Laguna 1555). Dioscorides also remarked the astringent properties of the root that was good (boiled in wax and oil) for skin problems, such as burns, ulcers, or erysipela (Osbaldeston 2000). As mentioned before, some of these uses have prevailed in Spain since the past century.

Food Composition Tables for raw tender leaves of *Anchusa azurea* (Tables 13.4, 13.5 and 13.6).

Table 13.4 Main constituents, per 100 g of fresh tender leaves of *A. azurea*

	Units	Average	Range	References
Energy (calculated value)	kcal	24	16–36	–
Moisture	g	91.2	89.7–92.7	1
Available carbohydrates	g	1.30	0.90–1.80	2
Dietary fibre	g	3.90	3.50–4.40	2
Proteins	g	1.45	1.00–2.80	2, 3
Lipids	g	0.54	0.07–0.93	2, 4
Ash	g	1.60	1.25–2.10	2, 3
K	mg	415	268–1172	2, 3
Na	mg	28.7	13.8–37.4	2
Ca	mg	164	126–219	2, 3
Mg	mg	17.4	9.60–35.1	2, 3
P	mg	66	–	3
Fe	mg	2.45	0.56–2.97	2, 3
Cu	µg	100	90–280	2
Mn	µg	469	150–699	2, 3
Zn	µg	307	185–860	2, 3

1 Morales et al. (2014) Spain (5 samples), *2* García-Herrera (2014) Spain (5 samples), *3* Ayan et al. (2006) Turkey

Table 13.5 Vitamins and other constituents, per 100 g of fresh tender leaves of *A. azurea*

	Units	Average	Range	References
Vitamin B_9 (total folates)	µg	278	256–299	1
Vitamin C	mg	12.1	5.41–18.11	2
Ascorbic acid	mg	0.67	0.57–1.39	2
Dehydroascorbic acid	mg	11.74	4.81–18.15	2
Vitamin E				
α-tocopherol	mg	0.36	0.28–0.44	2
β-tocopherol	mg	0.05	0.05–0.05	2
γ-tocopherol	mg	0.11	0.09–0.13	2
δ-tocopherol	mg	0.01	0.01–0.01	2
Organic acids				
Oxalic acid	mg	378	110–640	2
Glutamic acid	mg	191	181–211	2
Malic acid	mg	42.9	24.7–61	2
Citric acid	mg	8.02	nd–13.2	2
Fumaric acid	mg	71.8	26–130	2
Phenolics (total)	mg	148[a]	144–150	2
	mg	≈179[b]	176–182	3
Flavonoids	mg	84.8[c]	80.8–84.8	2

1 Morales et al. (2015) Spain (5 samples), *2* Morales et al. (2014) Spain (5 samples), *3* Conforti et al (2011) Italy; *nd* non-detected values

[a] Expressed as gallic acid

[b] Expressed as chlorogenic acid

[c] Expressed as catechin

Table 13.6 Fatty acids profile (% of total fatty acids) of fresh tender leaves of *A. azurea*

Individual compounds	Average	Range	References
12:0	0.07	0.05–0.09	1
14:0	0.35	0.33–0.37	1
16:0	10.4	9.83–11.1	1
16:1	0.14	0.13–0.15	1
18:0	1.67	0.48–2.86	1
18:1*n*−9	2.20	2.20–2.20	1
18:2*n*−6	12.2	12.0–12.3	1
18:3*n*−3	64.7	64.5–65.0	1
18:3*n*−6	1.46	1.43–1.49	1
20:0	1.64	2.47–1.81	1
20:1*n*−9	0.17	0.16–0.18	1
22:0	1.25	1.19–1.31	1
24:0	0.72	0.63–0.81	1
24:1*n*−9	0.40	0.37–0.43	1
Categories (calculated values)			
SFA	16.6	14.7–18.3	–
MUFA	3.1	3.0–3.1	–
PUFA	80.4	78.7–82.2	–
n−3	67.9	66.5–69.5	–
n−6	14.0	13.8–14.2	–
n−9	3.1	3.0–3.1	–

1 Morales et al. (2012b) Spain (4 samples)*SFA* saturated fatty acids, *MUFA* monounsaturated fatty acids, *PUFA* polyunsaturated fatty acids

Remarks (tender leaves of *A. azurea*)

- Low energy value (<40 kcal/100 g).
- Source of dietary fibre (>3 g/100 g).
- High proportion of PUFA and *n*−3 fatty acids content (mainly α-linolenic acid, 18:3*n*−3), compared to other vegetables. Good *n*−3/*n*−6 ratio (>4).
- High content of oxalates.
- Very low Na content (<40 mg/100 g). It can be considered as a source of Ca and vitamin B_9 (folates); sometimes also a source of K, Fe, Mn, and vitamin C.

13.3.3 Apium nodiflorum (L.) Lag. (Apiaceae)

Common Names Fool's watercress, *rabaça* (pt), *berra* (es), *ache faux cresson* (fr), *sedanina d'acqua* (it), *arkoseleno* (el), *bendik* (tr), *qurra* (ar).

Description Perennial herb, glabrous, with hollow and long stems up to 1 m, prostrate or floating in the water surface, rooting at the lower nodes. The leaves are alternate and pinnate with 3–13 ovate-lanceolate leaflets. The inflorescence is an umbel of umbels, up to 2 cm, with 3–15 rays, and very small white, 5-petalled flowers. The fruits 2–2.5 mm are ovoid, glabrous, with stout ridges.

It could be confused with the also edible *Rorippa nasturtium-aquaticum* (L.) Hayek and with the supposedly poisonous *Berula erecta* (Huds.) Coville, but the leaf stalks of the latter have a distinct ring towards the base, and the inflorescence is bigger and with long bracts.

Ecology and Distribution It grows in wet places such as ditches or streams, usually associated with *R. nasturtium-aquaticum*, 0–1200 m. It flowers from April to July. It is common in Central and South Europe, more frequent in the southwest; also in the north of Africa and west and central Asia.

Food Uses This species has been traditionally consumed in the Mediterranean countries, at least in Portugal (Carvalho and Telo 2013), Spain (e.g. González et al. 2011a; Menendez-Baceta et al. 2012; Parada et al. 2011; Tardío et al. 2006), Italy

(e.g. Ghirardini et al. 2007; Guarrera 2006; Lentini and Venza 2007; Picchi and Pieroni 2005), Cyprus (Della et al. 2006; Hadjichambis et al. 2008), Turkey (Ertuğ 2014), Lebanon (Marouf 2005), Jordan (Al-Qura'n 2010; Tukan et al. 1998), and Morocco (Nassif and Tanji 2013). Its tender leaves and stems are collected in winter and spring (González et al. 2011a). As other wild vegetables that grow in aquatic environments, *A. nodiflorum* is mainly consumed raw without any preparation or in green salads (e.g. Al-Qura'n 2010; Carvalho and Telo 2013; Della et al. 2006; Marouf 2005; Nassif and Tanji 2013; Nebel et al. 2006; Tardío et al. 2005). It is also cooked in different recipes, such as boiled with legumes (Della et al. 2006), in omelettes (Verde et al. 2003), as a turnover filling (Tukan et al. 1998), as a dough stuffing (Al-Qura'n 2010), or included in the Moroccan *beqoul*, a mixture of several wild food plants used to prepare a springtime meatless dish—highly prized by rural people, especially in mountainous areas (Nassif and Tanji 2013).
The fool's watercress is also employed as a condiment to aromatize soups or other stews (Della et al. 2006; Lentini and Venza 2007).

Other Uses Its tender aerial parts are considered digestive (Guarrera 2006; Rivera et al. 2005), intestinal anti-inflammatory (Bonet and Vallès 2002; Guarrera 2006), depurative, and employed to treat respiratory and urinary diseases (Guarrera 2006). Other perceived health benefits of *A. nodiflorum* include liver protection and bile production (Marouf 2005). The decoction of the whole plant was topically employed to treat eczema in southern Spain (Benítez et al. 2010).

Additionally, the fresh aerial parts were used as animal feed for pigs and partridges (Aceituno-Mata 2010; González et al. 2011b) and in ethnoveterinary science (Guarrera 2006).

This plant has been traditionally compared with the watercress *(Rorippa nasturtium-aquaticum)*, as its common and rather pejorative names in several languages (at least in English, French, and Spanish) seem to point out. They share the same habitat and can be confused, especially when the watercress is close to blooming. Although *A. nodiflorum* is registered as a vegetable in many ethnobotanical studies, some people considered that it is toxic in contraposition with the edibility of the watercress (Tardío et al. 2002; Velasco et al. 2010).

Historical References This species has probably been eaten since antiquity, though it is difficult to find any clear references to it. For instance, some of the ancient mentions of the celery (*A. graveolens* L.), a closely related species might be attributed to *A. nodiflorum*. This could be the case of some references found in Theophrastus or Dioscorides. In the sixteenth-century Spanish translation of the later work, Laguna (1555), in the chapter of the watercress (Book II, chapter CXVII), describes another similar species supposed to be toxic and with the Castilian name of *berrazas*, the same registered in the recent ethnobotanical works.

Food Composition Tables for raw young leaves and stems of *Apium nodiflorum* (Tables 13.7, 13.8 and 13.9).

Table 13.7 Main constituents, per 100 g of fresh young leaves and stems of *A. nodiflorum*

	Units	Average	Range	References
Energy (calculated value)	kcal	21	11–25	–
Moisture	g	92.0	90.0–94.0	1, 2
Available carbohydrates	g	1.20	0.70–2.10	2
Dietary fibre	g	2.70	1.90–3.40	2
Proteins	g	1.60	1.10–2.10	2
Lipids	g	0.42	0.07–0.74	2, 3
Ash	g	1.70	1.00–3.30	2
K	mg	165	114–1054	2
Na	mg	244	137–379	2
Ca	mg	152	63.9–246	2
Mg	mg	28.0	16.4–44.9	2
Fe	mg	1.80	0.81–3.15	2
Cu	μg	80.0	40.0–150	2
Mn	μg	290	170–340	2
Zn	μg	500	420–700	2

1 Morales (2011) Spain (4 samples), *2* García-Herrera (2014) Spain (4 samples), *3* Morales et al. (2012b) Spain (4 samples)

Table 13.8 Vitamins and other constituents, per 100 g of fresh young leaves and stems of *A. nodiflorum*

	Units	Average	Range	References
Vitamin B_9 (total folates)	μg	125	97–138	1
Vitamin C	mg	20.3	10.3–30.2	2
Ascorbic acid	mg	9.11	4.96–13.25	2
Dehydroascorbic acid	mg	15.0	11.6–18.4	2
Vitamin E				
α-tocopherol	mg	2.59	2.56–2.62	3
β-tocopherol	mg	0.25	0.25–0.25	3
γ-tocopherol	mg	0.21	0.21–0.21	3
δ-tocopherol	mg	nd	–	3
Organic acids				
Oxalic acid	mg	534	189–879	2
Malic acid	mg	72.7	18.4–127	2
Citric acid	mg	28.0	6.63–49.3	2
Fumaric acid	mg	1.37	0.09–2.65	2
Phenolics (total)	mg	80.5[a]	76.1–84.9	3
Flavonoids	mg	45.5[b]	43.9–47.1	3

1 Morales et al. (2015) Spain (4 samples), *2* Morales (2011) Spain (4 samples), *3* Morales et al. (2012a) Spain (4 samples); *nd* non-detected values

[a] Expressed as gallic acid

[b] Expressed as catechin

Table 13.9 Fatty acids profile (% of total fatty acids) of fresh young leaves and stems of *A. nodiflorum*

Individual compounds	Average	Range	References
12:0	nd	–	1
14:0	0.69	0.60–0.78	1
16:0	16.3	15.2–17.3	1
16:1	1.10	0.78–1.42	1
18:0	1.77	1.44–2.10	1
18:1$n-9$	3.33	3.31–3.35	1
18:2$n-6$	24.6	23.8–25.4	1
18:3$n-3$	43.5	43.4–43.5	1
20:0	0.34	0.30–0.38	1
20:1$n-9$	nd	–	1
22:0	1.22	1.17–1.27	1
22:1$n-9$	nd	–	1
24:0	2.07	2.05–2.09	1
24:1$n-9$	nd	–	1
Categories (calculated values)			
SFA	23.6	22.7–24.5	–
MUFA	4.7	4.4–4.9	–
PUFA	71.7	70.6–72.9	–
$n-3$	45.8	44.6–47.1	–
$n-6$	25.9	25.8–26.0	–
$n-9$	4.7	4.4–4.9	–

1 Morales et al. (2012b) Spain (4 samples); *nd* non-detected values
SFA saturated fatty acids, *MUFA* monounsaturated fatty acids, *PUFA* polyunsaturated fatty acids

Remarks (young leaves and stems of *A. nodiflorum*)

- Very low energy value (<25 kcal/100 g).
- High proportion of PUFA, compared to other vegetables.
- Very high content of oxalates: people with altered renal function should avoid this vegetable; boiling is recommended for general population.
- It can be considered as a source of vitamins E and B_9 (folates); sometimes also a source of Ca and vitamin C.

13.3.4 Arbutus unedo L. (Ericaceae)

Common Names Strawberry tree, *medronho* (pt), *madroño* (es), *arbousier* (fr), *corbezzolo* (it), *kumariá* (el), *kocayemiş* (tr), *sasnou* (ar).

Description Evergreen shrub or small tree, commonly 3–5 m tall, that has a twisted and gnarled trunk, with grey-brown or reddish bark that flakes and peels in thin plates. The alternate leaves are oval to oblong, glossy and leathery, dark green above and pale green underneath, with a serrated margin. The flowers are bell shaped, white or pale pink, and grouped in hanging racemes. The fruits are red berries, 1–2 cm diameter, with a rough surface.

Ecology and Distribution It grows in sclerophyllous shrublands, rocky slopes, and evergreen forests, up to 800–1200 m, in different types of usually well-drained soils. Blooming begins in autumn and continues into the winter. Since the fruits take a year to ripen, mature fruits are present when blooming. It is widespread in the Mediterranean Basin, from southwest to east of Europe, north of Africa (excluding Egypt and Libya) and Asia Minor, with some Atlantic locations in western France and southwestern Ireland. In the East Mediterranean, it coexists with *A. andrachne* L., also with edible fruits.

Food Uses The fruits of this small tree have been traditionally consumed, raw as a snack, as a dessert or, sometimes, used for elaborating jam or liqueurs, in most of the Mediterranean countries. It has been widely used at least in Portugal (Carvalho 2010; Mendonça de Carvalho 2006), Spain (Molina et al. 2011; Tardío et al. 2006), Italy (Guarrera 2006; Hadjichambis et al. 2008; Leonti et al. 2006), Croatia (Łuczaj et al. 2013a), Bosnia-Herzegovina (Redžić 2006), Greece (Leonti et al. 2006), Turkey (Dogan et al. 2004; Ertuğ 2014), Tunisia and Algeria (Le Floc'h 1983), and Morocco (Nassif and Tanji 2013). It is one of the most important wild fruit species in Spain (Tardío et al. 2006) and probably also in other Mediterranean countries. In general, the strawberry tree fruits only reach a really pleasant flavour if consumed totally ripe or even slightly overripe, then being delicate to transport. Some preservation techniques for extending their availability have also been used throughout the Mediterranean region, such as home-made compotes or jams (e.g. Bonet and Vallès 2002; Guarrera 2006; Łuczaj et al. 2013a; Verde et al. 2003). Furthermore, aromatic alcoholic beverages have been locally elaborated in different areas fermenting and distilling the fruits, being known as *aguardente de medronho* in Portugal, *koumaro* in Greece, and "liquore di corbezzolo" in Italy (Espírito-Santo et al. 2012), and sometimes even commercialized (Alarcão-E-Silva et al. 2001). On other occasions and especially in recent times, the liqueur is elaborated without distillation, by macerating the fruits in liquor (Guzmán 1997; Mendonça de Carvalho 2006; Verde et al. 2001).

Although smaller than those of *A. unedo*, the fruits of the Greek strawberry tree (*A. andrachne*) are also consumed in the East Mediterranean (Ertuğ 2014; Tukan et al. 1998).

Other Uses Regarding medicinal uses, the leaves, bark, and roots are said to be astringent, antiseptic, and even antiparasitic. The decoction of leaves or roots has been traditionally employed against diarrhoea in Spain (Verde et al. 2001; Villar et al. 1987), Italy (Guarrera 2006), Libya (De Natale and Pollio 2012) and Tunisia (Le Floc'h 1983). In Algeria and Cyprus, the fruit liqueur is also considered anti-diarrhoeic (Le Floc'h 1983; Lentini and Venza 2007), whereas in Cyprus, it is also said to have aphrodisiac and sedative effects (Lentini and Venza 2007). It has antiseptic properties that seem to explain why it is widely used against respiratory problems (Guarrera 2006), wounds, and other skin infections (Criado et al. 2008; Guarrera 2006; Pellicer 2001; Villar et al. 1987) and specifically as antiparasitic for scabies (Guarrera 2006; Mendonça de Carvalho 2006).

Its hard wood has been used as firewood and for making spoons and other domestic tools (Blanco and Cuadrado 2000; Guarrera 2006; Mendonça de Carvalho 2006).

Nowadays, *A. unedo* is widely cultivated as an ornamental tree and has become very popular both in public and private gardens.

Historical References This Mediterranean small tree was already mentioned by Theophrastus (371–287 BC), who also described the differences with *A. andrachne* and even the existence of a hybrid form (Teofrasto 1988). There are also references from the first century AD, such as those of Dioscorides and Pliny the Elder (Bostock and Riley 1855; Osbaldeston 2000), though in their opinion, the consumption of the fruits was not very appreciated. Dioscorides said, "when eaten they are bad for the stomach and causes headaches", whereas Pliny directly commented, "this is a fruit held in no esteem, in proof of which it has gained its name of *unedo*, people being generally content with eating but one". The fruit does not seem to be neither very appreciated in the sixteenth century. Clusius (1576) said that "the fruit was eaten by the greedy and poor people" and that "in Lisbon were sold by poor women". He also referred that the water extracted from its leaves and flowers was used in Portugal against eyes disorders and as antidote for plague and poisoning. However, this tree is very popular in central Spain, where it makes up part of the coat of arms (*El oso y el madroño*, the bear and the strawberry tree) of the city of Madrid.

Food Composition Tables for raw fruits of **Arbutus unedo** (Tables 13.10, 13.11 and 13.12).

Table 13.10 Main constituents, per 100 g of fresh fruits of *A. unedo*

	Units	Average	Range	References
Energy (calculated value)	kcal	167	98–210	–
Moisture	g	57.5	42.7–72.1	1, 2, 3, 4, 5, 6
Available carbohydrates	g	30.4	16.9–37.9	2, 4
Soluble sugars	g	33.7	15.0–59.6	2, 4, 7, 8
Fructose	g	16.8	3.64–28.1	2, 4, 6, 7, 8
Glucose	g	8.73	2.34–21.7	2, 4, 6, 7, 8
Sucrose	g	5.79	Traces–6.37	2, 4, 7, 8
Maltose	g	1.49	Traces–3.47	2, 4, 7, 8
Dietary fibre	g	17.3	12.6–19.9	4, 6
Insoluble fibre	g	13.9	8.70–18.7	4, 6
Soluble fibre	g	3.44	1.90–4.86	4, 6
Proteins	g	1.06	0.65–1.62	1, 2, 3, 4, 6
Lipids	g	0.57	0.23–1.02	1, 2, 3, 4, 6
Ash	g	0.73	0.25–1.37	1, 2, 3, 4, 6
K	mg	334	115–768	1, 4, 6
Na	mg	20.0	5.5–36.1	1, 4, 6
Ca	mg	111	28.9–237	1, 4, 6
Mg	mg	9.66	8.38–10.9	6
P	mg	32.5	7.4–66.8	1, 4, 6
Fe	mg	0.92	0.38–1.47	1, 4, 6
Cu	μg	99	57–208	1, 4
Mn	μg	146	37–230	1, 4
Zn	μg	437	330–762	1, 4, 6

1 Özcan and Haciseferoğullari (2007) Turkey, *2* Barros et al. (2010b) Portugal, *3* Ganhão et al. (2010) Spain, *4* Ruiz-Rodríguez et al. (2011) Spain (6 samples), *5* Morales et al. (2013) Spain (6 samples), *6* Vidrih et al. (2013) Croatia, *7* Ayaz et al. (2000) Turkey, *8* Alarcão-E-Silva et al. (2001) Portugal

Table 13.11 Vitamins and other constituents, per 100 g of fresh fruits of *A. unedo*

	Units	Average	Range	References
Carotenoids				
β-carotene	mg	0.32	0.02–0.89	1, 2, 3
Lutein + zeaxhantin	mg	0.04	0.03–0.05	1
RAE (calculated value)	μg	27	1.5–45	–
Vitamin B_3 (niacin)	mg	1.97	1.28–2.65	4
Vitamin C	mg	185	6–431	1, 3, 5, 6, 7
Ascorbic acid	mg	139	5–389	1, 2, 3, 4, 5, 6, 7, 8
Dehydroascorbic acid	mg	33.3	nd–36	3, 5, 6, 7
Vitamin E				
α-tocopherol	mg	3.10	0.02–8.93	1, 2, 6
β-tocopherol	mg	0.11	0.02–0.18	2, 6
γ-tocopherol	mg	0.33	0.26–0.44	2, 6
δ-tocopherol	mg	0.10	0.03–0.17	2, 6
Organic acids				
Oxalic acid	mg	90.5	53.5–139	3, 6, 8
Quinic acid	mg	2277	2157–2396	4, 6
Malic acid	mg	620	219–1052	3, 4, 6, 8
Citric acid	mg	13.5	nd–27.0	6, 8
Fumaric acid	mg	0.36	0.47–3.24	3, 6, 8
Phenolics (total)	mg	542[a]	367–639	7, 9
	mg	≈584[b]	–	4
	mg	1235[c]	872–1598	5
Phenolic acids	mg	≈78.5[c]	78.3–78.7	10
	mg	668[a]	487–944	5
Hydroxybenzoic acids	mg	41.3[a]	37.8–44.8	9
Hydroxycinnamic acids	mg	0.37[d]	0.33–0.40	9
Flavonols	mg	1.14[c]	0.57–1.48	1
	mg	10.4[e]	1.14–51.7	5, 9
Anthocianins	mg	3.77[c]	2.91–4.63	1
	mg	≈40.3	39.9–40.7	4
	mg	567[f]	251–1154	5
Tanins	mg	≈74.4[b]	73.6–75.2	4
Proanthocyanidins	mg	27.5[c]	26.5–28.4	1

1 Pallauf et al. (2008) Spain, *2* Barros et al. (2010b) Portugal, *3* Ruiz-Rodríguez et al. (2011) Spain (6 samples), *4* Alarcão-E-Silva et al. (2001) Portugal, *5* Ruiz-Rodríguez et al. (2014b) Spain (6 samples), *6* Morales et al. (2013) Spain (6 samples), *7* Vidrih et al. (2013) Croatia, *8* Pereira et al (2013) Portugal, *9* Ganhão et al. (2010) Spain, *10* Ayaz et al. (2000) Turkey

RAE retinol activity equivalents; *nd* non-detected values

[a] Expressed as gallic acid

[b] Expressed as catechin

[c] Expressed as sum of individual compounds (gallic acid, quercetin, cyanidin or catechin derivatives as major components of different families)

[d] Expressed as chlorogenic acid

[e] Expressed as rutin

[f] Expressed as pelargonidin-3-glucoside

Table 13.12 Fatty acids profile (% of total fatty acids) of fresh fruits of *A. unedo*

Individual compounds	Average	Range	References
12:0	0.78	0.55–1.04	1, 2
14:0	1.48	1.19–1.70	1, 2
16:0	12.5	7.95–20.8	1, 2, 3
16:1	0.09	0.03–0.13	1, 2
18:0	2.78	Traces–4.55	1, 2, 3
18:1*n*−9	20.2	20.9–24.8	1, 2, 3
18:2*n*−6	26.8	21.4–37.2	1, 2, 3
18:3*n*−3	33.0	31.2–37.1	1, 2, 3
18:3*n*−6	0.68	0.66–0.70	2
20:0	0.30	nd–0.66	1, 2
20:1*n*−9	0.13	nd–0.29	1, 2
22:0	0.62	0.39–0.84	1, 2
24:0	0.92	0.14–1.45	1, 2
Categories (calculated values)			
SFA	19.5	17.7–18.4	–
MUFA	20.5	22.1–25.1	–
PUFA	59.9	56.5–60.0	–
n−3	33.0	31.5–37.8	–
n−6	26.9	22.1–25.1	–
n−9	20.4	22.0–25.0	–

1 Barros et al. (2010b) Portugal, *2* Morales et al. (2013) Spain (6 samples), *3* Vidrih et al. (2013) Croatia; *nd* non-detected values

SFA saturated fatty acids, *MUFA* monounsaturated fatty acids, *PUFA* polyunsaturated fatty acids

Remarks (fruits of *A. unedo*)

- High content of sugars.
- High in dietary fibre (>6 g/100 g).
- High *n*−3 fatty acids content (mainly α-linolenic acid, 18:3*n*−3), compared to other wild fruits.
- High amount of quinic and malic acids.
- Very low Na content (<40 mg/100 g). It may be considered as a source of vitamin C; sometimes also a source of K, Ca, and vitamin E.
- Considerable amount of anthocyanins (cyanidin-3-glucoside as major compound).

13.3.5 Asparagus acutifolius L. (Asparagaceae)

Common Names Wild asparagus, *espargo-bravo* (pt), *espárrago triguero* (es), *asperge sauvage* (fr), *asparago selvatico* (it), *šparoga* (hr), *sparangi* (el), *tilkişen* (tr), *halyoun* (ar).

Description Perennial plant, growing up to 1.5 m tall, with rhizomatous tubers, from which new tender shoots (asparagus) sprout in spring. Spiny stem formed by little and narrow false leaves (cladodes), gathered in fascicles, whereas the real leaves are like scales. Little flowers, with 6 yellow cream tepals 4 mm long, growing in groups of 2–4 in the nodes of the stems. The fruit is a black or blue poisonous berry, up to 8 mm, with 1 or 2 seeds.

Ecology and Distribution It grows in all types of soils, in scrublands and borders of woods, throughout the Mediterranean region, from the sea level up to 1300 m. The flowering time covers from spring to autumn.

Food Uses Its young shoots have been traditionally consumed in most Mediterranean countries, such as Portugal (Carvalho and Telo 2013; Mendonça de Carvalho 2006), Spain (e.g. Bonet and Vallès 2002; González et al. 2011a; Tardío et al. 2006), France (Chauvet 2003), Italy (e.g. Ghirardini et al. 2007; Guarrera 2006; Picchi and Pieroni 2005), Croatia (Łuczaj et al. 2013b), Cyprus (Della et al. 2006), Turkey (Dogan 2012; Ertuğ 2014), Lebanon (Marouf 2005), and Morocco (Nassif and Tanji 2013). Although they can be eaten raw (Ertuğ 2004; Picchi and Pieroni 2005;

Tardío et al. 2006), they are generally consumed cooked in different dishes, such as rice dishes, soups (Carvalho and Telo 2013; Criado et al. 2008; Hadjichambis et al. 2008), and more frequently with eggs, that is, omelettes, scrambled, or poached eggs (Carvalho and Telo 2013; Dogan 2012; Ertuğ 2004; González et al. 2011a; Guarrera 2006; Hadjichambis et al. 2008; Lentini and Venza 2007; Łuczaj et al. 2013b; Marouf 2005; Mendonça de Carvalho 2006; Nebel et al. 2006; Pieroni et al. 2005; Tardío et al. 2006). In Italy, they are also fried in the pan with oil and onion, to add aroma to the tomato sauce, as seasoning for *risottos*, or to fill *ravioli* (Lentini and Venza 2007; Picchi and Pieroni 2005). They are also used in some traditional recipes, such as the *migas de espargos* in Portugal (Carvalho and Telo 2013) or the *sagne larghe* in the Italian region of Abruzzos (Picchi and Pieroni 2005).

This wild asparagus is nowadays one of the most appreciated and widely consumed noncultivated plants throughout the Mediterranean region. Apart from rural communities, recreational harvesting of *A. acutifolius* attains new collectors from urban areas, such as retired people and Sunday excursionists (Catani et al. 2001; Picchi and Pieroni 2005). Bundles of wild asparagus are frequently sold in local markets and on roadsides (Chauvet 2003; Mendonça de Carvalho 2006; Nassif and Tanji 2013; Picchi and Pieroni 2005; Tardío 2010) and served in luxury restaurants (Łuczaj et al. 2012; Parada et al. 2011). They are highly valued and reach high market prices (Molina et al. 2012; Rosati et al. 2005).

Other Uses It is popularly considered a healthy food with diuretic and purifying effects (Carrió and Vallès 2012; Guarrera 2006; Lentini and Venza 2007; Parada et al. 2011; Rivera et al. 2005; Signorini et al. 2009; Vallejo et al. 2009; Vázquez et al. 1997). For this reason, its consumption is recommended in folk medicine as digestive (González-Tejero et al. 2008; Guarrera 2006), to treat several kidney- and liver-related disorders (Benítez et al. 2010; González-Tejero et al. 2008; Guarrera 2006; Picchi and Pieroni 2005), prostate illnesses (Vallejo et al. 2009), bile duct lithiasis or bile inflammation (Guarrera 2006; Rivera et al. 2005), and to lose weight (Guarrera 2006; Picchi and Pieroni 2005).
The decoction of the plant has also been used to aid the placental expulsion in animals, such as sheep and cows (Ferrández and Sanz 1993; Pellicer 2001).

Ethnobotanical studies refer that this plant usually grows in olive-tree groves (Benítez 2009; Rosati 2001). One informant of the first study assured that the wild asparagus plants are propagated by some birds that consume olives after having eaten asparagus fruits.

Historical References The spears of this species may have been consumed since antiquity. Dioscorides, in the first century AD, besides mentioning other medicinal properties of the plant, also pointed out that the young shoots, boiled and eaten, soothe the intestines and encourage urine (Osbaldeston 2000).

Food Composition Tables for tender shoots of *Asparagus acutifolius*, raw (Tables 13.13, 13.14 and 13.15) and boiled (Tables 13.16 and 13.17).

Table 13.13 Main constituents, per 100 g of fresh young shoots of *A. acutifolius*

	Units	Average	Range	References
Energy (calculated value)	kcal	40	23–56	–
Moisture	g	85.4	81.2–88.5	1,2
Available carbohydrates	g	3.56	1.03–4.67	1,2
Soluble sugars	g	0.70	0.68–0.72	1
Fructose	mg	190	180–200	1
Glucose	mg	150	150–150	1
Sucrose	mg	320	310–330	1
Trehalose	mg	40	40–40	1
Dietary fibre	g	4.83	4.71–6.63	2
Proteins	g	2.40	1.69–3.25	1,2
Lipids	g	0.61	0.32–0.99	1,2,3
Ash	g	2.23	0.93–3.70	1,2
K	mg	585	492–1370	2,4,5
Na	mg	18.5	12.6–45.6	2,4,5
Ca	mg	54.1	23.0–109	2,4,5
Mg	mg	36.6	17.0–109	2,4,5,
Fe	mg	0.66	0.33–1.11	2,5
Cu	µg	105	50–140	2
Mn	µg	410	70–880	2
Zn	µg	1059	110–1810	2,5

1 Martins et al. (2011) Portugal, *2* García-Herrera (2014) Spain (5 samples), *3* Morales et al. (2012b) Spain (5 samples) *4* Bianco et al. (1996) Italy, *5* Romojaro et al. (2013) Spain

Table 13.14 Vitamins and other constituents, per 100 g of fresh young shoots of *A. acutifolius*

	Units	Average	Range	References
Carotenoids (total)				
β-carotene	mg	0.69	0.16–1.19	1, 2, 3
Lutein	mg	1.33	0.41–2.74	1, 3
Neoxanthin	mg	0.33	0.17–0.98	1, 3
Violaxanthin	mg	0.27	0.16–0.57	1, 3
RAE (calculated value)	µg	57.5	13.3–99.0	–
Vitamin B9 (total folates)	µg	217	194–239	4
Vitamin C	mg	37.8	26.0–49.9	5
Ascorbic acid	mg	22.7	13.9–39.3	5, 6
Dehydroascorbic acid	mg	14.4	9.58–19.2	5
Vitamin E				
α-tocopherol	mg	8.30	1.86–14.6	2, 7
β-tocopherol	mg	0.09	0.04–0.10	2, 7
γ-tocopherol	mg	2.95	0.73–5.88	2, 7
δ-tocopherol	mg	0.09	0.03–0.19	2, 7
Organic acids				
Oxalic acid	mg	98.1	70–218	5, 6
Glutamic acid	mg	39.0	nd–110	5
Malic acid	mg	112	15.7–260	5, 6

Table 13.14 (continued)

	Units	Average	Range	References
Citric acid	mg	308	30–388	5, 6
Fumaric acid	mg	Traces	–	5, 6
Succinic acid	mg	47.5	nd–93.0	5, 6
Phenolics (total)	mg	17.6[a]	17.3–17.9	7
	mg	132[b]	33.3–230	1, 8
Phenolic acids	mg	4.31[c]	–	1
Flavonoids	mg	6.09[d]	5.82–6.36	7
	mg	30.0[b]	28.2–32.3	8
	mg	226[e]	–	1, 2

1 Salvatore et al. (2005) Italy (3 samples), *2* Martins et al. (2011) Portugal, *3* García-Herrera (2013) Spain (5 samples), *4* Morales et al. (2015) Spain (5 samples), *5* Morales (2011) Spain (5 samples), *6* Pereira et al. (2013) Portugal, *7* Morales et al. (2012a) Spain (4 samples), *8* Barros et al. (2011b) Portugal

RAE retinol activity equivalents; *nd* non-detected values

[a] Expressed as gallic acid

[b] Expressed as sum of individual compounds (quercetin derivatives as major components)

[c] Expressed as caffeic acid

[d] Expressed as catechin

[e] Expressed as quercetin

Table 13.15 Fatty acids profile (% of total fatty acids) of fresh young shoots of *A. acutifolius*

Individual compounds	Average	Range	References
12:0	0.38	0.03–0.45	1, 2
14:0	0.37	0.03–0.72	1, 2
16:0	23.1	17.5–28.6	1, 2
16:1	0.09	nd–0.10	1
18:0	2.13	1.56–3.22	1, 2
18:1$n-9$	4.91	4.87–4.94	1, 2
18:2$n-6$	43.4	18.9–44.5	1, 2
18:3$n-3$	18.8	13.4–24.6	1, 2
20:0	0.44	0.42–0.45	1, 2
20:1$n-9$	0.26	0.26–0.26	1
22:0	2.22	2.20–2.24	1, 2
24:0	2.07	1.02–2.26	1, 2
Categories (calculated values)			
SFA	31.3	25.1–37.5	–
MUFA	5.2	5.0–5.4	–
PUFA	63.5	57.5–69.5	–
$n-3$	19.2	14.3–24.1	–
$n-6$	44.3	43.2–45.3	–
$n-9$	5.2	5.0–5.4	–

1 Martins et al. (2011) Portugal; *2* Morales et al. (2012b) Spain (5 samples)

SFA saturated fatty acids, *MUFA* monounsaturated fatty acids, *PUFA* polyunsaturated fatty acids

Remarks (young shoots of *A. acutifolius*)

- Source of dietary fibre (>3 g/100 g).
- Unusual ratio $n-3/n-6$ favourable to $n-6$.
- Citric acid as major organic acid.
- Low Na content (<120 mg/100 g). It may be considered as a source of K and vitamins C, E, and B_9; sometimes also a source of Mn.

Table 13.16 Main constituents, per 100 g of boiled young shoots of *A. acutifolius*

	Units	Average	Range	References
Energy (calculated value)	kcal	39	37–41	–
Moisture	g	89.6	89.6–89.6	1, 2
Available carbohydrates	g	5.00	4.60–5.40	2
Dietary fibre	g	3.60	3.60–3.60	2
Proteins	g	2.50	2.50–2.50	2
Lipids	g	0.12	0.12–0.12	2
Ash	g	0.60	0.60–0.60	2
K	mg	177	159–195	2
Na	mg	21.0	19.1–23.9	2
Ca	mg	31.5	31.2–31.8	2
Mg	mg	10.9	9.1–12.5	2
Fe	mg	0.22	0.13–0.31	2
Cu	µg	380	300–460	2
Mn	µg	120	110–130	2
Zn	µg	423	410–427	2

1 Morales (2011) Spain, *2* García-Herrera (2014) Spain

Table 13.17 Vitamins and other constituents, per 100 g of boiled young shoots of *A. acutifolius*

	Units	Average	Range	References
Vitamin B9 (total folates)	µg	283	273–290	1
Vitamin C	mg	47.9	46.1–49.7	1
Ascorbic acid	mg	24.2	23.7–24.7	1
Dehydroascorbic acid	mg	23.6	22.2–24.9	1
Organic acids				
Oxalic acid	mg	80	80–80	1
Glutamic acid	mg	30	30–30	1
Malic acid	mg	90	80–100	1
Citric acid	mg	200	130–270	1
Fumaric acid	mg	Traces	–	1

1 Morales (2011) Spain.

Remarks (boiled young shoots of *A. acutifolius*)

- Low energy value and still good source of dietary fibre and vitamins B_9 and C after boiling.
- Reduced oxalic acid levels and very low Na content.

13.3.6 Beta maritima L. (Amaranthaceae)

Common Names Sea beet, *acelga-brava* (pt), *acelga silvestre* (es), *barbabietola* (it), *divlja blitva* (bs, hr), *agrioteftlo, agriolachano* (el), *yabani pancar* (tr), *silq barri* (ar).

Description Perennial herb up to 80 cm, usually glabrous, that has long petiolate basal leaves, ovate to lanceolate, cuneate, the upper ones smaller. The greenish flowers grow in spikes, with 3 or more flowers in each axil. They are very small, 3 mm and with 5 sepals. Some botanists consider this species as subspecies of *B. vulgaris*, being usually thought to be the ancestor of the cultivated chard and sugar beet.

Ecology and Distribution It grows in saline soils, usually near the coast, in waste places as a ruderal plant in the Mediterranean region and south of Asia. It flowers from April to October.

Food Uses It has been traditionally used as a wild vegetable in several Mediterranean countries, as reported in Portugal (Mendonça de Carvalho 2006), Spain (Rivera et al. 2005; Tardío et al. 2005; Tardío et al. 2006), Italy (Di Tizio et al. 2012; Ghirardini et al. 2007; Guarrera 2006; Lentini and Venza 2007; Leonti et al. 2006), Slovenia (Cerne 1992), Croatia (Łuczaj et al. 2013b), Bosnia-Herzegovina (Redžić

2006), Greece (Leonti et al. 2006), Cyprus (Della et al. 2006), Turkey (Dogan 2012), Jordan (Al-Qura'n 2010; Tukan et al. 1998), and Tunisia (Le Floc'h 1983).

The young leaves of the basal rosette are gathered in spring, before the development of the flowering stems, and consumed cooked in a similar way as the cultivated chard (*B. vulgaris* subsp. *cicla* (L.) Schübl. & G. Martens). They are usually considered softer and tastier than those of the cultivated species (Tardío et al. 2002). This vegetable is generally boiled, drained, and then seasoned with olive oil and lemon or vinegar (Lentini and Venza 2007; Piera 2006; Tardío et al. 2002), fried with other ingredients (Di Tizio et al. 2012; Lentini and Venza 2007; Rivera et al. 2005), prepared in stews with legumes (Della et al. 2006; Mendonça de Carvalho 2006; Verde et al. 2003), in mixed soups (Cerne 1992; Lentini and Venza 2007; Mendonça de Carvalho 2006; Redžić 2006; Tukan et al. 1998) or in omelettes (Tardío et al. 2002; Verde et al. 2003). They are also employed for preparing the filling of some regional dishes such as the Italian *focacce scacciate* or the *cuddiruni* (Lentini and Venza 2007) and the traditional home-made pies composed of different blends of wild vegetables, such as those reported in southeastern Spain (Oltra 1998; Rabal 2000; Tardío 2010), Cyprus (Della et al. 2006), and Jordan (Tukan et al. 1998). Although wild beets are generally collected for domestic consumption, they can be occasionally found in vegetables mixes sold in the markets of southern Croatia (Łuczaj et al. 2013b).

The roots of this species were also used in Tunisia to prepare soups, and in scarcity times, they were also dried, ground, and mixed with wheat or barley flour (Le Floc'h 1983).

Other wild species of the same genus (*B. macrocarpa* Guss., known as *selg*) is also consumed as vegetable in Morocco (Nassif and Tanji 2013).

Other Uses This wild vegetable is used in Italy as a medicinal food against constipation when consumed in soups (Lentini and Venza 2007). It is also consumed for medicinal purposes in Spain, as stomachic, anti-inflammatory, laxative, and against dyspepsia (Martínez-Lirola et al. 1997; Rivera et al. 2005; Verde et al. 2003). Their leaves are occasionally employed as animal fodder (Moll 2005).

Historical References Its food use probably dates from prehistoric times. Although Theophrastus (third century BC) and Dioscorides (first century AD) did not mention the use of the wild forms, the beet was already mentioned by Theophrastus among the cultivated vegetables with fleshy roots and leaves (Teofrasto 1988) and by Dioscorides for its medicinal properties (Laguna 1555).

Food Composition Tables for raw tender leaves of *Beta maritima* (Tables 13.18, 13.19 and 13.20).

Table 13.18 Main constituents, per 100 g of fresh tender leaves of *B. maritima*

	Units	Average	Range	References
Energy (calculated value)	kcal	31	16–59	–
Moisture	g	87.3	75.4–91.4	1, 2, 3, 4, 5
Available carbohydrates	g	1.71	0.75–4.30	2, 6
Dietary fibre	g	4.38	3.29–9.50	2, 6
Proteins	g	3.10	1.80–3.91	2, 6
Lipids	g	0.34	0.18–0.70	2, 5, 6
Ash	g	2.68	2.00–5.60	2, 3, 6
K	mg	988	540–2356	3, 6
Na	mg	201	45–288	3, 6
Ca	mg	67.1	19–250	3, 6
Mg	mg	66.9	13.2–135	3, 6
P	mg	49.2	44.2–54.2	3
Fe	mg	2.88	1.42–4.31	3, 6
Cu	μg	190	90–330	3, 6
Mn	μg	730	640–1260	3, 6
Zn	μg	845	570–1230	3, 6

1 Guil-Guerrero et al. (1997a) Spain (5 samples), *2* Guil-Guerrero et al. (1997b) Spain, (5 samples), *3* Guil-Guerrero et al. (1999a) Spain (5 samples), *4* Sánchez-Mata et al. (2012) Spain (2 samples), *5* Morales et al. (2014) Spain (4 samples), *6* García-Herrera (2014) Spain (4 samples)

Table 13.19 Vitamins and other constituents, per 100 g of fresh tender leaves of *B. maritima*

	Units	Average	Range	References
Carotenoids (total)	mg	14.3[a]	11.3–17.3	1
Vitamin B_9 (total folates)	μg	302	250–354	2
Vitamin C	mg	36.4	18.3–66.0	1, 3, 4
Ascorbic acid	mg	27.5	7.82–48.0	1, 3, 4
Dehydroascorbic acid	mg	12.8	8.1–18.0	1, 3, 4
Vitamin E				
α-tocopherol	mg	0.51	0.49–0.53	4
β-tocopherol	mg	0.01	0.01–0.01	4
γ-tocopherol	mg	0.14	0.14–0.14	4
δ-tocopherol	mg	Traces	–	4
Organic acids				
Oxalic acid	mg	581	50.0–1070	1, 3, 4
Glutamic acid	mg	nd	–	4
Malic acid	mg	51.4	3.18–59.0	3, 4
Citric acid	mg	107	12.4–138	3, 4
Fumaric acid	mg	4.80	nd–10.0	3, 4
Phenolics (total)	mg	61.9[b]	54.4–69.4	4
Flavonoids	mg	21.5[c]	20.7–22.4	4
Nitrate	mg	200	170–230	1

1 Guil-Guerrero et al. (1997a) Spain (5 samples), *2* Morales et al. (2015) Spain (4 samples), *3* Sánchez-Mata et al. (2012) Spain (2 samples), *4* Morales et al. (2014) Spain (4 samples); *nd* non-detected values

[a] Expressed as β-carotene

[b] Expressed as gallic acid

[c] Expressed as catechin

Table 13.20 Fatty acids profile, (% of total fatty acids) of fresh tender leaves of *B. maritima*

Individual compounds	Average	Range	References
12:0	0.06	0.06–0.06	1
14:0	0.61	0.34–0.89	1, 2
16:0	16.8	10.8–22.6	1, 2
16:1	4.39	–	2
18:0	2.06	1.66–2.46	1, 2
18:1*n*−7	0.38	–	2
18:1*n*−9	4.58	3.50–5.66	1, 2
18:2*n*−6	20.2	18.8–21.7	1, 2
18:3*n*−3	43.6	29.4–57.8	1, 2
18:3*n*−6	0.17	0.17–0.17	2
18:4*n*−3	0.50	–	2
20:0	0.37	0.28–0.50	1, 2
20:1*n*−9	0.30	0.23–0.38	1, 2
22:0	0.97	0.89–1.04	1, 2
22:1	0.06	0.06–0.06	1
24:0	0.98	0.43–1.54	1, 2
Categories (calculated values)			
SFA	23.8	16.2–31.5	–
MUFA	8.1	4.0–12.3	–
PUFA	68.0	56.2–79.8	–
n−3	46.1	33.8–58.4	–
n−6	21.8	21.5–22.1	–
n−9	5.4	3.9–6.9	–

1 Morales et al. (2012b) Spain (4 samples), *2* Guil-Guerrero et al. (1996b) Spain (various samples)
SFA saturated fatty acids, *MUFA* monounsaturated fatty acids, *PUFA* polyunsaturated fatty acids

Remarks (tender leaves of *B. maritima*)

- Source of dietary fibre (>3 g/100 g).
- Very high content of oxalates: People with altered renal function should avoid this vegetable; boiling is recommended for general population.
- Often high Na content and nitrate levels; not recommended for infants.
- It can be considered as a source of K, Mn, and vitamins C and B_9 (folates); sometimes also a source of Mg, Fe, Cu, and Zn.
- High amount of carotenoids.

13.3.7 *Borago officinalis* L. (Boraginaceae)

Common Names Borage, *borrage* (pt), *borraja* (es), *bourrache* (fr), *borragine* (it), *buránza* (el), *borač* (hr), *hodan* (tr), *lisan al thawr* (ar).

Description Annual herb, up to 60 cm tall, entirely covered with white, stiff, and prickly hairs, especially abundant near the inflorescence. The basal leaves are large and wrinkled, with a long stalk of up to 15 cm; the upper ones without petiole. Their purplish-blue flowers (white in some cultivated forms) have a characteristic conical group of prominent dark purple anthers.

Ecology and Distribution It grows in nitrified areas near dwellings, in all kinds of soils. Native to the Mediterranean region and southwest of Asia, it has been cultivated and naturalized in most parts of Europe.

Food Uses The basal leaves (sometimes only their midribs) and, to a lesser extent, the tender stems, have been used as a vegetable throughout the Mediterranean region, at least in Portugal (Carvalho 2005), Spain (Tardío et al. 2006), Italy (Picchi and Pieroni 2005), Slovenia (Cerne 1992), Croatia (Łuczaj et al. 2013b), Greece (Leonti et al. 2006), Cyprus (Della et al. 2006), Turkey (Ertuğ 2014), Lebanon (Marouf 2005), Tunisia (Le Floc'h 1983), and Morocco (Nassif and Tanji 2013). They are usually boiled, sometimes with potatoes and other ingredients, and used to prepare vegetable soups (Carvalho 2005; Hadjichambis et al. 2008; Le Floc'h 1983; Lentini and Venza 2007), the stock being considered to be very nice, nutritive, and with medicinal properties (Ferrández and Sanz 1993; Font Quer 1962). Other times they are drained after boiling and simply seasoned with olive oil, lemon, and salt (Fajardo et al. 2007; Font Quer 1962; Lentini and Venza 2007), or lightly fried in olive oil, with garlic and a bit of ham (Verde et al. 1998). On some occasions, this species is included in traditional recipes of mixed vegetables, such as *collejas* (Verde et al. 1998), *preboggion* (Picchi and Pieroni 2005) and *beqoul* (Nassif and Tanji 2013). It is also used for omelettes (Hadjichambis et al. 2008; Lentini and Venza 2007) or as filling for pasta, such as *agnolotti* (ravioli), *frittelles* (buñuelos), calzone (Picchi and Pieroni 2005) or even to prepare the green *tagliatelle* (Lentini and Venza 2007), or couscous (Nassif and Tanji 2013).

This plant is especially well known and appreciated in some Spanish regions, such as Aragon and Navarre (Ferrández and Sanz 1993), where it is frequently cultivated and widely naturalized. There, the part usually consumed are the midribs of the basal leaves. However, there are also other recipes in several Spanish regions, especially those of the East, that use the whole leaves batter-fried in olive oil. Some of these preparations are finally sweetened with sugar or honey. For instance, the *crespillos* is a typical dessert from some Aragonese areas that is prepared in the catholic festivity of the Incarnation (25th of March). They are prepared frying the tender leaves, previously battered with a mix made mainly of eggs, flour, milk, baking powder, and sugar. It is believed that making *crespillos* on this day will favour a good harvest of olives next year (Ferrández and Sanz 1993).

In spite of the hairiness of the whole plant, the slightly fuzzy young leaves can be eaten raw in salads (Hadjichambis et al. 2008; Picchi and Pieroni 2005). The tender shoots are sometimes eaten in vinegar as pickles (Della et al. 2006; Picchi and Pieroni 2005).

In Spain, the beautiful flowers of this species were sucked for their sweet nectar, especially by children (Benítez 2009; González et al. 2011a; Pellicer 2001). The flowers have also been used to decorate and season salads (Hadjichambis et al. 2008; Picchi and Pieroni 2005) as already mentioned by Quer, a Spanish Botanist of the eighteenth century (Quer 1762).

Other Uses Borage has been traditionally considered a healthy vegetable or a medicinal food, thanks to its attributed digestive, laxative, purifying, and diuretic properties (e.g. Benítez 2009; Carvalho 2010; Font Quer 1962; Lentini and Venza 2007; Picchi and Pieroni 2005). Moreover, an infusion or decoction of the leaves or of the aerial part is also locally taken for treating respiratory diseases or rheumatism (Ferrández and Sanz 1993; Villar et al. 1987). It has also been used as a veterinary medicine and considered a honey plant (Carvalho 2010).

Historical References According to the opinion of several translators (Laguna 1555; Osbaldeston 2000), Dioscorides (first century AD) mentioned that the aerial part of this species, boiled in wine, is good for the chills of acute fevers, even for those produced by different types of malaria, for treating abscesses, and also as a cordial plant that causes merriment.

Food Composition Tables for raw tender leaves of *Borago officinalis* (Tables 13.21, 13.22 and 13.23).

Table 13.21 Main constituents, per 100 g of fresh tender leaves of *B. officinalis*

	Units	Average	Range	References
Energy (calculated value)	kcal	44	34–50	–
Moisture	g	87.2	86.4–88.8	1, 2
Available carbohydrates	g	9.45	7.23–10.7	2
Soluble sugars (total)	g	0.32	0.30–0.34	2
Glucose	mg	80.2	70.2–90.4	2
Fructose	mg	18.4	14.5–22.3	2

Table 13.21 (continued)

	Units	Average	Range	References
Sucrose	mg	205	183–222	2
Proteins	g	1.17	0.97–1.37	2
Lipids	g	0.16	0.13–0.19	2
Ash	g	2.35	1.93–2.91	1, 2
K	mg	567	508–626	1
Na	mg	43.5	31.0–56.0	1
Ca	mg	344	164–524	1
Mg	mg	8.50	7.00–10.0	1

1 Bianco et al. (1998) Italy, *2* Pereira et al. (2011) Portugal

Table 13.22 Vitamins and other constituents, per 100 g of fresh tender leaves of *B. officinalis*

	Units	Average	Range	References
Carotenoids	mg	0.31[a]	0.30–0.32	1
β-carotene	mg	2.86	2.50–32.2	2
Lutein	mg	3.81	2.69–4.93	2
Neoxanthin	mg	1.13	1.08–1.18	2
Violaxanthin	mg	1.15	1.03–1.28	2
RAE (calculated value)	μg	238	208–268	–
Vitamin C (ascorbic acid)	mg	1.83	1.16–2.73	1, 3
Vitamin E				
α-tocopherol	mg	≈1.15	1.08–1.22	1
β-tocopherol	mg	≈0.04	0.04–0.04	1
γ-tocopherol	mg	≈0.30	0.28–0.32	1
δ-tocopherol	mg	≈0.02	0.02–0.02	1
Organic acids				
Oxalic acid	mg	≈565	561–569	3
Quinic acid	mg	nd	–	3
Malic acid	mg	≈34.0	30.5–49.5	3
Shikimic acid	mg	≈1.58	1.58–1.58	3
Citric acid	mg	≈93.9	89.8–98.0	3
Fumaric acid	mg	≈304	299–308	3
Succinic acid	mg	≈82.5	80.4–84.6	3
Phenolics (total)	mg	39.6[b]	25.3–58.9	2, 4
Phenolic acids	mg	6.40[c]	–	2
Flavonoids	mg	18.9[d]	–	2
Nitrate	mg	69.2	–	5

1 Pereira et al. (2011) Portugal, *2* Salvatore et el. (2005) Italy, *3* Pereira et al. (2013) Portugal, *4* Gatto et al. (2011) Italy, *5* Bianco et al. (1998) Italy

RAE retinol activity equivalents; *nd* non-detected values

[a] Expressed as β-carotene

[b] Expressed as sum of individual compounds (procyanidin and caffeic acid derivatives as major components)

[c] Expressed as caffeic acid

[d] Expressed as quercetin

Table 13.23 Fatty acids profile (% of total fatty acids) of fresh tender leaves of *B. officinalis*

Individual compounds	Average	Range	References
12:0	0.92	0.73–1.11	1
14:0	2.80	2.11–3.49	1
16:0	15.9	11.3–19.8	1, 2, 3
16:1	0.13	0.12–0.14	4
18:0	1.99	0.70–3.00	1, 2, 3
18:1*n*−9	4.16	1.90–8.50	1, 2, 3
18:2*n*−6	10.9	8.2–12.3	1, 2, 3
18:3*n*−3	22.8	10.3–30.2	1, 2, 3
18:3*n*−6	8.92	5.00–12.0	1, 2, 3
18:4*n*−3	17.3	15.2–19.5	1, 2, 3
20:0	11.9	11.6–12.2	1
22:0	12.0	10.9–13.1	1
22:1	0.44	0.38–0.50	1
24:0	4.55	4.12–4.98	1
24:1	0.17	0.15–0.19	1
Categories (calculated values)			
SFA	43.4	40.8–50.0	–
MUFA	4.5	3.4–6.8	–
PUFA	52.1	46.6–52.4	–
n−3	34.9	18.4–35.2	–
n−6	17.1	15.9–17.2	–
n−9	4.5	3.4–6.8	–

1 Pereira et al. (2011) Portugal, *2* Del Río-Celestino et al. (2008) Spain, *3* Mhamdi et al. (2007) Tunisia (2 samples)

SFA saturated fatty acids, *MUFA* monounsaturated fatty acids, *PUFA* polyunsaturated fatty acids

Remarks (tender leaves of *B. officinalis*)

- High available carbohydrate content, compared to other vegetables; very low lipid content (<0.5 g/100 g).
- Unusual fatty acids profile with high γ-linolenic acid proportion compared to other vegetables.
- Very high content of oxalates: People with altered renal function should avoid this vegetable; boiling is recommended for general population.
- Low Na content (<120 mg/100 g). It can be considered as a source of K, Ca, and vitamin A (provitamin A as β-carotene).

13.3.8 Bryonia dioica Jacq. (Cucurbitaceae)

Common Names White bryony, *norça* (pt), *espárrago de nuez* (es), *bryone* (fr), *zucca matta* (it).

Description Climbing and herbaceous plant, whose annual and vigorous stems sprout in spring from perennial tuber roots, scrambling up to other plants by means of long and coiled tendrils. The slightly rough leaves are palmate, with 5 lobes. Dioecious species, that is, the male and female flowers are found in different individuals. The flowers are greenish-white with green veins and 5 petals. The fruit is a toxic berry that turns red when ripe.

Ecology and Distribution It grows in hedgerows, scrubs, woodland borders, and it is also frequent in ditches and waste lands, moist and cool environments with well-drained soils. It can be found in Western Mediterranean and Western Europe.

Food Uses According to recent ethnobotanical sources, its use as food is not widespread in the Mediterranean region, the tender young shoots being traditionally consumed in some parts of Portugal (Carvalho 2010), Spain (e.g González et al. 2011a; Tardío et al. 2006) and Italy (Guarrera 2006; Picchi and Pieroni 2005; Pieroni 1999). The edible portion includes the tips of their young shoots—of about 20 cm in length—with the first 6–8 immature leaves not yet fully expanded (Barros et al. 2011b; Fajardo 2008). They are always consumed cooked. In Italy, they are utilized in pancakes, stuffing for *tortelli* or simply eaten boiled in mixtures with others wild greens (Pieroni 1999). In Spain and Portugal, they are also boiled and then fried or sautéed, and prepared generally in omelettes or soups, although they can be eaten simply boiled and seasoned with garlic and olive oil (e.g. Carvalho 2010; González et al. 2011a).

The toxicity of this species is mentioned in most treaties on toxic plants (e.g. Couplan 1990; Mulet 1997), describing the toxic compounds of its tuberous root and, above all, of the fruits (see Other Uses). However, some of these studies mention the poisonousness of the young shoots. The presence of some toxicity in the spears in certain circumstances has been noticed popularly and registered in some ethnobotanical studies. For instance, in the province of Madrid, our informants advised that "when the plant flowers and fruits, they cannot be eaten because they produce diarrhoea" (Tardío et al. 2002). Other Spanish informants have mentioned the existence of some plants with bitter asparaguses that must not be collected, being selected after tasting them raw (Blanco 1995; Tardío et al. 2002). One of these could be the cause of a mild intoxication reported in Barcelona (Nogué 2010) by eating an omelette made with the young shoots of this plant. For those reasons, most of the ethnobotanical information remarked that the young shoots must be only gathered in spring before the development of the immature floral buds and always consumed cooked (e.g. Carvalho 2010; Couplan 1990; Pieroni 1999).

Other Uses Several toxic compounds have been found in its roots and fruits, such as cucurbitacins (bitter substances with cytotoxic and antitumoral activity), glucosides (brionin and brionidin), an alkaloid (brionicin), and some resins (Couplan 1990; Morán et al. 2011; Mulet 1997). It is also said that the consumption of 40 fruits is enough to kill an adult and 15 a child (Couplan 1990). Nevertheless, both plant parts have been employed in folk medicine in external use. Poultices of the grinded root were used to alleviate headache, nervousness, and rheumatism pain (Villar et al. 1987); the root fried in olive oil for healing wounds (Fajardo et al. 2007) or boiled against scorpions bites (Verde et al. 2001). Its fruits crushed and applied topically were used to mitigate rheumatism pain in Portugal and Spain (Aceituno-Mata 2010; Carvalho 2010). Nevertheless, some medicinal and veterinary internal uses have been described with the cold maceration of the tubers. It is used as diuretic in Italy (Pieroni 1999) or as antihypertensive in Spain (Ferrández and Sanz 1993). The consumption of its young shoots is also considered purgative (Bonet and Vallès 2002).

Historical References Besides other ancient uses of the root and fruits of this species, such as a hair remover for preparing leather, mentioned by Theophrastus in the third century BC (Scarborough 1978), the consumption of the young shoots was considered to be diuretic and have other medicinal properties, both by Dioscorides and Pliny the Elder, in the first century AD (Bostock and Riley 1855; Laguna 1555). For that reason, its current edible use could be a reminder of their ancient medicinal use (Tardío 2010). The consumption of plants that were first appreciated for their medicinal qualities, and today have lost their medicinal function and are simply regarded as a food, have been recorded in several species (Etkin 1996).

Food Composition Tables for young shoots of *Bryonia dioica*, raw (Tables 13.24, 13.25 and 13.26) and boiled (Tables 13.27 and 13.28).

Table 13.24 Main constituents, per 100 g of fresh young shoots of *B. dioica*

	Units	Average	Range	References
Energy (calculated value)	kcal	55	14–141	–
Moisture	g	85.9	70.9–90.8	1, 2, 3, 4
Available carbohydrates	g	4.21	0.80–10.37	1, 4
Soluble sugars (total)	g	1.20	1.12–1.22	1
Fructose	mg	591	582–605	1
Glucose	mg	513	500–525	1
Sucrose	mg	102	100–104	1
Dietary fibre	g	4.60	3.40–10.7	4
Proteins	g	3.97	1.00–11.9	1, 4
Lipids	g	1.12	0.10–2.90	1, 4, 5
Ash	g	1.48	1.00–3.30	1, 4
K	mg	487	338–1029	4
Na	mg	27.6	12.0–56.8	4
Ca	mg	53.3	30.5–141	4
Mg	mg	28.8	13.8–70.5	4
Fe	mg	0.70	0.24–1.78	4
Cu	μg	220	10.0–460	4
Mn	μg	250	110–610	4
Zn	μg	890	270–2630	4

1 Martins et al. (2011) Portugal, *2* Morales (2011) Spain (5 samples), *3* Sánchez-Mata et al (2012) Spain (2 samples), *4* García-Herrera (2014) Spain (5 samples), *5* Morales et al. (2012b) Spain (4 samples)

Table 13.25 Vitamins and other constituents, per 100 g of fresh young shoots of *B. dioica*

	Units	Average	Range	References
Carotenoids				
β-carotene	mg	0.57	0.15–1.95	1, 2, 3
Lutein	mg	1.63	0.68–3.70	1, 3
Neoxanthin	mg	0.73	0.17–3.83	3
Violaxanthin	mg	1.14	0.03–2.15	3
RAE (calculated value)	μg	47.5	12.5–162	–
Vitamin B_9 (total folates)	μg	43.2	39.3–47.0	4
Vitamin C	mg	21.4	16.0–27.7	1, 5, 6
Ascorbic acid	mg	12.1	4.27–20.8	2, 5, 6, 7
Dehydroascorbic acid	mg	10.7	3.99–14.9	5, 6
Vitamin E				
α-tocopherol	mg	2.50	0.69–6.39	1, 2, 8
β-tocopherol	mg	0.13	0.11–0.14	1, 8
γ-tocopherol	mg	1.29	0.41–3.18	1, 2, 8
δ-tocopherol	mg	0.10	0.06–0.13	6, 8
Vitamin K	μg	95.0	–	1
Organic acids				
Oxalic acid	mg	311	70–630	5, 6

Table 13.25 (continued)

	Units	Average	Range	References
Quinic acid	mg	≈245	233–252	7
Glutamic acid	mg	83	70–120	5
Malic acid	mg	1044	Traces–1710	5, 6, 7
Citric acid	mg	46.2	3.36–60	5, 6, 7
Fumaric acid	mg	5.07	2.00–15.0	5
Succinic acid	mg	nd	–	5, 6
Phenolics (total)	mg	99.5[a]	18.7–276	1, 2, 8
Flavonoids	mg	8.80[b]	1.31–17.0	2, 8
	mg	241[c]	229–253	9

1 Vardavas et al. (2006b) Crete, *2* Martins et al. (2011) Portugal, *3* García-Herrera et al. (2013) Spain (5 samples), *4* Morales et al. (2015) Spain (4 samples), *5* Morales (2011) Spain (5 samples), *6* Sánchez-Mata et al (2012) Spain (2 samples), *7* Pereira et al. (2013) Portugal, *8* Morales et al. (2012a) Spain (4 samples), *9* Barros et al. (2011b) Portugal

RAE retinol activity equivalents; *nd* non-detected values

[a] Expressed as gallic acid

[b] Expressed as catechin

[c] Expressed as sum of individual compounds (apigenin derivatives as major components)

Table 13.26 Fatty acids profile (% of total fatty acids) of fresh young shoots of *B. dioica*

Individual compounds	Average	Range	References
12:0	0.08	0.01–0.12	1
14:0	0.43	0.05–0.90	1, 2, 3
16:0	19.0	13.6–26.4	1, 2, 3
18:0	1.98	1.00–2.66	1, 2, 3
18:1*n*−9	2.53	1.21–4.85	1, 2, 3
18:2*n*−6	15.5	6.38–33.6	1, 2, 3
18:3*n*−3	54.5	25.4–70.3	1, 2, 3
18:3*n*−6	0.25	–	2
20:0	0.39	0.31–0.63	1, 2, 3
20:1*n*−9	0.13	nd–0.15	1, 2
22:0	0.76	0.60–1.09	1, 2, 3
22:1*n*−9	0.13	nd–0.26	1, 2
24:0	1.85	1.47–2.50	1, 2, 3
Categories (calculated values)			
SFA	26.1	21.4–32.8	–
MUFA	3.8	1.6–8.0	–
PUFA	70.1	59.3–76.7	–
n−3	54.5	25.5–70.3	–
n−6	15.6	6.4–33.9	–
n−9	2.8	1.5–5.3	–

1 Morales et al. (2012b) Spain (4 samples), *2* Vardavas et al. (2006a) Crete, *3* Martins et al. (2011) Portugal; *nd* non-detected values

SFA saturated fatty acids, *MUFA* monounsaturated fatty acids, *PUFA* polyunsaturated fatty acids

Remarks (young shoots of *B. dioica*)

- It may be considered as a source of fibre (>3 g/100 g).
- High lipid content, PUFA proportion and $n-3$ fatty acids content (mainly α-linolenic acid, 18:3$n-3$), compared to other vegetables. Good $n-3/n-6$ ratio (>3).
- High levels of malic acid, as major organic acid. High content of oxalates; boiling is recommended.
- Low Na content (<120 mg/100 g). It can be considered as a source of K and vitamin B_9 (folates); sometimes also a source of Cu and vitamins E and K.
- Caution must be taken to avoid poisoning: just tender shoots after boiling are edible (see section Food Uses and Other Uses of this monograph).

Table 13.27 Main constituents, per 100 g of boiled young shoots of *B. dioica*

	Units	Average	Range	References
Energy (calculated value)	kcal	20	18–26	–
Moisture	g	93.4	93.2–93.6	1,2
Available carbohydrates	g	1.20	1.02–1.44	2
Dietary fibre	g	2.92	2.90–2.93	2
Proteins	g	2.00	1.65–2.40	2
Lipids	g	0.15	0.13–0.17	2
Ash	g	0.65	0.65–0.65	2
K	mg	171	161–181	2
Na	mg	18.1	17.9–18.3	2
Ca	mg	16.0	14.5–17.5	2
Mg	mg	9.05	8.70–10.7	2
Fe	mg	0.20	0.20–0.20	2
Cu	µg	100	90.4–110	2
Mn	µg	80.3	80.0–80.5	2
Zn	µg	30.0	27.0–33.5	2

1 Morales (2011) Spain, *2* García-Herrera (2014) Spain

Table 13.28 Vitamins and other constituents, per 100 g of boiled young shoots of *B. dioica*

	Units	Average	Range	References
Vitamin B_9 (total folates)	µg	120	114–126	1
Vitamin C (total)	mg	6.42	6.17–6.67	1
Ascorbic acid	mg	6.39	6.04–6.71	1
Dehydroascorbic acid	mg	0.22	0.21–0.23	1
Organic acids				
Oxalic acid	mg	130	120–140	1
Glutamic acid	mg	nd	–	1
Malic acid	mg	98	92–1050	1
Citric acid	mg	40	40–40	1
Fumaric acid	mg	Traces	–	1

1 Morales (2011) Spain; *nd* non-detected values

Remarks (boiled young shoots of *B. dioica*)

- Very low lipid content (<0.5 g/100 g), low energy value (<40 g/100 g), very low Na amount (<40 mg/100 g), and source of dietary fibre and folates after boiling.
- Oxalic acid content reduces after boiling.
- Caution must be taken to avoid poisoning: Just tender shoots after boiling are edible (see sections Food Uses and Other Uses of this monograph).

13.3.9 Capsella bursa-pastoris (L.) Medik. (Brassicaceae)

Common Names Shepherd's-purse, *bolsa de pastor* (pt, es), *capselle* (fr), *borsa del pastore* (it), *agriokardamo* (el), *çobançantası* (tr), *kees rai* (ar).

Description Annual herb, with a basal rosette of deeply lobed leaves, that develops erect and slender flower stalks up to 0.3–0.5 m. The small and cross-shaped flowers have 4 white petals and give way to the characteristic heart-shaped (or purse-like) pods, with numerous tiny seeds.

Ecology and Distribution It is a ruderal plant that grows in disturbed and waste places, being a common weed in cultivated soils. It flowers almost all year round, producing several generations each year. Native to Europe, it is now naturalized all around the world, except for areas with tropical climates.

Food Uses The traditional consumption of this wild vegetable has been registered in several Mediterranean countries, such as Spain (Tardío et al. 2006; Verde et al. 1998), Italy (Guarrera 2006; Nebel et al. 2006; Picchi and Pieroni 2005), Slovenia (Cerne 1992), Croatia (Łuczaj et al. 2013a,b), Bosnia-Herzegovina (Łuczaj et al. 2013b), Cyprus (Della et al. 2006; Hadjichambis et al. 2008), Turkey (Dogan 2012; Ertuğ 2014; Kültür 2008), Palestine (Ali-Shtayeh et al. 2008), and Morocco (Nassif and Tanji 2013).

The tender rosettes of young leaves are collected in spring and generally consumed raw in salads (Della et al. 2006; Dogan et al. 2004; Dogan 2012; Guarrera 2006; Hadjichambis et al. 2008; Picchi and Pieroni 2005) or raw with bread (Ali-Shtayeh et al. 2008). It is also common to prepare them boiled (Guarrera 2006; Łuczaj et al. 2013a), in soups (Dogan et al. 2004; Dogan 2012; Guarrera 2006; Łuczaj et al. 2013a; Picchi and Pieroni 2005), with *pasta* or rice (Guarrera 2006). In some cases, the leaves are mixed with other wild vegetables, as in the Moroccan dish *beqoul*, a springtime meatless dish composed of up to 20 wild vegetables (Nassif and Tanji 2013). Other examples of wild vegetable mixes that include *Capsella bursa-pastoris* are the Italian dish *erbucci* (Picchi and Pieroni 2005), the Croatian *svakober* (Łuczaj et al. 2013a), and the Spanish *caldo verde* (Verde et al. 1998).

Additional food uses include the consumption of its flowers and fruits raw (Tardío et al. 2005; Tardío et al. 2006), and the use of the plant to curdle milk, as recorded in the Italian region of Basilicata (Guarrera 2006).

Outside the Mediterranean region, the food and medicinal uses of this plant are well known in China and other Asian countries (Picchi and Pieroni 2005; Zhou 1998).

Other Uses Medicinal uses of this species were registered in some Mediterranean folk pharmacopoeias. For example, it has been employed to treat kidney-related disorders in Algeria (González-Tejero et al. 2008), as digestive, cardiovascular, and to treat skin ailments in Portugal (Carvalho 2010), Italy (Guarrera 2006; Picchi and Pieroni 2005) and Cyprus (González-Tejero et al. 2008). It is also considered a hypotensive and hypocholesterolemic plant in southeastern Spain (Rivera et al. 2005). The infusion of the aerial parts was employed to control the menstrual flow in Portugal (Mendonça de Carvalho 2006), Spain (Akerreta 2009; Villar et al. 1987), and Italy (Guarrera 2006; Picchi and Pieroni 2005).

Historical References Although it is often said that it was one of the medicinal plants described by Dioscorides (Picchi and Pieroni 2005), it is surely due to an incorrect identification of the related species *thlaspi* in the annotated translation of Mattioli, in the sixteenth century (Font Quer 1962; Laguna 1555; Osbaldeston 2000). However, many of the medicinal uses registered in the ethnobotanical works for this plant probably come from these ancient works, being even included in recent herbal products marketed in several European countries (European Medicines Agency 2011).

No clear historical references about its food use in the Mediterranean region have been found.

Food Composition Tables for raw tender leaves of *Capsella bursa-pastoris* (Tables 13.29, 13.30 and 13.31).

Table 13.29 Main constituents, per 100 g of fresh tender leaves of *C. bursa-pastoris*

	Units	Average	Range	References
Energy (calculated value)	kcal	52	38–72	–
Moisture	g	82.1	79.1–94.4	1, 2, 3
Available carbohydrates	g	6.42	5.54–7.30	3
Dietary fibre	g	6.31[a]	–	2
	g	2.52[b]	2.04–3.00	3
Proteins	g	3.92	2.04–6.19	2, 3, 4
Lipids	g	0.59	0.45–0.67	2, 3
Ash	g	2.13	1.24–3.81	2, 3, 4
K	mg	395	315–564	3, 4
Na	mg	34.0	25.0–43.0	3, 4
Ca	mg	292	115–426	3, 4
Mg	mg	52.0	20.7–101	3, 4
P	mg	53.9	42.9–81.0	3, 4
Fe	mg	4.80	3.50–6.14	3
Cu	μg	90	70–110	3, 4
Mn	μg	830	670–1110	3, 4
Zn	μg	452	355–690	3, 4

1 Bianco et al. (1998) Italy, *2* Rodríguez-Berrocal et al. (1987) Spain, *3* Guil-Guerrero et al. (1999b) Spain (5 samples), *4* Ayan et al. (2006) Turkey

[a] Acid-detergent fibre

[b] Neutral-detergent fibre

Table 13.30 Vitamins and other constituents, per 100 g of fresh tender leaves of *C. bursa-pastoris*

	Units	Average	Range	References
Carotenoids	mg	7.40[a]	5.40–9.40	1
Vitamin C	mg	144	91–169	1, 2
Organic acids				
Oxalic acid	mg	89.0	64.0–114	1
Nitrate	mg	256	–	3

1 Guil-Guerrero et al. (1999b) Spain (5 samples), *2* Kiliç and Coşkun (2007) Turkey, *3* Bianco et al. (1998) Italy

[a] Expressed as β-carotene

Table 13.31 Fatty acids profile (% of total fatty acids) of fresh tender leaves of *C. bursa-pastoris*

Individual compounds	Average	Range	References
14:0	2.97	–	1
16:0	15.1	–	1
16:1	0.89	–	1
18:0	2.60	–	1
18:1*n*−7	0.60	–	1
18:1*n*−9	1.38	–	1
18:2*n*−6	11.1	–	1
18:3*n*−3	50.7	–	1
18:4*n*−3	3.35	–	1
20:0	3.93	–	1
20:1*n*−9	0.55	–	1
22:0	1.18	–	1
22:1*n*−9	0.34	–	1
24:0	1.60	–	1
Categories (calculated values)			
SFA	28.4	–	–
MUFA	3.9	–	–
PUFA	67.7	–	–
n−3	56.1	–	–
n−6	11.5	–	–
n−9	2.4	–	–

1 Guil-Guerrero et al. (1999b) Spain (various samples)
SFA saturated fatty acids, *MUFA* monounsaturated fatty acids, *PUFA* polyunsaturated fatty acids

Remarks (tender leaves of *C. bursa-pastoris*)

- High protein content compared to other wild plants.
- Good *n*−3/*n*−6 ratio (>3).
- It may reach high ash level presenting low Na content (<120 mg/100 g).
- It may be considered as a source of K, Ca, Fe, Mn, and vitamin C; often also a source of Mg.
- Very high nitrate content; not recommended for infants.

13.3.10 *Chenopodium album* L. (Amaranthaceae)

Common Names Fat hen, *ansarina-branca* (pt), *cenizo* (es), *ansérine blanche* (fr), *farinello comune* (it), *loboda* (bs, hr), *nena e egër* (sq), *aksirken* (tr), *alklab* (ar).

Description Annual herb, up to 1.5 m tall, that has a green striate stem, sometimes reddish, and grey branches. The leaves are 1–8 cm, alternate, peciolate, rhombic, ovate or lanceolate, usually irregularly dentate. Greenish small flowers, 3 mm in diameter, with 5 sepals and 5 stamens, grouped in an inflorescence in panicle. The fruit is an achene with a very little, glossy black, and rounded seed, around 1 mm in diameter.

Ecology and Distribution This ruderal weedy plant grows in disturbed habitats, such as groves, roadsides, and hedgerows in all types of rich in nitrogen soils, up to 1000 m. Flowering time from May to November. It occurs in almost all the regions of the world, being more common in temperate and subtropical zones.

Food Uses This wild vegetable has been traditionally consumed in most Mediterranean countries, being reported at least in Spain (Rigat et al. 2009; Rivera et al. 2005; Tardío et al. 2006), Italy (Ghirardini et al. 2007; Guarrera 2006; Lentini and Venza 2007; Pieroni et al. 2005), Slovenia (Cerne 1992), Croatia (Łuczaj et al. 2013a; Łuczaj et al. 2013b), Bosnia-Herzegovina (Redžić 2006), Albania (Hadjichambis et al. 2008; Pieroni 2008), and Turkey (Dogan 2012; Ertuğ 2014; Kültür 2008).

The young leaves or the tender tips are collected in spring, during March–April (Łuczaj et al. 2013a), but gathering can be prolonged until October in some areas (Cerne 1992). They are consumed raw (Pieroni 2008; Verde et al. 2003) but are more usually consumed after cooking (Dogan 2012; Hadjichambis et al. 2008; Lentini and Venza 2007; Łuczaj et al. 2013a; Pieroni et al. 2005; Rigat et al. 2009; Tardío et al. 2006). It is sometimes included in different traditional recipes with mixes of wild vegetables, such as the Italian *preboggion* of Liguria and the *pistic* of Friuli (Picchi and Pieroni 2005), and also the Croatian *svakober* whose name can be translated as "pick all" (Łuczaj et al. 2013a). Other cooked preparations include boiled and mixed with butter or cream (Hadjichambis et al. 2008), soups (Cerne 1992), omelettes (Verde et al. 2003), *risotto* (Lentini and Venza 2007) or fried in olive oil with other ingredients (Dogan 2012; Lentini and Venza 2007; Pieroni et al. 2005). It is also used as stuffing for vegetables pies, such as the Albanian *byrek* (Hadjichambis et al. 2008; Pieroni 2008), or to prepare other fillings (Lentini and Venza 2007), such as that for *tortelli* (Picchi and Pieroni 2005).

Although with some regional differences, this wild plant, so popular in the past, is at the present time scarcely used (Łuczaj et al. 2013a; Picchi and Pieroni 2005), even almost abandoned (Verde et al. 2003).

Other Uses This wild vegetable has been often considered a medicinal food, since the boiled leaves are regarded as a mild laxative (Picchi and Pieroni 2005; Rivera et al. 2005), digestive (Polat et al. 2013), tonic, and good for treating anaemia (Guarrera 2006; Picchi and Pieroni 2005). The high levels of Fe found in the plant (see Table 13.32 and remarks section) seem to confirm the last application.

Some people consider that its excessive consumption is harmful, even toxic (Guarrera 2006; Picchi and Pieroni 2005), probably due to its high levels of oxalic acid (see Table 13.33).

This species, widely considered as a weed (e.g. Fajardo et al. 2007; Moll 2005), is also used as animal feed for pigs, rabbits, and hens (Fajardo et al. 2007; Guarrera 2006; Picchi and Pieroni 2005).

Historical References This food plant has been gathered from the wild since ancient times, its remains appearing in Neolithic archaeological sites in Greece and the Balkans, northern and central Europe (Andrews 1948; Dalby 2003; Kirleis and Klooß 2014). According to some authors (Andrews 1948; Osbaldeston 2000), its use was reported by Dioscorides and Pliny the Elder. They recognized two closely related plants: *Atraphaxis* and *Atriplex*. The former being a cultivated species, the orache (*Atriplex hortensis* L.) and the latter likely being *Chenopodium album*. According to Dioscorides, both species were eaten as vegetables being considered as mild laxative, besides also being externally applied against skin problems (Laguna 1555). As for the alleged toxicity of the plant, Pliny, mentioning the estates of some Greek authors, accused it of being extremely difficult of digestion and of producing dropsy, jaundice, and paleness of the complexion. In their opinion, it should never be boiled without changing the water repeatedly (Bostock and Riley 1855).

Food Composition Tables for raw tender leaves of *Chenopodium album* (Tables 13.32, 13.33 and 13.34).

Table 13.32 Main constituents, per 100 g of fresh tender leaves of *C. album*

	Units	Average	Range	References
Energy (calculated value)	kcal	53	41–81	–
Moisture	g	79.8	70.6–96.3	1, 2, 3, 4, 5
Available carbohydrates	g	5.89	5.51–6.27	2
Dietary fibre	g	6.38	5.97–6.79	2
Proteins	g	2.74	0.43–8.83	2, 4, 5
Lipids	g	0.63	0.44–0.75	2, 6, 7
Ash	g	3.40	2.13–4.39	2, 3
K	mg	1155	855–1444	2, 3, 4
Na	mg	9.04	4.14–138	2, 3, 4
Ca	mg	307	236–438	2, 3, 4
Mg	mg	207	112–393	2, 3, 4
Fe	mg	5.29	4.79–5.80	2, 4
P	mg	25.0	3.70–79.0	2, 4, 5
Cu	μg	185	40–330	2, 4
Mn	μg	1070	550–1590	2, 4
Zn	μg	1300	750–1850	2, 4

1 Guil-Guerrero et al. (1997a) Spain (5 samples), *2* Guil-Guerrero and Torija-Isasa (1997) Spain (5 samples), *3* Bianco et al. (1998) Italy (2 samples), *4* Yildrim et al. (2001) Turkey, *5* Özbucak et al. (2007) Turkey, *6* Guil-Guerrero et al. (1996b) Spain (various samples), *7* Guil-Guerrero and Rodriguez-García (1999) Spain (5 samples)

Table 13.33 Vitamins and other constituents, per 100 g of fresh tender leaves of *C. album*

	Units	Average	Range	References
Carotenoids (total)	mg	10.6[a]	6.16–12.5	1, 2, 3, 4, 5[e]
β-carotene	mg	≈22.8	–	5[e]
Lutein	mg	≈37.4	–	5[e]
Neoxanthin	mg	nd	–	5[e]
Violaxanthin	mg	≈28.4	–	5[e]
Zeaxanthin	mg	≈1.00	–	5[e]
RAE (calculated value)	μg	≈1900	–	–
Vitamin B_2	μg	20.0	–	6[e]
Vitamin B_9 (total folates)	μg	96.0	–	7[e]
Vitamin C	mg	155	137–171	3
Ascorbic acid	mg	113	42.4–171	1, 2, 3, 8
Dehydroascorbic acid	mg	38.0	29.0–47.0	3
Organic acids				
Oxalic acid	mg	1162	361–2027	2, 3, 9
Phenolics (total)	mg	861[b]	785–937	10[e]
Flavonoids	mg	80[c]	74–87	10[e]

Table 13.33 (continued)

	Units	Average	Range	References
Proanthocyanidins	mg	378[d]	342–414	10[e]
Nitrate	mg	197	44.9–397	3, 9

1 Aliotta and Pollio (1981) Italy, *2* Guil-Guerrero and Torija-Isasa (1997) Spain (5 samples), *3* Guil-Guerrero et al. (1997a) Spain (5 samples), *4* Guil-Guerrero and Rodriguez-García (1999) Spain (5 samples), *5* Raju et al. (2007), India (2 samples), *6* Kuhnlein (1990) Canada (4 samples), *7* Pehrsson et al. (2011) USA (3 samples), *8* Yildrim et al. (2001) Turkey, 9 Bianco et al. (1998) Italy (2 samples), *10* Afolayan and Jimoh (2009) South Africa

RAE retinol activity equivalents; *nd* non-detected values

[a] Expressed as β-carotene

[b] Expressed as tannic acid

[c] Expressed as quercetin

[d] Expressed as catechin

[e] Plants gathered from locations different from Mediterranean or surrounding areas

Table 13.34 Fatty acids profile (% of total fatty acids) of fresh tender leaves of *C. album*

Individual compounds	Average	Range	References
14:0	0.66	–	1,2
16:0	15.7	–	1,2
18:0	1.69	–	1,2
18:1*n*−9	2.90	–	1,2
18:2*n*−6	15.9	–	1,2
18:3*n*−3	44.8	–	1,2
18:4*n*−3	0.17	–	1,2
20:0	0.23	–	1,2
22:0	1.00	–	1,2
24:0	0.61	–	1,2
Categories (calculated values)			
SFA	23.8	–	–
MUFA	3.5	–	–
PUFA	72.8	–	–
n−3	53.5	–	–
n−6	19.0	–	–
n−9	3.5	–	–

1 Guil-Guerrero et al. (1996b) Spain (various samples), *2* Guil-Guerrero and Torija-Isasa (1997) Spain (5 samples)

SFA saturated fatty acids, *MUFA* monounsaturated fatty acids, *PUFA* polyunsaturated fatty acids

Remarks (tender leaves of *C. album*)

- High in dietary fibre (>6 g/100 g).
- Very high content of oxalates: People with altered renal function should avoid this vegetable; boiling is recommended for general population.
- High ash level; good source of K, Ca, Mg, and Fe; sometimes also a source of Cu , Mn and Zn.
- Source of vitamin C; it may be also considered as a source of vitamins A (provitamin A as β-carotene) and B_9 (folates).

13.3.11 Chondrilla juncea L. (Asteraceae)

Common Names Skeleton weed, *gingeira* (pt), *ajonjera* (es), *chicorée à la bûche* (fr), *ginestrella* (it), *abèloradiko* (el), *karakavuk* (tr).

Description Perennial herbaceous plant, up to 1 m tall, with a milky sap when broken. Basal rosette of glabrous and sharply lobed leaves that wither when the stiff flowering stems build up. Base of the stems with characteristic dense, bristly, downward-pointing hairs. Linear stem leaves and small flower heads, with a 7–12 yellow ligulate flowers.

Ecology and Distribution It grows in dry open habitats: roadsides, croplands, and semiarid pastures, in well-drained soils. Eurasian and Mediterranean in origin, it has been introduced in North America, Chile, Australia, and New Zealand.

Food Uses The human consumption of this species, both of the young basal leaves and the tender flowering shoots, has been mentioned in many Mediterranean countries, such as Portugal (Carvalho and Telo 2013), Spain (e.g. Fajardo et al. 2007; Parada 2008; Tardío et al. 2006), France (Couplan 1989b), Italy (Guarrera 2006; Lentini and Venza 2007; Picchi and Pieroni 2005), Slovenia (Cerne 1992), Croatia (Łuczaj et al. 2013a), Greece (Couplan 2009b; Leonti et al. 2006), and Turkey (Dogan 2012; Ertuğ 2014). It is very appreciated in some regions, such as Marche, in central Italy, where its collection is object of a real cult, being included in *mesticanza* (or *misticanza*), a salad with a mix of several species (Guarrera 2006; Picchi and Pieroni 2005) or in a salad with sliced roots of *Campanula rapunculus* L. (Picchi and Pieroni 2005). However, in Spain, the spring salads of the long and subterranean sprouts were typically consumed in the past, dressed with salt, vinegar, and olive oil. The blanched shoots with leaves, collected when the crops were hand-weeded, were produced from any part of the roots that were previously severed, due to the remarkable powers of regeneration of this weed (Blanco 1998; Parada 2008; Tardío et al. 2002; Velasco et al. 2010). In opinion of the informants, these sprouts were delicious, very tender, juicy, and non-bitter, especially if gathered in sandy and removed soils because they are longer and more tender. Sometimes the plants were even covered with ground to obtain blanched shoots (Ferrández and Sanz 1993).

The basal leaves of *Chondrilla juncea* have also been consumed cooked, either boiled and then seasoned with olive oil and lemon, or sautéed with garlic and olive oil (Picchi and Pieroni 2005), in vegetable soups (Carvalho and Telo 2013; Picchi and Pieroni 2005), pastries (Dogan 2012; Picchi and Pieroni 2005), or included in mixed wild vegetable recipes (Verde et al. 1998). The tender flowering shoots were consumed cooked in the south of Italy, included in omelettes, or simply boiled and seasoned with olive oil and lemon (Lentini and Venza 2007; Picchi and Pieroni 2005), whereas in Spain they were only eaten raw (Tardío et al. 2002).

The latex of the plant has been used as chewing gum for children (Velasco et al. 2010) or even to curdle milk and make cheese (Aceituno-Mata 2010).

Other Uses This species has been considered a medicinal food that lowers blood pressure (Parada 2008), helps cure insomnia (Lentini and Venza 2007), purifies the blood, and is good against vitamin deficiency (Bonet 2001). The roots, boiled and macerated in olive oil, have been employed as an antiseptic to heal wounds in humans and animals (Bonet 2001). The root latex has also been used to cure hand wounds and warts (Picchi and Pieroni 2005; Velasco et al. 2010).

The root latex has also been utilized in many Spanish regions to prepare *liga*, a kind of glue used for trapping birds (e.g. Pellicer 2001; Tardío et al. 2002).

This species is also eaten by livestock or even collected to feed them, being very appreciated by sheep and goats, donkeys, rabbits, hens, and pigs (e.g. Aceituno-Mata 2010; Ferrández and Sanz 1993; Picchi and Pieroni 2005).

The plants, after flowering, were commonly used to make home-made brooms in rural areas, at least in central Spain (Blanco 1998; Tardío et al. 2002; Velasco et al. 2010) and Portugal (Carvalho and Telo 2013). Though not registered in a recent review (Nedelcheva et al. 2007), the Italian name *ginestrella* (little broom) could also suggest this use. These big and hard brooms without stick were employed to sweep the road, the stable, and the threshing floor.

This species is a weed of winter cereal crops, especially in the sandy soils of the Mediterranean region. Women groups, who used to hand-weeding these crops in the spring, kept this useful species for dinner (Picchi and Pieroni 2005).

Historical References According to Picchi and Pieroni (2005), its human consumption was described together with the lettuce by Galeno (second century AD). These authors also mentioned that Mattioli, a famous Italian doctor of the sixteenth century, remarked that peasants from Tuscany were fond of this delight for the palate. However, Clusius (1576), the coetaneous Flemish botanist who travelled through Spain and Portugal, only mentioned its use for making brooms and a kind of glue for trapping goldfinches and other small birds in Salamanca.

Food Composition Tables for raw tender leaves of *Chondrilla juncea* (Tables 13.35, 13.36 and 13.37).

Table 13.35 Main constituents, per 100 g of fresh tender leaves of *C. juncea*

	Units	Average	Range	References
Energy (calculated value)	kcal	44	22–104	–
Moisture	g	83.4	65.9–89.7	1, 2, 3, 4
Available carbohydrates	g	3.58	1.49–9.69	2, 3
Dietary fibre	g	7.70	4.10–13.4	2, 3
Proteins	g	2.50	1.83–6.13	2, 3
Lipids	g	0.80	0.09–1.50	2, 3, 5
Ash	g	2.41	1.39–4.35	2, 3
K	mg	1015	433–1277	2, 3
Na	mg	16.5	3.85–58.0	2, 3
Ca	mg	230	22.0–472	2, 3
Mg	mg	70.4	2.70–100	2, 3
P	mg	12.7	–	2
Fe	mg	4.43	1.47–6.57	2, 3
Cu	μg	430	120–900	3
Mn	μg	970	570–1450	3
Zn	μg	1630	530–3810	3

1 Sánchez-Mata et al. (2012) Spain (2 samples), *2* Ranfa et al. (2014) Italy (various samples), *3* García-Herrera et al. (2014b) Spain (4 samples), *4* Morales et al. (2014) Spain (4 samples), *5* Morales et al. (2012b) Spain (4 samples)

Table 13.36 Vitamins and other constituents, per 100 g of fresh tender leaves of *C. juncea*

	Units	Average	Range	References
Carotenoids (total)	mg	2.13[a]	–	1
Vitamin B_9 (total folates)	μg	90.2	74.5–106	2
Vitamin C	mg	16.5	11.1–22.1	3, 4
Ascorbic acid	mg	1.88	Traces–3.55	3, 4
Dehydroascorbic acid	mg	14.6	10.5–22.1	3, 4
Vitamin E	mg	2.73	–	1
α-tocopherol	mg	0.57	0.39–0.75	4
β-tocopherol	mg	0.01	nd–0.02	4
γ-tocopherol	mg	0.12	0.11–0.13	4
δ-tocopherol	mg	0.05	0.04–0.06	4
Organic acids				
Oxalic acid	mg	50.0	6.71–104	3, 4
Glutamic acid	mg	nd	–	4
Malic acid	mg	51.0	8.14–190	3, 4
Citric acid	mg	24.5	0.83–50.5	3, 4
Fumaric acid	mg	0.78	0.38–1.31	3, 4
Phenolics (total)	mg	37.7[b]	35.3–40.1	4
Flavonoids	mg	7.43[c]	7.15–7.71	4

1 Ranfa et al. (2014) Italy (various samples), *2* Morales et al. (2015) Spain (4 samples), *3* Sánchez-Mata et al. (2012) Spain (2 samples), *4* Morales et al. (2014) Spain (4 samples); *nd* non-detected values

[a] Expressed as β-carotene

[b] Expressed as gallic acid

[c] Expressed as catechin

Table 13.37 Fatty acids profile (% of total fatty acids) of fresh tender leaves of *C. juncea*

Individual compounds	Average	Range	References
12:0	0.05	0.03–0.07	1
14:0	0.58	0.53–0.63	1
16:0	13.0	12.5–13.4	1
16:1	0.11	0.11–0.11	1
18:0	2.17	2.14–2.20	1
18:1*n*−9	1.91	1.90–1.92	1
18:2*n*−6	19.9	19.7–20.1	1
18:3*n*−3	56.3	56.1–56.4	1
18:3*n*−6	0.13	0.13–0.13	1
20:0	2.31	2.16–2.46	1
20:1*n*−9	0.04	0.04–0.04	1
22:0	1.54	1.33–1.75	1
24:0	0.81	0.66–0.96	1
Categories (calculated values)			
SFA	20.7	19.9–21.5	–
MUFA	2.1	2.1–2.1	–
PUFA	77.2	76.0–78.2	–
n−3	57.1	56.4–57.8	–
n−6	20.3	20.2–20.4	–
n−9	2.0	2.0–2.0	–

1 Morales et al. (2012b) Spain (4 samples)
SFA saturated fatty acids, *MUFA* monounsaturated fatty acids, *PUFA* polyunsaturated fatty acids

Remarks (tender leaves of *C. juncea*)

- It can reach high available carbohydrates content.
- Good source of dietary fibre (>3 g/100 g).
- High PUFA proportion, with higher *n*−3 fatty acids content (mainly α-linolenic acid, 18:3*n*−3), compared to other vegetables.
- Low Na content (<120 mg/100 g). It can be considered as a source of K, Mn, and vitamin B_9 (folates), sometimes also a source of Ca, Mg, Fe, Cu, Zn, and vitamin C.
- Very good oxalic acid/Ca ratio.

13.3.12 Cichorium intybus L. (Asteraceae)

Common Names Chicory, *almeirão* (pt), *achicoria* (es), *chicorée* (fr), *cicoria, radicchio* (it), *cikorija* (hr), *agrioradikia* (gr), *hindiba* (tr), *hindbeh* (ar).

Description Perennial herb, with a deep taproot, and a rosette of basal lanceolate leaves. They are lobed to deeply toothed near the bottom, with conspicuous hairs along the central vein of the lower leaf surface, though they can show a wide variation in shape and hairiness. Erect and branched flowering stems up to 1.5 m, with alternate and smaller leaves. Bright blue and petal-like flowers, with 5 tiny teeth at the tip, clustered in heads widely spaced along the flowering stalks, short-stalked or stalkless. The fruit is an achene, without pappus (feathery hairs).

Ecology and Distribution It can be found in humanized environments, such as roadsides, arable lands, pastures, waste grounds, and other disturbed open habitats, mainly on nitrified and slightly moist soils. Native to Eurasia, it is a widespread weed all over the world, with a cosmopolitan distribution.

Food Uses Its use as a wild vegetable is widespread in the Mediterranean region, as recorded in many ethnobotanical studies from Spain (e.g. Parada et al. 2011; Rivera et al. 2005; Tardío et al. 2006), Italy (e.g. Guarrera 2006; Lentini and Venza 2007; Picchi and Pieroni 2005), Slovenia (Cerne 1992), Croatia (Łuczaj et al. 2013b), Bosnia-Herzegovina (Redžić 2006), Greece (Hadjichambis et al. 2008; Leonti et al. 2006), Cyprus (Della et al. 2006; Hadjichambis et al. 2008), Turkey (Dogan 2012; Ertuğ 2014), Lebanon (Batal and Hunter 2007; Marouf 2005), Tunisia (Le Floc'h 1983), and Morocco (Nassif and Tanji 2013). The tender basal leaves are collected before blooming and consumed either raw or more frequently cooked, since it has a characteristic bitter taste that must be reduced by boiling several times and discarding the water. The raw consumption of the plant was associated with its gathering in tilled soils when hand-weeding cereal crops, so common in the past. When they grow underground, they produce a tight whorl of blanched leaves that are much less bitter, being a traditional version of the present cultivated witloof chicory, whose compact and elongated head is the result of being forced to sprout in complete darkness (Tardío 2010). The plant was eaten raw with olives, onions, and bread (Della et al. 2006), in salads (e.g. Nassif and Tanji 2013; Picchi and Pieroni 2005; Tardío et al. 2006), either with yogurt (Batal and Hunter 2007; Dogan 2012) or in mixed salads with other species such as the traditional Italian *misticanza* (Guarrera 2006). Regarding its cooked consumption, there are more recipes, such as soups (e.g. Cerne 1992; Hadjichambis et al. 2008; Le Floc'h 1983; Lentini and Venza 2007)—including the Italian soup *fuje amnesteche* o *foje* (Guarrera 2006)—stewed with legumes (Della et al. 2006; Hadjichambis et al. 2008; Picchi and Pieroni 2005), boiled and seasoned with oil, salt, and lemon (Dogan 2012; Lentini and Venza 2007), boiled and fried or sautéed with onion or garlic and oil (Batal and Hunter 2007; Dogan 2012; Tardío et al. 2006), or even pickled in vinegar (Della et al. 2006). Due to its bitter taste, it was often included in several recipes or blends of wild vegetables in Spain and Italy (Tardío 2010). Today, it is still possible to find it in local markets of Dalmatia, in Croatia (Łuczaj et al. 2013b).

The roots of the wild chicory were also traditionally used as a coffee substitute, as those of the cultivated varieties. They were used mainly in scarce times, as reported in Portugal, Spain, and Italy (Guarrera 2006; Mendonça de Carvalho 2006; Tardío et al. 2006). The roots were dried, roasted, and ground, and sometimes mixed with roasted barley or coffee. For the same purpose, the dried basal leaves were sometimes used, generally mixed with coffee (Tardío et al. 2005; Velasco et al. 2010).

Other Uses Bitter herbs, such as chicory, are commonly perceived as healthy in many Mediterranean folk cuisines. Its consumption as a vegetable or drinking the decoction water is considered appetizer, depurative, diuretic, laxative, blood cleanser, and with positive effects on kidneys and liver (Batal and Hunter 2007; Bonet and Vallès 2002; Guarrera 2006; Le Floc'h 1983; Lentini and Venza 2007; Picchi and Pieroni 2005; Tardío et al. 2005). It is also used to treat gastrointestinal illnesses (Guarrera 2006; Redžić 2006), kidney and liver problems (Batal and Hunter 2007; Lentini and Venza 2007; Tardío et al. 2005), high blood pressure (Guarrera 2006), and anaemia (Batal and Hunter 2007; Guarrera 2006; Picchi and Pieroni 2005) among others. Medicinal properties were also attributed to the boiled roots (Villar et al. 1987).

Its leaves are also used as animal fodder, mainly for rabbits, pigs, and poultry (e.g. Aceituno-Mata 2010; Benítez 2009; Ferrández and Sanz 1993).

Historical References This wild vegetable is known since antiquity, being mentioned by Theophrastus (third century BC; Teofrasto 1988). Its consumption and medicinal properties are also referred by Dioscorides and Pliny the Elder (first century AD; Bostock and Riley 1855; Osbaldeston 2000). It was regarded as astringent, cooling, and good for the stomach, especially when boiled and taken with vinegar (Osbaldeston 2000). Its juice, mixed with rose oil and vinegar, was employed against headache; and taken with wine, for pains in the liver and bladder (Bostock and Riley 1855).

Food Composition Tables for raw tender leaves of *Cichorium intybus* (Tables 13.38, 13.39 and 13.40).

Table 13.38 Main constituents, per 100 g of fresh tender leaves of *C. intybus*

	Units	Average	Range	References
Energy (calculated value)	kcal	33	10–58	–
Moisture	g	87.9	75.0–94.5	1, 2, 3, 4, 5, 6
Available carbohydrates	g	3.50	1.80–4.7	5
Dietary fibre	g	3.6	1.20–6.70	1, 5
Proteins	g	1.83	0.20–4.30	1, 3, 5
Lipids	g	0.46	Traces–0.92	1, 5, 7
Ash	g	1.65	1.25–2.10	1, 2, 5
K	mg	299	50–1085	2, 5
Na	mg	70.8	21.6–170	2, 5
Ca	mg	153	45.5–276	2, 5
Mg	mg	19.8	12.0–33.9	2, 5
P	mg	10	–	3
Fe	mg	1.29	0.41–2.00	5
Cu	μg	110	60.2–210	5
Mn	μg	260	170–470	5
Zn	μg	370	80–510	5

1 Cowan et al. (1963), Lebanon, *2* Bianco et al. (1998) Italy (2 samples), *3* Özbucak et al. (2007) Turkey, *4* Sánchez-Mata et al (2012) Spain (2 samples), *5* García-Herrera et al.(2014b) Spain (4 samples), *6* Morales et al. (2014) Spain (4 samples), *7* Morales (2012b) Spain (4 samples)

Table 13.39 Vitamins and other constituents, per 100 g of fresh tender leaves of *C. intybus*

	Units	Average	Range	References
Carotenoids				
β-carotene	mg	2.94	1.44–4.45	1, 2, 3
Lutein	mg	3.65	3.05–4.54	2, 3
Neoxanthin	mg	1.44	1.15–1.73	2
Violaxanthin	mg	1.70	1.39–2.01	2
RAE (calculated value)	μg	245	120–371	–
Vitamin B_9 (total folates)	μg	253	244–262	4
Vitamin C	mg	19.7	11.5–23.0	3, 5, 6
Ascorbic acid	mg	2.72	0.34–7.00	5, 6
Dehydroascorbic acid	mg	15.2	11.5–18.9	5, 6
Vitamin E				
α-tocopherol	mg	0.99	0.88–1.10	3, 6
β-tocopherol	mg	0.04	0.03–0.05	6
γ-tocopherol	mg	1.23	0.58–1.97	3, 6
δ-tocopherol	mg	0.08	0.08–0.08	6
Vitamin K	mg	173	–	3
Organic acids				
Oxalic acid	mg	40.0	3.94–76.1	5, 6
Glutamic acid	mg	70.2	56.3–84.1	6
Malic acid	mg	92.5	11.1–150	5, 6
Citric acid	mg	30.3	4.62–55.9	5, 6
Fumaric acid	mg	2.40	Traces–3.38	5, 6

Table 13.39 (continued)

	Units	Average	Range	References
Phenolics (total)	mg	96.3[a]	73.0–119	3, 6
	mg	122[b]	–	2
	mg	306[c]	99.4–396	7, 8
Phenolic acids	mg	15.8[d]	–	2
Flavonoids	mg	20.8[c]	20.5–21.0	7
	mg	31.3[e]	30.3–32.3	6
	mg	106[f]	–	2
Nitrate	mg	12.9	–	9

1 Cowan et al. (1963) Lebanon, *2* Salvatore et al. (2005) Italy (3 samples), *3* Vardavas et al. (2006b) Crete (3 samples), *4* Morales et al. (2015) Spain (4 samples), *5* Sánchez-Mata et al (2012) Spain (2 samples), *6* Morales et al. (2014) Spain (4 samples), *7* Conforti et al. (2009) Italy, *8* Conforti et al. (2011) Italy (4 samples), 9 Bianco et al. (1998) Italy (2 samples)

RAE retinol activity equivalents

[a] Expressed as gallic acid

[b] Expressed as sum of individual compounds

[c] Expressed as chlorogenic acid

[d] Expressed as caffeic acid

[e] Expressed as catechin

[f] Expressed as quercetin

Table 13.40 Fatty acids profile (% of total fatty acids) of fresh tender leaves of *C. intybus*

Individual compounds	Average	Range	References
12:0	0.05	0.04–0.06	1
14:0	0.99	0.29–1.60	1, 2
16:0	15.8	10.0–21.1	1, 2
16:1	0.40	0.09–0.67	1, 2
18:0	2.13	1.33–2.87	1, 2
18:1$n-9$	1.57	1.53–1.61	1, 2
18:2$n-6$	18.1	15.1–21.1	1, 2
18:3$n-3$	48.8	37.1–60.9	1, 2
18:3$n-6$	0.33	0.33–0.33	1, 2
20:0	0.82	0.63–1.00	1, 2
20:1$n-9$	0.16	0.05–0.27	1
22:0	0.50	0.10–1.13	1, 2
24:0	4.38	1.12–7.60	1, 2
Categories (calculated values)			
SFA	26.0	20.5–30.2	–
MUFA	2.5	1.8–4.9	–
PUFA	69.2	68.0–76.9	–
$n-3$	50.3	50.2–54.4	–
$n-6$	19.0	17.7–22.6	–
$n-9$	2.2.1	1.7–2.5	–

1 Morales (2012b) Spain (4 samples), *2* Vardavas et al. (2006a) Crete (3 samples)

SFA saturated fatty acids, *MUFA* monounsaturated fatty acids, *PUFA* polyunsaturated fatty acids

Remarks (tender leaves of *C. intybus*)

- In some cases it may be considered as a source of dietary fibre (>3 g/100 g).
- Good oxalic acid/Ca ratio. Low oxalate levels (<100 mg/100 g).
- Low Na content (<120 mg/100 g); sometimes may be a source of K and Ca.
- It can be considered as a source of vitamins C, B_9 (folates), and A (provitamin A as β-carotene).

13.3.13 Crataegus monogyna Jacq. (Rosaceae)

Common Names Common hawthorn, *pilriteiro* (pt), *majuelo* (es), *aubépine* (fr), *biancospino comune* (it), *glog* (hr), *trikokkiá* (el), *yemişen* (tr), *zaarour bari* (ar).

Description Shrub or small tree, up to 7 m, spiny, deciduous that usually has a rounded top. The leaves are petiolate, obovate, or spatulate, cuneate, with up to 7 acute to rounded lobes. The flowers are grouped in 5–11 corymbiform groups, long pedicelate, with 5 sepals and 5 rounded petals, white or cream 20 stamens, with purple anthers. The fruits are spherical pomes up to 1 cm, usually red, fleshy, with 1 seed or pyrene. It flowers from March to July.

Ecology and Distribution It grows in open forests, edges and thickets, frequently on fresh and deep soils. It lives in almost all of Europe, northwest of Africa and west Asia, up to 2000 m. It is highly variable and six subspecies are considered in Europe.

Food Uses The fruits were commonly eaten in Portugal (Carvalho 2010), Spain (e.g. Aceituno-Mata 2010; Benítez 2009; Menendez-Baceta et al. 2012; Tardío et al. 2006), Italy (Guarrera 2006; Leonti et al. 2006), Slovenia (Cerne 1992), Bosnia-Herzegovina (Redžić 2006), Cyprus (Hadjichambis et al. 2008), Turkey (Ertuğ 2004, Ertuğ 2014), Tunisia (Le Floc'h 1983), and Morocco (Nassif and Tanji 2013). They ripen in autumn and were commonly eaten raw directly from the field (e.g. Blanco 1998; Cobo and Tijera 2011) and also consumed as a dessert or even stored for the winter (Aceituno-Mata 2010; Carvalho and Telo 2013), and sometimes even sold (Guzmán 1997; Tardío et al. 2002; Velasco et al. 2010). The fruits are usually small, with little flesh, and therefore they were mainly a children food and also used in famine times (Blanco and Diez 2005). However, there are trees with bigger and tastier fruits that were preferably exploited (Tardío et al. 2002), Nevertheless, they have been traditionally believed to be very rich in vitamins and, for instance, in some regions of Portugal, children were forced to eat them (Carvalho 2010). The pomes have been also used for making jam in Spain (Benítez 2009; Sánchez-Romero 2003), Italy (Lentini and Venza 2007) and Cyprus (Della et al.

2006). A liqueur of the fruits is prepared in Spain (e.g. González et al. 2011a; Verde et al. 2001), Italy (Guarrera 2006), and Slovenia (Cerne 1992), macerating them in brandy or liquor, alone or with other herbs (Parada 2008).
The raw flowers, young shoots, and leaves have been eaten in Spain (e.g. Tardío et al. 2006; Velasco et al. 2010) and Italy (Guarrera 2006), and the young leaves are also eaten in salads in Cyprus (Hadjichambis et al. 2008).

Other Uses It is a very important medicinal plant. The infusion of the flowers is commonly taken as sedative (e.g. Benítez 2009; Blanco and Cuadrado 2000; Pellicer 2001), considered hypotensive (e.g. Bonet and Vallès 2002; Guzmán 1997; Pardo-de-Santayana 2003), cardiotonic (e.g. Fajardo et al. 2007; Sánchez-Romero 2003) and used to cure circulatory (Fajardo et al. 2007; García Jiménez 2007) or respiratory ailments (Carvalho 2010). The leaves (Velasco et al. 2010) or fruits are sometimes used against such complaints (Villar et al. 1987). The fruits are considered a medicinal food, having relaxing (Lentini and Venza 2007), laxative (Villar et al. 1987), and digestive properties (García Jiménez 2007), and the jam is used as anticatarrhal (e.g. Benítez 2009; Fajardo et al. 2007).

The young shoots, leaves, and fruits are appreciated for feeding animals (Benítez 2009; Guarrera 2006), especially goats, pigs, and turkeys (e.g. Aceituno-Mata 2010; Fajardo et al. 2007; Velasco et al. 2010). It is also consumed by wild animals (Criado et al. 2008; Parada 2008), and the bees make good honey from its flowers (Fajardo et al. 2007; Lastra 2003; Molina 2001).

Its wood is used for manufacturing tool handles, musical instruments, bowls or spoons in Spain (e.g. Aceituno-Mata 2010; Fajardo et al. 2007), Portugal (Carvalho and Telo 2013), Italy (Guarrera 2006), and Turkey (Ertuğ 2004).

It has been used as rootstock for grafting trees, such as pear and apple trees (Aceituno-Mata 2010; Fajardo et al. 2007; Velasco et al. 2010), medlar (Guarrera 2006), service tree (Ferrández and Sanz 1993; Pardo-de-Santayana 2003), plum tree (Moll 2005), or climbing rosetrees (Mesa 1996). This plant is also used for hedges (Ferrández and Sanz 1993; Guarrera 2006) and as firewood since it has a high calorific power (Velasco et al. 2010).

The fruits were employed in several children games, such as throwing the seeds with a peashooter (e.g. Ferrández and Sanz 1993; Tardío et al. 2002), or elaborating bracelets and necklaces (Aceituno-Mata 2010; Benítez 2009).

The flowering and fruiting branches are used as ornamentals (Carvalho 2010; Parada 2008) and for popular bunches for sweetheart (Criado et al. 2008). This small tree was also employed for crop yield prediction: In the years with an abundant flowering, a good crop yield was expected by the peasants (Benítez 2009).

Historical References The fruits of plants of this genus have been used since antiquity, and their remains have been found in many archaeological sites (e.g. Font Quer 1962; Heiss 2014; Rottoli 2014). They were mentioned by Theophrastus in the third century BC and by Dioscorides in the first century (Osbaldeston 2000; Teofrasto 1988). The Spanish botanist of the sixteenth century, Laguna (1555) reported that he commonly ate the fruits as a child in his homeland, Segovia, Spain.

Food Composition Tables for raw fruits of *Crataegus monogyna* (Tables 13.41, 13.42 and 13.43).

Table 13.41 Main constituents, per 100 g of fresh fruits of *C. monogyna*

	Units	Average	Range	References
Energy (calculated value)	kcal	139	80–206	–
Moisture	g	61.8	32.1–77.8	1, 2, 3, 4, 5, 6, 7, 8
Available carbohydrates	g	25.9	9.40–37.5	5, 7
Soluble sugars (total)	g	14.4	7.7–17.7	1, 5, 7
Fructose	g	2.82	1.54–4.14	1, 5, 7
Glucose	g	7.90	1.05–14.3	1, 5, 7
Sucrose	g	0.21	0.05–0.44	1, 5, 7
Dietary fibre	g	11.8	7.23–16.4	7
Insoluble fibre	g	8.90	5.50–15.3	7
Soluble fibre	g	3.20	2.10–6.80	7
Proteins	g	1.43	0.39–2.48	4, 5, 7, 8
Lipids	g	0.63	0.38–1.22	1, 4, 5, 7, 8
Ash	g	2.32	1.07–4.26	1, 2, 4, 5, 7, 8
K	mg	459	113–603	1, 2, 3, 7, 8
Na	mg	15.6	3.14–31.2	1, 7, 8
Ca	mg	235	115–304	1, 2, 7, 8
Mg	mg	83.3	11.8–150	1, 2, 7, 8
P	mg	86.4	25.9–147	1, 8
Fe	mg	1.82	0.43–3.27	1, 2, 7, 8
Cu	μg	179	84.0–233	1, 2, 7
Mn	μg	742	46.0–1141	1, 2, 7
Zn	μg	311	187–540	1, 2, 7

1 Doležal et al. (2001) Czech Republic, *2* Boudraa et al. (2010) Algeria, *3* Egea et al. (2010) Spain, *4* Ganhão et al. (2010) Spain, *5* Barros et al. (2011a) Portugal, *6* Morales et al. (2013) Spain (6 samples), *7* Ruiz-Rodríguez (2014) Spain (6 samples), *8* Özcan et al. (2005) Turkey

Table 13.42 Vitamins and other constituents, per 100 g of fresh fruits of *C. monogyna*

	Units	Average	Range	References
Carotenoids				
β-carotene	mg	1.32	Traces–1.59	1, 2, 3, 4
RAE (calculated value)	μg	110	90–130	–
Vitamin B_1	μg	5	5–5	2
Vitamin C	mg	27.8	9.62–38.5	1, 5, 6
Ascorbic acid	mg	8.85	nd–25.2	2, 3, 5, 6, 7, 8
Dehydroascorbic acid	mg	21.6	9.31–36.3	5, 6
Vitamin E				
α-tocopherol	mg	1.88	0.70–3.02	2, 5
β-tocopherol	mg	0.15	0.14–0.16	5
γ-tocopherol	mg	0.17	0.16–0.18	5
δ-tocopherol	mg	0.16	0.13–0.19	5
Organic acids				
Oxalic acid	mg	70.7	30.5–84.3	5, 8
Shikimic acid	mg	9.2	9.2–9.2	8
Malic acid	mg	463	146–608	5, 8
Citric acid	mg	48.9	29.2–72.0	5, 8

Table 13.42 (continued)

	Units	Average	Range	References
Fumaric acid	mg	0.15	nd–0.40	5, 8
Succinic acid	mg	3.08	nd–10.3	5, 8
Phenolics (total)	mg	681[a]	347–1375	6
	mg	745[b]	312–2525	9, 10
Phenolic acids	mg	467[b]	245–911	6
Hydroxibenzoic	mg	0.55[b]	0.36–0.84	10
Hydroxycinnamic acids	mg	23.5[c]	18.5–28.4	10
Flavonols	mg	110[d]	32.3–468	6, 10
Anthocyanins	mg	11.1[e]	0.88–21.4	9, 10
	mg	27.3[f]	9.54–56.0	6
Sorbitol	g	4.34	–	1

1 Doležal et al. (2001) Czech Republic, *2* Boudraa et al. (2010) Algeria, *3* Egea et al. (2010) Spain, *4* Ruiz-Rodríguez (2014) Spain (6 samples), *5* Morales et al. (2013) Spain (6 samples), *6* Ruiz-Rodríguez et al. (2014a) Spain (6 samples), *7* Barros et al. (2011a) Portugal, *8* Pereira et al. (2013) Portugal, *9* Froehlicher et al. (2009) France; *10* Ganhão et al. (2010) Spain

RAE retinol activity equivalents; *nd* non-detected values

[a] Expressed as sum of individual compounds (gallic acid derivatives as major components)

[b] Expressed as gallic acid

[c] Expressed as chlorogenic acid

[d] Expressed as rutin

[e] Expressed as cyanidin-3-glucoside

[f] Expressed as pelargonidin-3-glucoside

Table 13.43 Fatty acids profile (% of total fatty acids) of fresh fruits of *C. monogyna*

Individual compounds	Average	Range	References
12:0	2.42	2.36–2.48	1
14:0	1.21	1.21–1.21	1
16:0	30.6	25.2–31.22	1, 2
16:1	0.40	0.38–0.42	1
18:0	5.24	3.81–6.61	1, 2
18:1$n-9$	17.8	10.1–24.9	1, 2
18:2$n-6$	10.3	10.2–10.8	1, 2
18:3$n-3$	16.1	15.7–16.4	1
20:0	nd	–	1
20:1$n-9$	3.06	2.93–3.19	1
22:0	4.67	4.64–4.70	1
24:0	13.3	11.5–15.1	1
Categories (calculated values)			
SFA	54.6	52.2–62.9	–
MUFA	20.2	13.5–28.6	–
PUFA	25.1	20.6–27.2	–
$n-3$	15.3	14.9–15.6	–
$n-6$	10.2	9.8–10.3	–
$n-9$	21.3	13.4–28.5	–

1 Morales et al. (2013) Spain (6 samples), *2* Doležal et al. (2001) Czech Republic; *nd* non-detected values

SFA saturated fatty acids, *MUFA* monounsaturated fatty acids, *PUFA* polyunsaturated fatty acids

Remarks (fruits of *C. monogyna*)

- High content of sugars.
- High in dietary fibre (>6 g/100 g).
- High ash level but low Na content (<120 mg/100 g).
- It can be considered as a source of Ca and vitamin A (provitamin A as β-carotene); sometimes also a source of K, Mg, Cu, Mn, and vitamins C and E.
- High phenolics content.

13.3.14 Crithmum maritimum L. (Apiaceae)

Common Names Rock samphire, *funcho do mar* (pt), *hinojo marino* (es), *criste marine* (fr), *finocchio marino* (it), *motar* (hr), *kritamo* (el), *denizteresi* (tr), *shanar bahariya* (ar).

Description Perennial herb, woody at the base, that bears glabrous, fleshy, and pinnate leaves, with 2–5 cm lanceolate leaflets, usually acute. Flowers mostly hermaphrodite, with minute sepals and pale yellow petals grouped in a compound umbel. The fruits are 3–6 mm, ovoids, glabrous, with 10 ridges.

Ecology and Distribution It grows on maritime rocks and hanging on cliffs, sometimes on sand, 0–150 m. It flowers usually from May to July. Very common in the coasts of almost all of Europe, Mediterranean Asia, and the north of Africa.

Food Uses The tender leaves and stems, collected both in spring and summer, and even in autumn, have been traditionally consumed as a salad vegetable in most Mediterranean countries, such as Spain (e.g. Carrió 2013; Parada 2008; Pellicer 2004a), Italy (e.g. Guarrera 2006; Picchi and Pieroni 2005), Slovenia (Cerne 1992), Croatia (Łuczaj et al. 2013b), Bosnia-Herzegovina (Redžić 2006), Greece (Hadjichambis et al. 2008), Cyprus (Della et al. 2006), Turkey (Dogan 2012; Ertuğ 2014), Tunisia (Le Floc'h 1983), and Morocco (Nassif and Tanji 2013). Although it can be eaten raw, it is frequently scalded in boiling water to soften, before using in salads. The tender leaves and stems have been traditionally preserved as pickles or brines throughout the Mediterranean region (e.g. Carrió 2013; Della et al. 2006; Guarrera 2006; Hadjichambis et al. 2008; Le Floc'h 1983; Moll 2005; Parada 2008; Pellicer 2004a; Picchi and Pieroni 2005). In a mode of preparation described in Alicante (Spain), the young leaves are cut into pieces and cleaned, slightly boiled, and let to cool completely. They are put in a wide-mouth recipient and covered with brine made with water, vinegar, and salt in a proportion of 5:1:0.5, or with a bit higher proportion of water if the brine is preferred softer. Following Picchi and Pieroni (2005), in Italy, the leaves are exposed to the sun, before being introduced in vinegar. In Cyprus, it is eaten like an appetizer with several kinds of food (Della et al.

2006); in the Balearic Islands, they are typically eaten with bread and olive oil (Carrió 2013), or prepared with capers (Moll 2005).
It is also used as a cooked vegetable, usually mixed with other species, so it can be also considered as a condiment, depending on the proportions of use. It is sold included in wild vegetable mixes, in the markets of Dalmatia, Croatia (Łuczaj et al. 2013b), and in Menorca, Spain, where it is cooked with rice (Moll 2005).

As a condiment, it has been used for seasoning olives, together with other aromatic plants, such as thyme and oregano in Catalonia, Spain, and also for homemade anchovies in brine, being covered with the leaves of the plant that confer them an special scent (Font Quer 1962). In Italy, the plant is used for the elaboration of sauces (Guarrera 2006), or as a fish condiment, being also employed in modern cuisine (Picchi and Pieroni 2005). Besides the leaves, the seeds are also used as a condiment, in Sicily (Lentini and Venza 2007).

In Mallorca, Spain, some people used this plant, instead of fennel (*Foeniculum vulgare*) in the elaboration of the typical herb liqueur (Carrió 2013).

In the past, this plant used to be sold on the streets of the Isle of Mallorca, while shouting *fonoll marí!* (sea fennel), being an interesting economic supplement for the collectors (Carrió 2013). Nowadays, some supermarkets sell jars with the pickled leaves.

Other Uses The medicinal properties of the consumption of this species are mentioned in many ethnobotanical works, as digestive, diuretic, and against scurvy (Carrió 2013; Guarrera 2006). Sailors have traditionally brought it in the boats for fighting against scurvy and renal and bladder stones. That is probably the reason of one of its popular names (St Peter's herb) in several European languages, such the Italian *erba di San Pietro* or the French *Saint Pierre*, whose derivation produce the English name samphire.

Historical References This species has a long history of use, both as a vegetable and as a medicinal plant, so it is possible to find lots of references. For instance, in the first century, both Dioscorides and Pliny the Elder referred to its consumption, eaten either raw or boiled, and also preserved in brine (Bostock and Riley 1855; Osbaldeston 2000). Besides describing some other medicinal properties, they also agreed in the effectiveness as diuretic of a drink made with its seed, root, and leaves boiled in wine, especially to help frequent painful urination. Many other historical sources refer its use. Laguna (1555) mentioned that the preserved plant was brought to Spain from Sicily. Outside the Mediterranean region, the English Nicholas Culpeper describes the plant in his Complete Herbal (1653) as having a pleasant, hot and spicy taste and being a great "digestive". The Spanish botanist Gómez-Ortega (1784) said that the part of Spain where the plant was more appreciated was Catalonia. There, the biggest sea fennel individuals were collected and prepared in vinegar to eat them, in salad or other several ways, all the year round, especially in winter.

Food Composition Tables for raw tender leaves of *Crithmum maritimum* (Tables 13.44, 13.45 and 13.46).

Table 13.44 Main constituents, per 100 g of fresh leaves of *C. maritimum*

	Units	Average	Range	References
Energy (calculated value)	kcal	30	23–38	–
Moisture	g	86.8	47.7–88.6	1, 2, 3
Available carbohydrates	g	2.08	1.90–2.7	1
Dietary fibre	g	4.68	3.04–5.60	1
Proteins	g	2.23	1.7–2.8	1
Lipids	g	0.43	0.28–0.58	1, 4, 5
Ash	g	2.90	2.54–3.48	1
K	mg	252	198–343	1, 3, 6
Na	mg	464	185–636	1, 3, 6
Ca	mg	224	85–414	1, 3, 6
Mg	mg	76.6	57.4–97.0	1, 3, 6, 7
P	mg	21.5	16.0–24.0	3, 7
Fe	mg	2.29	1.09–3.70	1, 3, 6, 7
Cu	μg	120	95.0–151	1, 3
Mn	μg	990	432–1080	1, 3
Zn	μg	665	334–870	1, 3, 6

1 Guil-Guerrero et al. (1996a) Spain (10 samples), *2* Guil-Guerrero et al. (1997a) Spain (5 samples), *3* Guil-Guerrero et al. (1998b) Spain (5 samples), *4* Guil-Guerrero et al. (1996b) Spain (various samples), *5* Guil-Guerrero and Rodríguez García (1999) Spain (5 samples), *6* Romojaro et al (2013) Spain, *7* Guil-Guerrero and Torija-Isasa (2002) Spain

Table 13.45 Vitamins and other constituents, per 100 g of fresh leaves of *C. maritimum*

	Units	Average	Range	References
Carotenoids (total)	mg	4.46[a]	3.30–5.60	1, 2, 3
Vitamin C	mg	62.0	39.0–76.6	1, 2, 4
Ascorbic acid	mg	39.0	22.0–50.7	1, 2, 4
Dehydroascorbic acid	mg	23.0	17.0–26.3	2, 4
Organic acids				
Oxalic acid	mg	53.0	16.0–77.0	1, 2
Phenolics (total)	mg	≈1260	≈715–1090	5
Flavonoids	mg	≈35.6[b]	21.1–50.2	5
Taninns	mg	≈193	35.6–350	5
Nitrate	mg	63.0	41.0–85.0	2

1 Guil-Guerrero et al. (1996a) Spain (10 samples), *2* Guil-Guerrero et al. (1997a) Spain (5 samples), *3* Guil-Guerrero and Rodríguez García (1999) Spain (5 samples), *4* Franke (1982) Italy, *5* Males et al. (2003) Croatia (10 samples)

[a] Expressed as β–carotene

[b] Expressed a quercetin

Table 13.46 Fatty acids profile (% of total fatty acids) of fresh leaves of *C. maritimum*

Individual compounds	Average	Range	References
14:0	1.75	–	1
16:0	9.41	–	1
16:1*n*−7	0.62	–	1
18:0	1.49	–	1
18:1*n*−7	nd	–	1
18:1*n*−9	16.1	–	1
18:2*n*−6	12.0	–	1
18:3*n*−3	9.98	–	1
18:3*n*−6	0.25	–	1
18:4*n*−3	0.42	–	1
20:0	1.38	–	1
20:1*n*−9	0.54	–	1
Categories (calculated values)			
SFA	26.0	–	–
MUFA	32.0	–	–
PUFA	42.0	–	–
n−3	19.0	–	–
n−6	22.7	–	–
n−9	30.8	–	–

1 Guil-Guerrero et al. (1996b) Spain (various samples);
SFA saturated fatty acids; *nd* non-detected values
MUFA monounsaturated fatty acids, *PUFA* polyunsaturated fatty acids

Remarks (leaves of *C. maritimum*)

- Low energy value (<40 kcal/100 g).
- Source of dietary fibre (>3 g/100 g).
- High MUFA proportion (oleic acid).
- Low oxalate levels (<100 mg/100 g).
- High ash levels and Na content.
- It can be considered as a source of Mg, Mn, and vitamin C, sometimes also a source of Ca and Fe.
- Very high amount of phenolics.

13.3.15 Eruca vesicaria (L.) Cav. (Brassicaceae)

Common Names Rocket, *oruga* (es), *roquette* (fr), *rucola* (it), *roca* (el), *yabani roka* (tr), *jarjeer* (ar).

Description This annual herb has a rosette of basal pinnatifid leaves with several lateral lobes and a large terminal lobe. The flowers have 4 erect sepals and perpendicular to these appear 4 white petals, with characteristic purple or brown veins. The fruit is a siliqua, more than 2 cm long, with seeds growing in two rows and a prominent flattened beak.

The cultivated form is named *E. vesicaria* subsp. *sativa* (Mill.) Thell and usually have bigger leaves and fruits and early deciduous sepals.

Ecology and Distribution It is a ruderal plant that grows along roads, in waste places, and cultivated grounds. It flowers in late winter and spring. Native to the Mediterranean region, it occurs in south of Europe, north Africa, and west Asia.

Food Uses This plant has been traditionally consumed throughout the Mediterranean countries, at least in Spain (e.g. Parada et al. 2011; Pellicer 2004a; Tardío et al. 2006), Italy (Guarrera 2006; Pieroni et al. 2005), Slovenia (Cerne 1992), Croatia (Łuczaj et al. 2013b), Greece (Hadjichambis et al. 2008), Cyprus (Della et al. 2006), Turkey (Dogan 2012; Ertuğ 2014), Lebanon (Marouf 2005), Jordan (Al-Qura'n 2010; Tukan et al. 1998), Palestine (Ali-Shtayeh et al. 2008), Tunisia (Le Floc'h 1983), and Morocco (Nassif and Tanji 2013). Its basal leaves are collected in spring, being mainly consumed raw in salads (Al-Qura'n 2010; Ali-Shtayeh et al.

2008; Cerne 1992; Della et al. 2006; Dogan 2012; Guarrera 2006; Hadjichambis et al. 2008; Marouf 2005; Piera 2006; Tukan et al. 1998), or raw without preparation (Al-Qura'n 2010; Dogan 2012; Tukan et al. 1998). Additionally, they are consumed cooked (Tardío et al. 2006; Triano et al. 1998), in soups (Piera 2006), used to make sauces (Guarrera 2006; Nassif and Tanji 2013; Piera 2006), or as garnish for couscous (Nassif and Tanji 2013). The leaves have a strong flavour, generally used in small proportions, mixed with other vegetables. Therefore, it is sometimes considered as a spice (Cerne 1992).

Its seeds are also consumed, either eaten raw or used as seasoning for cakes, as registered in Spain (Tardío et al. 2006).

This species is collected from the wild, but it is also cultivated as a crop (Della et al. 2006; Dogan 2012), although its domestication cannot be considered complete (Paludosi and Pignone 1997). It can be purchased throughout the year from local markets and stores. Due to its strong and sometimes aggressive flavour (even that of the cultivated forms), in some places of Italy, it has been substituted by two perennial species of the same family, *Diplotaxis tenuifolia* (L.) DC and *D. muralis* (L.) DC Picchi (Picchi and Pieroni 2005). The three species are included in the collective name rocket and both *Eruca* and *D. tenuifolia* are cultivated (Paludosi and Pignone 1997).

Other Uses This vegetable has been traditionally considered as a medicinal food with several medicinal properties attributed to its consumption, such as digestive and depurative (Guarrera 2006; Piera 2006). Modern studies have shown some of these properties and even some antitumoral activity has been suggested (Esiyok et al. 2004).

Historical References The leaves have been surely consumed since ancient times. It is generally agreed that the vegetable mentioned in the Bible (2 King 4:39–40) as *oroth* is this plant (Wright 2012). It has a long history of cultivation, being cited by Theophrastus among the cultivated vegetables with vigorous seeds which germinate quickly (Teofrasto 1988). Pliny the Elder, in the first century, mentioned that this plant may be grown with the greatest facility either in summer or winter, being able to stand the cold (Bostock and Riley 1855). He also compared this vegetable with lettuce, pointing out that their properties are quite different, the rocket being a great provocative of lust. For this reason, he said that these two plants should be mixed in dishes, the excess of cold in the one being compensated by the equal degree of heat in the other. Dioscorides also stated that eaten in great amount encourages the pursuit of sexual pleasure (aphrodisiac), both the leaves and the seeds, also being diuretic, digestive, and good for the bowels. He mentioned the use of the seeds as a condiment for other foods, specifying the existence in Spain of a wild form whose seeds were used instead of mustard, being more diuretic and far sharper than the cultivated (Laguna 1555; Osbaldeston 2000). The supposed aphrodisiac quality of rocket was the reason why Roman Catholic Church banned its cultivation in monastic gardens (Wright 2012).

Food Composition Tables for raw tender leaves of *Eruca vesicaria* (Tables 13.47, 13.48 and 13.49).

Table 13.47 Main constituents, per 100 g of fresh leaves of *E. vesicaria*

	Units	Average	Range	References
Energy (calculated value)	kcal	28	–	–
Moisture	g	89.6	–	1
Available carbohydrates	g	2.10	–	2
Dietary fibre	g	1.60	–	2
Proteins	g	2.60	–	2
Lipids	g	0.70	–	2
Ash	g	2.12	–	1
K	mg	413	369–449	1, 2, 3, 4
Na	mg	14.1	6.64–27.0	1, 2, 3, 4
Ca	mg	250	160–327	1, 2, 3, 4
Mg	mg	33.7	22–48.1	1, 3, 4
Fe	mg	1.81	1.04–2.91	1, 3, 4
Cu	μg	≈63.7	44.5–83.0	3, 4
Mn	μg	≈389	336–443	3, 4
Zn	μg	429	349–538	1, 4

1 Bianco et al. (1998) Italy (2 samples), *2* Souci et al. (2008) unknown origin, *3* Barlas et al. (2011) Turkey (30 samples), *4* Villatoro-Pulido et al. (2012), various origins (27 samples)

Table 13.48 Vitamins and other constituents, per 100 g of fresh leaves of *E. vesicaria*

	Units	Average	Range	References
Carotenoids				
β-carotene	mg	1.28	1.15–1.40	1, 2
Lutein	mg	2.48	–	1
RAE (calculated value)	μg	110	90–120	–
Vitamin B_1	μg	40	–	2
Vitamin B_2	μg	90	–	2
Vitamin B_3 (nicotinamide)	μg	300	–	2
Vitamin B_6	μg	70	–	2
Vitamin C	mg	125	–	1
Vitamin E				
α-tocopherol	mg	3.07	–	1
γ-tocopherol	mg	0.09	–	1
Vitamin K	μg	31	–	1
Organic acids				
Oxalic acid	mg	7.00	–	3
Phenolics (total)	mg	211[a]	–	1
Nitrate	mg	257	–	3

1 Vardavas et al. (2006b) Crete, *2* Souci et al. (2008) unknown origin, *3* Bianco et al. (1998) Italy (2 samples)

RAE retinol activity equivalents

[a] Expressed as gallic acid

Table 13.49 Fatty acids profile (% of total fatty acids) fresh leaves of *E. vesicaria*

Individual compounds	Average	Range	References
14:0	3.4	–	1
16:0	29.5	–	1
16:1	1.3	–	1
18:0	3.3	–	1
18:1*n*−9	2.9	–	1
18:2*n*−6	11.7	–	1
18:3*n*−3	44.3	–	1
18:3*n*−6	0.4	–	1
24:0	0.4	–	1
Categories (calculated values)			
SFA	37.7	–	–
MUFA	4.3	–	–
PUFA	58.0	–	–
n−3	46.0	–	–
n−6	12.4	–	–
n−9	3.0	–	–

1 Vardavas et al. (2006a) Greece *SFA* saturated fatty acids, *MUFA* monounsaturated fatty acids, *PUFA* polyunsaturated fatty acids

Remarks (leaves of *E. vesicaria*)

- Low energy value (<40 kcal/100 g).
- Good *n*−3/*n*−6 ratio (>3).
- Low oxalic acid level (<100 mg/100 g).

13.3.16 Foeniculum vulgare Mill. (Apiaceae)

Common Names Wild fennel, *funcho* (pt), *hinojo* (es), *fenouil* (fr), *finocchio selvatico* (it), *komorač* (hr), *marathos* (el), *rezene* (tr), *shumar* (ar).

Description Perennial herb, glabrous, with erect and striate stems up to 1–2.5 m. The basal leaves are long petiolate, 3 or 4 pinnate, with filiform terminal lobes. Small yellow flowers found in large flat compound umbels. Fruits 3–8 mm, ovoid, glabrous, with 10 ridges, and, like the whole plant, with anise scent.

Ecology and Distribution In dry, waste places, roadsides, and shrublands, on basic or acid soils, 0–1200 m. It flowers from June to November. Native to the Mediterranean region, it has become naturalized in worldwide temperate areas.

Food Uses It has been widely employed in the Mediterranean countries, cited in Portugal (e.g. Carvalho 2010; Mendonça de Carvalho 2006), Spain (e.g. Parada et al. 2011; Tardío et al. 2006), France (Marco et al. 2003), Italy (e.g. Guarrera 2006; Picchi and Pieroni 2005), Greece (e.g. Leonti et al. 2006; Zeghichi et al. 2003), Croatia (Łuczaj et al. 2013b), Bosnia-Herzegovina (Redžić 2006), Cyprus (Della et al. 2006), Turkey (Dogan 2012; Ertuğ 2014), Lebanon (Batal and Hunter 2007; Marouf 2005), Palestine (Ali-Shtayeh et al. 2008), Jordan (Al-Qura'n 2010; Tukan et al. 1998), Egypt (Hadjichambis et al. 2008), Tunisia (Le Floc'h 1983), and Morocco (Nassif and Tanji 2013). It is a very versatile species which has been used as vegetable, for seasoning and food preservation, and as herbal tea. Both the vegetative aerial parts and the fruits are aromatic and have food uses. It is commonly marketed (Łuczaj et al. 2013b; Picchi and Pieroni 2005).

Its tender leaves and stems are collected in spring and eaten raw, sometimes only chewed and sucked as a snack, or in salads (e.g. Al-Qura'n 2010; Guarrera 2006; Nassif and Tanji 2013; Tardío et al. 2006), It is also cooked in several dishes, either with meat (Nassif and Tanji 2013), boiled and seasoned with oil and lemon (Della et al. 2006; Picchi and Pieroni 2005; Zeghichi et al. 2003), cooked with rice and other vegetables (Hadjichambis et al. 2008), or in soups (e.g. Ali-Shtayeh et al. 2008; Picchi and Pieroni 2005; Tardío et al. 2006), and also in omelettes, tomato sauces, pasta (Picchi and Pieroni 2005), and couscous (Le Floc'h 1983), sometimes mixed with other wild vegetables (Hadjichambis et al. 2008; Tardío 2010). In Portugal and Spain (Carvalho 2010; Pardo-de-Santayana 2008), its consumption as a stewed green vegetable seems to be more common among gypsy communities, as shown by the traditional Spanish recipe called *olla gitana* (gypsy pot). Due to its strong flavour, this wild vegetable is commonly employed as a condiment. For example, the traditional Cyprus recipes called *eliotes* (olive pies), *kolokotes* (pumpkin pies) and *spanakopites* (spinach pies) are seasoned with its tender leaves and stems (Della et al. 2006), while in Palestine, the dried aerial parts are added to cakes (Ali-Shtayeh et al. 2008).

The small fruits (popularly known as seeds), collected in summer or autumn, are used for seasoning soups, stews (e.g. Batal and Hunter 2007; Carvalho 2010; Hadjichambis et al. 2008; Lentini and Venza 2007), sausages or pork meat (Picchi and Pieroni 2005), olives (Picchi and Pieroni 2005; Tardío et al. 2006), boiled chestnuts (Carvalho 2010; Hadjichambis et al. 2008; Parada 2008), figs (Mendonça de Carvalho 2006; Tardío et al. 2005) and as flavouring and stuffing for pies (Batal and Hunter 2007; Nassif and Tanji 2013). The fruits are boiled and added to bread in many countries (e.g. Ali-Shtayeh et al. 2008; Guarrera 2006; Le Floc'h 1983; Nassif and Tanji 2013), and sometimes even the fresh leaves (Tukan et al. 1998). Both the vegetative parts and the fruits have been frequently employed with the double purpose of flavouring and food preservation in pickles (Della et al. 2006) and dried figs (Tardío et al. 2006).

In some countries, the fruits are used to prepare digestive liqueurs or herbal teas (Ali-Shtayeh et al. 2008; Carvalho 2010; Guarrera 2006; Tardío et al. 2006).

Other Uses The folk medicinal uses are also very popular in the Mediterranean region, being considered aperitive, refreshing, digestive, carminative, bechic, antispasmodic, and sedative (e.g. Batal and Hunter 2007; Carvalho 2010; Guarrera 2006; Le Floc'h 1983; Menendez-Baceta et al. 2014; Rivera et al. 2005; Vanzani et al. 2011), being widely employed for cold treatment, flatulence, and stomach disorders. Other applications include as diuretic (e.g. Guarrera 2006; Mendonça de Carvalho 2006), for weight loss (Batal and Hunter 2007), as galactogenous, decrease menstrual symptoms (e.g. Batal and Hunter 2007; Guarrera 2006), protect the liver (Carvalho 2010; Lentini and Venza 2007), treat kidney stones (Batal and Hunter 2007; Guarrera 2006), and so on. The mode of application is drinking the infusion of the aerial tender parts or fruits.

Historical References This aromatic plant is well known since antiquity, being already mentioned by Theophrastus in the third century BC (Teofrasto 1988). Its use as a condiment is cited in the Roman period (first century AD) by Pliny the Elder and Columella (Bostock and Riley 1855; Columela 1824), the second describing its use for seasoning olives and other food products prepared as pickles and brines. In the same century, Dioscorides mentioned its medicinal properties, many of them identical as those reported in recent ethnobotanical studies (Laguna 1555).

Food Composition Tables for raw tender leaves of *Foeniculum vulgare* (Tables 13.50, 13.51 and 13.52).

Table 13.50 Main constituents, per 100 g of fresh leaves of *F. vulgare*

	Units	Average	Range	References
Energy (calculated value)	kcal	63	14–130	–
Moisture	g	82.4	72.9–90.1	1, 2, 3, 4, 5, 6
Available carbohydrates	g	9.67	1.40–22.4	2, 3, 6
Total soluble sugars	g	6.57	6.40–7.01	2
Fructose	mg	1510	1450–1570	2
Glucose	mg	4710	4560–4860	2
Sucrose	mg	350	290–410	2
Dietary fibre	g	3.87	2.70–6.20	1, 2, 6
Proteins	g	2.76	0.60–4.20	1, 2, 3, 6
Lipids	g	0.42	0.08–0.80	1, 2, 3, 6, 7
Ash	g	2.34	1.50–2.41	1, 2, 3, 6
K	mg	753	237–2774	6, 8, 9
Na	mg	89.6	35.2–138	2, 6, 8, 9, 10
Ca	mg	365	196–822	2, 6, 8, 9, 10
Mg	mg	85.0	30.4–331	2, 6, 8, 9, 10
P	mg	46.2	45.1–47.2	2, 8
Fe	mg	2.21	0.07–10.2	2, 6, 8, 9, 10
Cu	μg	255	10–577	6, 8, 9
Mn	μg	881	300–2090	6, 8, 9
Zn	μg	799	250–1840	2, 6, 8, 9, 10

1 Cowan et al. (1963) Lebanon, *2* Trichopoulou et al. (2000) Greece, *3* Barros et al. (2010c) Portugal, *4* Morales (2011) Spain (4 samples), *5* Sánchez-Mata et al. (2012) Spain (2 samples), *6* García-Herrera (2014) Spain (4 samples), *7* Morales et al. (2012b) Spain (4 samples), *8* Zeghichi et al. (2003) Crete, *9* Özcan et al. (2008) Turkey, *10* Romojaro et al. (2013) Spain

Table 13.51 Vitamins and other constituents, per 100 g of fresh leaves of *F. vulgare*

	Units	Average	Range	References
Carotenoids				
β-carotene	mg	1.19	–	1
Lutein	mg	3.66	–	1
RAE (calculated value)	μg	100	–	–
Vitamin B_9 (total folates)	μg	271	255–286	2
Vitamin C	mg	66.7	18–101	1, 3, 4
Ascorbic acid	mg	12.5	5.34–16.9	3, 4
Dehydroascorbic acid	mg	25.8	16.1–37.0	3, 4
Vitamin E				
α-tocopherol	mg	1.66	0.63–4.21	1, 5, 6
β-tocopherol	mg	0.59	0.4–0.78	6
γ-tocopherol	mg	0.33	0.14–0.81	1, 6
δ-tocopherol	mg	nd	–	6
Vitamin K	μg	239	–	1
Organic acids				
Oxalic acid	mg	278	40–402	3, 4, 7
Malic acid	mg	366	13.5–1047	3, 4, 7
Citric acid	mg	146	4.66–387	3, 4, 7
Fumaric acid	mg	5.10	1.08–8.78	3, 4, 7

Table 13.51 (continued)

	Units	Average	Range	References
Shikimic acid	mg	≈1.48	0.49–2.47	7
Quinic acid	mg	≈278	277–279	7
Phenolics (total)	mg	133[a]	47.2–170	1, 5, 6
	mg	≈195[b]	157–235	8,9
Hydroxycinnamic acids	mg	495[b]	–	10
Flavonoids	mg	9.72[c]	9.02–10.4	6
	mg	≈27.5[b]	27.1–27.9	8
	mg	82.5[d]	–	11
Nitrate	mg	24.0	23.0–25.0	5

1 Vardavas et al. (2006b) Crete, *2* Morales et al. (2015) Spain (4 samples), *3* Morales (2011) Spain (4 samples), *4* Sánchez-Mata et al. (2012) Spain (2 samples), *5* Zeghichi et al. (2003) Crete, *6* Morales et al. (2012a) Spain (4 samples), *7* Pereira et al. (2013) Portugal, *8* Conforti et al. (2009) Italy, *9* Conforti et al. (2011) Italy, *10* Vanzani et al. (2011) Italy (4 samples), *11* Trichopoulou et al. (2000) Greece *RAE* retinol activity equivalents; *nd* non-detected values

[a] Expressed as gallic acid

[b] Expressed as chlorogenic acid

[c] Expressed as catechin

[d] Expressed as sum of individual compounds (myricetin and quercetin derivatives as major components)

Table 13.52 Fatty acids profile (% of total fatty acids) of fresh leaves of *F. vulgare*

Individual compounds	Average	Range	References
12:0	0.14	0.07–0.23	1, 2
14:0	1.56	0.72–3.10	1, 2, 3
16:0	17.9	12.7–23.7	1, 2, 3
16:1	0.82	0.14–1.50	1, 2, 3
18:0	1.42	1.20–1.58	1, 2, 3
18:1$n-9$	2.00	1.30–2.88	1, 2, 3
18:2$n-6$	35.2	28.7–40.7	1, 2, 3
18:3$n-3$	34.9	32.3–37.3	1, 2, 3
18:3$n-6$	0.10	0.10–0.10	3
20:0	0.87	0.80–1.15	1, 2, 3
20:1$n-9$	0.03	nd–0.12	1, 2, 3
22:0	1.10	0.99–1.14	1, 2, 3
24:0	1.13	0.64–1.50	1, 2, 3
24:1$n-9$	0.30	–	3
Categories (calculated values)			
SFA	29.4	22.9–32.9	
MUFA	3.3	2.5–4.6	–
PUFA	67.3	63.1–74.6	–
$n-3$	31.4	23.6–36.5	–
$n-6$	35.9	30.2–39.5	–
$n-9$	2.7	1.4–4.6	–

1 Barros et al. (2010c) Portugal, *2* Morales et al. (2012b) Spain (4 samples), *3* Vardavas et al. (2006a) Crete; *nd* non-detected values; *SFA* saturated fatty acids, *MUFA* monounsaturated fatty acids, *PUFA* polyunsaturated fatty acids

Remarks (leaves of *F. vulgare*)

- High available carbohydrate content, compared to other vegetables.
- In some cases, it may be considered as a source of dietary fibre (>3 g/100 g).
- Despite wide variability, it can be considered as a source of Ca, Mn, and vitamins C and B_9 (folates); sometimes also a source of K, Mg, and vitamin K.

13.3.17 Humulus lupulus L. (Cannabaceae)

Common Names Hop, *lúpulo* (es, pt), *houblon* (fr), *luppolo* (it), *hmelj* (hr, bs), *şerbetçiotu* (tr), *hachiche addinar* (ar).

Description Perennial, climbing, and herbaceous plant up to 5–10 m long, which every spring sends up new twining shoots from a subterranean rhizome. The stems, with stout hooked hairs to aid grip, have opposite and palmate leaves with 3(5) lobes and serrate margins, both with a rough texture. It is dioecious, with unisexual flowers on separate plants. Male plants produce drooping panicles of small staminate flowers, and female plants have pistillate flowers grouped in cone-like inflorescences of 2.5–5 cm long, with yellow glands on the bracteoles of the cone. They produce a capsule with a single seed that is resinous and aromatic.

Ecology and Distribution It mostly grows in moist thickets, slopes, river banks, alluvial woods, and other moist and cool environments, often in sandy soils. Native to Europe and western Asia, cultivated in North and South America, Africa, Asia, and Australia, and naturalized in many areas or the world.

Food Uses Its young shoots or asparagus—approximately 20 cm long—are gathered in spring. Its consumption has been documented in several Mediterranean countries, including Spain (Tardío et al. 2006), Italy (Ghirardini et al. 2007; Guarrera 2006; Hadjichambis et al. 2008; Picchi and Pieroni 2005; Pieroni 2001), Slovenia (Cerne 1992), Bosnia-Herzegovina (Redžić 2006), Turkey (Ertuğ 2014), and Morocco (Hadjichambis et al. 2008). The asparaguses are generally boiled and then fried in omelettes (Guarrera 2006; Hadjichambis et al. 2008; Tardío et al. 2005) and also used in soups (Cerne 1992; Guarrera 2006; Tardío et al. 2002), such as the *acquacotta* in the Italian city of Viterbo (Picchi and Pieroni 2005), and in stews with potatoes (Picchi and Pieroni 2005; Tardío et al. 2002). In northern and central Italy they are also prepared with pasta, risotto, or gnocchi, used to fill meatballs, crepes, and pies, such as the *pasqualine* in Piamonte, or just scalded and served with butter or olive oil and vinegar (Guarrera 2006; Picchi and Pieroni 2005). The sprouts are occasionally consumed raw, as recorded in Morocco (Hadjichambis et al. 2008) and Spain (Tardío et al. 2002).

The female inflorescences, collected in August–September, are extensively used to aromatize beer (Guarrera 2006; Picchi and Pieroni 2005; Tardío et al. 2006), as well as in the preparation of some local spirits (Carvalho 2005; Parada et al. 2011), including the Italian *grappa* (Picchi and Pieroni 2005). Although nowadays it is widely cultivated for use in the brewing industry, in the past it was collected and sold to beer manufacturers, such as in some regions of northern Spain (Pardo-de-Santayana 2003).

Other Uses The medicinal use of hop as sedative is extensively documented. The infusion of the female inflorescences is considered digestive and sedative, as recorded in Spain, Portugal, and Italy (Calvo et al. 2011; Carvalho 2005; Neves et al. 2009; Picchi and Pieroni 2005). They were also put inside pillows to sleep better (Calvo et al. 2011) and used against headaches (Picchi and Pieroni 2005). The water obtained from boiling the roots was consumed to improve blood circulation in some Spanish regions (Velasco et al. 2010). Additionally, the alcoholic beverages prepared with its flowers (spirits and beer) are considered anti-enuretic (Parada et al. 2011).

In ethnoveterinary, the external use of the aerial parts of hop was recorded in northern Spain to treat pig diseases (Akerreta et al. 2010).

Historical References The food use of this species by the Romans was mentioned by the naturalist Pliny the Elder in the first century (Bostock and Riley 1855). In the sixteenth century, the Spanish doctor Laguna (1555) reported the consumption of its young shoots in Castile, comparing it with the real asparagus. Other than mentioning some medicinal properties of the plant, he also pointed out that with its strongly scented flowers beer was prepared in Germany and Flanders.

Food Composition Tables for young shoots of *Humulus lupulus*, raw (Tables 13.53, 13.54 and 13.55) and boiled (Tables 13.56 and 13.57).

Table 13.53 Main constituents, per 100 g of fresh young shoots of *H. lupulus*

	Units	Average	Range	References
Energy (calculated value)	kcal	39	29–55	–
Moisture	g	85.8	85.2–93.2	1, 2, 3
Available carbohydrates	g	1.85	1.40–2.20	3
Dietary fibre	g	4.85	4.35–6.42	3
Proteins	g	4.25	3.13–5.10	3
Lipids	g	0.37	0.10–1.08	3, 4
Ash	g	1.35	0.90–2.01	3
K	mg	469	314–675	3
Na	mg	43.5	25.3–65.8	3
Ca	mg	73.5	50.3–134	3
Mg	mg	32.5	27.6–48.2	3
Fe	mg	0.74	0.37–1.32	3
Cu	μg	140	100–170	3
Mn	μg	306	100–550	3
Zn	μg	975	710–1510	3

1 Morales (2011) Spain (5 samples), *2* Sánchez-Mata et al. (2012) Spain (2 samples), *3* García-Herrera (2014) Spain (5 samples), *4* Morales et al. (2012b) Spain (4 samples)

Table 13.54 Vitamins and other constituents, per 100 g of fresh young shoots of *H. lupulus*

	Units	Average	Range	References
Carotenoids				
β-carotene	mg	0.37	0.16–1.26	1
Lutein	mg	0.55	0.19–1.72	1
Neoxanthin	mg	0.73	0.13–1.57	1
Violaxanthin	mg	0.21	0.04–0.67	1
RAE (calculated value)	μg	30.8	13.0–105	–
Vitamin B_9 (total folates)	μg	144	108–179	2
Vitamin C	mg	51.5	28.6–61.1	3, 4
Ascorbic acid	mg	30.2	11.6–37.4	3, 4
Dehydroascorbic acid	mg	23.2	12.8–33.7	3, 4
Vitamin E				
α-tocopherol	mg	4.51	4.05–4.97	5
β-tocopherol	mg	0.24	0.22–0.26	5
γ-tocopherol	mg	8.98	8.95–9.01	5
δ-tocopherol	mg	0.69	0.68–0.70	5
Organic acids				
Oxalic acid	mg	92.6	50.8–190	3, 4
Glutamic acid	mg	nd	–	3
Malic acid	mg	619	216–1040	3, 4
Citric acid	mg	83.8	nd–170	3, 4

Table 13.54 (continued)

	Units	Average	Range	References
Fumaric acid	mg	traces	–	3, 4
Succinic acid	mg	nd	–	3
Phenolics (total)	mg	55.8[a]	54.5–57.1	5
Flavonoids	mg	9.56[b]	8.91–10.21	5

1 García-Herrera et al. (2013) Spain (4 samples), *2* Morales et al. (2015) Spain (4 samples), *3* Morales (2011) Spain (5 samples), *4* Sánchez-Mata et al. (2012) Spain (2 samples), *5* Morales et al. (2012a) Spain (4 samples)

RAE retinol activity equivalents; *nd* non-detected values

[a] Expressed as gallic acid

[b] Expressed as catechin

Table 13.55 Fatty acids profile (% of total fatty acids) of fresh young shoots of *H. lupulus*

Individual compounds	Average	Range	References
12:0	0.05	0.04–0.06	1
14:0	0.49	0.46–0.52	1
16:0	19.5	18.9–20.1	1
16:1	0.46	0.45–0.47	1
18:0	1.60	1.55–1.65	1
18:1$n-9$	1.88	1.78–1.98	1
18:2$n-6$	29.7	28.9–30.6	1
18:3$n-3$	38.2	38.1–38.2	1
18:3$n-6$	0.36	0.35–0.37	1
20:0	2.34	2.19–2.49	1
20:1	0.13	0.12–0.14	1
22:0	2.86	2.63–3.09	1
24:0	0.89	0.35–1.43	1
Categories (calculated values)			
SFA	28.2	27.3–29.0	–
MUFA	2.5	2.5–2.6	–
PUFA	69.3	68.4–70.3	–
$n-3$	39.2	38.2–40.1	–
$n-6$	30.5	30.5–30.6	–
$n-9$	2.0	2.0–2.1	–

1 Morales et al. (2012b) Spain (4 samples); *nd* non-detected values

SFA saturated fatty acids, *MUFA* monounsaturated fatty acids, *PUFA* polyunsaturated fatty acids

Remarks (young shoots of *H. lupulus*)

- Source of dietary fibre (>3 g/100 g).
- High protein content compared to other wild plants.
- High levels of malic acid, as major organic acid.
- Low Na content (<120 mg/100 g). It can be considered as a source of K, Mn and vitamins C, E, and B_9 (folates).

Table 13.56 Main constituents, per 100 g of boiled young shoots of *H. lupulus*

	Units	Average	Range	References
Energy (calculated value)	kcal	21	20–22	–
Moisture	g	90.7	90.5–90.9	1, 2
Available carbohydrates	g	1.90	1.7–2.1	2
Dietary fibre	g	3.80	3.8–3.8	2
Proteins	g	0.60	0.6–0.6	2
Lipids	g	0.15	0.14–0.16	2
Ash	g	0.70	0.70–0.70	2
K	mg	171	156–186	2
Na	mg	7.10	6.40–7.80	2
Ca	mg	40.4	34.8–45.9	2
Mg	mg	17.0	16.4–17.6	2
Fe	mg	0.57	0.50–0.60	2
Cu	µg	80	80–80	2
Mn	µg	180	160–200	2
Zn	µg	400	380–420	2

1 Morales (2011) Spain (5 samples), *2* García-Herrera (2014) Spain (5 samples)

Table 13.57 Vitamins and other constituents, per 100 g of boiled young shoots of *H. lupulus*

	Units	Average	Range	References
Vitamin B_9 (total folates)	µg	101	90.0–112	1
Vitamin C	mg	17.4	16.6–18.9	1
Ascorbic acid	mg	11.3	10.8–11.8	1
Dehydroascorbic acid	mg	5.56	5.37–5.75	1
Organic acids				
Oxalic acid	mg	80	80–80	1
Glutamic acid	mg	nd	–	1
Malic acid	mg	470	450–490	1
Citric acid	mg	120	90–150	1
Fumaric acid	mg	traces	–	1

1 Morales (2011) Spain (5 samples); *nd* non-detected values

Remarks (boiled young shoots of *H. lupulus*)

- Very low lipid content (<0.5 g/100 g), very low energy value (<25 kcal/100 g), and still good source of dietary fibre and vitamins B_9 and C after boiling.
- Oxalic acid levels reduce after boiling. Very low Na content (<40 mg/100 g).

13.3.18 *Malva sylvestris* L. (Malvaceae)

Common Names Common mallow, *malva* (pt, es), *mauve* (fr), *malva selvatica* (it), *sljez* (bs, hr), *molocha* (el), *ebegümeci* (tr), *khubbayzeh* (ar).

Description Biennial or perennial herb, up to 1.5 m, with erect or decumbent stems, whose long petiolate leaves are 5–10 cm, roundish, with 3–7 lobes, usually slightly dentate. The flowers are purple, with 5 dark-veined petals. Stamens numerous with their filaments joined surrounding the ovary and styles. The fruit is a rounded and reticulate schizocarp with 10–12 mericarps, brownish when ripe.

Ecology and Distribution This nitrophilous plant grows in waste grounds and roadsides, at altitudes of up to 1500 m. It flowers usually from January to October. It lives in almost all of Europe, the Mediterranean region, and the Macaronesian Islands; also introduced in North and Middle America, Asia, and Australia.

Food Uses This species has been widely consumed in the Mediterranean region, as recorded in Portugal (Mendonça de Carvalho 2006), Spain (e.g. Parada et al. 2011; Tardío et al. 2006; Velasco et al. 2010), Italy (e.g. Ghirardini et al. 2007; Guarrera 2006; Picchi and Pieroni 2005), Slovenia (Cerne 1992), Croatia (Łuczaj et al. 2013a), Bosnia-Herzegovina (Redžić 2006), Greece (Hadjichambis et al. 2008), Cyprus (Della et al. 2006), Turkey (Dogan 2012; Ertuğ 2014), Lebanon (Batal and Hunter 2007; Marouf 2005), Palestine (Ali-Shtayeh et al. 2000), Jordan (Al-Qura'n 2010), Egypt (Hadjichambis et al. 2008), Tunisia (Le Floc'h 1983), and Morocco (Nassif and Tanji 2013; Tanji and Nassif 1995).

The tender leaves and stems are gathered in spring, although the collection can be prolonged until summer in shady places with water (Marouf 2005). They are mainly eaten boiled (e.g. Al-Qura'n 2010; Della et al. 2006; Hadjichambis et al. 2008; Le Floc'h 1983; Łuczaj et al. 2013a; Parada et al. 2011; Picchi and Pieroni 2005) and then seasoned with oil (Lentini and Venza 2007), cooked with rice and meat stews (Hadjichambis et al. 2008; Marouf 2005), in soups (e.g. Batal and Hunter 2007; Cerne 1992; Della et al. 2006; Dogan 2012; Guarrera 2006; Hadjichambis et al. 2008), with legumes (Marouf 2005; Mendonça de Carvalho 2006), fried or sautéed in olive oil and served with other ingredients (Ali-Shtayeh et al. 2008; Della et al. 2006; Dogan 2012; Hadjichambis et al. 2008; Marouf 2005), and used as stuffing for pies (Dogan 2012; Picchi and Pieroni 2005). They can also be eaten raw in salads (e.g. Marouf 2005; Parada et al. 2011; Picchi and Pieroni 2005). Some examples of traditional recipes with mallow leaves are the Lebanese *khubbayzeh and labneh* (strained yogurt), the Cyprian soup called *molochosoupa* (Hadjichambis et al. 2008), or the Moroccan *beqoula*, a dish prepared chiefly using species of the genus *Malva* (Nassif and Tanji 2013).

The immature fruits are collected in early summer and eaten raw as a pastime during country walks, especially by children (e.g. Lentini and Venza 2007; Mendonça de Carvalho 2006; Parada et al. 2011; Tardío et al. 2005). They are called *panuzzi du Signuri* in Sicily (Lentini and Venza 2007) and *pan y quesito* in Spain (Aceituno-Mata 2010; Tardío et al. 2006), after their bread-like form.

Additionally, the flowers are used to prepare herbal teas and alcoholic beverages, such as *ratafia*, a liqueur traditionally elaborated in Catalonia (NE Spain) that includes many wild species (Bonet and Vallès 2002; Rigat et al. 2009).

The mallow is one of the most frequently gathered wild edible plants in countries such as Lebanon (Marouf 2005), Turkey (Dogan 2012), and Bosnia-Herzegovina (Redžić 2006), and it can be found in some local markets in Italy (Picchi and Pieroni 2005) and Morocco (Tanji and Nassif 1995).

Other Uses It is one of the medicinal plants most widely used in the Mediterranean popular phytotherapy. The leaves, flowers, and even the roots were employed, both in internal and external use, for a wide range of diseases, such as digestive problems (constipation, diarrhoea, stomach ache), respiratory (cough, sore throat), circulatory (anaemia), skin (acne, boils), and excretory system ailments (e.g. Batal and Hunter 2007; Bonet and Vallès 2002; Guarrera 2006; Le Floc'h 1983; Marouf 2005; Menendez-Baceta et al. 2014; Picchi and Pieroni 2005; Redžić 2006; Rigat et al. 2009). Since one of the ways of administration is its consumption as a vegetable, it is often considered a medicinal food (Tardío 2010).

It also has been used in ethnoveterinary (Guarrera 2006; Mendonça de Carvalho 2006) and animal feed (Guarrera 2006; Latorre 2009).

Historical References This species was already mentioned by Theophrastus in the third century BC among the wild vegetables that need to be cooked before being consumed (Teofrasto 1988). Dioscorides mentioned a domestic mallow that is better to be eaten than the wild one, referring to also the several medicinal properties, especially for the digestive system and in external use for different skin problems (Laguna 1555). In the opinion of Pliny the Elder both kinds of mallows are held in very general esteem (Bostock and Riley 1855). The cultivated mallow was also mentioned in Arabic references, such as that of Ibn al-Awwam, a Spanish Arab agriculturalist of the later twelfth century, who cited that this plant is more nutritious and more blood productive than all other vegetables (Cubero 2001).

Food Composition Tables for raw tender leaves of *Malva sylvestris* (Tables 13.58, 13.59 and 13.60).

Table 13.58 Main constituents, per 100 g of fresh tender leaves of *M. sylvestris*

	Units	Average	Range	References
Energy (calculated value)	kcal	35	23–50	–
Moisture	g	81.0	75.7–86.9	1, 2, 3, 4, 5, 6
Available carbohydrates	g	2.23	1.93–2.44	2
Soluble sugars (total)	g	0.72	–	7
Fructose	mg	248	–	7
Glucose	mg	292	–	7
Sucrose	mg	177	–	7
Dietary fibre	g	4.76	4.18–5.34	2
Proteins	g	3.00	0.83–5.70	2, 5, 6
Lipids	g	0.56	0.40–0.76	2, 6
Ash	g	3.21	2.32–5.44	2, 3, 4, 6
K	mg	692	547–836	3, 4, 8
Na	mg	106	68.0–134	3, 4, 8
Ca	mg	226	122–361	3, 4, 8, 9
P	mg	77.1	60.0–110	3, 5
Mg	mg	283	20.9–368	3, 4, 8, 9
Fe	mg	3.61	0.76–6.29	3, 8, 9
Cu	μg	208	100–330	3, 9
Mn	μg	498	203–760	3, 9
Zn	μg	1580	38–2665	3, 8, 9

1 Guil-Gerrero et al. (1997a) Spain (5 samples), *2* Guil-Guerrero et al. (1997b) Spain (5 samples), *3* Guil-Gerrero et al. (1999a), Spain (5 samples), *4* Bianco et al. (1998) Italy (2 samples), *5* Özbucak et al. (2007) Turkey, *6* Barros et al. (2010a) Portugal, *7* Souci et al. (2008) unknown origin, *8* Romojaro et al. (2013) Spain, *9* Hiçsomenez et al. (2009) Turkey (15 samples)

Table 13.59 Vitamins and other constituents, per 100 g of fresh leaves of *M. sylvestris*

	Units	Average	Range	References
Carotenoids	mg	4.33[a]	0.83–10.1	1, 2
Vitamin C	mg	133	72.0–178	1, 3
Ascorbic acid	mg	84.5	Traces–171	1, 2, 3
Dehydroascorbic acid	mg	7.50	5.00–11.0	1, 3
Vitamin E				
α-tocopherol	mg	≈20.1	19.6–20.6	2
β-tocopherol	mg	≈0.36	0.35–0.37	2
γ-tocopherol	mg	≈4.80	4.56–5.04	2
δ-tocopherol	mg	≈0.31	0.30–0.32	2
Organic acids				
Oxalic acid	mg	118	7.00–284	1, 4
Phenolics (total)	mg	≈1692[b]	1659–1733	2
Flavonoids	mg	≈925[c]	890–960	2
Nitrate	mg	87.6	84.8–106	1, 4

1 Guil-Guerrero et al. (1997a) Spain (5 samples), *2* Barros et al. (2010a) Portugal, *3* Franke and Kensbock (1981), Germany, *4* Bianco et al. (1998) Italy (2 samples)

[a] Expressed as β-carotene

[b] Expressed as gallic acid

[c] Expressed as catechin

Table 13.60 Fatty acids profile (% of total fatty acids) of fresh tender leaves of *M. sylvestris*

Individual compounds	Average	Range	References
12:0	0.09	0.08–0.10	1
14:0	0.62	0.44–0.77	1, 2
16:0	12.7	8.72–15.6	1, 2
16:1	2.03	1.99–2.07	1, 2
18:0	1.64	1.15–2.07	1, 2
18:1*n*−7	1.14	–	2
18:1*n*−9	1.73	–	2
18:2*n*−6	11.2	10.4–12.4	1, 2
18:3*n*−3	67.8	66.8–68.7	1
18:3*n*−6	0.50	–	2
18:4*n*−3	0.29	–	2
20:0	0.78	0.50–1.04	1, 2
20:1*n*−9	0.02	nd–0.04	1, 2
22:0	0.37	nd–0.83	1, 2
22:1	1.91	–	2
24:0	2.79	0.74–4.84	1, 2
Categories (calculated values)			
SFA	17.9	12–25.7	–
MUFA	6.4	5.3–7.4	–
PUFA	75.5	73.2–81.1	–
n−3	64.5	60.8–68.7	–
n−6	11.1	10.5–12.6	–
n−9	4.4	3.4–5.2	–

1 Barros et al. (2010a) Portugal, *2* Guil-Guerrero et al. (1996b) Spain (various samples); *nd* non-detected values

SFA saturated fatty acids, *MUFA* monounsaturated fatty acids, *PUFA* polyunsaturated fatty acids

Remarks (tender leaves of *M. sylvestris*)

- Source of dietary fibre (>3 g/100 g).
- High proportion of PUFA and *n*−3 fatty acids
- Good *n*−3/*n*−6 ratio (>5).
- High ash level. Despite wide variability, it can be considered as a source of Ca, K and vitamins C and E; sometimes also a source of Mg, Fe, Cu, Mn, and Zn.
- Low oxalic acid/Ca ratio.
- Very high phenolics levels compared to other vegetables.

13.3.19 Montia fontana L. (Montiaceae)

Common Names Water-blinks, *merujas* (pt), *borujas* (es), *montie des fontaines* (fr), *cetonchiello* (it).

Description Annual or perennial, glabrous and fleshy herb with thin and branching stems up to 50 cm or more when it grows in water and shorter when it appears in land. Small, opposite, and spatulate leaves (3–20 mm). Inconspicuous flowers of 2 mm in terminal cymes, with 3 sepals, 5 white petals, 3 stamens and 3 ovules. The fruit is a globose capsule, dehiscent by 3 valves and usually with 3 black, reticulated, and reniform seeds.

Ecology and Distribution It grows in clumps in damp places, frequently in water, in springs, streams, wet places among rocks, and moist pastures, especially on non-calcareous soils. It can be found in many temperate regions throughout the world including some European areas, but it is rare in East and South Europe.

Food Uses The young and tender stems and leaves of this plant are traditionally consumed raw in salads in the regions of the Iberian Peninsula where it grows. There are many modern ethnobotanical references from western and central Spain that report its current consumption (e.g. Criado et al. 2008; González et al. 2011a; Tardío et al. 2006). Ethnobotanical studies in the north of Portugal have also recorded the same use (Alves Ribeiro et al. 2000; Carvalho 2010). It is one of the most valued wild vegetables in these regions, especially in the Spanish province of Salamanca (e.g. González et al. 2011a; Velasco et al. 2010). A recent ethnobotanical study in the north of Madrid (Aceituno-Mata 2010) found that *Montia fontana* was one of the most important wild vegetables and still consumed by 64 % of the informants that cited its use, higher than that of other salad wild vegetables. Despite being worldwide distributed and mentioned in some comprehensive references (Couplan 1989b; PFAF 2014), we only have found another clear ethnobotanical reference (besides Iberian ones) about its traditional use at the end of the nineteenth century in Central Europe (Jage 1979). Following this author, water-blinks were sold in markets as a salad vegetable, especially in winter. Interestingly its use is revealed by several German folk names registered in Central Europe (*Flachssalat, Winzerlsalat, Schnippchensalat*).

Water-blinks are always eaten raw in salads after being meticulously cleaned. These salads are prepared by dressing the plants with salt, vinegar, and olive oil. Other ingredients, such as garlic, olives, preserved tomatoes, dry pepper, paprika, hard-boiled eggs, and tuna were sometimes added. It is consumed even nowadays, being still possible to find it in some greengrocers and even in some restaurants (Tardío et al. 2011).

The plants must be harvested before the flowering period, prior to their nice flavour turn bitter (Díaz-Fernández et al. 2008; PFAF 2014), but they must reach enough size to be collected. If scarcely developed, they carry a lot of mud when they are cut and then it takes a long time to clean them (Tardío et al. 2002). Therefore, in Spain they are collected from winter, in the milder areas, such as many places of Extremadura (Blanco and Cuadrado 2000), to spring, in the more continental areas, such as those of the regions of Madrid and Castile and Leon (e.g. Díaz-Fernández et al. 2008). Some informants report that plants collected in highlands are better than those from the lowlands as they are smaller but tastier (Blanco 1998).

Other Uses Regarding its ethnopharmacological use in the Mediterranean region, we only could find one reference in Montesinho, NE Portugal (Carvalho 2010). A very hot poultice of the plant fried in olive oil is applied over the bladder as a diuretic to treat difficulty urinating and pain caused by cystitis. Other external medicinal use has also been described in Scotland (Allen and Hatfield 2004). The heated plant was placed on the affected area to treat suppurating sores and rheumatism.

Historical References Apart from the above reference of Jage (1979), mentioning the traditional use of this plant in Central Europe at the end of the nineteenth century, we have not found another historical reference.

Food Composition Tables for raw tender leaves and stems of *Montia fontana* (Tables 13.61, 13.62 and 13.63).

Table 13.61 Main constituents, per 100 g of fresh young leaves and stems of *M. fontana*

	Units	Average	Range	References
Energy (calculated value)	kcal	36	17–50	–
Moisture	g	93.3	90.3–95.9	1, 2, 3
Available carbohydrates	g	2.54	1.26–3.28	1, 2
Soluble sugars (total)	g	0.14	0.12–0.15	1
Fructose	mg	36.3	35.1–40.0	1
Glucose	mg	48.5	47.0–49.2	1
Sucrose	mg	21.3	19.4–23.1	1
Trehalose	mg	18.1	15.5–21.0	1
Rafinose	mg	13.0	10.0–16.0	1
Dietary fibre	g	4.44	4.10–4.78	2
Proteins	g	1.20	0.64–1.90	1, 2
Lipids	g	1.33	0.13–2.07	1, 2, 3
Ash	g	0.92	0.72–1.26	1, 2
K	mg	385	304–407	2
Na	mg	90.7	86.3–95.0	2
Ca	mg	31.4	26.0–36.8	2
Mg	mg	31.7	27.9–35.4	2
Fe	mg	1.30	1.04–1.56	2
Cu	µg	50.0	40.5–60.0	2
Mn	µg	1070	800–1340	2
Zn	µg	380	310–450	2

1 Pereira et al. (2011) Portugal, *2* Tardío et al. (2011) Spain (5 samples), *3* Morales et al. (2012b) Spain (5 samples)

Table 13.62 Vitamins and other constituents, per 100 g of fresh young leaves and stems of *M. fontana*

	Units	Average	Range	References
Carotenoids				
β-carotene	mg	0.18[a]	0.17–0.19	1
RAE (calculated value)	µg	15.3	14.0–16.5	–
Vitamin B_9 (total folates)	µg	41.8	41.6–41.9	2
Vitamin C	mg	34.3	28.9–39.7	1, 3
Ascorbic acid	mg	28.9	21.9–37.0	3
Dehydroascorbic acid	mg	9.19	5.29–13.10	3
Vitamin E				
α-tocopherol	mg	4.5	2.90–6.08	1, 4
β-tocopherol	mg	0.18	0.11–0.25	1, 4
γ-tocopherol	mg	0.99	0.51–1.41	1, 4
Organic acids				
Oxalic acid	mg	388	40–990	5, 6
Glutamic acid	mg	nd	–	5, 6
Quinic acid	mg	65.8	54.8–76.8	5, 6
Malic acid	mg	36.5	nd–140	5, 6
Citric acid	mg	85.3	nd–148	5, 6
Fumaric acid	mg	nd	–	5, 6
Succinic acid	mg	Traces	–	5, 6
Phenolics (total)	mg	75.5[b]	68.4–82.5	4

Table 13.62 (continued)

	Units	Average	Range	References
Flavonoids	mg	16.7[c]	16.0–17.3	4

1 Pereira et al. (2011) Portugal, *2* Morales et al. (2015) Spain (5 samples), *3* Tardío et al. (2011) Spain (5 samples), *4* Morales et al. (2012a), Spain (5 samples), *5* Morales (2011) Spain (5 samples), *6* Pereira et al. (2013) Portugal

RAE retinol activity equivalents, *nd* non-detected values

[a] Expressed as β-carotene

[b] Expressed as gallic acid

[c] Expressed as catechin

Table 13.63 Fatty acids profile (% of total fatty acids) of young leaves and stems of *M. fontana*

Individual compounds	Average	Range	References
12:0	0.18	0.11–0.25	1, 2
14:0	0.51	0.43–0.74	1, 2
16:0	14.3	10.3–18.3	1, 2
16:1	0.24	0.27–0.39	1, 2
18:0	0.54	0.07–1.03	1,2
18:1$n-9$	4.21	1.99–6.47	1, 2
18:2$n-6$	18.5	17.8–19.2	1, 2
18:3$n-3$	51.6	47.4–56.4	1, 2
18:3$n-6$	0.04	0.04–0.04	1
20:0	10.7	7.73–12.2	1, 2
20:1$n-9$	0.15	0.04–0.31	1, 2
22:0	1.01	0.46–1.86	1, 2
22:1$n-9$	0.42	0.11–0.68	1, 2
24:0	0.41	0.18–0.64	1, 2
24:1$n-9$	0.01	0.01–0.01	1
Categories (calculated values)			
SFA	26.1	22.2–29.3	–
MUFA	5.3	3.3–7.3	–
PUFA	68.2	63.2–74.5	–
$n-3$	50.2	49.0–56.6	–
$n-6$	18.0	17.0–19.1	–
$n-9$	4.9	2.9–6.9	–

1 Pereira et al. (2011) Portugal, *2* Morales et al. (2012b) Spain (5 samples)

SFA saturated fatty acids, *MUFA* monounsaturated fatty acids, *PUFA* polyunsaturated fatty acids

Remarks (young leaves and stems of *M. fontana*)

- High moisture content and low energy value (<40 kcal/100 g).
- Source of dietary fibre (>3 g/100 g).
- High lipid content. It is among the highest $n-3$ fatty acids content, compared to other wild vegetables (estimated value around 0.56 g 18:3$n-3$/100 g edible portion).
- Low Na content (<120 mg/100 g); it can be also considered as a source of K, vitamins C, E, and B_9 (folates), standing out Mn content.
- High content of oxalates in some samples: People with altered renal function should avoid this vegetable.

13.3.20 *Myrtus communis* L. (Myrtaceae)

Common Names Myrtle, *murta* (pt), *mirto* (pt, es, it), *myrte* (fr), *mrča* (bs), *myrssini*, (el), *mersin* (tr), *rayhan* (ar).

Description Shrub, up to 5 m, erect, much-branched, having glandular-hairy twigs when young. The leaves are coriaceous with aromatic oil glands, ovate-lanceolate, acute, and usually opposite. The scented white flowers are solitary in the leaf axils, with 5 sepals and petals, and numerous stamens. The fruits are round berries, usually blue-black when ripe, in autumn.

Ecology and Distribution It grows in Mediterranean woodlands and shrublands, in south of Europe, north of Africa, and west Asia, up to 1000 m.

Food Uses This aromatic plant has been widely used in the Mediterranean, at least in Portugal (Mendonça de Carvalho 2006), Spain (Blanco and Cuadrado 2000; Carrió 2013; Cobo and Tijera 2011; Moll 2005; Pellicer 2001; Triano et al. 1998; Velasco et al. 1998), France (Couplan 1989b), Italy (Guarrera 2006; Hadjichambis et al. 2008; Lentini and Venza 2007), Bosnia-Herzegovina (Redžić 2006), Greece (Hadjichambis et al. 2008), Cyprus (Hadjichambis et al. 2008), Turkey (Dogan et al. 2004; Ertuğ 2014), Palestine (Ali-Shtayeh et al. 2008), and Morocco (Hadjichambis et al. 2008; Nassif and Tanji 2013). Almost all the references mentioned the raw consumption of the ripe fruits, being sometimes even sold in local markets (Ertuğ 2004). They are also used to prepare jam and marmalade (Hadjichambis et al. 2008; Lentini and Venza 2007).

Myrtle has also been traditionally used for elaborating digestive wines or liqueurs (Carrió 2013; Couplan 1989b; Font Quer 1962; Guarrera 2006; Hadjichambis et al. 2008; Mendonça de Carvalho 2006). The liqueur is elaborated by macerating the berries (and sometimes also the leaves) in liquor or alcohol for 2 months and adding a sugar solution (Carrió 2013). It is very popular in the islands of Corsica and Sardinia, where it is known as *myrte* or *mirto* (Barboni et al. 2010; Montoro et al. 2006). Other non-alcoholic beverages, such as herbal teas, have also been made with this plant (Ali-Shtayeh et al. 2008; Moll 2005; Nassif and Tanji 2013).

The leaves (Ali-Shtayeh et al. 2008), the flowering buds (Redžić 2006) and the sweet flavoured flowers (Lentini and Venza 2007) are consumed raw in salads in Palestine, Bosnia-Herzegovina, and Sicily, respectively.

Different aromatic parts of the plant, especially the leaves and the fruits, are used as highly valued condiments in the Mediterranean kitchen (e.g. Ali-Shtayeh et al. 2008; Hadjichambis et al. 2008; Nassif and Tanji 2013), to flavour grilled meats (Guarrera 2006) or to season olives in brine (Cobo and Tijera 2011; Galán 1993; Lentini and Venza 2007; Triano et al. 1998). In the latter case, some people mention its double role as condiment and preservative for avoiding the olives to soften (Molina 2001).

Other Uses Different parts of the plant have been used as medicine, mainly to treat digestive (Carrió 2013; Guarrera 2006; Moll 2005) and respiratory diseases (Carrió 2013; Guarrera 2006; Pellicer 2001; Redžić 2006) or in external use as anti-inflammatory and to treat skin diseases (Carrió 2013; Guarrera 2006; Pellicer 2001). The raw consumption of the fruit is popularly considered good against diarrhoea and a vitamin-rich tonic (Carrió 2013; Guarrera 2006).

The essential oil from the leaves, flowers, and bark is used in the elaboration of different cosmetics (Carrió 2013; Guarrera 2006), such as the traditional *aigua de murta* (myrtle water), made in the Balearic Islands and used as a perfume and a rejuvenating tonic for the skin.

Among other uses, myrtle has been used as an ornamental plant, both in gardens and in different ritual and religious ceremonies (Carrió 2013; Galán 1993; Guarrera 2006).

Historical References This plant was well known and cultivated since antiquity, occupying a prominent place in the writings of Greek, Roman, and Arabian writers (Sumbul et al. 2011). Following these authors, its ceremonial use was mentioned in the Bible (Nehemiah 8:15) and its medicinal properties cited by Hippocrates (both, c. fifth century BC). Theophrastus mentioned it among the cultivated plants, giving advices about its cultivation and describing its morphological characteristics, its fragrance, and its use in the confection of garlands (Teofrasto 1988). Dioscorides also mentioned a myriad of medicinal applications of myrtle (Osbaldeston 2000), explaining, for example, how to prepare myrtle oil with the leaves for treating different skin diseases and how to elaborate myrtle wine with the fruits for digestive problems. Interestingly many of these remedies have been registered in recent ethnobotanical surveys.

Alonso de Herrera (1513), a Spanish author of the sixteenth century, mentioned the use of myrtle as a condiment, besides treating how to cultivate it and describing some of the medicinal properties already mentioned by Dioscorides.

Food Composition Tables for raw fruits of *Myrtus communis* (Tables 13.64, 13.65 and 13.66).

Table 13.64 Main constituents, per 100 g of fresh fruits of *M. communis*

	Units	Average	Range	References
Energy (calculated value)	kcal	89	80–107	–
Moisture	g	70.3	61.7–75.7	1, 2, 3, 4, 5
Available carbohydrates	g	7.88	7.50–8.26	3
Reducing sugars	g	7.56	6.69–8.64	1, 3
Dietary fibre	g	17.4	–	1
Proteins	g	2.55	1.66–4.17	1, 5
Lipids	g	1.31	0.87–2.37	1, 2, 4, 5
Ash	g	0.68	0.60–0.73	1, 5
K	mg	514	478–549.9	5
Na	mg	79.3	77.7–81.0	5
Ca	mg	142	122–163.2	5
Mg	mg	44.4	36.8–52.1	5
Fe	mg	2.08	1.60–2.56	5
Cu	μg	279	248–310	5
Mn	μg	872	723–1021	5
Zn	μg	1008	962–1053	5

1 Aydın and Ozcan (2007) Turkey, *2* Wannes et al. (2009) Tunisia, *3* Fadda and Mulas (2010) Italy (20 samples), *4* Wannes et al. (2010) Tunisia, *5* Hacıseferoğulları et al. (2012) Turkey (2 samples)

Table 13.65 Other constituents, per 100 g of fresh fruits of *M. communis*

	Units	Average	Range	References
Organic acids				
Malic acid	mg	60	40–70	1
Citric acid	mg	229	191–268	1
Tartaric acid	mg	0.07	0.07–0.08	1
Total phenolics	mg	≈1201[a]	531–2411	2, 3
Phenolic acids	mg	≈29.1[a]	14.4–60.1	2
Flavonols	mg	≈298[a]	114–577	2
Flavanols	mg	≈251[a]	115–548	2

1 Hacıseferoğulları et al. (2012) Turkey (2 samples), *2* Barboni et al. (2010) France (10 samples), *3* Serio et al. (2014) Italy

[a] Expressed as sum of individual compounds (myricetin derivatives as major components)

Table 13.66 Fatty acids profile (% of total fatty acids) of fresh fruits of *M. communis*

Individual compounds	Average	Range	References
12:0	1.46	0.36–4.30	1, 2, 3
14:0	1.08	0.35–3.00	1, 2, 3
16:0	13.4	10.2–15.7	1, 2, 3, 4
16:1	0.36	0.30–0.53	1, 2, 3
18:0	4.12	2.20–8.19	1, 2, 3, 4
18:1*n*−9	32.0	6.49–72.1	1, 2, 3, 4
18:2*n*−6	43.3	1.70–71.3	1, 2, 3, 4
18:3*n*−3	3.75	3.25–4.70	2, 3
20:0	0.64	0.4–1.04	2, 3, 4
20:1*n*−9	0.45	0.2–0.93	2, 3, 4
20:2	0.01	–	4
22:0	0.31	–	4
24:0	0.01	–	4
Categories (calculated values)			
SFA	21.0	17.9–53.6	–
MUFA	32.5	27.1–40.4	–
PUFA	46.2	19.2–47.1	–
n−3	4.7	3.7–12.6	–
n−6	41.4	6.6–43.3	–
n−9	32.2	26.0–40.1	–

1 Cakir (2004) Turkey, *2* Wannes et al. (2009), Tunisia, *3* Wannes et al. (2010), Tunisia, *4* Serce et al. (2010) Turkey (8 samples)
SFA saturated fatty acids, *MUFA* monounsaturated fatty acids, *PUFA* polyunsaturated fatty acids

Remarks (fruits of *M. communis*)

- High in dietary fibre (>6 g/100 g).
- High MUFA proportion (oleic acid).
- Low Na content (<120 mg/100 g). It can be considered as a source of K, Ca, Cu, and Mn; sometimes also a source of Fe.

13.3.21 Papaver rhoeas L. (Papaveraceae)

Common Names Poppy, *papoila* (pt), *amapola* (es), *coquelicot* (fr), *rosolaccio* (it), *mak* (hr), *paparouna* (el), *gelincik* (tr), *bennaâman* (ar).

Description Annual herb, up to 60 cm, with pinnate and usually petiolate leaves that has showy flowers with 2 green deciduous sepals and 4 red petals, often with a black spot at the base, and many bluish stamens. The fruit is an obovoid capsule of 1–1.5 cm, covered by a disk with 8–12 rays.

Ecology and Distribution It grows abundantly on ruderal and waste grounds, usually associated and mixed in cereal crops. It flowers in spring, but sometimes also in summer. It seems to be native to the south of Europe, although some authors consider it as a plant introduced with the cultivation of cereals. It also grows in Asia and North Africa.

Food Uses This wild species has been traditionally used as a vegetable in several Mediterranean countries. The consumption of its young leaves and shoots has been registered in Spain (e.g. Parada et al. 2011; Rigat et al. 2009; Tardío et al. 2006; Velasco et al. 2010), France (Couplan 2009a; Marco et al. 2003), Italy (e.g. Ghirardini et al. 2007; Guarrera 2006; Lentini and Venza 2007; Picchi and Pieroni 2005), Croatia (Łuczaj et al. 2013b), Greece (Hadjichambis et al. 2008; Leonti et al. 2006; Zeghichi et al. 2003), Cyprus (Della et al. 2006), Turkey (Dogan et al. 2004; Dogan 2012; Ertuğ 2000; Ertuğ 2014), Lebanon (Marouf 2005), and Morocco (Nassif and Tanji 2013). They are consumed raw in salads (e.g. Couplan 2009a; Dogan 2012; Guarrera 2006; Hadjichambis et al. 2008; Marouf 2005; Picchi and Pieroni 2005; Tardío et al. 2005), but more frequently boiled and used in vegetables mixes with other wild herbs like side dish (Hadjichambis et al. 2008; Marco et al. 2003; Nassif and Tanji 2013; Picchi and Pieroni 2005), eaten in soups (Dogan 2012; Guarrera 2006; Hadjichambis et al. 2008; Lentini and Venza 2007; Picchi and Pieroni 2005), or used as pastry stuffing in vegetable pies, *ravioli*, etc. (Couplan 2009a; Della et al. 2006; Dogan 2012; Picchi and Pieroni 2005; Tardío 2010; Zeghichi et al. 2003). Some of these recipes include more than 15–20 different wild species. Additionally, they are also consumed in omelettes (Couplan 2009a; Guarrera 2006; Verde et al. 2003).

The consumption of this wild vegetable was common in the past, mainly in shortage periods (Mesa 1996). However, it can be bought nowadays in some traditional markets, generally in vegetable mixes (Picchi and Pieroni 2005), such as the *mišanca*, a wild vegetable mix sold in every market of the Dalmatian Coast, in southern Croatia (Łuczaj et al. 2013b).

Apart from the young leaves and shoots, petals and seeds are also edible. For instance, it is traditional to chew and suck the petals (Tardío et al. 2006), to add them in salads in some parts of Spain (Rivera and Obón 1991), or even to prepare dishes based on them in Morocco (Nassif and Tanji 2013). In some Italian villages, the poppy seeds are spread on the bread and on some sweets offered to the bride in order to stimulate fertility and bring the couple happiness (Lentini and Venza 2007). These traditional uses are nowadays incorporated in modern cuisines, such as the use of the poppy petals in Italian luxury restaurants (Łuczaj et al. 2012) or of poppy seeds in bread.

Other Uses This species contain alkaloids with sedative and narcotic effects and has been employed in Mediterranean folk medicines mainly against nervousness and insomnia (Benítez et al. 2010; Guarrera 2006; Le Floc'h 1983; Marouf 2005; Picchi and Pieroni 2005), stomach ache (Guarrera 2006), toothache (Aceituno-Mata 2010; Guarrera 2006), eye infections (Benítez et al. 2010; Guarrera 2006), and sore throat or cough (Guarrera 2006; Picchi and Pieroni 2005; Pieroni et al. 2004; Rigat et al. 2013). Other uses include the treatment of baldness (Benítez et al. 2010) and measles (Rivera et al. 2005). The most frequent mode of use is to drink the decoction or tisane of the petals or seeds, and, sometimes, it is also applied topically. As warned in some Italian studies, the excessive use of seeds as a soporific for babies may cause convulsions (Signorini et al. 2009).

The consumption of its leaves and shoots can also be slightly toxic and cause, in exceptional cases, mild poisoning in children, adults, and animals (Couplan 1990), such as vomit or nausea (Guarrera 2006).

The aerial parts were employed as fodder, mainly for rabbits and pigs, in Spain (Aceituno-Mata 2010; Blanco and Cuadrado 2000; Ferrández and Sanz 1993; Martínez-Lirola et al. 1997), Italy (Guarrera 2006), and Turkey (Ertuğ 2000). They were also used in ethnoveterinary as a soft sedative for pigs (Guarrera 2006; Picchi and Pieroni 2005).

Historical References The edibility of this species and its laxative properties are already mentioned by Theophrastus in the third century BC (Teofrasto 1988). Pliny the Elder, in the first century, stated that the wild poppy, boiled in honey, was used as a remedy for diseases in the throat (Bostock and Riley 1855).

Food Composition Tables for raw tender leaves and stems of *Papaver rhoeas* (Tables 13.67, 13.68 and 13.69).

Table 13.67 Main constituents, per 100 g of fresh young leaves and stems of *P. rhoeas*

	Units	Average	Range	References
Energy (calculated value)	kcal	42	24–78	–
Moisture	g	88.3	68.5–91.2	1, 2, 3
Available carbohydrates	g	3.35	2.90–5.30	2, 4
Dietary fibre	g	4.40	2.50–11.10	2, 4
Proteins	g	3.50	1.50–5.90	2, 4
Lipids	g	0.64	0.15–1.03	4, 5
Ash	g	2.50	1.45–5.20	1, 4
K	mg	562	188–1672	1, 2, 4, 6
Na	mg	61.0	21.0–112	1, 2, 4, 6
Ca	mg	197	50–545	1, 2, 4, 6
Mg	mg	35.9	8.60–74.1	1, 2, 4, 6
P	mg	55.9	26.5–85.3	2, 6
Fe	mg	3.46	0.90–6.04	2, 4, 6
Cu	μg	364	130–1070	4, 6
Mn	μg	727	390–1060	4, 6
Zn	μg	978	200–2640	2, 4, 6

1 Bianco et al. (1998) Italy (2 samples), *2* Trichopulou et al. (2000) Greece, *3* Morales et al. (2014) Spain (4 samples), *4* García-Herrera (2014) Spain (4 samples), *5* Morales et al. (2012b) Spain (4 samples), *6* Zeghichi et al. (2003) Crete

Table 13.68 Vitamins and other constituents, per 100 g of fresh young leaves and stems of *P. rhoeas*

	Units	Average	Range	References
Carotenoids				
β-carotene	mg	0.75	–	1
Lutein	mg	1.15	–	1
RAE (calculated value)	μg	62.5	–	–
Vitamin B_9 (total folates)	μg	152	151–154	2
Vitamin C	mg	24	18.7–47.6	3, 4
Ascorbic acid	mg	14.1	11.9–17.1	1, 3, 4
Dehydroascorbic acid	mg	16.9	6.80–30.1	3, 4
Vitamin E				
α-tocopherol	mg	1.00	0.52–1.37	1, 4, 5
β-tocopherol	mg	0.66	0.53–0.79	4
γ-tocopherol	mg	0.04	0.03–0.05	1, 4
δ-tocopherol	mg	0.02	0.01–0.03	4
Vitamin K	μg	145	–	1
Organic acids				
Oxalic acid	mg	456	124–850	3, 4
Glutamic acid	mg	95.7	nd–118	4
Malic acid	mg	89.0	22.9–280	3, 4
Citric acid	mg	115	11.2–410	3, 4
Fumaric acid	mg	4.40	Traces–8.26	3, 4
Phenolics (total)	mg	119[a]	22.3–269	1, 4, 5
	mg	≈239[b]	219–256	6, 7
Flavonoids	mg	12.0[c]	11.5–12.5	4
	mg	31.0[d]	–	8

Table 13.68 (continued)

	Units	Average	Range	References
	mg	≈143[b]	141–145	6
Nitrate	mg	252	219–307	5, 9

1 Vardavas et al. (2006b) Crete, *2* Morales et al. (2015) Spain (4 samples), *3* Sánchez-Mata et al. (2012) Spain (2 samples), *4* Morales et al. (2014) Spain (4 samples), *5* Zeghichi et al. (2003) Crete, *6* Conforti et al. (2009) Italy, *7* Conforti et al. (2011) Italy, *8* Trichopulou et al. (2000) Greece, *9* Bianco et al. (1998) Italy

RAE retinol activity equivalents; *nd* non-detected values

[a] Expressed as gallic acid

[b] Expressed as chlorogenic acid

[c] Expressed as catechin

[d] Expressed as sum of compounds (quercetin derivatives as major components)

Table 13.69 Fatty acids profile (% of total fatty acids) of fresh young leaves and stems of *P. rhoeas*

Individual compounds	Average	Range	References
12:0	0.14	0.10–0.18	1
14:0	1.58	0.56–2.53	1, 2
16:0	15.2	9.27–20.6	1, 2
16:1	0.45	0.16–0.73	1, 2
18:0	1.93	1.45–2.40	1, 2
18:1$n-9$	1.88	1.36–2.40	1, 2
18:2$n-6$	15.5	14.5–16.5	1, 2
18:3$n-3$	53.6	42.1–65.0	1, 2
18:3$n-6$	0.20	–	2
20:0	1.65	0.97–2.27	1, 2
20:1$n-9$	0.05	0.04–0.06	1
22:0	2.01	1.59–3.50	1, 2
24:0	1.35	1.10–1.60	1, 2
24:1$n-9$	1.00	–	2
Categories (calculated values)			
SFA	28.2	15.9–34.4	–
MUFA	3.3	1.6–4.1	–
PUFA	68.5	61.5–82.5	–
$n-3$	52.5	45.8–65.8	–
$n-6$	16.2	15.9–16.7	–
$n-9$	2.7	1.4–3.3	–

1 Morales et al. (2012b) Spain (4 samples), *2* Vardavas et al. (2006a) Crete

SFA saturated fatty acids, *MUFA* monounsaturated fatty acids, *PUFA* polyunsaturated fatty acids

Remarks (young leaves and stems of *P. rhoeas*)

- It may represent a source of dietary fibre (>3 g/100 g).
- Low Na content (<120 mg/100 g); it can be considered as a source of K, Cu, Mn, and vitamins C and B_9 (folates); sometimes also a source of Ca, Fe, and vitamin K.
- High content of oxalates: People with altered renal function should avoid this vegetable; boiling is recommended for general population.
- Very high nitrate content, not recommended for infants.

13.3.22 Plantago lanceolata L. and *P. major* L. (Plantaginaceae)

Common Names *P. lanceolata*: narrowleaf plantain, *lingua de ovelha* (pt), *llantén menor* (es), *plantain lancéolé* (fr), *piantaggine* (it), *damarlıca* (tr), *adhan al kabsh* (ar). *P. major*: broadleaf plantain, *tanchagem* (pt), *llantén mayor* (es), *grand plantain* (fr), *piantaggine* (it), *trputac* (hr), *lithospasto* (el), *sinirotu* (tr), *lisan al hamal* (ar).

Description Both are perennial herbs with one or more basal leaves rosettes growing from the same root. The leaves of *P. lanceolata* are lanceolate, acute, entire, with 3–5 longitudinal veins, up to 20 cm, usually villous. In the case of *P. major*, the leaves are ovate or elliptical, acute, and with a smooth margin, up to 20 × 9 cm, with 3–9 longitudinal veins. The inflorescences are spikes, much longer and narrower in *P. major* than in *P. lanceolata*, whose flowers have white and longer stamens.

Ecology and Distribution They are common in waste places along roads and meadows and pastures in several types of soils. In southern droughty areas, *P. major* is restrained to humid soil, such as river flood-lands. Both *P. major* and *P. lanceolata* flower from spring to autumn. They grow in Europe, Asia, and Africa, in temperate and cold zones, now naturalized all around the world, except in tropical areas.

Food Uses Both species have been traditionally consumed as vegetables in the Mediterranean countries, at least in Spain (Menendez-Baceta et al. 2012; Parada 2008; Piera 2006; Velasco et al. 2010), Italy (Guarrera 2006; Hadjichambis et al.

2008; Picchi and Pieroni 2005), Slovenia (Cerne 1992), Croatia (Łuczaj et al. 2013a), Greece (Hadjichambis et al. 2008), Turkey (Dogan 2012; Ertuğ 2014; Kültür 2008), and Lebanon (Marouf 2005).

Their tender basal leaves are collected in spring and eaten raw in salads (Dogan 2012; Hadjichambis et al. 2008; Marouf 2005; Menendez-Baceta et al. 2012; Parada 2008; Picchi and Pieroni 2005; Piera 2006; Velasco et al. 2010) or cooked (Guarrera 2006; Hadjichambis et al. 2008; Kültür 2008; Łuczaj et al. 2013a; Marouf 2005; Menendez-Baceta et al. 2012), sometimes in soups (Guarrera 2006; Picchi and Pieroni 2005). They can also be mixed with other wild vegetables, as in the traditional Italian dishes *acquacotta* and *pistic* (Guarrera 2006), the latter being a vegetable plate elaborated with a mix of a large number of wild vegetables (Paoletti et al. 1995). The leaves of *P. lanceolata* are also used as stuffing to prepare vegetable pies in Turkey (Dogan 2012).

The inflorescences of *P. major* are consumed raw in the southwest of Spain (Hadjichambis et al. 2008) and in the north of Italy, close to the Switzerland border, to elaborate a soup with a curious and pleasant fungus flavour (Picchi and Pieroni 2005). As also reported by these authors, its consumption has become very popular in the French *nouvelle cuisine* of the 90s.

In northeastern Spain their leaves are employed in the elaboration of a traditional liqueur called *ratafia* prepared with green walnuts and many wild herbs that sometimes include the basal leaves of these species (Bonet and Vallès 2002; Tardío et al. 2006).

Other Uses The consumption of the leaves of *P. major* is considered healthy for the stomach function (Marouf 2005). Different parts of both species (mainly the leaves, sometimes the seeds and the roots) have also been traditionally employed to treat several digestive, respiratory, urinary, and skin disorders, as registered in Portugal, Spain, Italy, Albania, Turkey, Cyprus, Lebanon, and Algeria (e.g. Bonet et al. 1999; Camejo-Rodrigues et al. 2003; González-Hernández et al. 2004; González-Tejero et al. 2008; Guarrera 2006; Marouf 2005; Menendez-Baceta et al. 2014; Neves et al. 2009; Picchi and Pieroni 2005; Tetik et al. 2013). It has also been used in ethnoveterinary (Carrió and Vallès 2012; Carvalho 2010).

These species are usually considered weeds and their leaves and seeds have been employed as animal food, for ruminants, rabbits, and birds (Parada 2008; Velasco et al. 2010).

Historical References Theophrastus, in the third century BC, cites the narrowleaf plantain among the wild vegetables sprouting in spring (Teofrasto 1988). Apart from describing many medicinal applications of the leaves and roots of both species, Dioscorides mentioned their consumption as cooked vegetables, with lentils (Osbaldeston 2000). Some of these medicinal prescriptions include the consumption of the cooked leaves, being therefore described as a medicinal food.

Food Composition Tables for raw tender leaves of *Plantago lanceolata* (Tables 13.70, 13.71 and 13.72).

Table 13.70 Main constituents, per 100 g of fresh tender leaves of *P. lanceolata*

	Units	Average	Range	References
Energy (calculated value)	kcal	28	17–26	–
Moisture	g	86.2	83.4–88.8	1
Available carbohydrates	g	2.81	2.37–3.25	1
Soluble sugars	g	1.06	–	2
Fructose	mg	211	–	2
Glucose	mg	740	–	2
Sucrose	mg	112	–	2
Dietary fibre	g	3.71	–	1
Insoluble fibre	g	2.70	–	2
Proteins	g	1.72	1.36–2.39	1,3
Lipids	g	0.33	0.27–0.39	1
Ash	g	2.07	1.43–3.26	1, 3
K	mg	361	263–415	1, 3, 4
Na	mg	21.3	13.6–33.4	1, 4
Ca	mg	304	57–660	1, 3, 4
Mg	mg	52.6	20.7–88.0	1, 3, 4
P	mg	≈28.0	–	3
Fe	mg	3.91	1.11–5.12	1, 3, 4
Cu	µg	159	90–190	1, 4
Mn	µg	661	310–1012	1, 3, 4
Zn	µg	548	295–770	1, 3, 4

1 Guil Guerrero (2001) Spain (10 samples), *2* Souci et al. (2008) unknown origin, *3* Ayan et al. (2006) Turkey, *4* Queralt et al. (2005) Bulgaria

Table 13.71 Vitamins and other constituents, per 100 g of fresh tender leaves of *P. lanceolata*

	Units	Average	Range	References
Carotenoids (total)	mg	6.77[a]	4.07–9.47	1
Vitamin C (ascorbic acid)	mg	13.6	12.3–14.9	1
Organic acids				
Oxalic acid	mg	88.0	–	1
Phenolics (total)	mg	≈480[b]	298–662	2
	mg	≈629[c]	309–950	3, 4
Hydroxycinnamic acids	mg	509[d]	–	5
Flavonoids	mg	≈49.6[e]	46.0–53.2	4
Anthocyanins	mg	0.54[f]	–	5

1 Guil Guerrero (2001) Spain (10 samples), *2* Gatto et al. (2011) Italy, *3* Gálvez et al. (2005) Spain, *4* Vârvan and Vârvan (2012) Romania, *5* Vanzani et al. (2011) Italy (4 samples)

[a] Expressed as β-carotene

[b] Expressed as sum of different compounds (verbascoside as major component)

[c] Expressed as gallic acid

[d] Expressed as chlorogenic acid

[e] Expressed as catechin

[f] Expressed as cyanidin-3-glucoside

Table 13.72 Fatty acids profile (% of total fatty acids) of tender leaves of *P. lanceolata*

Individual compounds	Average	Range	References
12:0	0.97	–	1
14:0	0.49	–	1
16:0	14.6	–	1
16:1	1.78	–	1
16:2*n*−6	0.38	–	1
16:3*n*−3	0.98	–	1
18:0	0.87	–	1
18:1*n*−9	1.45	–	1
18:2*n*−6	18.9	–	1
18:3*n*−3	45.1	–	1
20:0	1.98	–	1
22:0	0.87	–	1
22:1*n*−9	3.89	–	1
24:0	0.65	–	1
Categories (calculated values)			
SFA	21.8	–	–
MUFA	7.8	–	–
PUFA	70.4	–	–
n−3	49.6	–	–
n−6	20.8	–	–
n−9	5.9	–	–

1 Guil Guerrero (2001) Spain (10 samples)
SFA saturated fatty acids, *MUFA* monounsaturated fatty acids, *PUFA* polyunsaturated fatty acids

Remarks (tender leaves of *P. lanceolata*)

- Very low energy value (<40 kcal/100 g).
- Very low lipid content (<0.5 g/100 g), with high PUFA proportion.
- It may be a source of dietary fibre (>3 g/100 g).
- Low oxalate (<100 mg/100 g) and very low Na (<40 mg/100 g) contents.
- It may be considered as a source of Mn and vitamin C; often also a source of K, Ca, Mg, Fe, and Cu.

Food Composition Tables for raw tender leaves of *Plantago major* (Tables 13.73, 13.74 and 13.75).

Table 13.73 Main constituents, per 100 g of fresh tender leaves of *P. major*

	Units	Average	Range	References
Energy (calculated value)	kcal	27	21–32	–
Moisture	g	87.7	85.6–89.8	1, 2, 3
Available carbohydrates	g	2.03	1.35–2.71	3
Dietary fibre	g	3.88	3.19–4.57	3
Proteins	g	2.30	1.98–2.60	3
Lipids	g	0.20	0.14–0.22	3
Ash	g	3.19	2.86–3.82	3, 4
K	mg	318	283–357	3, 4
Na	mg	124	70–137	3, 4
Ca	mg	108	86.0–134	3, 4
Mg	mg	95.0	81.0–109	3, 4
P	mg	23.4	20.0–26.8	3, 4
Fe	mg	2.25	1.20–2.80	3, 4, 5
Cu	μg	150	100–230	3, 4, 5
Mn	μg	450	300–520	3, 4, 5
Zn	μg	250	100–520	3, 4, 5

1 Guil-Guerrero et al. (1996b) Spain (various samples), *2* Guil-Guerrero et al. (1997a) Spain (5 samples), *3* Guil-Guerrero (2001) Spain (10 samples), *4* Guil-Guerrero et al. (1998b) Spain (5 samples), *5* Stef et al. (2010), Romania

Table 13.74 Vitamins and other constituents, per 100 g of fresh tender leaves of *P. major*

	Units	Average	Range	References
Carotenoids (total)	mg	6.42[a]	4.87–9.18	1, 2, 3, 4, 5
Vitamin C	mg	55.0	24.3–92.0	3, 5, 6
Ascorbic acid	mg	32.5	17.0–40.0	1, 3, 6
Dehydroascorbic acid	mg	41.5	27.0–52.0	3, 6
Organic acids				
Oxalic acid	mg	67.3	54.4–136	3, 5
Nitrate	mg	101	94–108	3, 5

1 Aliotta and Pollio (1981) Italy, *2* Guil-Guerrero et al. (1996b) Spain (various samples), *3* Guil-Gerrero et al. (1997a) Spain (5 samples), *4* Guil-Guerrero and Rodriguez-García (1999) Spain (5 samples), *5* Guil-Guerrero (2001) Spain (10 samples), *6* Franke and Kensbock (1981) Germany

[a] Expressed as β-carotene

Table 13.75 Fatty acids profile (% of total fatty acids) of fresh tender leaves of *P. major*

Individual compounds	Average	Range	References
12:0	3.57	–	1
14:0	1.99	1.79–2.20	1, 2
16:0	16.2	15.9–16.6	1, 2
16:1	2.77	1.47–3.84	1, 2
18:0	2.36	2.12–2.61	1, 2
18:1*n*−9	2.09	1.86–2.32	1, 2
18:2*n*−6	12.5	11.2–13.8	1, 2
18:3*n*−3	36.7	33.2–40.0	1, 2
18:4*n*−3	2.02	–	2
20:0	1.46	1.31–1.61	1, 2
22:0	2.07	–	1
22:1	3.66	3.45–3.87	1, 2
24:0	1.09	0.98–1.21	1, 2
Categories (calculated values)			
SFA	30.6	29.1–32.0	–
MUFA	10.0	9.7–10.3	–
PUFA	59.5	57.7–61.2	–
n−3	43.4	42.9–43.9	–
n−6	14.1	12.8–14.8	–
n−9	10.0	9.7–10.3	–

1 Guil-Guerrero (2001) Spain (10 samples), *2* Guil-Guerrero et al. (1996b) Spain (various samples)
SFA saturated fatty acids, MUFA monounsaturated fatty acids, PUFA polyunsaturated fatty acids

Remarks

- Low energy value (<40 kcal/100 g).
- Source of dietary fibre (>3 g/100 g).
- Very low lipid content (<0.5 g/100 g), with high PUFA proportion.
- Low oxalate content (<100 mg/100 g).
- High ash level, low Na content (<40 mg/100 g).
- It may be considered as a source of Mg, Mn, and vitamin C; often also a source of K, Ca, Fe, and Cu.
- High content of nitrate; not recommended for infants.

13.3.23 *Portulaca oleracea* L. (Portulacaceae)

Common Names Purslane, *beldroega* (pt), *verdolaga* (es), *poupier* (fr), *portulaca* (it), *tušt* (hr), *andrakla* (el), *semizotu* (tr), *baqleh* (ar).

Description Annual fleshy herb, glabrous, with prostrate and branched stems up to 50 cm. The obovate and succulent leaves are sessile and up to 2 cm. The yellow flowers have 2 sepals and 4–6 petals. The fruit is a little capsule up to 1 cm, dehiscing by a transverse line, containing numerous little seeds up to 1 mm, black and reniform. Five different subspecies are known.

Ecology and Distribution It grows around cultivated places, in all type of soils. It flowers from February to September. It is common in the entire Mediterranean region and in all the temperate areas throughout the world.

Food Uses The leaves and stems have been traditionally consumed as raw or cooked vegetables in most Mediterranean countries. There are references from Portugal (Carvalho 2010; Carvalho and Telo 2013; Mendonça de Carvalho 2006), Spain (e.g. Blanco and Cuadrado 2000; Fajardo et al. 2007; Pellicer 2004a; Rigat et al. 2009; Tardío et al. 2005; Velasco et al. 2010), France (Marco et al. 2003), Italy (Guarrera 2006; Lentini and Venza 2007; Leonti et al. 2006; Picchi and Pieroni 2005), Slovenia (Cerne 1992), Croatia (Łuczaj et al. 2013a; Łuczaj et al. 2013b), Albania (Hadjichambis et al. 2008), Greece (Hadjichambis et al. 2008; Leonti et al. 2006), Cyprus (Della et al. 2006), Turkey (Dogan 2012; Ertuğ 2014), Lebanon (Batal and Hunter 2007; Marouf 2005), Jordan (Al-Qura'n 2010; Tukan et al. 1998), Palestine (Ali-Shtayeh et al. 2008), Egypt (Hadjichambis et al. 2008), Tunisia (Le Floc'h 1983), and Morocco (Hadjichambis et al. 2008; Nassif and Tanji 2013). The leaves, and sometimes also the tender stems, have been widely consumed raw in salads, either simply dressed with olive oil and lemon or vinegar (e.g. Ali-Shtayeh et al. 2008; Pellicer 2004a), or with more ingredients, such as tomatoes, capers, and cucumbers (Lentini and Venza 2007), anchovies (Picchi and Pieroni 2005), or yogurt (Dogan 2012). Purslane has been also broadly eaten cooked in different recipes, such as soups (Carvalho 2010; Cerne 1992; Dogan 2012; Guarrera 2006; Lentini and Venza

2007; Marco et al. 2003; Marouf 2005; Mendonça de Carvalho 2006), as a side dish, simply boiled, strained, and sprinkled with olive oil (Łuczaj et al. 2013a), boiled and fried with garlic and olive oil (Picchi and Pieroni 2005), or for accompanying legume dishes, such as the Spanish *cocido* or *potaje* (Blanco and Cuadrado 2000; Cobo and Tijera 2011; Mesa 1996; Sánchez-Romero 2003; Triano et al. 1998; Verde et al. 2003), and even included in mixes with other wild vegetables (Marouf 2005; Nassif and Tanji 2013). It is also used for omelettes (Benítez 2009; Mesa 1996; Molero Mesa et al. 2001; Obón 2006; Tardío et al. 2002; Verde et al. 2003) and as filling for vegetable pies like the Lebanese *fatayer*, triangle-shaped pies filled with vegetables, meat, or cheese (Batal and Hunter 2007; Marouf 2005). It has been traditionally preserved by pickling throughout the Mediterranean region (Carvalho and Telo 2013; Galán 1993; Guarrera 2006; Picchi and Pieroni 2005).

The seeds have been also eaten, at least in the north of Africa, both directly as a snack and for quenching thirst and hunger (Le Floc'h 1983) or mixed with cereals in porridge or with dried powdered dates in a sort of *sfouf* (Moroccan cake) from the Sahara (Nassif and Tanji 2013).

In Spain, the consumption of this plant is often associated with famine or scarcity times (Aceituno-Mata 2010; Benítez 2009; González Turmo 1997; Moll 2005).

Other Uses Purslane has been also traditionally used in folk medicine. It is eaten to improve the digestion or against constipation and stomach ache (Batal and Hunter 2007; Guarrera 2006; Velasco et al. 2010). It has been also used against urinary problems (González-Tejero 1989; Guarrera 2006), skin diseases (Guarrera 2006; Parada et al. 2002; Picchi and Pieroni 2005), to cure insomnia (Picchi and Pieroni 2005; Velasco et al. 2010), or to lower cholesterol (Batal and Hunter 2007).

It has been also widely employed as animal feed, especially for pigs, hens, or rabbits (e.g. Aceituno-Mata 2010; Carvalho 2010; Guarrera 2006; Pellicer 2004a). Today it is considered a typical summer weed in vegetable gardens (e.g. Aceituno-Mata 2010; Carvalho 2010).

Historical References This edible plant has been used and even cultivated since antiquity. In the third century BC, Theophrastus named purslane as one of the several summer cultivated vegetables that must be sown in April (Teofrasto 1988). Dioscorides, in his *Materia Medica*, describes a myriad of medicinal applications (Osbaldeston 2000), most of them still registered in modern ethnobotanical studies. Also in the first century, the Spanish-Roman Columella describes the way of preserving pickled purslane (Columela 1824). In the sixteenth century, Ruel and Daléchamps mention the existence of a cultivated form and say that purslane was preserved with salt and vinegar for winter and Lent (Marco et al. 2003).

Food Composition Tables for raw tender leaves of *Portulaca oleracea* (Tables 13.76, 13.77 and 13.78).

Table 13.76 Main constituents, per 100 g of fresh tender leaves of *P. oleracea*

	Units	Average	Range	References
Energy (calculated value)	kcal	25	19–32	–
Moisture	g	92.6	90.0–94.3	1, 2, 3, 4, 5
Available carbohydrates	g	1.98	1.11–2.70	4
Dietary fibre	g	1.20	0.90–1.80	1, 4
Proteins	g	3.00	2.50–3.50	1, 4
Lipids	g	0.35	0.30–0.40	1, 4, 6
Ash	g	1.88	1.25–2.95	1, 2, 4
K	mg	464	280–611	2, 5, 7
Na	mg	52.0	9.00–105	2, 5, 7
Ca	mg	116	51.0–181	2, 5, 7
Mg	mg	166	56–276	5, 7
P	mg	33.0	20.1–46.0	2, 7
Fe	mg	4.29	2.90–5.68	2, 7
Cu	μg	390	360–420	7
Mn	μg	590	540–640	7
Zn	μg	740	570–900	2, 7

1 Cowan et al. (1963), Lebanon, *2* Gurses and Artik (1984) Turkey (10 samples), *3* Guil-Gerrero et al. (1997a) Spain (5 samples), *4* Guil-Guerrero et al. (1997b) Spain (5 samples), *5* Bianco et al. (1998) Italy (2 samples), *6* Guil-Guerrero et al. (1996b) Spain (various samples), *7* Guil-Guerrero et al. (1999a) Spain (5 samples)

Table 13.77 Vitamins and other constituents, per 100 g of fresh edible portion of *P. oleracea*

	Units	Average	Range	References
Carotenoids	mg	6.2[a]	1.3–9.8	1, 2, 3, 4
β-carotene	mg	≈3.50	–	4
Lutein	mg	≈5.40	–	4
Neoxanthin	mg	≈0.50	0.10–0.90	5[d], 6[d]
Violaxanthin	mg	≈2.50	1.10–3.90	6[d]
Zeaxanthin	mg	≈0.20	–	4
RAE (calculated value)	μg	≈229	–	–
Organic acids				
Oxalic acid	mg	681	519–869	3, 7
Vitamin C	mg	101.5	29–109	1, 3, 8
Ascorbic acid	mg	54.7	4.82–105	2, 3, 9, 10
Dehydroascorbic acid	mg	9.0	4.0–18.0	3, 9
Phenolics	mg	270[b]	267–272	11
Phenolic acids	mg	6.32[c]	–	12[d]
Flavonoids	mg	0.30[c]	–	12[d]
Anthocyanins	mg	0.24	–	12[d]
Nitrate	mg	42.1	32–52.1	3, 7

1 Cowan et al. (1963), Lebanon, *2* Aliotta and Pollio (1981) Italy, *3* Guil-Guerrero et al. (1997a) Spain (5 samples), *4* Dias et al. (2009) Portugal, *5* Mercadente and Rodríguez Amaya (1990) Brazil, *6* Raju et al. (2007) India (2 samples), *7* Bianco et al. (1998) Italy (2 samples), *8* Bruno et al. (1980) Italy, *9* Franke and Kensbock (1981) Germany, *10* Gurses and Artik (1984) Turkey (10 samples), *11* Conforti et al. (2011) Italy, *12* Andarwulan et al. (2012) Indonesia; *RAE* retinol activity equivalents

[a] Expressed as β-carotene or as sum of compounds

[b] Expressed as chlorogenic acid

[c] Expressed as sum of compounds (chlorogenic acid and quercetin as major component)

[d] Plants gathered from locations different from Mediterranean or surrounding areas

Table 13.78 Fatty acids profile (% of total fatty acids) of fresh tender leaves of *P. oleracea*

Individual compounds	Average	Range	References
14:0	0.71	–	1
16:0	17.4	–	1
16:1*n*−7	20.9	–	1
18:0	3.46	–	1
18:1*n*−7	0.74	–	1
18:1*n*−9	5.89	–	1
18:2*n*−6	16.8	–	1
18:3*n*−3	32.6	–	1
18:3*n*−6	0.27	–	1
20:0	0.87	–	1
22:0	3.33	–	1
24:0	0.49	–	1
Categories (calculated values)			
SFA	25.4	–	–
MUFA	26.6	–	–
PUFA	48.0	–	–
n−3	31.7	–	–
n−6	16.5	–	–
n−9	5.9	–	–

1 Guil-Guerrero et al. (1996b) Spain (various samples); *SFA* saturated fatty acids, *MUFA* monounsaturated fatty acids, *PUFA* polyunsaturated fatty acids

Remarks (tender leaves of *P. oleracea*)

- Low energy value (<40 kcal/100 g).
- Very low lipid content (<0.5 g/100 g).
- Low Na content (<120 mg/100 g); it can be considered as a source of Mg, Fe, Cu, Mn, vitamin C; it might be also a source of K, Ca, and vitamin A (provitamin A as β-carotene).
- Very high content of oxalates: People with altered renal function should avoid this vegetable; boiling is recommended for general population.

13.3.24 Prunus spinosa L. (Rosaceae)

Common Names Blackthorn, *abrunheiro* (pt), *endrino* (es), *prunellier* (fr), *prugnolo selvatico* (it), *trnina* (hr), *tspourna* (el), *dağ eriği* (tr), *berqouq el ouach* (ar).

Description Shrub up to 3 m, with intricate spiny branches, deciduous. Leaves usually obovate, serrate, with a 1 cm petiole. The numerous flowers grow solitary before the leaves appear; with 5 glabrous sepals and 5 white petals. Blue or black globular fruits, pruinose, 10–15 mm in diameter, acid and astringent when raw.

Ecology and Distribution It grows in almost all types of soils, with preference for basic and rich ones, in hedges, forest edges, roads, or waysides. It lives in almost all of Europe, Siberia, Caucasus, southwest Asia, and northwest Africa.

Food Uses The fruits (sloes) were traditionally consumed at least in Portugal (Carvalho 2010; Carvalho and Telo 2013), Spain, (e.g. Rigat et al. 2009; Tardío et al. 2006; Tejerina 2010), Italy (Guarrera 2006; Lentini and Venza 2007), Slovenia (Cerne 1992), Croatia (Łuczaj et al. 2013a), Bosnia-Herzegovina (Redžić 2006), and Turkey (Ertuğ 2000; Ertuğ 2014). They can be eaten raw, directly from the plant, when they are ripe in autumn or, even better, when overripe after the first frosts, and the fruits become sweeter (e.g. Aceituno-Mata 2010; Fajardo 2008; Menendez-Baceta et al. 2012; Pardo-de-Santayana 2003; Tardío et al. 2002). The mature fruits are very astringent and sour, so they were usually left to overripe in a bag introduced into straw, put in the shade, buried for a week (Benítez 2009; Molina 2001) or were dried in the sun like raisins (Blanco 2002; Criado et al. 2008; Sánchez-Romero 2003; Verde et al. 2001). That way, the sloes could be stored at home for being consumed during the long winter (Aceituno-Mata 2010; Tardío et al. 2002), being even a typical Christmas dessert in some Spanish regions (Fajardo et al. 2007). The fruits have been also used to elaborate marmalades (Bonet 1993; Guarrera 2006; Parada 2008) and other kind of sweet preserves (Cofradía Extremeña de Gastronomía 1985).

As in other European countries, sloes are also traditionally gathered to elaborate liqueurs in the Mediterranean region, at least in Portugal (Carvalho 2010), Spain (e.g. Akerreta 2009; Benítez 2009; González et al. 2011a; Tardío et al. 2006), and Italy (Guarrera 2006). In Spain, a liqueur called *pacharán* is widely popular. It originated in Navarre (north of Spain), but it is now highly appreciated throughout the country. Many people prepare it at home, but it is also widely marketed and there are many commercial brands. It is made by soaking the fruits in anisette with some coffee beans and cinnamon (e.g. Akerreta 2009; Fajardo 2008; Menendez-Baceta et al. 2012; Tardío et al. 2002). But there are also many variants of these traditional sloe liqueurs. The fruits can be macerated in anisette, liquor (Pardo-de-Santayana 2003), or brandy (Verde et al. 2001). Some people also add chamomile, green walnuts, or *Jasonia glutinosa* (Aceituno-Mata 2010; Pardo-de-Santayana 2003; Villar et al. 1987). It is also mixed with a great number of herbs to elaborate liqueurs like the *ratafia* of Catalonia in northeastern Spain (Parada 2008). In the south of Spain, the boiled fruits were also mixed with wine (Fernández Ocaña 2000). These liqueurs were traditionally taken as a medicinal drink, especially for their digestive properties, as detailed in the following section.

Other Uses Given their astringent properties, the raw fruits were eaten against diarrhoea (Parada et al. 2002), although when overripe they were considered laxative (Carvalho 2010; Pardo-de-Santayana 2003). The fruits are also eaten against stomach ache (Sánchez-Romero 2003). As mentioned above, the sloe liqueurs were commonly drunk to heal or prevent several conditions. It is especially considered as digestive, laxative, and stomach-ache reliever (e.g. Akerreta 2009; Carvalho 2010; Guarrera 2006; Mulet 1991; Villar et al. 1987). Many other medicinal properties have been attributed to the plant (Guarrera 2006). For instance, the infusion or decoction of the aerial parts (leaves, flowers, and fruits) has been used as an aid to reduce high blood glucose levels (Benítez 2009) and high blood pressure (Akerreta 2009).

Goats browse their twigs and young branches, and pigs and foxes like the fruits (Guarrera 2006; Pardo-de-Santayana 2003). In Montesinho, NE Portugal, the fruits were given to the pigs as diuretics (Carvalho 2010).

In Italy and Bosnia-Herzegovina, the leaves have been used as tobacco surrogate (Guarrera 2006; Redžić 2006).

The wood is hard and the branches are used to elaborate tool-handles (Fajardo et al. 2007) or beautiful walking sticks (Guarrera 2006; Pardo-de-Santayana 2003). It is a good stock for grafting some fruit trees, especially plum trees (e.g Aceituno-Mata 2010; Guarrera 2006; Tardío et al. 2002). It is also planted or protected for creating living hedges (e.g. Guarrera 2006; Velasco et al. 2010).

Historical References Several prehistoric archaeological sites reveal that the fruits have been eaten since ancient times. The fruit remains that have been found in a northern Italy Neolithic site seem to have even been cooked or roasted (Rottoli 2014). There are references of the plum tree in the Greek and Roman classical authors but not clear references of this species.

Food Composition Tables for raw fruits of *Prunus spinosa* (Tables 13.79, 13.80 and 13.81).

Table 13.79 Main constituents, per 100 g of fresh fruits of *P. spinosa*

	Units	Average	Range	References
Energy (calculated value)	kcal	114	61–190	–
Moisture	g	60.2	43.1–72.8	1, 2, 3, 4, 5, 6, 7
Available carbohydrates	g	19.7	8.64–35.5	3, 7
Soluble sugars (total)	g	14.1	7.30–19.0	3, 7
Fructose	g	2.43	1.40–3.90	1, 3, 7
Glucose	g	9.27	4.95–14.9	1, 3, 7
Sucrose	g	0.24	Traces–0.50	1, 3, 7
Dietary fibre	g	12.3	10.7–13.9	2, 7
Insoluble fibre	g	10.3	8.90–11.8	7
Soluble fibre	g	1.95	0.70–2.70	1, 7
Proteins	g	1.23	0.08–3.08	1, 2, 3, 5, 7
Lipids	g	0.52	0.30–0.90	1, 2, 3, 5, 7
Ash	g	1.69	1.09–3.39	1, 2, 3, 5, 7
K	mg	453	216–590	1, 2, 7
Na	mg	22.8	5.62–66.6	1, 2, 7
Ca	mg	45.3	32.6–67.8	1, 2, 7
Mg	mg	22.6	11.4–29.9	1, 2, 7
P	mg	42.9	–	1
Fe	mg	0.76	0.26–1.98	1, 2, 7
Cu	µg	166	102–360	1, 7
Mn	µg	117	77–171	1, 2, 7
Zn	µg	263	160–529	1, 2, 7

1 Doležal et al. (2001), Czech Republic, *2* Marakoğlu et al. (2005) Turkey, *3* Barros et al. (2010b) Portugal, *4* Egea et al. (2010) Spain, *5* Ganhão et al. (2010) Spain, *6* Morales et al. (2013) Spain (6 samples), *7* Ruiz-Rodríguez (2014) Spain (6 samples)

Table 13.80 Vitamins and other constituents, per 100 g of fresh fruits of *P. spinosa*

	Units	Average	Range	References
Carotenoids				
β-carotene	mg	0.33	0.28–0.44	1, 2
RAE (calculated value)	µg	27.0	23–37	–
Vitamin C	mg	10.5	2.00–30.7	1, 3, 4, 5, 6
Ascorbic acid	mg	3.20	0.10–30.7	2, 5, 6, 7, 8
Dehydroascorbic acid	mg	9.16	4.7–14.6	5, 6
Vitamin E				
α-tocopherol	mg	5.23	5.05–5.41	5
β-tocopherol	mg	0.08	0.07–0.09	5
γ-tocopherol	mg	0.09	0.07–0.11	5
δ-tocopherol	mg	nd	nd	5
Organic acids				
Oxalic acid	mg	66.8	38.2–150	5, 8
Quinic acid	mg	11.4	7.5–15.3	8
Shikimic acid	mg	1.80	1.5–2.1	8
Malic acid	mg	291	249–333	5, 8
Citric acid	mg	33.5	12.6–64.7	5, 8

Table 13.80 (continued)

	Units	Average	Range	References
Fumaric acid	mg	1.59	Traces–0.28	5, 8
Succinic acid	mg	4.8	nd–9.9	5, 8
Phenolics (total)	mg	2294[a]	1584–3005	6
	mg	590[b]	123–2062	2, 3, 4, 7, 9
Phenolic acids	mg	729[b]	544–1019	6
Hydroxybenzoic acids	mg	1.51[b]	1.36–1.66	9
Hydroxycinnamic acids	mg	32.2[c]	29.5–34.9	9
Flavonoids	mg	208[d]	180–235	2
Flavonols	mg	77.8[e]	19.1–180	6, 9
Anthocyanins	mg	1431[f]	1044–2736	6
Sorbitol	g	4.07	–	1

1 Doležal et al. (2001), Czech Republic, *2* Barros et al. (2010b) Portugal, *3* Ertuk et al. (2009) Turkey (6 samples), *4* Jabłońska-Ryś et al. (2009) Poland; *5* Morales et al. (2013) Spain (6 samples), *6* Ruiz-Rodríguez et al. (2014a) Spain (6 samples), *7* Egea et al. (2010) Spain, *8* Pereira et al (2013) Portugal, *9* Ganhão et al. (2010) Spain

RAE retinol activity equivalents; *nd* non-detected values

[a] Expressed as sum of individual compounds (cyanidin derivatives as major component)

[b] Expressed as gallic acid

[c] Expressed as chlorogenic acid

[d] Expressed as catechin

[e] Expressed as rutin

[f] Expressed as pelargonidin-3-glucoside

Table 13.81 Fatty acids profile (% of total fatty acids) of fresh fruits of *P. spinosa*

Individual compounds	Average	Range	References
12:0	1.13	0.1–2.23	1, 2
14:0	0.75	0.09–1.48	1
16:0	14.9	6.2–23.6	1, 3
16:1	0.88	0.63–1.20	1
18:0	8.30	2.36–15.0	1, 3
18:1$n-9$	32.8	8.03–57.9	1, 3
18:2$n-6$	18.9	13.3–23.9	1, 3
18:3$n-3$	6.69	2.59–11.4	1, 3
20:0	3.93	0.52–7.44	1
20:1$n-9$	0.49	0.06–1.34	1
22:0	1.26	0.26–2.42	1
24:0	1.09	0.46–1.76	1
Categories (calculated values)			
SFA	35.5	11.1–60.0	–
MUFA	36.4	11.6–61.2	–
PUFA	28.1	27.7–28.4	–
$n-3$	7.5	2.9–12.2	–
$n-6$	20.5	16.3–24.8	–
$n-9$	35.4	10.3–60.5	–

1 Barros et al. (2010b) Portugal, *2* Morales et al. (2013) Spain (6 samples), *3* Doležal et al. (2001) Czech Republic; *SFA* saturated fatty acids, *MUFA* monounsaturated fatty acids, *PUFA* polyunsaturated fatty acids

Remarks (fruits of *P. spinosa*)

- It can reach high available carbohydrates content.
- Often high in dietary fibre (>6 g/100 g).
- High MUFA proportion (oleic acid) compared to other fruits.
- Low Na content (<120 mg/100 g); it can be considered as a source of vitamin E; sometimes also source of K and Cu.
- Very high anthocyanins level, with cianidin-3-rutinoside as the major compound (up to 1964 mg/100 g fresh matter).

13.3.25 Rorippa nasturtium-aquaticum (L.) Hayek (Brassicaceae)

Common Names Watercress, *agrião* (pt), *berro* (es), *cresson* (fr), *crescione* (it), *dragušac* (bs, hr), *kardama* (el), *suteresi* (tr), *karsun mehi* (ar).

Description Perennial herb, with hollow stems, the roots growing in the nodes, procumbent or floating in water. The leaves are pinnate, the younger with round leaflets and the end leaflet usually larger than the others. However, the leaflets of the leaves close to the flowers are lanceolate. Small white flowers, with 4 petals up to 4 mm, in a racemose inflorescence. The fruit is a siliqua 1.5 cm, with the seeds in two rows.

Ecology and Distribution It grows in wet places or floating in the water in flowing streams. It flowers almost all year-round. Native to Europe and Asia, it is now naturalized all around the world, except in tropical zones.

Food Uses The watercress has a widespread tradition of use all over the Mediterranean region. Its consumption has been registered in Portugal (Carvalho 2010; Mendonça de Carvalho 2006), Spain (e.g. González et al. 2011a; Menendez-Baceta et al. 2012; Parada et al. 2011; Tardío et al. 2006), Italy (e.g. Guarrera 2006; Picchi and Pieroni 2005), Slovenia (Cerne 1992), Bosnia-Herzegovina (Redžić 2006), Turkey (Dogan 2012; Ertuğ 2014), Cyprus (Della et al. 2006), Lebanon (Batal and Hunter 2007), Jordan (Al-Qura'n 2010; Tukan et al. 1998), Egypt (Hadjichambis et al. 2008), Tunisia (Le Floc'h 1983), and Morocco (Hadjichambis et al. 2008; Nassif and Tanji 2013). The young, tender, and pungent leaves are collected in spring, before flowering. They are mainly consumed raw in salads (e.g. Carvalho and Telo 2013; Cerne 1992; Della et al. 2006; Dogan 2012; Hadjichambis et al. 2008; Lentini and Venza 2007; Nassif and Tanji 2013; Redžić 2006; Tardío et al. 2005), including the Jordanian yogurt salads (Al-Qura'n 2010; Tukan et al. 1998), the Lebanese *korra tabbouleh*, a salad with cracked wheat (Batal and Hunter 2007), salads of lettuce and watercress (Carvalho 2010), salads with other wild herbs (Picchi and Pieroni 2005), or only watercress salads dressed with olive oil and other ingredients (Carvalho and Telo 2013; Guarrera 2006).

To a lesser extent, it is also consumed cooked in soups (Carvalho 2010; Guarrera 2006; Hadjichambis et al. 2008; Mendonça de Carvalho 2006), in omelettes (Guarrera 2006; Parada et al. 2011), fried with onion (Dogan 2012), or other recipes such as the Moroccan *beqoul* (a mix of up to 20 wild vegetables), goat meat *tajine*, or as sauce for couscous (Nassif and Tanji 2013). The seeds can be sprouted and eaten in salads as well (Batal and Hunter 2007).

The raw consumption of contaminated watercress gathered in streams of cattle-rearing areas may cause fasciolosis, a parasitic disease cause by the flatworm *Fasciola hepatica*.

Other Uses Several medicinal properties are attributed to this species in Mediterranean folk medicines, both consuming the aerial parts as a vegetable (González et al. 2010; Guarrera 2006), preparing an infusion or as a syrup (Carvalho 2010). It is considered a diuretic, refreshing, and depurative plant (Bonet and Vallès 2002; Guarrera 2006; Le Floc'h 1983; Rivera et al. 2005), high in iodine, iron, and proteins, and it was used to treat anaemia (Guarrera 2006; Mendonça de Carvalho 2006), high blood pressure (Guarrera 2006; Menendez-Baceta et al. 2014), heart and vein diseases (Redžić 2006), and other circulatory illnesses since it fortifies the blood and lowers fat in blood (Batal and Hunter 2007). It is recommended as well to treat respiratory illnesses such as pneumonia, asthma (Carvalho 2010; Guarrera 2006; Mendonça de Carvalho 2006), to facilitate digestion and improve appetite (Batal and Hunter 2007; Guarrera 2006). Its consumption is considered beneficial for diabetics (Batal and Hunter 2007; Guarrera 2006), against scurvy (Guarrera 2006; Le Floc'h 1983), and as an aphrodisiac (Le Floc'h 1983). The leaves and/or roots were used topically to treat sciatica (Guarrera 2006), alopecia, and eczemas (Benítez et al. 2010; Guarrera 2006), whereas the roots boiled in water with sugar were drunk to stimulate the liver (Carvalho 2010).

The leaves have also been used in ethnoveterinary (Benítez 2009; Guarrera 2006) and as animal fodder of hens, pigs, and rabbits (González et al. 2011b; Mendonça de Carvalho 2006).

Historical References This aquatic plant has been surely consumed since ancient times and also known by its medicinal properties. Dioscorides, in the first century, said that it is eaten raw being warming and diuretic, and that externally applied at night and wiped away in the morning, it takes away freckles and sunburn (Osbaldeston 2000). Laguna (1555) commented that the plant of Dioscorides is called *berros* in Castile, confirming its admirable ability to induce urine and to shatter kidney stones.

Food Composition Tables for raw leaves of *Rorippa nasturtium-aquaticum* (Tables 13.82, 13.83 and 13.84).

Table 13.82 Main constituents, per 100 g of fresh leaves of *R. nasturtium-aquaticum*

	Units	Average	Range	References
Energy (calculated value)	kcal	25	24–25	–
Moisture	g	93.2	92.2–94.2	1, 2, 3
Available carbohydrates	g	4.51	4.49–4.53	1
Soluble sugars (total)	g	0.21	0.19–0.22	1
Fructose	mg	59.0	56.0–62.0	1
Glucose	mg	54.0	51.0–57.0	1
Sucrose	mg	66.0	62.0–70.0	1
Dietary fibre	g	1.75	1.50–2.00	2, 3
Insoluble fibre	g	0.81	–	3
Soluble fibre	g	0.66	–	3
Proteins	g	0.91	0.85–0.97	1
Lipids	g	0.19	0.18–0.21	1
Ash	g	1.13	1.05–1.20	1
K	mg	276	–	3
Na	mg	12.0	–	3
Ca	mg	175	170–180	2, 3
Mg	mg	24.5	15.0–34.0	2, 3
P	mg	58.0	52.0–64.0	2,3
Fe	mg	1.65	2.2–3.1	2, 3
Zn	μg	93.2	92.2–94.2	2

1 Pereira et al. (2011) Portugal, *2* Holland et al. (1991) unknown origin, *3* Souci et al. (2008) unknown origin

Table 13.83 Vitamins and other constituents, per 100 g of fresh leaves of *R. nasturtium-aquaticum*

	Units	Average	Range	References
Carotenoids	mg	4.90	–	1
β-carotene	mg	3.71	2.52–4.90	1, 2
RAE (calculated value)	μg	309	210–408	–
Vitamin B_1	mg	0.12	0.08–0.16	1, 2
Vitamin B_2	mg	0.11	0.06–0.17	1, 2
Vitamin B_3	mg	0.47	0.30–0.65	1, 2
Vitamin C	mg	53.0	0.95–96.0	1, 2, 3
Vitamin E				
α-tocopherol	mg	≈1.49	1.48–1.50	3
β-tocopherol	mg	≈0.04	0.04–0.04	3
γ-tocopherol	mg	≈0.14	0.14–0.14	3
δ-tocopherol	mg	≈0.09	0.08–0.11	3
Vitamin K	μg	250	–	1
Organic acids				
Oxalic acid	mg	≈252	247–258	4
Quinic acid	mg	≈23.9	16.1–31.7	4
Malic acid	mg	≈79.6	76.1–83.1	4
Shikimic acid	mg	≈1.02	1.02–1.02	4
Citric acid	mg	≈383	360–406	4
Succinic acid	mg	≈35.1	34.6–35.6	4
Fumaric acid	mg	≈0.95	0.84–1.09	4

1 Souci et al. (2008) unknown origin, *2* Holland et al. (1991) unknown origin, *3* Pereira et al. (2011) Portugal, *4* Pereira et al. (2013) Portugal; *RAE* retinol activity equivalents

Table 13.84 Fatty acids profile (% of total fatty acids) of fresh leaves of *R. nasturtium-aquaticum*

Individual compounds	Average	Range	References
12:0	0.28	0.22–0.34	1
14:0	0.64	0.58–0.70	1
16:0	13.2	12.9–13.5	1
16:1	0.49	0.48–0.50	1
18:0	0.92	0.90–0.94	1
18:1*n*−9	0.70	0.67–0.73	1
18:2*n*−6	11.8	11.7–12.0	1
18:3*n*−3	68.4	68.2–68.7	1
18:3*n*−6	0.08	0.08–0.08	1
20:0	0.14	0.12–0.16	1
20:1*n*−9	0.01	0.01–0.01	1
22:0	0.22	0.22–0.22	1
24:0	0.80	0.78–0.82	1
24:1*n*−9	0.36	0.32–0.40	1
Categories (calculated values)			
SFA	16.5	16.2–16.8	–
MUFA	1.6	1.5–1.7	–
PUFA	81.9	81.5–82.3	–
n−3	69.8	69.4–70.3	–
n−6	12.1	12.1–12.2	–
n−9	1.1	1.0–1.2	–

1 Pereira et al. (2011) Portugal
SFA saturated fatty acids, *MUFA* monounsaturated fatty acids, *PUFA* polyunsaturated fatty acids

Remarks (leaves of *R. nasturtium-aquaticum*)

- High moisture content, very low lipid content (<0.5 g/100 g), low energy value (<40 kcal/100 g), and very low Na content (<40 mg/100 g).
- High PUFA and *n*−3 proportion; very good *n*−3/*n*−6 ratio (>5).
- It could be considered as a source of vitamin A (provitamin A as β-carotene), C, K, and Ca.

13.3.26 Rubus ulmifolius Schott (Rosaceae)

Common Names Elmleaf blackberry, *silva* (pt), *zarzamora* (es), *ronce* (fr), *rovo* (it), *'ulliq* (ar).

Description Shrub, up to 2(6) m, that forms dense thickets with prickly and intricate stems and digitate compound leaves, with 3 or 5 obovate leaflets. The flowers grow in globose inflorescences have a calyx of 5 sepals and a corolla with 5 ovate and pink to white petals. The fruit that ripens at the end of summer and autumn is a blackberry, a head of little, one-seeded druplets.

Ecology and Distribution It grows on well-drained humid, acid or basic soils, both in sunny and full-shade habitats, usually at stream sides, near ways, hedges, and in the borders of woods up to 1500 m. It is common in Central and Western Europe and North Africa and introduced in America and Oceania. Other similar widespread species of *Rubus* in Central and Eastern Mediterranean countries are *R. plicatus* Weihe & Nees (=*R. fruticosus* L.) and *R. sanctus* Shreb.

Food Uses Blackberries are highly valued fruits in many Mediterranean countries. Those of this species are referred in Portugal (Carvalho and Telo 2013; Mendonça de Carvalho 2006), Spain (e.g. Menendez-Baceta et al. 2012; Pardo-de-Santayana 2008; Tardío et al. 2006), Italy (Guarrera 2006; Hadjichambis et al. 2008; Lentini and Venza 2007), Morocco (Nassif and Tanji 2013), and Tunisia (Le Floc'h 1983). They are usually eaten raw (Carvalho and Telo 2013; Lentini and Venza 2007; Parada et al. 2002), both directly in the field or as a dessert when brought home, and also smashed with sugar (Pardo-de-Santayana 2008; Tardío et al. 2002), and used to prepare jams, jellies, and desserts (Le Floc'h 1983; Tardío et al. 2006). Blackberries are commonly used to elaborate beverages as well. The fruits are macerated in liquor or anisette (e.g. Blanco 1998; Guarrera 2006; Hadjichambis et al. 2008; San Miguel 2004). They can be also smashed and mixed with wine and sugar to elaborate the so-called blackberry wine (Ferrández and Sanz 1993; Pardo-de-Santayana 2004; Tardío et al. 2002) or with water and sugar to prepare a refreshing beverage (Menendez-Baceta et al. 2012). In the north of Portugal, blackberries were added to grape juice to improve the colour and fermentation of wine, as grapes ripening is difficult in this region (Carvalho 2010). The sour peeled young shoots are consumed as a vegetable in spring when they sprout. They are usually eaten raw in the field (e.g. Carrió 2013; González et al. 2011a; Tardío et al. 2002) but can be also prepared in salads (e.g. Aceituno-Mata 2010; Bonet and Vallès 2002; Verde et al. 1998), boiled like asparagus (Mulet 1991; Triano et al. 1998), and included in omelettes (e.g. Guarrera 2006; Tardío et al. 2002). The shoots and young leaves are sometimes employed to elaborate liqueurs (Bonet and Vallès 2002; Lentini and Venza 2007).

Other Uses Besides its importance as a food plant, it has many other uses. Regarding medicine, some of the medicinal applications are related to its food use, and therefore it is sometimes considered a food-medicine, at least in Spain and Portugal. For instance, the consumption of the boiled or raw young shoots (Mulet 1991; Parada 2008; Pellicer 2001; Villar et al. 1987), the raw fruits (González-Tejero 1989), and the wine prepared with them (Blanco 1996) are considered good remedies to stop diarrhoea. The liqueur of the fruits is also drunk against digestive ailments (Arauzo et al. 2004; Blanco 1998; Velasco et al. 2010). The consumption of the young shoots is believed to have a depurative effect (Villar et al. 1987) and the jam to fight against iron deficiency (Akerreta 2009). Other remedies against diarrhoea, not so related with food use, are to drink the infusion or decoction of flowers (Carvalho 2010), leaves (Pardo-de-Santayana 2008), young shoots (Velasco et al. 2010; Villar et al. 1987) or fruits (González-Tejero 1989). Given its astringent properties, it is also used against skin disorders, such as healing wounds, applying the raw leaves (Aceituno-Mata 2010; Parada 2008), the dried leaves powder, or the decoction of the bark (Sánchez López et al. 1994). It has been also used against fever (Mulet 1991), diabetes (Sánchez López et al. 1994), haemorrhoids (Ferrández and Sanz 1993; Molina 2001; Fajardo et al. 2007), and many other complaints.

The bark is very flexible and was used for weaving baskets and tying brooms or many other things (e.g. Blanco 1998; Carvalho 2010; Guarrera 2006). In some localities the canes were used as fuel (Carvalho 2010; Velasco et al. 2010) and the fruits for making writing ink (Akerreta 2009; Verde et al. 1998).

Though brambles are considered weeds since they propagate easily in pastures roads or walls and are difficult to eradicate, they are also welcome or even planted to form thorny hedges (e.g. Blanco and Cuadrado 2000; Carvalho 2010).

Historical References The food use of different species of this genus is surely known since antiquity. Moreover, some of the medicinal uses registered in modern ethnobotanical works are already mentioned by Dioscorides in the first century AD (Osbaldeston 2000).

Food Composition Tables for raw fruits of *Rubus ulmifolius* (Tables 13.85, 13.86 and 13.87).

Table 13.85 Main constituents, per 100 g of fresh fruits of *R. ulmifolius*

	Units	Average	Range	References
Energy (calculated value)	kcal	99	51–145	–
Moisture	g	72.2	57.8–83.8	1, 2, 3, 4
Available carbohydrates	g	16.1	8.28–23.8	4
Soluble sugars (total)	g	11.4	4.25–23.6	4
Fructose	g	7.02	2.35–17.8	4
Glucose	g	6.70	2.11–17.5	4
Sucrose	g	0.17	0.14–0.67	4
Dietary fibre	g	11.5	7.05–16.0	4
Insoluble fibre	g	9.45	7.00–17.7	4
Soluble fibre	g	2.04	0.80–4.50	4
Proteins	g	1.74	0.56–2.32	2, 4
Lipids	g	0.52	0.19–1.41	2, 3, 4
Ash	g	0.79	0.60–1.1	2, 4
K	mg	196	122–270	4
Na	mg	32	6.8–57.2	4
Ca	mg	78.4	45.6–111.2	4
Mg	mg	43	16.8–69.2	4
Fe	mg	1.16	0.69–1.64	4
Cu	μg	268	132–404	4
Mn	μg	1560	270–2850	4
Zn	μg	409	279–539	4

1 Egea et al. (2010) Spain, *2* Ganhão et al. (2010) Spain, *3* Morales et al. (2013) Spain (6 samples), *4* Ruiz-Rodríguez (2014) Spain (6 samples)

Table 13.86 Vitamins and other constituents, per 100 g of fresh fruits of *R. ulmifolius*

	Units	Average	Range	References
Carotenoids				
β-carotene	mg	0.38	0.36–0.40	1
RAE (calculated value)	μg	31	30–33	–
Vitamin C	mg	23.0	10.0–33.8	2, 3, 4
Ascorbic acid	mg	15.7	3.66–29.3	1, 3, 4
Dehydroascorbic acid	mg	7.51	3.39–14.3	3, 4
Vitamin E				
α-tocopherol	mg	3.38	1.92–4.84	3
β-tocopherol	mg	0.24	0.14–0.34	3
γ-tocopherol	mg	3.70	2.07–5.39	3
δ-tocopherol	mg	3.69	2.04–5.34	3
Organic acids				
Oxalic acid	mg	98.1	66.1–130	3
Malic acid	mg	179	56.0–302	3
Citric acid	mg	46.6	17.8–75.4	3
Fumaric acid	mg	0.56	0.31–0.81	3
Succinic acid	mg	64.1	47.3–81.0	3
Phenolics (total)	mg	478[a]	247–951	1, 2, 5
	mg	608[b]	373–1345	4
Phenolic acids	mg	414[a]	156–967	4

Table 13.86 (continued)

	Units	Average	Range	References
Hydroxybenzoic acids	mg	2.80[a]	2.47–3.13	5
Hydroxycinnamic acids	mg	14.5[c]	11.9–16.9	5
Flavonols	mg	29.5[d]	4.2–84.7	1, 3
Anthocyanins	mg	101[e]	89–123	5
	mg	141[f]	68–301	4

1 Egea et al. (2010) Spain, *2* Jabłońska-Ryś et al. (2009), Poland, *3* Morales et al. (2013) Spain (6 samples), *4* Ruiz-Rodríguez et al. (2014b) Spain (6 samples), *5* Ganhão et al. (2010) Spain
RAE retinol activity equivalents
[a] Expressed as gallic acid
[b] Expressed as sum of individual compounds (gallic acid and cyanidin derivatives as major components)
[c] Expressed as chlorogenic acid
[d] Expressed as rutin
[e] Expressed as cyanidin-3-glucoside
[f] Expressed as pelargonidin-3-glucoside

Table 13.87 Fatty acids profile (% of total fatty acids) of fresh fruits of *R. ulmifolius*

Individual compounds	Average	Range	References
12:0	0.69	0.64–0.74	1
14:0	0.25	0.25–0.25	1
16:0	7.02	6.89–7.15	1
16:1	0.15	0.15–0.15	1
18:0	3.39	3.2–3.58	1
18:1*n*−9	22.6	21.6–23.6	1
18:2*n*−6	48.6	47.6–49.5	1
18:3*n*−3	13.3	13.1–13.5	1
20:0	0.36	0.34–0.38	1
20:1*n*−9	0.08	0.06–0.10	1
22:0	0.73	0.73–0.73	1
24:0	0.47	0.43–0.51	1
Categories (calculated values)			
SFA	13.8	13.1–13.3	–
MUFA	23.4	23.0–23.8	–
PUFA	63.4	62.9–63.9	–
n−3	13.6	13.5–13.8	–
n−6	49.8	49.4–50.1	–
n−9	23.2	22.8–23.7	–

1 Morales et al. (2013) Spain (6 samples)
SFA saturated fatty acids, *MUFA* monounsaturated fatty acids, *PUFA* polyunsaturated fatty acids

Remarks (fruits of *R. ulmifolius*)

- High in dietary fibre (>6 g/100 g).
- Low Na content (<120 mg/100 g); it can be considered as a very good source of Mn and vitamin E; sometimes also a source of Cu and vitamin C.

13.3.27 *Rumex acetosa* L. and *R. papillaris* Boiss. & Reut. (Polygonaceae)

Common Names *Rumex acetosa*: common sorrel, *azeda* (pt), *acedera* (es), *oseille* (fr), *acetosa* (it), *hummaid* (ar). *R. papillaris* is also known as *acedera* (es).

Description Both are perennial herbs up to 1–1.2 m, growing of a stout root, with basal leaves, long peciolate, lanceolate, and frequently arrow shaped. The leaves of *R. papillaris* are larger and narrower, with a papillose surface and usually divergent basal lobes. Little greenish to reddish flowers, grouped in branched inflorescences, bigger and more dense in the second species. The fruit is a trigonous nut, covered by rounded valves.

Ecology and Distribution *R. acetosa* typically grows in grasslands throughout Europe, though in drier Mediterranean regions it is restricted to wetter areas, such as river banks. *R. papillaris* grows in roadsides, pastures, and meadows in continental areas of the Iberian Peninsula, that is central Spain and northwest of Portugal.

Food Uses *R. acetosa* L. and other closely related sorrels less widely distributed in the Mediterranean region, such as *R. intermedius* DC, *R. thyrsiflorus* Fingerh, and *R. papillaris* (also known as *R. thyrsiflorus* subsp. *papillaris* (Boiss. & Reut.) Sagredo & Malag.), have a similar acid taste and consequently are used in the same way, though the majority of the ethnobotanical studies report the use of the first one. This species has been consumed in many Mediterranean countries, such as Portugal (Carvalho 2010; Mendonça de Carvalho 2006), Spain (e.g. Pardo-de-Santayana 2008; San Miguel 2004), Italy (Guarrera 2006; Hadjichambis et al. 2008; Picchi and

Pieroni 2005), Slovenia (Cerne 1992), Bosnia-Herzegovina(Redžić 2006), Albania (Hadjichambis et al. 2008), Palestine (Ali-Shtayeh et al. 2008), and Morocco (Nassif and Tanji 2013). Its leaves and sometimes the tender stems were sucked or eaten raw, being typically consumed by children and shepherds (Carvalho 2010; Pardo-de-Santayana 2008; Picchi and Pieroni 2005; San Miguel 2004). The leaves were also consumed raw in salads and cooked in soups (Carvalho 2010; Cerne 1992; Hadjichambis et al. 2008), with rice or pasta (Guarrera 2006), fried in olive oil (Ali-Shtayeh et al. 2008), like side dishes (Cerne 1992), or even in vegetables mixes, such as the Italian *pistic* (Guarrera 2006).

The Iberian endemism *R. papillaris* has been traditionally consumed in continental areas of central Spain (Aceituno-Mata 2010; Blanco 1998; Tardío et al. 2005; Velasco et al. 2010). The tender and basal leaves were collected at the end of winter and in springtime, mainly in ploughed lands because the plants were more tender (Aceituno-Mata 2010). They were usually consumed raw as a snack, with bread, or in salads (Tardío et al. 2005; Tardío et al. 2006). Because of the acid-lemon flavour, they were mixed in salads with other wild species, such as *Chondrilla juncea* and *Montia fontana* (Aceituno-Mata 2010).

Other Uses The leaves of the common sorrel have been also used as fodder, especially for pigs (Carvalho 2005; González et al. 2011b; Guarrera 2006). The plant is considered as lightly toxic, surely due to its oxalic acid content (González et al. 2011b; Guarrera 2006), and so it was previously cooked and mixed with other ingredients (flour and bran) for feeding pigs (Carvalho 2005; González et al. 2011b) and not recommended for people with renal calculus.

Concerning medicinal properties, *R. acetosa* has been traditionally used against diarrhoea (Guarrera 2006; Velasco et al. 2010), heart diseases (Redžić 2006), and regarded as a blood purifier (Moll 2005; Villar et al. 1987). Similarly, it was used in veterinary as a digestive remedy and antihelminthic (González et al. 2011b; Velasco et al. 2010).

The common sorrel has been also used for dying (Guarrera 2006) and as a cosmetic for removing spots and washing and caring the hands (Velasco et al. 2010).

The scarce references found for the Iberian sorrel seem to agree with some of the mentioned for the common sorrel, such as its use in animal feed (for pigs, chickens, goats, and sheep), and as a kind of cosmetic for washing the farmers hands with leaves after weeding (Aceituno-Mata 2010).

Historical References No historical reference can be attributed without any doubt to *R. papillaris*. However, some of them can be assigned to *R. acetosa* and other related taxa of this genus. Common sorrel has been used since antiquity and even cultivated in gardens, as referred by Dioscorides and Pliny the Elder, in the first century AD (Bostock and Riley 1855; Osbaldeston 2000). They also pointed out the medicinal properties of this cultivated plant and other wild related species, some being similar to those registered in modern ethnobotanical works.

Food Composition Tables for raw tender leaves of *Rumex papillaris* (Tables 13.88, 13.89 and 13.90).

Table 13.88 Main constituents, per 100 g of fresh tender leaves of *R. papillaris*

	Units	Average	Range	References
Energy (calculated value)	kcal	33	22–46	–
Moisture	g	89.1	87.8–90.7	1, 2
Available carbohydrates	g	2.00	1.60–2.70	3
Dietary fibre	g	4.40	4.00–5.00	3
Proteins	g	2.40	1.60–3.50	3
Lipids	g	0.66	0.16–1.11	2, 3
Ash	g	1.00	0.40–1.30	3
K	mg	351	255–468	3
Na	mg	25.6	15.7–37.9	3
Ca	mg	60.3	36.9–89.7	3
Mg	mg	45.0	33.7–50.8	3
Fe	mg	1.00	0.40–1.20	3
Cu	μg	80	30–150	3
Mn	μg	750	410–1180	3
Zn	μg	360	240–420	3

1 Sánchez-Mata et al. (2012) Spain (2 samples), *2* Morales et al. (2014) Spain (5 samples), *3* García-Herrera (2014) Spain (5 samples)

Table 13.89 Vitamins and other constituents, per 100 g of fresh tender leaves of *R. papillaris*

	Units	Average	Range	References
Vitamin B_9 (total folates)	μg	187	177–197	1
Vitamin C	mg	25.8	18.9–32.3	2, 3
Ascorbic acid	mg	14.3	3.15–22.5	2, 3
Dehydroascorbic acid	mg	11.0	2.85–16.9	2, 3
Vitamin E				
α-tocopherol	mg	1.29	0.85–1.73	3
β-tocopherol	mg	0.05	0.03–0.07	3
γ-tocopherol	mg	0.34	0.21–0.47	3
δ-tocopherol	mg	nd	–	3
Organic acids				
Oxalic acid	mg	251	80.5–271	2, 3
Glutamic acid	mg	253	185–321	3
Malic acid	mg	117	9.87–320	2, 3
Citric acid	mg	30.3	nd–100	2, 3
Fumaric acid	mg	nd	–	2, 3
Phenolics (total)	mg	104[a]	99.8–108	3
Flavonoids	mg	39.5[b]	36.2–42.7	3

1 Morales et al. (2015) Spain (5 samples), *2* Sánchez-Mata et al. (2012) Spain (2 samples), *3* Morales et al. (2014) Spain (5 samples); *nd* non-detected values

[a] Expressed as gallic acid

[b] Expressed as catechin

Table 13.90 Fatty acids profile (% of total fatty acids) of fresh tender leaves of *R. papillaris*

Individual compounds	Average	Range	References
12:0	0.10	0.08–0.12	1
14:0	0.51	0.16–0.86	1
16:0	11.2	10.9–11.5	1
16:1	0.20	0.11–0.29	1
18:0	0.80	0.74–0.86	1
18:1*n*−9	5.80	5.66–5.94	1
18:2*n*−6	22.8	22.6–23.2	1
18:3*n*−3	51.8	51.6–51.9	1
20:0	0.27	0.12–0.42	1
20:1*n*−9	0.19	0.17–0.21	1
22:0	0.76	0.72–0.80	1
24:0	1.18	1.14–1.22	1
24:1*n*−9	0.05	0.04–0.06	1
Categories (calculated values)			
SFA	15.5	14.5–15.4	–
MUFA	6.6	6.4–6.8	–
PUFA	77.9	77.8–79.1	–
n−3	54.1	53.9–55.0	–
n−6	23.8	23.9–24.1	–
n−9	6.4	6.3–6.5	–

1 Morales et al. (2012b) Spain (4 samples)
SFA saturated fatty acids, *MUFA* monounsaturated fatty acids, *PUFA* polyunsaturated fatty acids,

Remarks (tender leaves of *R. papillaris*)

- Low energy value (<40 kcal/100 g).
- Source of dietary fibre (>3 g/100 g).
- High proportion of PUFA and *n*−3 fatty acids content, compared to other vegetables.
- Very low Na content (<40 mg/100 g); it can be considered as a source of Mn and vitamins C and B_9 (folates); sometimes also a source of K.
- It may reach high oxalates levels; boiling is recommended for general population.

13.3.28 Rumex pulcher L. (Polygonaceae)

Common Names Fiddle dock, *romaza* (es), *patience élégante* (fr), *romice* (it), *štavljak* (hr), *xinidia* (el), *ekşilik* (tr), *hummaid* (ar).

Description Perennial and tap-rooted herb with basal leaves up to 14 cm long, dark green, peciolate, oblong, rounded above, and often with a narrow middle in the shape of a fiddle. Open inflorescence with many branches up to 70 cm, with little greenish flowers in clusters. The fruit is a trigonous brown nut, covered by thick valves, reticulate and toothed, with up to 9 teeth almost spiny.

Ecology and Distribution Ruderal plant that grows in almost all types of soils, in waste places, roadsides, 0–2000 m. Flowering time usually from April to July. Mediterranean region, centre and west of Europe, west of Asia, widespread as a weed in all the temperate regions of the world.

Food Uses It is a wild vegetable traditionally consumed in some Mediterranean countries, as recorded in Spain (e.g. González et al. 2011a; Rivera et al. 2005; Tardío et al. 2006), France (Marco et al. 2003), Italy (Bianco et al. 2009; Picchi and Pieroni 2005), Croatia (Łuczaj et al. 2013a), Greece (Hadjichambis et al. 2008), Turkey (Kültür 2008; Ertuğ 2014), Cyprus (Della et al. 2006), and Jordan (Al-Qura'n 2010). Its tender basal leaves are collected mainly in springtime, although they can also be gathered during mild winters (González et al. 2011a; Tardío 2010). Because of its strong taste, this species is generally consumed cooked, and it is often sweetened through repeated boiling and discarding the boiling water (Tardío 2010). Traditional recipes of fiddle dock with legumes have been reported both in Cyprus—*xinopoureka*—(Della et al. 2006; Hadjichambis et al. 2008), and Spain—*potaje*, popularly consumed during Lent—(Tardío 2010). This association of its consumption with Lent is also found in Italy (Picchi and Pieroni 2005). They are also prepared in wild vegetables mixtures at least in Greece, Italy, and Croatia (Hadjichambis et al. 2008; Łuczaj et al. 2013a; Picchi and Pieroni 2005), such as the Italian *pistic* (a vegetable dish from the region of Friuli elaborated with a mix of up to 54 wild vegetables), and the Croatian *svakober*, which translates as "pick all", where the plants are boiled for 20–30 min, strained and sprinkled with olive oil, and served as a side dish. These mixes are sometimes used as stuffing in vegetable pies (Al-Qura'n 2010; Della et al. 2006). In Cyprus these pies, which are called *pittes*, include up to 11 plants that are boiled or fried along with rice or *pourgouri* (like couscous) and spices (Della et al. 2006). This species has been also consumed in omelettes (Benítez 2009; Picchi and Pieroni 2005), used for preparing a sauce to accompany meat (Verde et al. 2003), and more occasionally eaten raw in salads (Al-Qura'n 2010; Bonet and Vallès 2002).

Other Uses Although this plant has been mainly used as a wild green, it has other minor uses in medicine and ethnoveterinary. The leaves, both eaten as a vegetable or in infusion, and also the flowers and fruits, have been traditionally considered as digestive, anti-diarrhoeic, and anti-catarrhal in several Spanish regions (Fajardo et al. 2007; Molero Mesa et al. 2001; Molina 2001; Rivera et al. 2005; Sánchez López et al. 1994; Verde et al. 2003). As registered by Blanco (2002), poultices made with the leaves were used against boils and infected grains, along with the leaves of the white henbane (*Hyoscyamus albus* L.). In ethnoveterinary it has been employed to treat urogenital myiasis (González et al. 2011b) and respiratory affections (Velasco et al. 2010).

It is also considered a lightly toxic species (Benítez 2009), which can cause drastic hypotension when abusively used (Agelet and Vallès 2003).

Its leaves were also used as animal fodder (Aceituno-Mata 2010; Fajardo et al. 2007) and, in times of scarcity, as a tobacco substitute (Criado et al. 2008; Sánchez López et al. 1994; Tardío et al. 2002; Velasco et al. 2010; Verde et al. 2001).

Historical References Some species of this genus (with the name of *lapathum* and other related ones), were described by Dioscorides and Pliny the Elder (first century AD), pointing out their medicinal properties and their use as a vegetable-medicine (Bostock and Riley 1855; Osbaldeston 2000). Some of these attributed properties are very similar to those registered in modern ethnobotanical works. The consumption like spinach of this wild vegetable in central Spain, especially in times of scarcity, is mentioned, together with other species, by Cienfuegos (1627).

Food Composition Tables for basal leaves of *Rumex pulcher*, raw (Tables 13.91, 13.92 and 13.93) and boiled (Tables 13.94 and 13.95).

Table 13.91 Main constituents, per 100 g of fresh tender leaves of *R. pulcher*

	Units	Average	Range	References
Energy (calculated value)	kcal	42	22–75	–
Moisture	g	86.4	69.0–89.1	1, 2
Available carbohydrates	g	3.28	1.50–4.50	3
Dietary fibre	g	5.45	4.00–12.6	3
Proteins	g	3.00	1.90–5.50	3
Lipids	g	0.61	0.10–1.03	3, 4
Ash	g	1.70	1.10–3.10	3
K	mg	591	382–955	3
Na	mg	106	33.4–132	3
Ca	mg	62.6	2.30–124	3
Mg	mg	40.6	2.20–71.0	3
Fe	mg	1.50	0.82–2.70	3
Cu	μg	110	50–240	3
Mn	μg	261	200–400	3
Zn	μg	563	370–1610	3

1 Sánchez-Mata et al. (2012) Spain (2 samples), *2* Morales et al. (2014) Spain (4 samples), *3* García-Herrera (2014) Spain (5 samples), *4* Morales et al. (2012b) Spain (4 samples)

Table 13.92 Vitamins and other constituents, per 100 g of fresh tender leaves of *R. pulcher*

	Units	Average	Range	References
Vitamin B_9 (total folates)	μg	478	440–529	1
Vitamin C	mg	40.8	28.7–46.5	2, 3
Ascorbic acid	mg	25.5	16.6–30.8	2, 3
Dehydroascorbic acid	mg	15.9	4.27–22.3	2, 3
Vitamin E				
α-tocopherol	mg	0.44	0.42–0.46	3
β-tocopherol	mg	0.03	0.02 − 0.01	3
γ-tocopherol	mg	0.07	0.07–0.07	3
δ-tocopherol	mg	nd	–	3
Organic acids				
Oxalic acid	mg	523	57.7–730	2, 3
Glutamic acid	mg	nd	–	3
Malic acid	mg	17.5	3.21–34.3	2, 3
Citric acid	mg	79.5	11.0–121	2, 3
Fumaric acid	mg	50.0	Traces–100	2, 3
Phenolics (total)	mg	73.4[a]	68.1–78.8	3
Flavonoids	mg	26.1[b]	25.2–26.9	3

1 Morales et al.(2015) Spain (4 samples), *2* Sánchez-Mata et al. (2012) Spain (2 samples), *3* Morales et al. (2014) Spain (4 samples)

nd non-detected values

[a] Expressed as gallic acid

[b] Expressed as catechin

Table 13.93 Fatty acids profile (% of total fatty acids) of fresh tender leaves of *R. pulcher*

Individual compounds	Average	Range	References
12:0	0.23	0.21–0.25	1
14:0	0.39	0.38–0.40	1
16:0	9.30	9.19–9.41	1
16:1	0.21	0.20–0.22	1
18:0	1.73	1.70–1.76	1
18:1*n*−9	4.22	4.21–4.23	1
18:2*n*−6	17.0	16.9–17.2	1
18:3*n*−3	62.9	62.9–63.0	1
20:0	0.44	0.37–0.51	1
20:1*n*−9	0.14	0.13–0.15	1
22:0	0.91	0.84–0.98	1
24:0	0.90	0.84–0.96	1
24:1*n*−9	0.06	0.03–0.09	1
Categories (calculated values)			
SFA	14.1	13.8–14.4	–
MUFA	4.8	4.7–4.8	–
PUFA	81.1	80.8–81.5	–
n−3	63.9	63.5–64.3	–
n−6	17.3	17.2–17.3	–
n−9	4.5	4.5–4.6	–

1 Morales et al. (2012b) Spain (4 samples)
SFA saturated fatty acids, *MUFA* monounsaturated fatty acids, *PUFA* polyunsaturated fatty acids

Remarks (tender leaves of *R. pulcher*)

- Good source of dietary fibre (>3 g/100 g).
- Good *n*−3/*n*−6 ratio (>3).
- High proportion of PUFA and *n* − 3 fatty acids content, compared to other vegetables.
- It can be considered as a source of K and vitamins B_9 (folates) and C.
- Very high content of oxalates: people with altered renal function should avoid this vegetable; boiling is recommended for general population.

Table 13.94 Main constituents of 100 g of boiled of tender leaves of *R. pulcher*

	Units	Average	Range	References
Energy (calculated value)	kcal	18	17–19	
Moisture	g	90.9	90.0–91.8	1
Available carbohydrates	g	2.00	1.90–2.10	1
Dietary fibre	g	4.40	4.30–4.50	1
Proteins	g	0.30	0.30–0.30	1
Lipids	g	0.03	0.03–0.03	1
Ash	g	0.80	0.80–0.80	1
K	mg	206	186–226	1
Na	mg	29.1	28.9–29.3	1
Ca	mg	107	97.4–115.8	1
Mg	mg	26.9	25.1–28.7	1
Fe	mg	1.10	0.92–1.28	1
Cu	µg	80	70–90	1
Mn	µg	200	180–220	1
Zn	µg	180	150–210	1

1 García-Herrera (2014) Spain

Table 13.95 Vitamins and other constituents, per 100 g of boiled of tender leaves of *R. pulcher*

	Units	Average	Range	References
Vitamin C	mg	21.6	20.7–22.47	1
Ascorbic acid	mg	20.4	19.7–21.03	1
Dehydroascorbic acid	mg	1.25	0.95–1.55	1
Vitamin B_9 (total folates)	µg	191	178–204	1
Organic acids				
Oxalic acid	mg	290	288–292	1
Glutamic acid	mg	nd	–	1
Malic acid	mg	nd	–	1
Citric acid	mg	30	30–30	1
Fumaric acid	mg	Traces	–	1

1 Morales (2011) Spain; *nd* non-detected values

Remarks (tender leaves of *R. pulcher*)

- Very low energy value (<25 kcal/100 g).
- Very low Na amount (<40 mg/100 g).
- Still good source of dietary fibre and vitamin C after boiling.
- Boiling reduces the high level of oxalic acid.

13.3.29 Scolymus hispanicus L. (Asteraceae)

Common Names Golden thistle, *cangarinha* (pt), *cardillo* (es), *chardon d'Espagne* (fr), *cardoncello silvatico* (it), *ascolimpros* (gr), *şevketi bostan* (tr), *garnina* (ar).

Description Prickly perennial herb, up to 80 cm, that has a rosette of basal leaves deeply divided, with spiny margins, pale or red-coloured veins, and a thickened midrib. The stems have prickly dentate wings and rigid leaves. The flower heads (capitula) appear in the axils of the upper leaves, surrounded by an involucre of bracts ending in a sharp point. All the flowers are ligulate, golden-yellow. The fruits (achenes) are 2–3 mm length, with a pappus of hairs forming a short corona.

Ecology and Distribution It grows in uncultivated agricultural fields, roadsides, waste places, and nitrified lands, on almost all types of soils, the bigger specimens obtained on sandy soils with some moisture. It has a circum-Mediterranean distribution, extending also to northwestern France and the Canary Islands.

Food Uses Different parts of this plant have been traditionally consumed, especially the midribs of the tender basal leaves and the outer part of the root. The whitish and fleshy midribs are eaten either raw as a snack, raw in salads, or much more frequently cooked, in most of the Mediterranean countries, such as Portugal (Barão and Soveral 2010; Mendonça de Carvalho 2006), Spain (e.g. Cobo and Tijera 2011; Tardío et al. 2006), Italy (e.g. Guarrera 2006; Nebel et al. 2006), Greece (Hadjichambis et al. 2008; Leonti et al. 2006), Cyprus (Della et al. 2006), Turkey (Dogan 2012; Ertuğ 2014), Tunisia (Le Floc'h 1983), and Morocco (Nassif and Tanji 2013). It is one of the most appreciated wild vegetable in many of these countries, being even considered as a part of the Mediterranean culinary heritage (e.g. Picchi and Pieroni 2005). Bundles of peeled basal leaves of this thistle can be found in local markets throughout the Mediterranean basin. It is occasionally cultivated even nowadays, at least in the south of Spain and Italy (COAGRICO 2014; Laghetti 2009).

The process of peeling the leaves, that is, the elimination of their greenish and spiny parts, is usually carried out by hand, with a scraping movement of three fingers of one hand from the base to the apex of each leaf, while the other hand holds the bottom of the rosette. If done with determination and with a little practice, punctures are minimal (Tardío et al. 2002). The midribs are washed, sliced, and usually boiled in water with salt, being next prepared in different ways, such as sautéed with garlic and ham, fried in batter, with scrambled eggs or in omelettes, for stuffing pies like the South-Italian *verdhët*, etc. (Guarrera 2006; Hadjichambis et al. 2008; Picchi and Pieroni 2005; Tardío 2010). They have been also commonly used as a vegetable garnish for other dishes, like the Spanish *cocido* (Tardío et al. 2006) or the North African couscous (Le Floc'h 1983).

The consumption of the root, mentioned by Theophrastus in third century BC, is very rare nowadays in Spain, with only two recent ethnobotanical references (Tejerina 2010; Velasco et al. 2010). Neither it seems to be a common use in Italy, Turkey, and Greece, being almost reduced to the Greek island of Crete (Dogan 2012; Guarrera 2006; Hadjichambis et al. 2008), where the outer part of the root is consumed in a typical Crete recipe with lamb (Dogan 2012).

Other food uses have been described for the flowers of this species. The traditional use of the flowers as a saffron substitute, documented by Clusius in the sixteenth century in Spain (Clusius 1576), only has been recently registered in the province of Madrid (Aceituno-Mata 2010; Tardío et al. 2005), though completely abandoned. The flowers were also employed as preservative, to curdle milk for making cheese or curd in several parts of Spain (Aceituno-Mata 2010; Criado et al. 2008; Galán 1993; Pellicer 2001; Perera López 2005) and Italy (Picchi and Pieroni 2005). Besides curdling the milk, the dry flowers give the cheese a characteristic yellowish colour (Aceituno-Mata 2010).

Other Uses The infusion or decoction of the flowers to treat diarrhoea and other digestive problems have been widespread in Spain, both for humans or animals (e.g. Aceituno-Mata 2010; Benítez et al. 2010; Blanco and Cuadrado 2000; Velasco et al. 2010). The use of this plant against fever has been mentioned in Italy (Guarrera 2006) and Spain, where a decoction of the flowering plant was employed against malt fever (Benítez et al. 2010).

There are also many references of the use of this species as animal food (e.g. Guarrera 2006; Tejerina 2010; Velasco et al. 2010), being their leaves, as those of other thistles, highly prized by donkeys and also eaten by goats and sheep. The cooked basal leaves or the residuals from peeling them were also used to feed pigs (e.g. Aceituno-Mata 2010; Velasco et al. 2010).

Historical References The oldest known references are those of Theophrastus (371–287 BC) and Pliny the Elder (23–79 AD). Theophrastus told us that its root is eaten both raw and cooked, being juicier when the plant is flowering (Teofrasto 1988). Pliny the Elder mentioned that it is eaten in the East and used by the poor, and finally that it is said to possess diuretic properties in a very high degree (Bostock and Riley 1855). Clusius (1576) mentioned the edible use of the young plants with their root, either raw or in meat stews in the Spanish province of Salamanca.

Food Composition Tables for raw and peeled basal leaves of *Scolymus hispanicus* (Tables 13.96, 13.97 and 13.98).

Table 13.96 Main constituents, per 100 g of fresh midribs of *S. hispanicus*

	Units	Average	Range	References
Energy (calculated value)	kcal	38	13–89	–
Moisture	g	84.1	79.2–92.7	1, 2, 3
Available carbohydrates	g	3.40	1.10–9.20	2
Dietary fibre	g	7.00	3.10–12.30	2
Proteins	g	1.70	0.30–5.20	2
Lipids	g	0.37	0.08–0.74	2, 4
Ash	g	3.20	1.70–5.20	2
K	mg	1040	559–1772	2
Na	mg	39.1	11.2–65.3	2
Ca	mg	235	124–410	2
Mg	mg	93.9	17.5–210	2
Fe	mg	2.36	1.39–3.11	2
Cu	µg	90.0	50.0–130	2
Mn	µg	370	160–570	2
Zn	µg	500	360–920	2

1 Sánchez-Mata et al. (2012) Spain (2 samples), *2* García-Herrera et al. (2014b) Spain (4 samples), *3* Morales et al. (2014) Spain (4 samples), *4* Morales et al. (2012b) Spain (4 samples)

Table 13.97 Vitamins and other constituents, per 100 g of fresh midribs of *S. hispanicus*

	Units	Average	Range	References
Carotenoids				
β-carotene	mg	0.10	–	1
Lutein	mg	0.33	–	1
RAE (calculated value)	µg	8.08	–	–
Vitamin B_9 (total folates)	µg	103	96.2–110	2
Vitamin C	mg	12.6	0.91–22.0	1, 3, 4
Ascorbic acid	mg	1.11	Traces–22.0	3, 4
Dehydroascorbic acid	mg	2.73	0.91–3.24	3, 4
Vitamin E				
α-tocopherol	mg	0.04	0.02–0.06	1, 4
β-tocopherol	mg	0.02	0.02–0.06	4
γ-tocopherol	mg	0.01	0.01–0.02	1, 4
δ-tocopherol	mg	Traces	–	4
Vitamin K	µg	38.0	–	1
Organic acids				
Oxalic acid	mg	485	14.6–955	3, 4
Glutamic acid	mg	nd	–	4
Malic acid	mg	54.9	26.9–93.2	3, 4
Citric acid	mg	31.3	3.16–64.0	3, 4
Fumaric acid	mg	1.29	nd–2.33	3, 4
Phenolics (total)	mg	38.7[a]	20.0–56.0	1, 4
Flavonoids	mg	8.39[b]	7.27–9.51	4

1 Vardavas et al. (2006b) Crete, *2* Morales et al. (2015) Spain (4 samples), *3* Sánchez-Mata et al. (2012) Spain (2 samples), *4* Morales et al. (2014) Spain (4 samples)
RAE retinol activity equivalents; *nd* non-detected values
[a] Expressed as gallic acid
[b] Expressed as catechin

Table 13.98 Fatty acids profile (% of total fatty acids) of fresh midribs of *S. hispanicus*

Individual compounds	Average	Range	References
12:0	0.24	0.23–0.25	1
14:0	0.89	0.79–1.00	1, 2
16:0	22.9	19.8–25.2	1, 2
16:1	1.63	1.28–1.83	1, 2
18:0	3.24	2.58–4.04	1, 2
18:1*n*−9	7.50	6.34–8.58	1, 2
18:2*n*−6	27.3	26.2–28.2	1, 2
18:3*n*−3	28.6	26.6–30.5	1, 2
20:0	1.54	1.34–1.82	1, 2
20:2	0.21	0.19–0.23	1
22:0	1.53	0.75–2.44	1, 2
24:0	1.07	0.42–1.80	1, 2
24:1*n*−9	0.27	0.10–0.33	1, 2
Categories (calculated values)			
SFA	32.5	32.4–32.5	–
MUFA	9.9	8.5–11.3	–
PUFA	57.7	56.3–59.1	–
n−3	29.5	27.4–31.7	–
n−6	28.2	26.9–28.9	–
n−9	7.9	7.0–9.4	–

1 Morales et al. (2012b) Spain (4 samples), *2* Vardavas et al. (2006a) Crete
SFA saturated fatty acids, *MUFA* monounsaturated fatty acids, *PUFA* polyunsaturated fatty acids

Remarks (midribs of *S. hispanicus*)

- It can reach high available carbohydrates content.
- Good source of dietary fibre (>3 g/100 g).
- High ash level, with low Na content (<120 mg/100 g).
- It can be considered as a source of K, Ca, and vitamin B_9 (folates); sometimes also a source of Mg, Fe, and Mn.
- High content of oxalates: People with altered renal function should avoid this vegetable; boiling is recommended for general population.

13.3.30 Silene vulgaris (Moench.) Garcke (Caryophyllaceae)

Common Names Bladder campion, *colleja* (es), *silène commun* (fr), *strigoli* (it), *stroufouli* (el), *ecibücü* (tr).

Description Perennial herbaceous plant, up to 80 cm tall, with woody rhizomes that sometimes produce a dense turf of stems. The leaves, always sessile and opposed, are bluish-green, glabrous, and lanceolate with tiny white teeth on edge. The flowers have white or pale pink petals with a slit in the end and a characteristic inflated calyx that persists in fruiting. Fruit in capsule with many kidney-shaped tiny seeds (1–2 mm).
It is a very plastic species, with several subspecies described, but the most common and frequently used is subsp. *vulgaris*.

Ecology and Distribution It can be found in humanized environments, such as roadsides, arable lands, grassy slopes, and even waste places, being much more frequent in lime soil. Native to most of Europe, especially in the Mediterranean region, nowadays it has been introduced in many other regions of the world, being even considered as a weed.

Food Uses The tender leaves and stems of this species have been widely consumed, either raw in salads or more frequently cooked, in many Mediterranean countries, such as Spain (e.g. Blanco and Cuadrado 2000; Fajardo et al. 2000; Mesa 1996; Tardío et al. 2006; Velasco et al. 2010), France (Couplan 1989a; Couplan 1989b; Marco et al. 2003), Italy (e.g. Ghirardini et al. 2007; Guarrera 2006; Laghetti et al. 1994; Picchi and Pieroni 2005), Slovenia (Cerne 1992), Croatia (Łuczaj et al. 2013b), Bosnia-Herzegovina (Redžić 2006), Greece (Hadjichambis et al. 2008; Leonti et al. 2006; Simopoulos 2004), Cyprus (Della et al. 2006), Turkey (Dogan 2012; Ertuğ 2014), Lebanon (Marouf 2005), Egypt (Hadjichambis et al. 2008), and Morocco (Hadjichambis et al. 2008; Nassif and Tanji 2013). Though not widely, it was also used until recent times in other northern countries, such as in Poland, but abandoned nowadays (Łuczaj 2010).

It has a soft flavour and can be consumed raw in salads. However, it is more frequently eaten after cooking, either as a main vegetable in omelettes and as a stuffing for different kind of pies or as an ingredient of other recipes, such as *arroz con collejas* (rice with bladder campion) or *potaje*, a typical Spanish dish with stewed chickpeas, beans, rice, and cod, often consumed during Lent. In several regions of Italy, it is also used in the elaboration of archaic recipes, such as the *pistic* of Friuli, the *preboggion* of Liguria, and the *minestrella di Gallicano* of Toscana, that are composed of a large number of wild vegetables (Picchi and Pieroni 2005). Other recipes with mixes of wild vegetables that include this species have also been registered in other Mediterranean countries, such as *ensalada de matas* and *herbes bullides* in Spain (Tardío 2010) or *beqoul* in Morocco (Nassif and Tanji 2013).

Other Uses This plant was well known by the children of rural areas of Spain and Italy because they used to enjoy playing with the characteristic inflated calyxes, blowing them up on their hands or foreheads (e.g. Blanco and Cuadrado 2000; Laghetti et al. 1994; Picchi and Pieroni 2005; Velasco et al. 2010). Though not widely extended, some applications in popular medicine have been described in Spain and Italy, especially to aid in digestive disorders (González-Tejero 1989; Martínez-Lirola et al. 1997; Muntané 1991; Picchi and Pieroni 2005).

Historical References This species might be consumed since ancient times, but the earliest references we have found are from the seventeenth century in Spain (Cienfuegos 1627) and the eighteenth century in France (Garidel 1715, cited in Marco et al. 2003) and England (Bryant 1783). Cienfuegos, in the second volume (mss 3358: 521) of his manuscript work conserved at the Spanish National Library, mentioned its consumption together with other wild vegetables, such as *Allium ampeloprasum* L., *Rumex pulcher* L., and *Anchusa azurea* Mill., particularly important in times of scarcity. Bryant, in his *Flora Diaetetica* (1783), mentioned this species as one of the better ones, when boiled, for salads, and even proposed that it should be cultivated.

Food Composition Tables for young leaves and stems of *Silene vulgaris*, raw (Tables 13.99, 13.100 and 13.101) and boiled (Tables 13.102 and 13.103).

Table 13.99 Main constituents, per 100 g of fresh young leaves and stems of *Silene vulgaris*

	Units	Average	Range	References
Energy (calculated value)	kcal	34	17–56	–
Moisture	g	85.9	86.6–88.5	1, 2, 3
Available carbohydrates	g	2.32	1.03–3.90	1, 3
Dietary fibre	g	4.36	2.60–6.63	1, 3
Proteins	g	2.47	1.31–3.60	1, 3, 4
Lipids	g	0.67	0.31–1.31	1, 3, 5
Ash	g	1.53	0.20–4.33	1, 3
K	mg	601	410–1583	3, 5, 6, 7
Na	mg	22.4	5.90–114	3, 6, 7
Ca	mg	160	70.7–254	3, 5, 6
Mg	mg	50.4	24.2–109	3, 5, 6
P	mg	44.2	28.0–60.5	5, 6
Fe	mg	1.93	0.21–3.49	3, 5, 6
Cu	μg	144	10.0–367	3, 6
Mn	μg	709	540–1010	3, 5, 6
Zn	μg	408	110–538	3, 5, 6

1 Alarcón et al. (2006) Spain (10 samples), *2* Sánchez-Mata et al. (2012) Spain (2 samples), *3* García-Herrera (2014) Spain (5 samples), *4* Ayan et al. (2006) Turkey, *5* Morales et al. (2012b) Spain (4 samples), *6* Zeghichi et al. (2003) Crete, *7* Egea-Gilabert et al. (2013) Spain (14 samples)

Table 13.100 Vitamins and other constituents, per 100 g of fresh young leaves and stems of *S. vulgaris*

	Units	Average	Range	References
Carotenoids				
β-carotene	mg	1.03	–	1
Lutein	mg	2.01	–	1
RAE (calculated value)	μg	85.7	–	–
Vitamin B_9 (total folates)	μg	267	214–320	2
Vitamin C	mg	25.5	10.8–27.7	1, 3, 4
Ascorbic acid	mg	17.1	8.4–27.7	1, 3, 4
Dehydroascorbic acid	mg	7.34	nd–13.9	3, 4
Vitamin E				
α-tocopherol	mg	10.1	8.96–11.3	1, 5, 6
β-tocopherol	mg	0.31	0.26–0.36	6
γ-tocopherol	mg	0.40	0.09–0.76	1, 6
δ-tocopherol	mg	0.51	0.51–0.51	6
Vitamin K	μg	172	–	1
Organic acids				
Oxalic acid	mg	390	20.5–800	3, 4, 7
Glutamic acid	mg	nd	–	3
Malic acid	mg	25.7	nd–100	3, 4
Citric acid	mg	41.8	nd–130	3, 4
Fumaric acid	mg	1.13	nd–1.57	3, 4
Succinic acid	mg	nd	–	3

Table 13.100 (continued)

	Units	Average	Range	References
Phenolics (total)	mg	61.9[a]	26.7–117.4	1, 5, 6
	mg	68.7[b]	12.7–124	7, 8
	mg	≈88.6[c]	–	9
Hydroxycinnamic acids	mg	230[b]	–	10
Flavonoids	mg	21.6[d]	16.1–27.1	6
Anthocyanins	mg	1.12[e]	–	10
Nitrate	mg	178	84.0–357	5, 7

1 Vardavas et al. (2006b) Crete, *2* Morales et al. (2015) Spain (5 samples), *3* Morales (2011) Spain (5 samples), *4* Sánchez-Mata et al. (2012) Spain (2 samples), *5* Zeghichi et al. (2003) Crete, *6* Morales et al. (2012a) Spain (4 samples), *7* Egea-Gilabert et al. (2013) Spain (14 samples), *8* Conforti et al. (2011) Italy, *9* Gatto et al. (2011) Italy, *10* Vanzani et al. (2011) Italy

RAE retinol activity equivalents; *nd* non-detected values

[a] Expressed as gallic acid

[b] Expressed as chlorogenic acid

[c] Expressed as sum of individual compounds (apigenin derivatives as major component)

[d] Expressed as catechin

[e] Expressed as cyanidin-3-glucoside

Table 13.101 Fatty acids profile (% of total fatty acids) of fresh young leaves and stems of *S. vulgaris*

Individual compounds	Average	Range	References
12:0	0.44	0.30–0.58	1
14:0	1.41	0.92–1.78	1, 2
16:0	16.4	10.8–21.7	1, 2, 3
16:1	0.34	0.32–0.36	1
18:0	1.06	0.2–1.95	1, 2, 3
18:1*n*−9	2.19	1.22–2.84	1, 2, 3
18:2*n*−6	18.6	0.88–26.8	1, 2, 3
18:3*n*−3	48.1	39.6–56.9	1, 2, 3
18:4*n*−3	0.39	–	2
20:0	2.17	1.56–2.84	1
20:1*n*−9	0.06	0.04–0.08	1
22:0	1.03	1.02–1.04	1
22:1*n*−9	2.65	0.31–5.51	1, 3
24:0	2.06	1.61–2.39	1, 2
Categories (calculated values)			
SFA	28.7	15.1–29.7	–
MUFA	2.4	1.3–7.8	–
PUFA	68.9	68.7–77.1	–
n−3	43.5	41.0–54.1	–
n−6	25.4	23.0–27.8	–
n−9	2.2	1.3–7.8	–

1 Morales et al. (2012b) Spain (4 samples), *2* Vardavas et al. (2006a) Crete, *3* Alarcón et al. (2006) Spain (10 samples)

SFA saturated fatty acids, *MUFA* monounsaturated fatty acids, *PUFA* polyunsaturated fatty acids

Remarks (young leaves and stems of *S. vulgaris*)

- In some cases, it may represent considered as a source of dietary fibre (>3 g/100 g).
- Low Na content (<120 mg/100 g); it can be considered as a source of K, Mn, and vitamins K and B_9 (folates); sometimes also a source of Ca as well as vitamins C and E.
- High content of nitrate; not recommended for infants.
- High content of oxalates: People with altered renal function should avoid this vegetable; boiling is recommended for general population.

Table 13.102 Main constituents of 100 g of boiled young leaves and stems of *S. vulgaris*

	Units	Average	Range	References
Energy (calculated value)	kcal	21	21–21	–
Moisture	g	92.1	91.3–92.5	1, 2
Available carbohydrates	g	1.12	1.10–1.13	2
Dietary fibre	g	3.32	3.30–3.34	2
Proteins	g	2.00	2.00–2.00	2
Lipids	g	0.22	0.20–0.24	2
Ash	g	0.80	0.80–0.80	2
K	mg	226	223–230	2
Na	mg	11.8	9.4–13.3	2
Ca	mg	67.9	62.5–73.3	2
Mg	mg	21.1	19.9–21.3	2
Fe	mg	0.47	0.45–0.49	2
Cu	μg	60	40–80	2
Mn	μg	600	590–610	2
Zn	μg	200	190–210	2

1 Morales (2011) Spain, *2* García-Herrera (2014) Spain

Table 13.103 Vitamins and other constituents, per 100 g of boiled young leaves and stems of *S. vulgaris*

	Units	Average	Range	References
Vitamin B_9 (total folates)	μg	153	133–173	1
Vitamin C	mg	13.9	12.9–14.9	1
Ascorbic acid	mg	13.9	12.9–14.9	1
Dehydroascorbic acid	mg	nd	–	1
Organic acids				
Oxalic acid	mg	450	420–480	1
Glutamic acid	mg	nd	–	1
Malic acid	mg	nd	–	1
Citric acid	mg	60	40–80	1
Fumaric acid	mg	Traces	–	1

1 Morales (2011) Spain; *nd* non-detected values

Remarks (boiled young leaves and stems of *S. vulgaris*)

- Very low lipid content (<0.5 g/100 g), very low energy value (<25 kcal/100 g).
- Very low Na content (<40 mg/100 g).
- Still a source of dietary fibre (>3 g/100 g), Mn, vitamins B_9 and C after boiling, but it also keeps high oxalic acid content.

13.3.31 Silybum marianum (L.) Gaertner (Asteraceae)

Common Names Milk thistle, *cardo borriquero* (es), *chardon Marie* (fr), *cardo mariano* (it), *nerokavlos* (el), *devedikeni* (tr), *shouk el-diman* (ar).

Description Annual or biennial plant, depending on the climatic conditions, growing to more than 1.5 m. Large and thorny basal leaves, dark green usually marbled or streaked with characteristic white veins. Solitary and densely spiny heads located at the end of the stems, with purple and tubular flowers. The fruits (achenes) are black and smooth with a white pappus of simple hairs.

Ecology and Distribution It grows in fertile and well-drained soils, especially in disturbed areas. Native to the Mediterranean regions of Europe, North Africa, and the Middle East, it is also naturalized in other parts of the world.

Food Uses The consumption of several parts of this plant have been referred in many Mediterranean countries, such as Spain (e.g. Benítez 2009; Fajardo 2008; Tardío et al. 2006), Italy (Guarrera 2006; Lentini and Venza 2007; Picchi and Pieroni 2005), Greece (Couplan 2009b), Cyprus (Della et al. 2006), Turkey (Ertuğ 2014), Jordan (Tukan et al. 1998), Palestine (Ali-Shtayeh et al. 2008), Tunisia (Le Floc'h 1983), and Morocco (Nassif and Tanji 2013).
As a vegetable, one of the parts most widely used are the midribs of the basal leaves that must be collected before the flowering shoot begins to grow. The leaves have to be peeled with a knife and a good pair of gloves to remove the leaf blades with their very sharp thorns. Though sometimes eaten raw in salads (Akerreta 2009; Fernández Ocaña 2000; Guarrera 2006), they are usually boiled in water and then consumed in different manners: dressed with olive oil (e.g. Della et al. 2006; Fajardo 2008; Picchi and Pieroni 2005; Tejerina 2010), fried (Della et al. 2006; Pellicer 2004b; Picchi and Pieroni 2005), battered (Lentini and Venza 2007; Triano et al. 1998), included in soups (Ali-Shtayeh et al. 2008; Guarrera 2006) or in different stews with legumes (e.g. Fajardo et al. 2007; Tardío et al. 2002; Tukan et al. 1998; Verde et al. 1998). The tender sprouts of the main stem and the portion of the stem near the inflorescence, collected before or during the flowering period and with outer parts removed, are chiefly eaten raw as a snack, directly in the field (e.g. Ali-

Shtayeh et al. 2008; Couplan 2009b; Lentini and Venza 2007; Tardío et al. 2002) or even in salads (Guarrera 2006; Tardío et al. 2002). In addition, the immature and thorny inflorescences, after carefully removing their hardy thorns, were eaten as wild artichokes. They were mainly eaten raw by children (Mesa 1996; Rabal 2000; Tardío et al. 2002) and shepherds (Nassif and Tanji 2013), and, sometimes, also in salads (Tardío et al. 2002), cooked (Guarrera 2006; Lentini and Venza 2007), or in brine (Guarrera 2006). This wild vegetable was very well valued, especially when collected tender enough. It has a good flavour that resembles that of the artichokes. In Spain, it was often used in the past, especially in times of shortage after the Spanish Civil War, in the 1940s. Nowadays its consumption is almost abandoned.

The mature fruits, gathered in the summer, were also traditionally consumed, either raw as a snack in central Spain (Criado et al. 2008; Tardío et al. 2002), or roasted, in Jordan (Tukan et al. 1998). They are today used to produce silymarin (for medicinal purposes) and, as a by-product, an edible oil of good quality (Le Floc'h 1983) and rich in vitamin E (El-Mallah et al. 2003).

Finally, its use as a preservative has also been described, with the flowers employed in several Spanish regions to curdle milk for making cheese or curd (e.g. Ferrández and Sanz 1993; Moll 2005; Parada 2008).

Other Uses Its medicinal use can be found in many different ethnobotanical references from the Mediterranean countries. This species has been frequently considered a food-medicine, since its cooked or raw consumption is considered to have properties as laxative (Guarrera 2006), against liver problems and hypertension (Akerreta 2009), or blood circulation (Verde et al. 2003). However, the majority of the works referred the use of the fruits, mainly in decoction, against liver diseases, mentioned as choleretic and cholagogue (Akerreta 2009; Benítez 2009; Le Floc'h 1983; Picchi and Pieroni 2005). The decoction of the inflorescences has also been employed against diarrhoea (Aceituno-Mata 2010; Parada et al. 2002), Malta fever, and liver protection (Benítez 2009).

Most ethnobotanical references report its use as animal food, at least in Italy (Guarrera 2006) and Spain (e.g. Criado et al. 2008; Moll 2005; Pellicer 2004b). In Spain, it is well known that donkeys are very fond of this plant (Criado et al. 2008), being one of its Spanish name *cardo borriquero* (ass thistle). It has been also used to feed pigs (Aceituno-Mata 2010; Ferrández and Sanz 1993) and the fruits for cage birds (Benítez 2009; Sánchez López et al. 1994).

Historical References Although this species has been probably consumed since ancient times, a few clear references of it can be found in historical sources, surely duc to thc unclcar dcscriptions of these works. Dioscorides (first century AD) probably mentioned it when describing a broad and prickled plant, named *Silybo*, which was eaten newly sprung-up boiled with oil and salt (Osbaldeston 2000). Curiously, he did not mention its medicinal properties—only the emetic properties of the juice of the milk of thistle root.

Food Composition Tables for raw and peeled basal leaves of *Silybum marianum* (Tables 13.104, 13.105 and 13.106).

Table 13.104 Main constituents, per 100 g of fresh peeled basal leaves of *S. marianum*

	Units	Average	Range	References
Energy (calculated value)	kcal	13	9–18	–
Moisture	g	95.1	92.9–96.8	1, 2, 3, 4
Available carbohydrates	g	1.10	0.50–1.70	3
Dietary fibre	g	2.50	2.30–2.90	3
Proteins	g	0.60	0.50–0.80	3
Lipids	g	0.13	Traces–0.26	3, 5
Ash	g	1.50	1.00–1.90	3
K	mg	718	432–1300	3
Na	mg	80.9	24.7–128	3
Ca	mg	132	42.0–171	3
Mg	mg	17.3	10.3–22.6	3
Fe	mg	0.50	0.40–0.50	3
Cu	μg	80.5	10.0–170	3
Mn	μg	100	30–210	3
Zn	μg	260	210–350	3

1 Bianco et al. (1998) Italy (2 samples), *2* Sánchez-Mata et al. (2012) Spain (2 samples), *3* García-Herrera et al. (2014b) Spain (4 samples), *4* Morales et al. (2014) Spain (4 samples), *5* Morales et al. (2012b) Spain (4 samples)

Table 13.105 Vitamins and other constituents, per 100 g of fresh peeled basal leaves of *S. marianum*

	Units	Average	Range	References
Vitamin B_9 (total folates)	μg	41.7	11.7–71.6	1
Vitamin C	mg	2.74	1.87–3.32	2, 3
Ascorbic acid	mg	0.36	nd–1.04	2, 3
Dehydroascorbic acid	mg	2.39	1.80–3.05	2, 3
Vitamin E				
α-tocopherol	mg	0.04	0.04–0.04	3
β-tocopherol	mg	Traces	–	3
γ-tocopherol	mg	0.01	0.01–0.01	3
δ-tocopherol	mg	0.10	0.10–0.10	3
Organic acids				
Oxalic acid	mg	960	171–1889	2, 3
Glutamic acid	mg	nd	–	3
Malic acid	mg	2.08	nd–4.71	2, 3
Citric acid	mg	1.16	nd–2.09	2, 3
Fumaric acid	mg	68.0	0.90–220	2, 3
Phenolics (total)	mg	3.72[a]	3.36–4.08	3
Flavonoids	mg	1.13[b]	0.88–1.40	4
Nitrate	mg	113	–	1

1 Morales et al. (2015) Spain (4 samples), *2* Sánchez-Mata et al. (2012) Spain (2 samples), *3* Morales et al. (2014) Spain (4 samples), *4* Bianco et al. (1998) Italy (2 samples); *nd* non-detected values

[a] Expressed as gallic acid

[b] Expressed as catechin

Table 13.106 Fatty acids profile (% of total fatty acids) of fresh peeled basal leaves of *S. marianum*

Individual compounds	Average	Range	References
12:0	0.48	0.39–0.57	1
14:0	0.99	0.67–1.31	1
16:0	28.7	27.1–30.3	1
18:0	5.05	4.64–5.46	1
18:1$n-9$	3.86	3.76–3.96	1
18:2$n-6$	31.0	30.4–31.6	1
18:3$n-3$	21.6	21.3–21.8	1
22:0	2.43	1.16–3.70	1
24:0	3.36	3.06–3.66	1
Categories (calculated values)			
SFA	41.0	40.0–44.0	–
MUFA		3.9–4.1	–
PUFA	54.0	52.2–55.9	–
$n-3$	21.6	21.3–21.8	–
$n-6$	31.0	30.4–31.6	–
$n-9$	3.9	3.8–4.0	–

1 Morales et al. (2012b) Spain (4 samples)
SFA saturated fatty acids, *MUFA* monounsaturated fatty acids, *PUFA* polyunsaturated fatty acids

Remarks (peeled basal leaves of *S. marianum*)

- High moisture content. Very low lipid content (<0.5 g/100 g); very low energy value (<25 kcal/100 g).
- It can be considered as a source of K; sometimes also a source of Ca and vitamin B_9 (folates).
- Very high content of oxalates: People with altered renal function should avoid this vegetable; boiling is recommended for general population.

13.3.32 *Smilax aspera* L. (Smilacaceae)

Common Names Rough bindweed, *zarzaparrilla* (es), *salsepareille* (fr), *straccia-braghe* (it), *gıcırdikeni* (tr).

Description Perennial climbing herb, often woody below, dioecious. The stems are angled and more or less prickly, with alternate leaves; these are coriaceous and very variable in form, cordate, acute, ovate to sagittate, petiolate, with parallel veins and reticulate, usually with prickles in the petiole and principal veins. The little 6 tepals flowers grow in axillary or terminal umbels. The fruit is a round berry, 7 mm in diameter, red or black, with 3 seeds.

Ecology and Distribution It grows climbing in forests and in its substitution scrublands, or in hedges. Its range of distribution covers all of the south of Europe, north of Africa, the Macaronesian region, and also in west and south of Asia.

Food Uses The young tender shoots have been traditionally consumed raw or cooked in the Mediterranean countries, at least in Spain (Cobo and Tijera 2011; Pellicer 2001), Italy (Guarrera 2006; Lentini and Venza 2007; Picchi and Pieroni 2005), Slovenia (Cerne 1992), Bosnia-Herzegovina (Redžić 2006), Greece, Cyprus (Hadjichambis et al. 2008), and Turkey (Dogan 2012; Ertuğ 2014). As a vegetable, the tender shoots, sometimes with leaves, were eaten raw as a snack (Dogan 2012; Pellicer 2001), prepared in salads seasoned with oil and lemon (Lentini and Venza

2007), and more frequently cooked like asparagus (Cobo and Tijera 2011; Guarrera 2006; Hadjichambis et al. 2008; Redžić 2006), in soups (Cerne 1992), roasted with onions and mixed with eggs (Dogan et al. 2004; Ertuğ 2004), fried alone or with eggs (Hadjichambis et al. 2008), or cooked with rice (Dogan 2012).

The shoots are also consumed as pickles (Dogan et al. 2004), sometimes after boiling in vinegar and preserving in oil (Guarrera 2006; Picchi and Pieroni 2005).

Similar to several American species of this genus, the European sarsaparilla has been used to elaborate beverages, at least in Spain. In some parts of southern Spain, its root was added (together with other ingredients) to a typical home-made liqueur known as *resol* (Arauzo et al. 2004; Benítez 2009; Sánchez-Romero 2003). As in other parts of the world, the refreshing drink sarsaparilla, made with carbonic water and the root extract of this species (and overall other American species) was very popular in some Spanish regions (Fernández Ocaña 2000; Guzmán 1997; Villar et al. 1987).

Other Uses As a medicinal plant it is considered a panacea, being used for a great number of diseases, at least in Italy and Spain (e.g. Benítez 2009; Guarrera 2006; Guzmán 1997). The leaves and especially the roots have been used mainly as digestive, diuretic, blood depurative, hypotensive, and against respiratory and skin diseases (Benítez 2009; Guarrera 2006; Guzmán 1997; Moll 2005; Pellicer 2001; Sánchez-Romero 2003; Villar et al. 1987).

It has also been used in veterinary medicine for healing wounds (Guarrera 2006). Their fruits (red) and roots (yellow) were used as dyer and also to make ink (Guarrera 2006).

Historical References Dioscorides (first century AD) described this species and mentioned that a decoction of the leaves and fruits is an antidote for deadly poisons, taken as a drink beforehand (or afterwards), being used in mixes for antipoison medicines (Laguna 1555). The more widespread use of the root of this species begun in the sixteenth century after the introduction in Europe of the Asian and American species of *Smilax* as a general tonic for treating several diseases, and especially as a cure for syphilis and rheumatism (Monardes 1574).

Food Composition Tables for raw tender shoots and leaves of *Smilax aspera* (Tables 13.107 and 13.108).

Table 13.107 Main constituents, per 100 g of fresh tender shoots and leaves of *S. aspera*

	Units	Average	Range	References
Energy (calculated value)	kcal	100	86–113	–
Moisture	g	64.1	61.9–66.4	1
Available carbohydrates	g	9.92	8.76–11.1	1
Dietary fibre	g	18.8	15.6–22.1	1
Proteins	g	3.50	3.00–3.99	1
Lipids	g	0.95	0.90–1.00	1
Ash	g	2.44	2.27–2.61	1
K	mg	≈180	161–199	2
Ca	mg	≈353	17.5–689[a]	2

Table 13.107 (continued)

	Units	Average	Range	References
Mg	mg	≈61.2	45.5–77.0	2
Fe	mg	≈1.66	1.01–2.31	2
Mn	μg	≈551	262–840	2
Zn	μg	≈2130	665–3605[a]	2

1 Cabiddu and Decandia (2000) Italy (8 samples), *2* Poschenrieder et al. (2012) Spain

[a] The highest values were found in plants gathered near an ancient lead/barite mine

Table 13.108 Vitamins and other constituents, per 100 g of fresh tender shoots and leaves of *S. aspera*

	Units	Average	Range	References
Vitamin E				
α-tocopherol	mg	≈29.1	–	1
β-tocopherol	mg	≈4.5	–	1

1 Demo et al. (1998) Greece

Remarks (tender shoots and leaves of *S. aspera*)

- Low moisture content; high lipids and available carbohydrate content.
- High in dietary fibre (>6 g/100 g).
- It may be a source of vitamin E.

13.3.33 Sonchus asper (L.) Hill (Asteraceae)

Common Names Prickly sow thistle, *serralha áspera* (pt), *cerraja* (es), *laiteron epineux* (fr), *grespino spinoso* (it), *tsochos* (el), *eşekgevreği* (tr), *tifaf* (ar).

Description Annual herb up to 2 m, with glabrous and spiny leaves. The young ones are pinnatifid with triangular lobes and form a basal rosette, the leaves of the flowering stem are auriculate, with rounded lobes that clasp the stem. The inflorescences (capitula) grow at the end of the stems, up to 1.5 cm in diameter, yellow and ligulate flowers. Achenes around 3 mm, compressed and usually winged, with a deciduous pappus.

Ecology and Distribution It lives around cultivated grounds and in waste places, in all of Europe, north of Africa, and west Asia, now naturalized in other parts of the world, being considered a weed in agricultural systems.

Food Uses This wild vegetable has been traditionally consumed in the Mediterranean countries, at least in Spain (Mulet 1991; Tardío et al. 2005; Tardío et al. 2006; Velasco et al. 2010; Verde et al. 2003), Italy (Guarrera 2006; Hadjichambis et al. 2008; Lentini and Venza 2007; Leonti et al. 2006; Nebel et al. 2006; Picchi and Pieroni 2005), Croatia (Łuczaj et al. 2013b), Greece (Leonti et al. 2006), Turkey (Dogan 2012; Ertuğ 2004, Ertuğ 2014), and Morocco (Nassif and Tanji 2013). The edible parts are the basal leaves, and sometimes, the tender stems, collected in spring (Ertuğ 2004; Guarrera 2006; Hadjichambis et al. 2008; Lentini and Venza 2007; Tardío et al. 2006), although in some areas only the peeled basal leaves are used (Hadjichambis et al. 2008; Nassif and Tanji 2013; Tardío et al. 2005).

They are consumed raw in salads (Dogan 2012; Guarrera 2006; Hadjichambis et al. 2008; Lentini and Venza 2007; Nassif and Tanji 2013; Tardío et al. 2005; Verde et al. 2003) and also cooked for preparing soups (Dogan 2012; Guarrera

2006; Lentini and Venza 2007), omelettes (Guarrera 2006; Verde et al. 2003), fried with onion and eggs (Dogan 2012), or used as pastry stuffing (Dogan 2012). It is sometimes boiled mixed with other herbs (Hadjichambis et al. 2008), as in some traditional recipes such as the Moroccan *beqoul* (mix of up to 20 wild vegetables), the Italian *pistic* (Guarrera 2006; Paoletti et al. 1995), and the *pastissets de brosses*, traditional vegetable pies from eastern Spain made with several *Sonchus* species (Mulet 1991).

Other Uses This species has also been employed as fodder for pigs, cows, sheep, goats, and rabbits (Guarrera 2006; Tardío et al. 2002; Velasco et al. 2010).

Some medicinal properties have been also reported. Its consumption raw in salads is considered to have depurative and diuretic effects (Guarrera 2006; Guzmán 1997), and it has been recommended in western Turkey attributing antitumoral activity (Polat and Satil 2012). Other medicinal uses recorded in the Italian folk medicine include its use to treat high blood pressure, sore throat, wounds, sores and boils (or furuncles), and scorpion or insect bites (Guarrera 2006).

Historical References The edibility of *Sonchus* species is probably known since antiquity, being already mentioned by Dioscorides in the first century, who recognized two kinds of *sónkhos*, one more rough and prickly, the other more tender and edible (Laguna 1555), probably *Sonchus asper* and *S. oleraceus*, respectively (Osbaldeston 2000).

Food Composition Tables for raw leaves of *Sonchus asper*, (Tables 13.109, 13.110 and 13.111).

Table 13.109 Main constituents, per 100 g of fresh tender leaves of *S. asper*

	Units	Average	Range	References
Energy (calculated value)	kcal	37	34–40	–
Moisture	g	86.4	85.3–87.5	1, 2
Available carbohydrates	g	3.25	2.93–3.57	2
Dietary fibre	g	1.98	1.87–2.09	2
Proteins	g	3.56	3.34–3.78	2
Lipids	g	0.68	0.61–0.75	2
Ash	g	3.04	2.74–3.34	2
K	mg	511	440–633	1, 2
Na	mg	137	85.0–211	1, 2
Ca	mg	137	80.1–175	1, 2
Mg	mg	26.4	17.9–39.9	1, 2
P	mg	49.3	39.4–59.2	2
Fe	mg	2.98	2.20–3.76	2
Cu	µg	310	240–380	2
Mn	µg	880	780–980	2
Zn	µg	900	830–970	2

1 Bianco et al. (1998) Italy (2 samples), *2* Guil-Guerrero et al. (1998a) Spain (5 samples)

Table 13.110 Vitamins and other constituents, per 100 g of fresh tender leaves of *S.* asper

	Units	Average	Range	References
Carotenoids (total)	mg	8.00[a]	5.90–10.1	1
Vitamin B_1	μg	≈18	4–32	2[g]
Vitamin B_2	μg	≈86	71–101	2[g]
Vitamin B_3 (niacin)	μg	≈6	Traces–14	2[g]
Vitamin C (ascorbic acid)	mg	62.8	53.8–71.8	1
Organic acids				
Oxalic acid	mg	61.0	nd–141	1,3
Phenolics (total)	mg	≈102[b]	100–103	4
	mg	106[c]	93.6–119	5[g]
	mg	308[d]	279–337	6
Flavonoids	mg	9.92[e]	8.91–10.9	5[g]
Flavonols	mg	9.21[e]	6.58–11.8	5[g]
Proanthocyanidins	mg	22.7[f]	21.2–24.2	5[g]
Nitrate	mg	72.6	–	3

1 Guil-Guerrero et al. (1998a) Spain (5 samples), *2* Hussain et al. (2010) Pakistan, *3* Bianco et al. (1998) Italy (2 samples), *4* Conforti et al. (2011) Italy, *5* Afolayan and Jimoh (2009), South Africa, *6* Gatto et al. (2011) Italy; *nd* non-detected values

[a] Expressed as β-carotene

[b] Expressed as chlorogenic acid

[c] Expressed as tannic acid

[d] Expressed as sum of different compounds (cichoric acid as major component)

[e] Expressed as quercetin

[f] Expressed as catechin

[g] Plants from non-Mediterranean origin

Table 13.111 Fatty acids profile (% of total fatty acids) of fresh tender leaves of *S.* asper

Individual compounds	Average	Range	References
14:0	3.44	–	1
16:0	15.3	–	1
16:1	0.99	–	1
18:0	2.01	–	1
18:1*n*−9	3.34	–	1
18:2*n*−6	9.9	–	1
18:3*n*−3	39.9	–	1
18:3*n*−6	0.47	–	1
18:4*n*−3	0.09	–	1
20:0	3.39	–	1
20:1*n*−9	nd	–	1
22:0	2.96	–	1
24:0	1.42	–	1
Categories (calculated values)			
SFA	34.3	–	–
MUFA	5.2	–	–

Table 13.111 (continued)

Individual compounds	Average	Range	References
PUFA	60.5	–	–
n−3	48.2	–	–
n−6	12.5	–	–
n−9	4.0	–	–

1 Guil-Guerrero et al. (1998a) Spain (5 samples); *nd* non-detected values
SFA saturated fatty acids, *MUFA* monounsaturated fatty acids, *PUFA* polyunsaturated fatty acids

Remarks (tender leaves of *S. asper*)

- Low energy value (<40 kcal/100 g).
- High ash level.
- It may be considered as a source of K, Fe, Cu, Mn, and vitamin C; often also a source of Ca.

13.3.34 Sonchus oleraceus L. (Asteraceae)

Common Names Sow thistle, *serralha* (pt), *cerraja* (es), *laiteron potager* (fr), *grespino comune* (it), *kostriš* (hr), *tsochos* (el), *kuzugevreği* (tr), *tifaf* (ar).

Description Annual herb, up to 1 m. Basal leaves are glabrous, pinnatisect to pinnatifid, spineless, with dentate lobes; the stem leaves are usually simple and amplexicaul. The inflorescences (capitula), up to 1.5 cm, grow at the end of the branches in relatively numerous groups. Flowers are yellow and ligulate. The fruits are achenes, around 3 mm, rugose, with a persistent pappus.

Ecology and Distribution It lives around cultivated fields, pastures, roadsides, and waste places. Native to all of Europe, north of Africa, and west Asia, but introduced as a weed in almost all the world.

Food Uses This species has been commonly employed as a wild green throughout the Mediterranean region, as recorded in Portugal (Mendonça de Carvalho 2006), Spain (e.g. Parada et al. 2011; Rivera et al. 2005; Tardío et al. 2006), Italy (e.g. Guarrera 2006; Nebel et al. 2006; Picchi and Pieroni 2005; Signorini et al. 2009), Slovenia (Cerne 1992), Croatia (Łuczaj et al. 2013b), Bosnia-Herzegovina (Redžić 2006), Greece (Hadjichambis et al. 2008; Leonti et al. 2006; Zeghichi et al. 2003), Cyprus (Della et al. 2006), Turkey (Dogan 2012; Ertuğ 2014), Jordan (Al-Qura'n 2010), Egypt (Hadjichambis et al. 2008), Tunisia (Le Floc'h 1983), and Morocco (Nassif and Tanji 2013).

Its young leaves and stems are gathered in springtime before blooming. They are consumed raw in salads (e.g. Al-Qura'n 2010; Della et al. 2006; Dogan 2012; Guarrera 2006; Mendonça de Carvalho 2006; Nassif and Tanji 2013; Parada et al. 2011; Pieroni et al. 2005; Tardío et al. 2006; Zeghichi et al. 2003), and cooked, both in soups (Guarrera 2006; Hadjichambis et al. 2008; Lentini and Venza 2007; Pieroni et al. 2005), with legumes (Mendonça de Carvalho 2006; Velasco et al. 2010), fried with eggs or in omelettes (Dogan 2012; Guarrera 2006; Hadjichambis et al. 2008; Lentini and Venza 2007; Verde et al. 2003), used as pastry stuffing (Dogan et al. 2004; Dogan 2012; Zeghichi et al. 2003), boiled and prepared as a salad (Dogan 2012; Guarrera 2006), or boiled and then fried with olive oil, garlic, and other ingredients (Pieroni et al. 2005). This species is commonly prepared in mixtures with other wild species, as recorded in different Mediterranean recipes. Some examples are the Italian *misticanza*, a salad with a mix of several raw vegetables, the *foie* and the *chòrta vramena*, mixtures of several wild herbs previously boiled (Guarrera 2006; Nebel et al. 2006), the *pastissets de brosses* and the *minxos*, in eastern Spain, traditional vegetable pies of wild herbs (Mulet 1991; Tardío 2010), or the Moroccan *beqoul*, a mixture of up to 20 wild food plants used to prepare a springtime meatless dish (Nassif and Tanji 2013). Although it is generally collected for domestic consumption, it can also be found in some wild vegetable mixes sold in the markets of southern Croatia (Łuczaj et al. 2013b).

Its flowers have also been employed to curdle milk, as reported in some Italian regions (Guarrera 2006).

Other Uses This species has been used for medicinal purposes in a lesser extent. Its leaves are considered refreshing and depurative (Oltra 1998; Rivera et al. 2005), and they were directly consumed or prepared in infusion as a liver protector (Guarrera 2006; Mendonça de Carvalho 2006), or against pyrosis (Rivera et al. 2005). Its latex was topically used to treat warts (Benítez et al. 2010; Guarrera 2006; Menendez-Baceta et al. 2014; Tardío et al. 2002), and the decoction of the whole plant against haemorrhoids (Benítez et al. 2010). Other medicinal uses are mainly related to intestinal and skin disorders (Ferrández and Sanz 1993; Guarrera 2006).

It has also been used as animal fodder, especially for rabbits, pigs, and hens, as recorded in Spain (Molina 2001; Tardío et al. 2002; Velasco et al. 2010) and Portugal (Carvalho 2010; Mendonça de Carvalho 2006).

Historical References The species of this genus has been probably consumed since ancient times. Dioscorides in the first century, mentioned two kinds of *sónkhos*, one more rough and prickly, the other more tender and edible (Laguna 1555), probably *Sonchus asper* and *S. oleraceus* (Osbaldeston 2000).

Food Composition Tables for raw leaves of *Sonchus oleraceus* (Tables 13.112, 13.113 and 13.114).

Table 13.112 Main constituents, per 100 g of fresh tender leaves of *S. oleraceus*

	Units	Average	Range	References
Energy (calculated value)	kcal	33	16–56	–
Moisture	g	87.6	83.0–91.9	1, 2, 3, 4, 5, 6
Available carbohydrates	g	2.29	0.94–4.20	3, 4, 5
Dietary fibre	g	3.37	2.60–5.57	3, 4, 5
Proteins	g	2.22	1.11–3.48	3, 4, 5, 7
Lipids	g	0.60	0.20–1.28	3, 4, 5, 7, 8
Ash	g	2.17	1.58–3.00	2, 3, 5, 7
K	mg	481	308–790	2, 3, 4, 5, 7, 9
Na	mg	144	43.0–270	2, 3, 4, 5, 9
Ca	mg	131	32.0–280	2, 3, 4, 5, 9
Mg	mg	43.3	17.4–230	2, 3, 4, 5, 7, 9
P	mg	60.4	35.0–85.9	7, 9
Fe	mg	2.85	0.57–5.62	2, 3, 4, 5, 7, 9
Cu	μg	249	20–580	3, 5, 9
Mn	μg	881	370–1269	3, 5, 9
Zn	μg	597	176–840	3, 4, 5, 7, 9

1 Guil-Guerrero et al. (1997a) Spain (5 samples), *2* Bianco et al. (1998) Italy (2 samples), *3* Guil-Guerrero et al. (1998a) Spain (5 samples), *4* Trichopoulou et al. (2000) Greece, *5* García Herrera et al. (2014b) Spain (5 samples), *6* Morales et al. (2014) Spain (5 samples), *7* Ayan et al. (2006) Turkey, *8* Guil-Guerrero et al (1996b) Spain (various samples), *9* Zeghichi et al. (2003) Crete

Table 13.113 Vitamins and other constituents, per 100 g of fresh tender leaves of *S. oleraceus*

	Units	Average	Range	References
Carotenoids (total)	mg	15.7[a]	12.1–18.4	1, 2, 3
β-carotene	mg	1.05	–	4
Lutein	mg	1.83	–	4
RAE (calculated value)	μg	87.5	–	–
Vitamin B_9 (total folates)	μg	85.8	70.2–101	5
Vitamin C	mg	38.7	10.1–86.0	1, 2, 3, 4, 6, 7
Ascorbic acid	mg	23.9	0.71–66.0	2, 7
Dehydroascorbic acid	mg	9.50	7.87–13.0	2, 7
Vitamin E				
α-tocopherol	mg	0.74	0.29–1.75	4, 7, 8
β-tocopherol	mg	0.04	0.03–0.05	7
γ-tocopherol	mg	0.47	0.44–0.50	4, 7
δ-tocopherol	mg	0.01	0.01–0.01	7
Vitamin K	μg	175	–	4
Organic acids				
Oxalic acid	mg	436	98.0–840	2, 7
Glutamic acid	mg	43.7	41.2–46.0	7
Malic acid	mg	318	50.0–488	7
Citric acid	mg	13.2	12.1–14.3	7
Fumaric acid	mg	1.14	0.40–1.84	7
Phenolics (total)	mg	56.8[b]	47.2–53.1	4, 7, 8
	mg	≈217[c]	166–270	9, 10

Table 13.113 (continued)

	Units	Average	Range	References
	mg	≈220[d]	196–244	11
Hydroxycinnamic acids	mg	474[c]	–	12
Flavonoids	mg	14.8[e]	13.8–15.8	7
	mg	34.4[d]	–	13
	mg	≈90.6[c]	89.6–91.5	9
Anthocyanins	mg	3.59[f]	–	12
Nitrate	mg	100	51.5–191	2, 8, 14

1 Saleh et al. (1977), Egipt, *2* Guil-Guerrero et al. (1997a) Spain (5 samples), *3* Guil-Guerrero et al. (1998a) Spain (5 samples), *4* Vardavas et al. (2006b) Crete, *5* Morales et al. (2015) Spain (5 samples), *6* Bruno et al. (1980) Italy, *7* Morales et al. (2014) Spain (5 samples), *8* Zeghichi et al. (2003) Crete, *9* Conforti et al. (2009) Italy, *10* Conforti et al. (2011) Italy, *11* Gatto et al. (2011) Italy, *12* Vanzani et al. (2011) Italy (4 samples), *13* Trichopoulou et al. (2000) Greece, *14* Bianco et al. (1998) Italy (2 samples) *RAE* retinol activity equivalents

[a] Expressed as β-carotene

[b] Expressed as gallic acid

[c] Expressed as chlorogenic acid

[d] Expressed as sum of different compounds (cichoric acid and quercetin as major compound)

[e] Expressed as catechin

[f] Expressed as cyanidin-3-glucoside

Table 13.114 Fatty acids profile (% of total fatty acids) of fresh tender leaves of *S. oleraceus*

Individual compounds	Average	Range	References
12:0	0.15	0.12–0.18	1
14:0	2.00	1–2.67	1, 2, 3
16:0	15.67	9.53–17.5	1, 2, 3
18:0	1.87	1.43–2.20	1, 2, 3
18:1*n*−9	1.38	0.82–2.15	1, 2, 3
18:2*n*−6	11.2	8.10–14.3	1, 2, 3
18:3*n*−3	54.8	43.6–66.3	2, 3
18:3*n*−6	0.27	0.2–0.34	2, 3
18:4*n*−3	0.09	–	2
20:0	1.36	–	1, 2
20:1*n*−9	0.87	1.5–1.60	2, 3
22:0	2.21	nd–3.01	1, 2
24:0	1.04	0.83–1.42	1, 2
24:1*n*−9	1.40	–	3
Categories (calculated values)			
SFA	26.2	18.2–34.4	–
MUFA	2.5	0.9–3.8	–
PUFA	71.3	62.8–80.9	–
n−3	59.2	52.9–67.0	–
n−6	12.1	10.2–13.9	–
n−9	2.5	0.9–3.8	–

1 Morales et al. (2012b) Spain (4 samples), *2* Guil-Guerrero et al (1996b) Spain (various samples), *3* Vardavas et al. (2006a) Crete

SFA saturated fatty acids, *MUFA* monounsaturated fatty acids, *PUFA* polyunsaturated fatty acids

Remarks (tender leaves of *S. oleraceus*)

- In some cases, it may be considered as a source of dietary fibre (often >3 g/100 g).
- High proportion of PUFA and $n-3$ fatty acids, compared to other vegetables. Good $n-3/n-6$ ratio (>4).
- It can be considered as a source of K, Mn, and vitamin B9 (folates); sometimes also a source of Ca, Fe, and vitamin C.
- High content of oxalates: people with altered renal function should avoid this vegetable; boiling is recommended for general population.
- It also may reach high nitrate content.

13.3.35 Tamus communis L. (Dioscoreaceae)

Common Names Black bryony, *lupios* (es), *respounchous* (fr), *tamaro* (it), *bljušt* (hr, bs), *avronies* (el), *dolanbaç* (tr), *jarma'* (ar).

Description Climbing, herbaceous plant, with twining stems till 2–4 m tall that dies down in winter but with a perennial and large tuber. The leaves are smooth, shining, and heart shaped with a long petiole. Dioecious plant, with greenish-yellow and inconspicuous flowers produced in longer racemes in the male plant, while in shorter clusters in the female one. The fruits are bright berries.

Ecology and Distribution It grows in moist and shady places, such as woodland edges, hedges, or scrub areas, usually in riversides, frequently associated with brambles. It is widely distributed in much of Europe, northwest Africa, and western Asia.

Food Uses The young shoots are traditionally consumed and popularly considered as a kind of wild asparagus in many countries of the Mediterranean area, such as Portugal (Carvalho 2010), Spain (e.g. Cobo and Tijera 2011; Tardío et al. 2006; Tejerina 2010), France (Couplan 2009a), Italy (Lentini and Venza 2007; Picchi and Pieroni 2005), Slovenia (Cerne 1992), Bosnia-Herzegovina (Redžić 2006), Croatia (Łuczaj et al. 2013b), Greece (Couplan 2009a), Turkey (Dogan et al. 2004; Ertuğ 2004; Ertuğ 2014), and Palestine (Ali-Shtayeh et al. 2008). They are collected in spring, before the flowering period, from March to the mid of May in central Spain. The spears of this species have a quite bitter flavour, and hence some informants recommend to boil them immediately after gathering and to change the boiling water once or twice (Tardío et al. 2002). However, their bitter and characteristic flavour makes them very appreciated in many regions, such as central Spain, southwest of France or Pesaro in Italy, suggesting other additional reasons for this consumption.

As with other asparagus, they are usually boiled in water and subsequently prepared in different ways (e.g. Couplan 2009a; Criado et al. 2008; Lentini and Venza 2007; Picchi and Pieroni 2005; Tardío et al. 2002). The most common recipe is in omelettes or with scrambled eggs. However, several other recipes are also prepared:

boiled shoots in a salad seasoned with vinegar, olive or walnut oil, adding a hard-boiled egg (Couplan 2009a; Tardío et al. 2002); fried in olive oil with garlic (after being boiled), sometimes also with breadcrumb (Tardío et al. 2002); in *risottos* or cooked with tomato sauce (Lentini and Venza 2007). In other cases, the cooking water is not eliminated and used for making a stew with potatoes or a kind of soup, like in Crete (Couplan 2009a), the Italian *acquacotta* (Picchi and Pieroni 2005), or the Spanish *sopa de rabiacanes*. In the latter, the young shoots are directly fried in olive oil with a spoon of chopped onion, paprika, minced garlic, and parsley. All of them have to be boiled in water, adding salt, a beaten egg, and finally covering with sliced bread (Cofradía Extremeña de Gastronomía 1985).

The spears of *Tamus communis* are still much appreciated. Nowadays, it is possible to taste them in some luxury restaurants, such as the traditional omelette (*tortilla de rabiacanes*) in the "Parador Nacional de Jarandilla" (a state-run luxury hotel, built in a Castle), in the northern part of the Spanish region of Extremadura (Paradores 2014).

Other Uses Some informants tell that this plant should not be eaten in flowering or fruiting stage, because it is poisonous then. Certainly, the whole plant is lightly toxic due to its saponins content, though these compounds are poorly absorbed by the body and broken down by thorough cooking (PFAF 2014). The actually toxic parts are the red berries and the root tubers which also contain calcium oxalate crystals that if internally used can cause irritation of mucous membranes, digestive and respiratory disorders (Couplan 2009a). Because of this rubefacient effect, the fruits and roots have been externally used against musculo-eskeletic and rheumatic pain in several regions, such as Spain (Aceituno-Mata 2010; Tardío et al. 2002; Velasco et al. 2010), Portugal (Carvalho 2010) and Italy (Picchi and Pieroni 2005). The crushed fruits are either directly applied or macerated in alcohol, liquor, or lard. The informants warned that the treatment should be carefully done because it could be painful and cause blisters.

The decoction of the roots have been externally used in baths against feet and leg inflammations, both for humans and animals (Aceituno-Mata 2010).

Historical References The food use of this plant can be found in some references of the first century, such as those of Pliny the Elder (Bostock and Riley 1855) or Dioscorides (Laguna 1555). Following Dioscorides, their new stalks are eaten cooked as other garden vegetables, because of their medicinal properties. They are urinary, expel the menstrual blood, reduce the spleen, and are good for epilepsy, vertigo, and paralysis. Picchi and Pieroni (2005) comment that, according to the Italian herbalists of the sixteenth century, bundles of this asparagus were sold in the past in most of the Italian markets. The present consumption of this wild asparagus and its appreciation in many Mediterranean regions could be a reminder of its ancient medicinal use.

Food Composition Tables for young shoots of *Tamus communis*, raw (Tables 13.115, 13.116 and 13.117) and boiled (Tables 13.118 and 13.119).

Table 13.115 Main constituents, per 100 g of fresh young shoots of *T. communis*

	Units	Average	Range	References
Energy (calculated value)	kcal	46	25–85	–
Moisture	g	85.2	82.0–89.0	1, 2, 3
Available carbohydrates	g	5.20	1.80–11.7	1, 4
Soluble sugars (total)	g	1.28	1.26–1.30	1
Fructose	g	640	620–660	1
Glucose	g	300	280–325	1
Sucrose	mg	120	110–131	1
Trehalose	mg	223	73–373	1
Dietary fibre	g	4.35	3.50–6.00	4
Proteins	g	3.13	2.50–3.80	1, 4
Lipids	g	0.49	0.10–1.28	1, 2, 4
Ash	g	1.25	0.90–2.40	1, 2, 4
K	mg	371	178–562	4
Na	mg	18.5	12.1–32.0	4
Ca	mg	47.0	20.9–94.4	4
Mg	mg	22.4	14.8–42.0	4
Fe	mg	0.65	0.44–1.36	4
Cu	μg	130	60–170	4
Mn	μg	165	80–340	4
Zn	μg	745	630–990	4

1 Martins et al. (2011) Portugal, *2* Morales et al. (2012a). Spain (4 samples), *3* Sánchez-Mata et al. (2012) Spain (2 samples), *4* García-Herrera (2014) Spain (5 samples)

Table 13.116 Vitamins and other constituents, per 100 g of fresh young shoots of *T. communis*

	Units	Average	Range	References
Carotenoids				
β-carotene	mg	0.44	0.32–0.60	1
Lutein	mg	1.14	0.70–1.73	1
Neoxanthin	mg	1.19	0.77–1.90	1
Violaxanthin	mg	0.62	0.38–1.09	1
RAE (calculated value)	μg	36.6	26.6–49.9	–
Vitamin B_9 (total folates)	μg	38.1	37.2–39.0	2
Vitamin C	mg	65.4	58.6–79.4	3, 4
Ascorbic acid	mg	45.2	25.9–57.1	3, 4, 5
Dehydroascorbic acid	mg	10.0	1.45–28.3	3, 4
Vitamin E				
α-tocopherol	mg	1.41	0.12–3.00	6, 7
β-tocopherol	mg	1.42	0.09–2.75	6, 7
γ-tocopherol	mg	1.57	1.28–2.00	6, 7
δ-tocopherol	mg	0.23	0.08–0.44	6, 7
Organic acids				
Oxalic acid	mg	67.4	9.65–90.0	3, 4, 5
Quinic acid	mg	nd	–	5
Glutamic acid	mg	nd	–	3

Table 13.116 (continued)

	Units	Average	Range	References
Malic acid	mg	27.6	nd–80.0	3, 4, 5
Shikimic acid	mg	≈30.9	30.9–30.9	5
Citric acid	mg	274	nd–340	3, 4, 5
Phenolics (total)	mg	49.5[a]	44.5–53.6	7
	mg	220[b]	214–226	8
Flavonoids	mg	9.33[c]	7.89–10.77	7
	mg	201[b]	195–207	8

1 García-Herrera et al. (2013) Spain (5 samples), *2* Morales et al. (2015) Spain (4 samples), *3* Morales (2011) Spain (5 samples), *4* Sánchez-Mata et al. (2012) Spain (2 samples), *5* Pereira et al. (2013) Portugal, *6* Martins et al. (2011) Portugal, *7* Morales et al. (2012a). Spain (4 samples), *8* Barros et al. (2011b)

RAE retinol activity equivalents; *nd* non-detected values

[a] Expressed as gallic acid

[b] Expressed as sum of different compounds (chlorogenic acid, quercetin and kaempferol derivatives as major components)

[c] Expressed as catechin

Table 13.117 Fatty acids profile (% of total fatty acids) of fresh young shoots of *T. communis*

Individual compounds	Average	Range	References
12:0	0.09	0.06–0.12	1, 2
14:0	0.15	0.15–0.15	1, 2
16:0	15.9	14.8–17.0	1, 2
16:1	0.17	0.15–0.20	1, 2
18:0	0.97	0.96–0.98	1, 2
18:1$n-9$	6.04	4.56–7.51	1, 2
18:2$n-6$	42.2	42.0–42.5	1, 2
18:3$n-3$	29.4	27.5–31.4	1, 2
18:3$n-6$	0.45	nd–0.91	1, 2
20:0	0.23	0.21–0.24	1, 2
20:1$n-9$	0.33	0.19–0.41	1, 2
22:0	0.88	0.75–1.02	1, 2
22:1	0.01	nd–0.02	1, 2
24:0	2.03	1.92–2.21	1, 2
24:1$n-9$	0.05	0.04–0.06	2
Categories (calculated values)			
SFA	20.5	19.6–21.4	–
MUFA	6.6	5.1–8.2	–
PUFA	72.8	70.4–75.3	–
$n-3$	30.2	27.8–32.5	–
$n-6$	43.1	42.5–43.7	–
$n-9$	6.5	4.9–8.0	–

1 Martins et al. (2011) Portugal, *2* Morales et al. (2012b) Spain (4 samples); *nd* non-detected values

SFA saturated fatty acids, *MUFA* monounsaturated fatty acids, *PUFA* polyunsaturated fatty acids

Remarks (*raw young shoots of T. communis*)

- It can reach high available carbohydrate content.
- Source of dietary fibre (>3 g/100 g).
- High proportion of PUFA, compared to other vegetables. Unusual ratio $n-3/n-6$, favourable to $n-6$.
- Citric acid as major organic acid. Low oxalate levels (<100 mg/100 g).
- Very low Na amount (<40 mg/100 g).
- It can be considered as a source of vitamins C and B_9 (folates); sometimes also a source of vitamin E.
- Caution must be taken to avoid poisoning: Just tender shoots after boiling are edible (See comments in the section Other Uses).

Table 13.118 Main constituents of 100 g of boiled young shoots of *T. communis*

	Units	Average	Range	References
Energy (calculated value)	kcal	22	21–23	–
Moisture	g	90.6	90.6–90.6	1, 2
Available carbohydrates	g	3.10	2.90–3.30	2
Dietary fibre	g	3.30	3.30–3.30	2
Proteins	g	0.50	0.50–0.50	2
Lipids	g	0.08	0.08–0.08	2
Ash	g	0.70	0.70–0.70	2
K	mg	202	180–225	2
Na	mg	14.9	13.2–16.6	2
Ca	mg	47.3	44.0–50.6	2
Mg	mg	15.2	14.0–16.4	2
Fe	mg	0.43	0.39–0.47	2
Cu	μg	90.0	70.0–110	2
Mn	μg	130	120–140	2
Zn	μg	440	400–480	2

1 Morales (2011) Spain, *2* García-Herrera (2014) Spain

Table 13.119 Vitamins and other constituents, per 100 g of boiled young shoots of *T. communis*

	Units	Average	Range	References
Vitamin C	mg	40.9	39.3–41.5	1
Ascorbic acid	mg	36.2	35.4–37.0	1
Dehydroascorbic acid	mg	3.50	2.36–5.91	1
Organic acids				
Oxalic acid	mg	80	70–90	1
Glutamic acid	mg	nd	–	1
Malic acid	mg	nd	–	1
Citric acid	mg	240	230–250	1
Fumaric acid	mg	Traces	–	1

1 Morales (2011) Spain; *nd* non-detected values

Remarks (boiled young shoots of *T. communis*)

- Very low lipid content (<0.5 g/100 g), and energy value (<25 kcal/100 g).
- Still good source of dietary fibre and vitamin C after boiling.
- Low oxalate levels (<100 mg/100 g) and very low Na amount (<40 mg/100 g).
- Caution must be taken to avoid poisoning: Just tender shoots, after boiling are edible.

13.3.36 Taraxacum officinale Weber and *T. obovatum* (Willd.) DC. (Asteraceae)

Common Names Dandelion, *dente de leão* (pt), *diente de león* (es), *dente di leone* (it), *maslačak* (bs, hr), *agrioradiki* (el), *karahindiba* (tr), *tarakhon* (ar).

Description These species are perennial herbs with a taproot and a rosette of basal leaves, variable in shape, lanceolate to obovate, usually divided in acute and triangular lobes that are the origin of many of their common names. In the case of *T. obovatum* the leaves are in general broader, obovate, contracted to the petiole, with a shallow toothed at the margin. The red or brown stems that produce the flower heads, reaching up to 40 cm in the first species, are shorter in the second one. The yellow flowers are all ligulate and the fruits are tuberculate to rugose achenes with a silky pappus with white hairs growing in many rows.

Ecology and Distribution Both species grow in similar habitats, in woods, meadows, and pastures, roadsides and disturbed soils, the second being more frequent on calcareous soils. They flower from late winter to the beginning of summer. *Taraxacum officinale* is a plant of European origin, naturalized in temperate zones worldwide, whereas *T. obovatum* grows in the west of the Mediterranean region.

Food Uses The species of this taxonomically complicated genus are generally not differentiated by non-specialists and therefore used in the same way. In fact, most of the ethnobotanical references include the information about this genus under the quite ambiguous name of *T. officinale*. However, where both species are available, *T. obovatum* seems to be preferred by it less bitter taste and its shining appearance (Marco et al. 2003). The traditional consumption of the leaves, root, and inflorescences of *Taraxacum* species have been registered in many Mediterranean countries, such as Spain (e.g. González et al. 2011a; Menendez-Baceta et al. 2012; Parada et al. 2011; Tardío et al. 2006), France (Marco et al. 2003), Italy (e.g. Ghirardini et al. 2007; Lentini and Venza 2007; Picchi and Pieroni 2005), Slovenia (Cerne 1992), Croatia (Łuczaj et al. 2013b), Bosnia-Herzegovina (Redžić 2006), Turkey (Dogan 2012; Ertuğ 2004; Ertuğ 2014), Cyprus (Della et al. 2006), Jordania (Al-Qura'n 2010), and Morocco (Nassif and Tanji 2013). The consumption of

T. obovatum has been documented in Spain (Fernández Ocaña 2000; Verde et al. 2003), France (Marco et al. 2003), and Morocco (Nassif and Tanji 2013).

The basal rosette of tender leaves are collected in spring and consumed as a vegetable, usually raw as a snack or in salads (e.g. Al-Qura'n 2010; Cerne 1992; Della et al. 2006; Dogan 2012; Guarrera 2006; Nassif and Tanji 2013; Parada et al. 2011; Picchi and Pieroni 2005; Tardío et al. 2006). They are also eaten cooked, in the same way as spinach or sautéed (Guarrera 2006), browned in a frying pan (Lentini and Venza 2007), in omelettes (Lentini and Venza 2007; Verde et al. 2003), soups (Cerne 1992; Guarrera 2006), added to pastries (Dogan 2012), or cooked in the Moroccan *beqoul* (Nassif and Tanji 2013). They can be found in some markets of Croatia and Italy (Łuczaj et al. 2013b; Picchi and Pieroni 2005). The hollow and milky stems are also consumed raw as a snack (Menendez-Baceta et al. 2012; Tardío et al. 2005) and the root after boiling, seasoned with olive oil (Lentini and Venza 2007; Redžić 2006). Additionally, in Sicilia and other regions of Italy, the buds are preserved like capers, and are used to flavour main courses (Lentini and Venza 2007), though this is considered a modern use by some authors (Picchi and Pieroni 2005).

Some beverages are elaborated with these plants, such as leaf juices (Velasco et al. 2010) or herbal teas with different parts of the plant. These digestive or tonic teas were made with the inflorescences (González-Tejero 1989), the leaves or the whole plant (Villar et al. 1987), the dry stems (Guzmán 1997), or the roasted root (Guarrera 2006; Menendez-Baceta et al. 2012), the latter two taken as a coffee surrogate. A cold drink was also prepared with the flowers (Dogan et al. 2004).

Other Uses These species are traditionally considered as medicinal food. Their consumption as vegetable (or the infusion of its leaves and/or roots) is considered a liver and kidney protector (e.g. Benítez et al. 2010; Bonet and Vallès 2002; Guarrera 2006; Picchi and Pieroni 2005; Rivera et al. 2005), and appropriate to treat digestive or gastrointestinal diseases (Carvalho 2005; Guarrera 2006; Parada et al. 2011; Redžić 2006), because of its detoxifying and laxative effect (Calvo et al. 2011; Guarrera 2006; Lentini and Venza 2007). It has also been employed to treat other circulatory, respiratory, and urinary illnesses (Guarrera 2006; Velasco et al. 2010). In Portugal, the infusion of the dried inflorescences was employed to control cholesterol (Carvalho 2005). The latex is topically used to treat warts, wounds, and cuts (Guarrera 2006; Menendez-Baceta et al. 2014).

The leaves were also employed as fodder for pigs, chickens, and rabbits (Aceituno-Mata 2010; Velasco et al. 2010). The stems and the infrutescences were used in children games (Aceituno-Mata 2010; Guarrera 2006; Pardo-de-Santayana 2003; Picchi and Pieroni 2005).

Historical References The ancient knowledge and consumption of this group of species is evidenced by Theophrastus who mentioned them among the wild bitter vegetables (Teofrasto 1988). It could be also among the plants eaten in Egypt mentioned by Pliny the Elder (Bostock and Riley 1855).

Food Composition Tables for raw leaves of *Taraxacum officinale* (Tables 13.120, 13.121 and 13.122).

Table 13.120 Main constituents, per 100 g of fresh leaves of *T. officinale*

	Units	Average	Range	References
Energy (calculated value)	kcal	84	81–87	–
Moisture	g	83.8	77.1–88.6	1, 2[a]
Available carbohydrates	g	16.2	15.9–16.3	2[a]
Soluble sugars	g	1.36	1.27–1.46	2[a]
Fructose	mg	60.6	56.4–64.8	2[a]
Glucose	mg	434	395–474	2[a]
Sucrose	mg	763	710–815	2[a]
Dietary fibre	g	0.90	–	3
Proteins	g	2.26	1.79–2.84	3, 4
Lipids	g	0.62	–	2[a]
Ash	g	1.12	0.29–0.95	3, 4, 5
K	mg	378	–	3, 4
Na	mg	6.67	–	3
Ca	mg	75.8	–	3, 4
Mg	mg	25.5	–	3, 4
Fe	mg	5.25	0.34–14.4	3, 5
Cu	μg	80.0	63.0–90.0	3, 5
Mn	μg	699	–	3, 4
Zn	μg	85.2	10.2–215	3, 4, 5
Al	μg	3.96	0.39–7.53	3, 5
Si	μg	26.7	–	3
Sr	μg	0.34	0.26–0.42	3, 5
Rb	μg	0.06	–	3
Br	μg	0.19	–	3

1 Bianco et al. (1998) Italy (2 samples), *2* Dias et al. (2014) Portugal, *3* Bockholt and Schnittke (1996) Germany, *4* Ayan et al. (2006) Turkey, *5* Gjorgieva et al. (2010) Macedonia

Table 13.121 Vitamins and other constituents, per 100 g of fresh leaves of *T. officinale*

	Units	Average	Range	References
Carotenoids (total)	mg	3.59	3.00–4.18	1
Vitamin C (ascorbic acid)	mg	30.2	8.00–62.2	1, 2[a]
Vitamin E				
α-tocopherol	mg	3.52	3.25–3.78	2[a]
β-tocopherol	mg	0.13	0.11–0.16	2[a]
γ-tocopherol	mg	0.35	0.31–0.40	2[a]
Organic acids				
Oxalic acid	mg	995	986–1003	2[a]
Malic acid	mg	957	928–986	2[a]
Citric acid	mg	138	138–138	2[a]
Fumaric acid	mg	4.00	4.00–4.00	2[a]
Phenolics (total)	mg	272[b]	236–291	2[a], 3
Phenolic acids	mg	193[b]	191–195	2[a]
Flavonoids	mg	43.3[b]	42.3–44.2	2[a]
Nitrate	mg	29.6	–	4

1 Aliotta and Pollio (1981) Italy, *2* Dias et al. (2014) Portugal, *3* Gatto et al. (2011) Italy, *4* Bianco et al. (1998) Italy (2 samples) Italy (2 samples)

[a] *Taraxacum* sect. *Ruderalia*

[b] Expressed as sum of compounds (cichoric acid as major component)

Table 13.122 Fatty acids profile (% of total fatty acids) of fresh leaves of *T. officinale*

Individual compounds	Average	Range	References
16:0	11.3	8.03–14.1	1[a], 2[b]
18:0	2.40	2.10–2.70	1[a]
18:1$n-9$	0.90	0.69–1.11	1[a]
18:2$n-6$	19.4	13.4–26.1	1[a], 2[b]
18:3$n-3$	62.1	52.4–72.0	1[a], 2[b]
20:0	0.30	0.27–0.33	1[a]
22:0	0.30	0.24–0.36	1[a]
24:0	1.50	1.30–1.60	1[a]
Categories (calculated values)			
SFA	14.2	11.0–17.4	–
MUFA	0.9	0.7–1.1	–
PUFA	85.3	81.7–89.0	–
$n-3$	64.8	62.6–67.0	–
$n-6$	20.6	14.7–26.4	–
$n-9$	0.9	0.7–1.1	–

1 Liu et al. (2002) Australia, *2* Dias et al. (2014) Portugal
SFA saturated fatty acids, *MUFA* monounsaturated fatty acids, *PUFA* polyunsaturated fatty acids
[a] Plants gathered from locations different from Mediterranean or surrounding areas
[b] *Taraxacum* sect. *Ruderalia*

Remarks (leaves of *T. officinale*)

- High available carbohydrate content.
- High proportion of PUFA and $n-3$ fatty acids content (mainly α-linolenic acid, 18:3$n-3$), compared to other vegetables.
- Sometimes a source of Fe, Mn, and vitamin C.
- Very high content of oxalates in the leaves (lower values have been described in other parts such as flowers): people with altered renal function should avoid this vegetable; boiling is recommended for general population.

Food Composition Tables for raw leaves of *Taraxacum obovatum* (Tables 13.123, 13.124 and 13.125).

Table 13.123 Main constituents, per 100 g of fresh tender leaves of *T. obovatum*

	Units	Average	Range	References
Energy (calculated value)	kcal	40	23–58	–
Moisture	g	83.3	79.2–86.7	1, 2, 3
Available carbohydrates	g	3.34	1.63–5.39	2
Dietary fibre	g	7.01	5.37–8.65	2
Proteins	g	1.57	1.02–2.09	2
Lipids	g	0.69	0.19–1.16	2, 4

Table 13.123 (continued)

	Units	Average	Range	References
Ash	g	2.13	1.75–2.49	2
K	mg	566	375–685	2
Na	mg	34.9	5.10–61.7	2
Ca	mg	117	16.0–269	2
Mg	mg	18.2	2.30–34.6	2
Fe	mg	3.57	2.58–4.18	2
Cu	μg	150	80–220	2
Mn	μg	330	150–530	2
Zn	μg	500	220–900	2

1 Sánchez-Mata et al. (2012) Spain (2 samples), *2* García-Herrera et al. (2014b) Spain (4 samples), *3* Morales et al. (2014) Spain (4 samples), *4* Morales et al. (2012b) Spain (4 samples)

Table 13.124 Vitamins and other constituents, per 100 g of fresh tender leaves of *T. obovatum*

	Units	Average	Range	References
Vitamin B_9 (total folates)	μg	110	90.9–130	1
Vitamin C	mg	16.4	11.5–20.8	2, 3
Ascorbic acid	mg	1.99	1.15–2.83	2, 3
Dehydroascorbic acid	mg	14.5	9.66–19.3	2, 3
Vitamin E				
α-tocopherol	mg	0.51	0.49–0.53	3
β-tocopherol	mg	0.01	0.01–0.01	3
γ-tocopherol	mg	0.05	0.05–0.05	3
δ-tocopherol	mg	0.03	0.02–0.04	3
Organic acids				
Oxalic acid	mg	20.0	2.30–36.6	2, 3
Glutamic acid	mg	62.9	Traces–105	3
Malic acid	mg	106.2	23.1–161	2, 3
Citric acid	mg	27.6	Traces–38.1	2, 3
Fumaric acid	mg	3.28	Traces–3.78	2, 3
Phenolics (total)	mg	58.3[a]	57.4–59.1	3
Flavonoids	mg	30.0[b]	29.4–30.7	3

1 Morales et al. (2015) Spain (4 samples), *2* Sánchez-Mata et al. (2012) Spain (2 samples), *3* Morales et al. (2014) Spain (4 samples)

[a] Expressed as gallic acid

[b] Expressed as catechin

Table 13.125 Fatty acids profile, as percentage (%of total fatty acids) of fresh tender leaves of *T. obovatum*

Individual compounds	Average	Range	References
12:0	0.35	0.30–0.40	1
14:0	0.80	0.73–0.87	1
16:0	11.8	11.7–12.0	1
16:1	0.21	0.21–0.21	1
18:0	2.45	2.43–2.48	1

Table 13.125 (continued)

Individual compounds	Average	Range	References
18:1$n-9$	3.24	3.23–3.25	1
18:2$n-6$	17.6	16.6–18.7	1
18:3$n-3$	58.5	58.3–58.8	1
18:3$n-6$	0.16	0.16–0.16	1
20:0	0.56	0.44–0.68	1
20:1$n-9$	0.09	0.07–0.11	1
22:0	0.99	0.98–10.2	1
24:0	1.51	1.46–1.58	1
24:1$n-9$	0.08	0.08–0.08	1
Categories (calculated values)			
SFA	18.5	18.0–19.0	–
MUFA	3.6	3.6–3.7	–
PUFA	77.5	77.1–77.7	–
$n-3$	59.5	58.3–60.3	–
$n-6$	18.1	16.8–18.9	–
$n-9$	3.4	3.4–3.5	–

1 Morales et al. (2012b) Spain (4 samples)
SFA saturated fatty acids, *MUFA* monounsaturated fatty acids, *PUFA* polyunsaturated fatty acids

Remarks (tender leaves of *T. obovatum*)

- Good source of dietary fibre (often >6 g/100 g). High proportion of PUFA and *n*-3 fatty acids.
- Low Na content (<120 mg/100 g). It can be considered as a source of K and Fe; sometimes also a source of Ca, Cu, and vitamins C and B_9 (folates).
- Low oxalate levels (<100 mg/100 g).

13.3.37 Urtica dioica L. (Urticaceae)

Common Names Stinging nettle, *urtiga* (pt), *ortiga* (es), *ortie* (fr), *ortica* (it), *kopriva* (bs, hr), *hithër* (sq), *tsouknida* (el), *ısırgan* (tr), *qurais* (ar).

Description Perennial and dioecious herb, rhizomatous and covered with stinging hairs. The leaves are opposite and petiolate, with a cordate base, an acuminate tip, and a serrate margin. The small flowers, 1–2 mm, which grow in axilar branched racemes up to 10 cm, have 4 petals, and 4 stamens or 1 ovary and a simple style. The fruit is a brownish achene, 2 mm.

Ecology and Distribution This nitrophilous plant grows in waste places and roadsides, in moist soils, 100–2500 m. Flowering occurs from April to September. Its range of distribution covers all the temperate zones of the world.

Food Uses This well-known species has been used as a vegetable also in the Mediterranean countries, at least in Portugal (Carvalho 2010), Spain (Rigat et al. 2009; Tardío et al. 2006; Velasco et al. 2010), Italy (Guarrera 2006; Hadjichambis et al. 2008; Picchi and Pieroni 2005), Slovenia (Cerne 1992), Croatia (Łuczaj et al. 2013a; Łuczaj et al. 2013b), Bosnia-Herzegovina (Redžić 2006), Albania (Hadjichambis et al. 2008; Pieroni 2008), Turkey (Dogan 2012; Ertuğ 2014; Kültür 2008), Lebanon (Marouf 2005), and Morocco (Nassif and Tanji 2013). The tender tips with the leaves are collected mainly in spring before flowering and usually consumed after cooking. They are typically used to prepare different recipes, such as soups (Carvalho 2010; Cerne 1992; Guarrera 2006; Hadjichambis et al. 2008; Redžić 2006; Rigat et al. 2009), rice dishes (Guarrera 2006; Hadjichambis et al. 2008; Picchi and Pieroni 2005), in omelettes or with scrambled eggs (e.g. Bonet and Vallès 2002; Fajardo et al. 2007; García Jiménez 2007; Guarrera 2006; Hadjichambis et al. 2008; Verde et al. 2003), boiled and salted with garlic (Fajardo et al. 2007), onion and mallow (Dogan 2012), or with noodles (Hadjichambis et al. 2008). They are also included in stews with legumes (Fajardo et al. 2007; Rigat et al. 2009) and, sometimes as a filling for pasta dishes, such as lasagna, ravioli, or tortelli (Guarrera 2006; Hadjichambis et al. 2008; Picchi and Pieroni 2005), or for vegetable pies (Dogan 2012; Guarrera 2006; Hadjichambis et al. 2008; Picchi and Pieroni

2005; Pieroni 2008; Redžić 2006). It is one of the species added in different recipes of wild vegetables mixes (Guarrera 2006; Łuczaj et al. 2013a; Picchi and Pieroni 2005). Its delicate sweet flavour often mitigates the excessive bitterness of other species (Picchi and Pieroni 2005).

Although less common, nettles are also consumed raw in salads (Dogan 2012; Fajardo et al. 2007; García Jiménez 2007; Gil Pinilla 1995; Marouf 2005). The leaves have to be scalded in hot water and washed to remove their stinging hairs.

In Turkey, it is one of the wild edible greens most commonly eaten (Ertuğ 2004, Dogan 2012). It is sold in the vegetable markets of Dalmatia, in Croatia (Łuczaj et al. 2013b).

This plant has also been used to elaborate herbal liqueurs (Guarrera 2006; Rigat et al. 2009) and the young leaves to curdle milk (Guarrera 2006).

Other Uses Stinging nettle is one of the most widely used medicinal plants, being also employed in the Mediterranean region (e.g. Benítez 2009; Carvalho 2010; Ertuğ 2004; Guarrera 2006; Kültür 2008; Łuczaj et al. 2013a; Marouf 2005). Some of the more common remedies are referred to the circulatory system, being almost generally considered as a blood depurative. It is also widely employed for urinary, digestive, and respiratory disorders. In external use it has been employed against alopecia and dandruff (e.g. Akerreta 2009; Carvalho 2010).

It has also been extensively used as fodder for different animals (e.g. Benítez 2009; Pardo-de-Santayana 2003; Tardío et al. 2002), especially for pigs, turkeys, and hens, but also for horses and cows, being considered as improver of health and production. For example, it is considered good for improving milk production in mother pigs (Carvalho 2010), for increasing the quality of meat, eggs, milk, and butter (Picchi and Pieroni 2005), and to increase the brightness of the hair of horses (Guarrera 2006).

Other traditional uses of the stinging nettle include the extraction of textile fibre from their stems (García Jiménez 2007; Picchi and Pieroni 2005), a yellow dye from their roots (Guarrera 2006; Picchi and Pieroni 2005), or a biopesticide with the plant maceration (Benítez 2009; Guarrera 2006; Velasco et al. 2010).

Historical References A plant of this genus was already mentioned by Theophrastus in the third century BC among the wild vegetables that need to be cooked before being consumed (Teofrasto 1988). Dioscorides mentioned several medicinal properties of nettles, saying for example that the leaves (boiled together with small shellfish) soften the bowels, dissolve windiness, and induce urine (Osbaldeston 2000). Pliny, also in the first century AD, says that nettle first begins to grow in spring, at which period it is by no means a disagreeable food, being even considered as a preventive from diseases the whole year through (Bostock and Riley 1855). In Apicius, a collection of Roman cookery recipes, nettles are included among the vegetables, after field herbs, specifying that "the female nettle, when the sun is in the position of Aries, is supposed to render valuable services against ailments of various kinds" (Dommers Vehling 1936).

Food Composition Tables for raw leaves of *Urtica dioica* (Tables 13.126, 13.127 and 13.128).

Table 13.126 Main constituents, per 100 g of fresh leaves of *U. dioica*

	Units	Average	Range	References
Energy (calculated value)	kcal	51	24–70	–
Moisture	g	82.9	76.9–90.7	1, 2, 3, 4
Available carbohydrates	g	5.04	3.71–7.05	1
Glucose	mg	246	–	5
Fructose	mg	284	–	5
Sucrose	mg	317	–	5
Starch	mg	462	–	5
Dietary fibre	g	2.96	0.88–4.50	1, 2, 6
Insoluble fibre	g	1.32	–	5
Soluble fibre	g	1.79	–	5
Proteins	g	4.29	1.35–5.42	1, 2, 6, 7
Lipids	g	0.86	0.25–1.29	1, 2
Ash	g	1.54	1.33–3.41	1, 2, 6, 8
K	mg	391	19.3–740	1, 2, 6, 7, 8
Na	mg	14.9	1.92–27.0	1, 2, 6, 8
Ca	mg	625	246–982	1, 2, 6, 7, 8
Mg	mg	171	14–482	1, 2, 7, 8
P	mg	111	57.1–178	2, 6
Fe	mg	≈5.39	1.9–13.0	9
Cu	μg	350	100–502	1, 2, 7, 8
Mn	μg	1698	580–3000	1, 2, 7, 8
Zn	μg	839	80.0–1428	1, 2, 7, 8
Se	μg	3.00	–	2

1 Martínez-Para et al. (1979) Spain, *2* Wetherilt (1992) Turkey, *3* Bianco et al. (1998) Italy (2 samples), *4* Guil-Guerrero et al. (2003) Spain (5 samples), *5* Souci et al. (2008) unknown origin, *6* Bockholt and Schnittke (1996) Germany, *7* Adamski and Bieganska (1980) Poland, *8* Krstic-Pavlovic and Dzamic (1985) Yugoslavia, *9* Gjorgieva et al. (2011) Macedonia

Table 13.127 Vitamins and other constituents, per 100 g of fresh leaves of *U. dioica*

	Units	Average	Range	References
Carotenoids				
β-carotene	mg	5.72	1.18–10.9	1, 2, 3
Hydroxy-α-carotene	mg	≈0.90	–	1
Lutein (epoxide)	mg	5.61	5.25–5.97	1, 3
Luteoxanthin	mg	≈1.85	–	1
Neoxhantin	mg	≈0.43	0.40–0.46	4
Violaxanthin	mg	1.92	1.10–2.65	1, 3
RAE (calculated value)	μg	476	–	–
Vitamin B_1	μg	20.0	–	2
Vitamin B_2	μg	230	–	2
Vitamin B_3	μg	620	–	2
Vitamin B_6	μg	70.0	–	2
Vitamin C	mg	285	238–333	5

Table 13.127 (continued)

	Units	Average	Range	References
Vitamin E				
α-tocopherol	mg	14.4	–	2
Nitrate	mg	92.4	–	6

1 Kudritsata et al. (1987), Ukraine, *2* Wetherilt (1992) Turkey, *3* Guil-Guerrero et al. (2003) Spain (5 samples), *4* Gjorgieva et al. (2011) Macedonia, *5* Franke and Kensbock (1981) *RAE* retinol activity equivalents

Table 13.128 Fatty acids profile (% of total fatty acids) of fresh leaves of *U. dioica*

Individual compounds	Average	Range	References
12:0	0.37	–	1
14:0	0.20	0.20–0.20	1
16:0	18.2	16.3–21.0	1
16:1	2.40	0.90–4.20	1, 2
18:0	1.35	0.80–2.00	1, 2
18:1$n-9$	2.53	2.25–3.00	1, 2
18:2$n-6$	26.3	16.8–34.6	1, 2
18:3$n-3$	38.9	37.1–43.9	1
20:0	1.00	1.00–1.00	1
20:1$n-9$	0.70	0.70–0.70	2
22:1	0.50	0.40–0.60	2
Categories (calculated values)			
SFA	22.8	20.0–24.8	–
MUFA	6.7	3.4–8.9	–
PUFA	70.6	66.3–76.6	–
$n-3$	42.1	39.6–44.7	–
$n-6$	28.5	20.4–37.0	–
$n-9$	4.0	2.4–4.5	–

1 Souci et al. (2008) unknown origin, *2* Guil-Guerrero et al. (2003) Spain (5 samples)
SFA saturated fatty acids, *MUFA* monounsaturated fatty acids, *PUFA* polyunsaturated fatty acids

Remarks (leaves of *U. dioica*)

- High protein content compared to other wild plants.
- High lipid content and $n-3$ fatty acids content (mainly α-linoleic acid, 18:3$n-3$), compared to other vegetables.
- Very low Na content (<40 mg/100 g). Even taking account the wide variability, it can be considered as a source of Ca, Mn, and vitamin C; sometimes also a source of K, Mg, Fe, Cu, as well as vitamins B_2, E, and A (provitamin A as β-carotene).
- High content of nitrate; not recommended for infants.

13.3.38 Ziziphus lotus (L.) Lam. (Rhamnaceae)

Common Names Lotus tree, *arto* (es), *jujubier des lotophages* (fr), *zinzula sarvaggia* (it), *lotos* (el), *sarı hünnap* (tr), *shizaf* (he), *sidir, sedra* (ar).

Description This deciduous, spiny, and intricate shrub, up to 3 m, has grey, zigzagged, and frequently curved, down branches. The alternate leaves are 2–4 cm, elliptical, shallowly crenate, with three parallel longitudinal veins in their surface. There are two hard thorns in the insertions of each leaf; one is straight and the other is hooked. The small and greenish flowers, with 5 sepals, 5 petals, and 5 stamens, grow in axillary cymes. The fruit is a subglobose and brown-reddish drupe, 1–2 cm in diameter, with a large stone in the centre and surrounded by a dry fleshy pulp. It flowers in spring or summer and the fruits ripen in autumn.

Ecology and Distribution It grows in dry places in almost all types of soils, with preference for basic grounds, at low altitudes. It lives in some Southern European areas of Spain, Italy (Sicily), Greece, Cyprus, west Asia, and more abundantly along the north of Africa from Morocco to Egypt.

Food Uses The edible fruits, though not of high quality, have been consumed in some countries of the Mediterranean region, such as southeast Spain (Martínez-Lirola et al. 1997; Rabal 2000), Cyprus (Della et al. 2006), Turkey (Ertuğ 2014), Jordan (Al-Qura'n 2010; Tukan et al. 1998), Syria (Tanji and Nassif 1995), Libya (Louhaichi et al. 2011), Tunisia (Benammar et al. 2010; Tanji and Nassif 1995), Algeria (Benammar et al. 2010; Tanji and Nassif 1995), and Morocco (Abouri et al. 2012; Nassif and Tanji 2013; Tanji and Nassif 1995). They are mainly eaten as a dessert, both raw when ripe or after dehydration throughout the year (Abouri et al. 2012; Al-Qura'n 2010; Della et al. 2006; Martínez-Lirola et al. 1997; Nassif and Tanji 2013; Rabal 2000; Tanji and Nassif 1995; Tukan et al. 1998), though they are also used to prepare jams (Della et al. 2006; Tukan et al. 1998) or even an acid beverage in Lybia (Louhaichi et al. 2011).

In the north of Africa, the dried fruits are ground into flour for making bread and the seeds kernel to make *amlou* (Nassif and Tanji 2013), a Moroccan chunky paste usually made with argan oil, almonds or peanuts, and sometimes honey or sugar. The fruits are prized by nomads that take them as travel provisions and frequently sold in local markets and on roadsides, at least in some Moroccan localities (Nassif and Tanji 2013).

The fruits of other wild species of the same genus (all of them small trees) are also consumed and frequently considered of better quality. This is the case of *Z. spina-christi* (L.) Desf. or *Z. nummularia* (Burm. f.) Wight & Arn., in the Middle East (Al-Qura'n 2010; Dafni et al. 2005), or the more widely cultivated and sometimes naturalized *Z. jujuba* Mill., and *Z. mauritiana* Lam., the Chinese and Indian jujuba, respectively (Azam-Ali et al. 2006).

Other Uses Different parts of the plant (fruits, leaves, root) have been employed in North African traditional medicine for treating diverse ailments (Abouri et al. 2012; Benammar et al. 2010; El-Mokasabi 2014), such as diabetes, digestive, respiratory and cardiac diseases, or skin problems. The external use of the leaf powder for healing abscesses and wounds, or the fruit decoction for treating diabetes, digestive and respiratory problems is common in central Morocco (Abouri et al. 2012).

This plant has been also used as fodder and honey plant (Louhaichi et al. 2011), fuel (Al-Qura'n 2010; Louhaichi et al. 2011), and for hedges and erosion control (Louhaichi et al. 2011).

Historical References *Ziziphus* fruit remains, which could be assigned to *Z. lotus* and *Z. spina-christi*, have been found in a number of archaeological excavations, in west Asia and Egypt, dated from Neolithic and Bronze Age (Azam-Ali et al. 2006). Its abundance in Libya was mentioned by Herodotus (fifth century BC) and Theophrastus (third century BC), describing its edible fruits, with a sweetness similar to dates, and identifying them with those consumed by the mythological lotus-eaters (lotophages) referred to in the Homer's Odyssey (García-González 2008; Teofrasto 1988). They also mentioned that the fruits were used to make wine. The same was repeated by Pliny the Elder in the first century AD, also adding that those who eat its fruits are subject to no maladies of the stomach (Bostock and Riley 1855).

Food Composition Tables for raw fruits of *Ziziphus lotus* (Tables 13.129, 13.130 and 13.131).

Table 13.129 Main constituents, per 100 g of fresh fruits of *Z. lotus*

	Units	Average	Range	References
Energy (calculated value)	kcal	61	55–65	–
Moisture	g	12.3	12.3–12.3	1, 2, 3
Soluble sugars (total)	g	9.87	9.25–10.5	2, 3
Reducing sugars	g	2.27	–	2
Sucrose	g	7.05	–	2
Dietary fibre	g	5.12	4.29–5.42	2, 3
Pectins	g	2.07	1.74–2.40	3

Table 13.129 (continued)

	Units	Average	Range	References
Proteins	g	1.18	0.82–1.54	3
Lipids	g	0.79	0.77–0.81	3
Ash	g	3.01	2.81–3.35	1, 3
K	mg	126	118–141	1, 3
Na	mg	10.0	9.0–11.1	3
Ca	mg	490	–	1
Mg	mg	373	349–416	1, 3
P	mg	9.29	8.92–10.8	3
Fe	mg	1.25	1.17–1.37	1, 3
Mn	μg	1900	1030–2708	1, 3
Zn	μg	413	386–460	1, 3

1 Boudraa et al. (2010) Algeria (5 samples), *2* Saadoui et al. (2012) Algeria, *3* Abdeddaim et al. (2014) Algeria

Table 13.130 Vitamins and other constituents, per 100 g of fresh fruits of *Z. lotus*

	Units	Average	Range	References
Carotenoids (total)	mg	32.1[a]	1.47–62.8	1, 2
Vitamin B_1 (thiamin)	μg	39	34–44	2
Vitamin C (ascorbic acid)	mg	86.4	5.67–167	1, 2
Vitamin E (tocopherols)	mg	5.41[b]	0.97–9.85	1, 2
Total phenolics	mg	≈8.26[c]	–	3
Flavonoids	mg	≈4.61[d]	–	3
Tannins	mg	≈33.6[e]	–	3

1 Benammar et al. (2010) Algeria, *2* Boudraa et al. (2010) Algeria (5 samples), *3* Rsaissi et al. (2013) Morocco

[a] Expressed as β-carotene
[b] Expressed as α-tocopherol
[c] Expressed as gallic acid
[d] Expressed as quercetin
[e] Expressed as catechin

Table 13.131 Fatty acids profile (% of total fatty acids) of fresh raw fruits of *Z. lotus*

Individual compounds	Average	Range	References
16:0	27.6	–	1
18:0	11.2	–	1
18:1$n-9$	24.5	–	1
18:2$n-6$	36.6	–	1
Categories (calculated values)			
SFA	24.5	–	–
MUFA	38.8	–	–
PUFA	36.6	–	–
$n-6$	36.6	–	–
$n-9$	24.5	–	–

1 Benammar et al. (2010) Algeria
SFA saturated fatty acids, *MUFA* monounsaturated fatty acids, *PUFA* polyunsaturated fatty acids

Remarks (fruits of *Z. lotus*)

- Source of dietary fibre (>3 g/100 g).
- High ash level, with very low Na content (<40 mg/100 g). It can be considered as a source of Mg and Mn; sometimes also a source of Ca, Cu, and vitamin E.
- High proportion of unsaturated fatty acids.

References

Abdeddaim M, Lombarkia O, Bacha A, Fahloul D, Abdeddaim D, Farhat R, Saadoudi M, Noui Y, Lekbir A (2014) Biochemical characterization and nutritional properties of *Zizyphus lotus* L. fruits in Aures region, northeastern of Algeria. Ann Food Sci Technol 15(1):75–81

Abouri M, Mousadik A E, Msanda F, Boubaker H, Saadi B, Cherifi K (2012) An ethnobotanical survey of medicinal plants used in the Tata Province, Morocco. Int J Med Plant Res 1(7):99–123

Aceituno-Mata L (2010) Estudio etnobotánico y agroecológico de la Sierra Norte de Madrid. PhD Thesis. Universidad Autónoma de Madrid

Adamski R, Bieganska J (1980) Investigations on chemical substances present in *Urtica dioica* L. leaves.1. Trace-Elements. Herba Pol 26(3):177–180

Afolayan A, Jimoh F (2009) Nutritional quality of some wild leafy vegetables in South Africa. Int J Food Sci Nutr 60(5):424–431

Agelet A, Vallès J (2003) Studies on pharmaceutical ethnobotany in the region of Pallars (Pyrenees, Catalonia, Iberian Peninsula). Part II. New or very rare uses of previously known medicinal plants. J Ethnopharmacol 84(2–3):211–227

Akerreta S (2009) Etnobotánica farmaceútica en Navarra: del uso tradicional de las plantas medicinales a su evidencia científica. PhD Thesis. Universidad de Navarra

Akerreta S, Calvo MI, Cavero RY (2010) Ethnoveterinary knowledge in Navarra (Iberian Peninsula). J Ethnopharmacol 130:369–378

Alarcão-E-Silva MLCMM, Leitão AEB, Azinheira HG, Leitão MCA (2001) The *Arbutus* berry: studies on its color and chemical characteristics at two mature stages. J Food Compos Anal 14:27–35

Alarcón R, Ortiz LT, García P (2006) Nutrient and fatty acid composition of wild edible bladder campion populations [*Silene vulgaris* (Moench.) Garcke]. Int J Food Sci Technol 41:1239–1242

Aliotta G, Pollio A (1981) Vitamin A and C content in some edible wild plants in Italy. Riv Ital EPPOS 63:47–48

Ali-Shtayeh MS, Yaniv Z, Mahajna J (2000) Ethnobotanical survey in the Palestinian area: a classification of the healing potential of medicinal plants. J Ethnopharmacol 73:221–232. (1 Feb)

Ali-Shtayeh M, Jamous R, Al-Shafie' J, Elgharabah W, Kherfan F, Qarariah K, Khdair I, Soos I, Musleh A, Isa B, Herzallah H, Khlaif R, Aiash S, Swaiti G, Abuzahra M, Haj-Ali M, Saifi N, Azem H, Nasrallah H (2008) Traditional knowledge of wild edible plants used in Palestine (Northern West Bank): a comparative study. J Ethnobiol Ethnomed 4(1):13

Allen DE, Hatfield G (2004) Medicinal plants in folk tradition. An ethnobotany of Britain & Ireland. Timber Press, Cambridge

Alonso-de-Herrera G (1513) Agricultura general. (1818 edition). Real Sociedad Económica Matritense, Madrid

Al-Qura'n SA (2010) Ethnobotanical and ecological studies of wild edible plants in Jordan. Libyan Agric Res Cen J Int 1(4):231–243

Alves Ribeiro J, Monteiro AM, Fonseca da Silva ML (2000) Etnobotânica. Plantas bravias, comestíveis, condimentares e medicinais. Joâo Azevedo Editor, Mirandela

Andarwulan N, Kurniasih D, Apriady RA, Rahmat H, Roto AV, Bolling BW (2012) Polyphenols, carotenoids, and ascorbic acid in underutilized medicinal vegetables. J Funct Foods 4(1):339–347

Andrews AC (1948) Orach as the Spinach of the Classical Period. Isis 39(3):169–172

Arauzo MA, Fierro C, González A, Iribarren I, López L, Muñoz J, Palomo G, Revilla A (2004) Aproximación a la flora de las Tablas de Villarrubia de los Ojos del Guadiana. Parte del entorno de las Tablas de Daimiel. ARBA, Madrid

Ayan I, Acar Z, Mut H, Basaran U, Asci O (2006) Morphological, chemical and nutritional properties of forage plants in a natural rangeland in Turkey. Bangladesh J Bot 35(2):133–142

Ayaz FA, Kucukislamoglu M, Reunanen M (2000) Sugar, non-volatile and phenolic acids composition of strawberry tree (*Arbutus unedo* L. var. *ellipsoidea*) fruits. J Food Compos Anal 13(2):171–177

Aydin C, Özcan M (2007) Determination of nutritional and physical properties of myrtle (*Myrtus communis* L.) fruits growing wild in Turkey. J Food Eng 79(2):453–458

Azam-Ali S, Bonkoungou E, Bowe C, deKock C, Godara A, Williams JT (2006) Ber and other jujubes. International Centre for Underutilised Crops, Southampton

Barandiarán JM, Manterola A (1990) La alimentación doméstica en Vasconia. Atlas etnográfico de Vasconia, vol III. Gobierno Vasco & Etniker Euskalerria, Bilbao

Barão MJ, Soveral A (2010) The ecology and use of edible thistles in Évora, Alentejo, South-eastern Portugal. In: Pardo-de-Santayana M, Pieroni A, Puri R (eds) Ethnobotany in the New Europe: people, health and wild plant resources. Berghahn Books, Oxford, pp 189–210

Barboni T, Venturini N, Paolini J, Desjobert JM, Chiaramonti N, Costa J (2010) Characterisation of volatiles and polyphenols for quality assessment of alcoholic beverages prepared from Corsican *Myrtus communis* berries. Food Chem 122(4):1304–1312

Barlas NT, Irget ME, Tepecik M (2011) Mineral content of the rocket plant *(Eruca sativa)*. Afr J Biotechnol 10(64):14080–14082

Barros L, Carvalho AM, Ferreira ICFR (2010a) Leaves, flowers, immature fruits and leafy flowered stems of *Malva sylvestris*: a comparative study of the nutraceutical potential and composition. Food Chem Toxicol 48(6):1466–1472

Barros L, Carvalho AM, Morais JS, Ferreira ICFR (2010b) Strawberry-tree, blackthorn and rose fruits: detailed characterisation in nutrients and phytochemicals with antioxidant properties. Food Chem 120(1):247–254

Barros L, Carvalho AM, Ferreira ICFR (2010c) The nutritional composition of fennel (*Foeniculum vulgare*): Shoots, leaves, stems and inflorescences. LWT-Food Sci Technol 43(5):814–818

Barros L, Carvalho AM, Ferreira ICFR (2011a) Comparing the composition and bioactivity of *Crataegus monogyna* flowers and fruits used in folk medicine. Phytochem Anal 22:181–188

Barros L, Dueñas M, Ferreira ICFR, Carvalho AM, Santos-Buelga C (2011b) Use of HPLC–DAD–ESI/MS to profile phenolic compounds in edible wild greens from Portugal. Food Chem 127:169–173

Batal M, Hunter E (2007) Traditional Lebanese recipes based on wild plants: an answer to diet simplification? Food Nutr Bull 28(2):S303–S311

Benammar C, Hichami A, Yessoufou A, Simonin AM, Belarbi M, Allali H, Khan NA (2010) *Ziziphus lotus* L. (Desf.) modulates antioxidant activity and human T-cell proliferation. BMC Complement Altern Med 10:54

Benítez G (2009) Etnobotánica y Etnobiología del poniente granadino. PhD Thesis. Facultad de Farmacia. Universidad de Granada

Benítez G, González-Tejero MR, Molero Mesa J (2010) Pharmaceutical ethnobotany in the Western part of Granada province (Southern Spain): ethnopharmacological synthesis. J Ethnopharmacol 129:87–105

Bianco VV, Santamaria P, Elia A (1998) Nutritional value and nitrate content in edible wild species used in Southern Italy. Acta Hort (ISHS) 467:71–87

Bianco VV, Rocco M, Santamaria P (2009) Piante spontanee nella cucina tradizionale molese. Storie, curiosità e ricette. Edizioni Levante, Bari

Blanco E (1995) Investigaciones etnobotánicas en la Sierra del Caurel (Lugo) y en la Calabria Extremeña (Badajoz). PhD Thesis. Universidad Autónoma de Madrid

Blanco E (1996) El Caurel, las plantas y sus habitantes. Fundación Caixa Galicia, La Coruña

Blanco E (1998) Diccionario de etnobotánica segoviana. Ayuntamiento de Segovia, Segovia

Blanco E (2002) Etnobotánica en los Montes de Toledo. Asociación Cultural Montes de Toledo, Toledo

Blanco E, Cuadrado C (2000) Etnobotánica en Extremadura. Estudio de La Calabria y La Siberia extremeñas. Emilio Blanco y CEP Alcoba de los Montes, Madrid

Blanco E, Diez J (2005) Guía de la Flora de Sanabria, Carballeda y los Valles. Catálogo de Etnoflora. ADISAC-La Voz, Zamora

Bockolt R, Schnittke C (1996) Contents of nutrients and minerals in herb species of intensive managed peat soil meadows. Wirtsch Futter 42(3):209–216

Bonet MA (1993) Etnobotànica de la Vall del Tenes (Vallès Oriental). Publicaciones de L'Abadia de Montserrat, Barcelona

Bonet MA (2001) Estudi etnobotànic del Montseny. PhD Thesis. Universidad de Barcelona

Bonet MA, Vallès J (2002) Use of non-crop food vascular plants in Montseny biosphere reserve (Catalonia, Iberian Peninsula). Int J Food Sci Nutr 53:225–248

Bonet MA, Parada M, Selga A, Vallès J (1999) Studies on pharmaceutical ethnobotany in the regions of L' Alt Empordà and Les Guilleries (Catalonia, Iberian Peninsula). J Ethnopharmacol 68:145–168

Bostock J, Riley HT (1855) The natural history of Pliny. H. G. Bohn, London. http://www.perseus.tufts.edu/hopper/text?doc=Plin.+Nat.+toc

Boudraa S, Hambaba L, Zidani S, Boudraa H (2010) Composition minérale et vitaminique des fruits de cinq espèces sous exploitées en Algérie: *Celtis australis* L., *Crataegus azarolus* L., *Crataegus monogyna* Jacq., *Elaeagnus angustifolia* L. et *Zizyphus lotus* L. Fruits 65(02):75–84

Bruno S, Amico A, Stefanizzi L (1980) Contenuto in vitamina C di piante eduli e medicamentose della regione pugliese. Boll Soc Ital Biol Sper 56(20):2067–2070

Bryant C (1783) Flora dietetica or history of esculent plants. Benjamin White, London

Cabiddu A, Decandia M (2000) A note on the chemical composition and tannin content of some Mediterranean shrubs browsed by Sarda goats. Cahiers Options Méditerranéennes 52:175–178

Cakir A (2004) Essential oil and fatty acid composition of the fruits of *Hippophae rhamnoides* L. (Sea Buckthorn) and *Myrtus communis* L. from Turkey. Biochem Sys Ecol 32(9):809–816

Calvo MI, Akerreta S, Cavero RY (2011) Pharmaceutical ethnobotany in the Riverside of Navarra (Iberian Peninsula). J Ethnopharmacol 135:22–33

Camejo-Rodrigues J, Ascensão L, Bonet MA, Vallès J (2003) An ethnobotanical study of medicinal and aromatic plants in the Natural Park of "Serra de São Mamede" (Portugal). J Ethnopharmacol 89:199–209

Carrió E (2013) Contribució al coneixement etnobotànic de Mallorca. La biodiversitat vegetal i la seva gestió en una illa mediterrània. PhD Thesis. Universitat de Barcelona

Carrió E, Vallès J (2012) Ethnobotany of medicinal plants used in Eastern Mallorca (Balearic Islands, Mediterranean Sea). J Ethnopharmacol 141:1021–1040

Carvalho AM (2005) Etnobotánica del Parque Natural de Montesinho. Plantas, tradición y saber popular en un territorio del Nordeste de Portugal. PhD Thesis. Universidad Autónoma de Madrid

Carvalho AM (2010) Plantas y sabiduría popular del Parque Natural de Montesinho. Un estudio etnobotánico en Portugal. CSIC, Madrid

Carvalho AM, Telo M (2013) Etnoflora da Terra de Miranda. Cultibos, yerbas i saberes. FRAUGA-Istituto Politécnico de Bragança, Bragança

Castroviejo S (gen. coord.) (1986–2015) Flora iberica, vols. 1–16(I),17,18,20,21. Real Jardín Botánico, CSIC, Madrid. http://www.floraiberica.es

Catani M, Amaya S, Díaz-Aguilar AL (2001) Comer en Tentudía: aproximación etnográfica a la comida y a los hábitos de vida de las gentes de la comarca de Tentudía en los últimos setenta años. Centro de Desarrollo Comarcal de Tentudía, Monesterio
Cerne M (1992) Wild plants from Slovenia used as vegetables. Acta Hort (ISHS) 318:87–96
Chauvet M (2003) Inventorying food plants in France. In: Knüpffer H, Oschmann J (eds) Rudolf Mansfeld and plant genetic resources. Proceedings of a symposium dedicated to the 100th birthday of Rudolf Mansfeld. Gatersleben, Germany. 8–9 October, 2001. ZADI, Bonn, pp 98–107
Church SM (2006) The History of Food Composition Databases. Nutr Bull 31:15–20
Cienfuegos B (1627) Historia de las plantas. Manuscripts 3357–3363, conserved at Biblioteca Nacional de España (National Library), Madrid
Clusius C (1576) Rariorum aliquot stirpium per Hispanias observatarum Historia. Libris duobus expressa, Anvers
COAGRICO (2014) Cooperativa Ntra. Sra. de las Virtudes. http://www.coagrico.com. Accessed 4 March 2015
Cobo MP, Tijera RE (2011) Etnobotánica de Doñana. Mancomunidad de Desarrollo y Fomento del Aljarafe, Sevilla
Cofradía Extremeña de Gastronomía (1985) Recetario de cocina extremeña. Estudio de sus orígenes. Universitas Editorial, Badajoz
Columela LJM (1824) Los doce libros de Agricultura (*De Re Rustica*). Trad. J. M. Álvarez de Sotomayor. Imprenta de D. Miguel de Burgos, Madrid
Concon JM (1988) Toxicology. Principles and concepts. Marcel Dekker, NewYork
Conforti F, Sosa S, Marrelli M, Menichini F, Statti GA, Uzunov D, Tubaro A, Menichini F (2009) The protective ability of Mediterranean dietary plants against the oxidative damage: the role of radical oxygen species in inflammation and the polyphenol, flavonoid and sterol contents. Food Chem 112(3):587–594
Conforti F, Marrelli M, Carmela C, Menichini F, Valentina P, Uzunov D, Statti GA, Duez P, Menichini F (2011) Bioactive phytonutrients (omega fatty acids, tocopherols, polyphenols), in vitro inhibition of nitric oxide production and free radical scavenging activity of non-cultivated Mediterranean vegetables. Food Chem 129(4):1413–1419
Couplan F (1989a) La cuisine sauvage. Comment accomoder mille plantes oubliées.Enciclopédie des plantes comestibles de l'Europe, vol 2. Equilibres Aujourd'houi, Flers
Couplan F (1989b) Le regal vegetal. Plantes sauvages comestibles. Encyclopedie des plantes comestibles de l'Europe, vol 1. Equilibres Aujourd'houi, Flers
Couplan F (1990) Les belles veneneuses. Encyclopedie des plantes comestibles de l'Europe, vol 3. Equilibres Aujourd'houi, Flers
Couplan F (2009a) Le regal vegetal. Plantes sauvages comestibles. Encyclopedie des plantes comestibles de l'Europe. Éditions Sang de la Terre, Paris
Couplan F (2009b) Le véritable régime crétois. Les éditions Ellébore, Paris
Cowan JW, Sabry ZI, Shadarev SB, Sakr AH (1963) Composition of edible wild plants of Lebanon. J Sci Food Agric 14(7):484–488
Criado J, Fernández López JM, Leocadio G, Núñez RM, Blanco E (2008) Uso tradicional de las plantas en Toledo. Diputación de Toledo, Toledo
Cubero JI (2001) El libro de agricultura de Al Awan. Junta de Andalucía. Consejería de Agricultura y Pesca, Sevilla
Dafni A, Levy S, Lev E (2005) The ethnobotany of Christ's Thorn Jujube (*Ziziphus spina-christi*) in Israel. J Ethnobiol Ethnomed 1(1):8
Dalby A (2003) Food in the ancient world from A–Z. Routledge, London
De Natale A, Pollio A (2012) A forgotten collection: the Libyan ethnobotanical exhibits (1912–14) by A. Trotter at the Museum O. Comes at the University Federico II in Naples, Italy. J Ethnobiol Ethnomed 8(1):4
Del Rio-Celestino M, Font R, de Haro-Bailon A (2008) Distribution of fatty acids in edible organs and seed fractions of borage (*Borago officinalis* L.). J Sci Food Agric 88(2):248–255

Della A, Paraskeva-Hadjichambi D, Hadjichambis AC (2006) An ethnobotanical survey of wild edible plants of Paphos and Larnaca countryside of Cyprus. J Ethnobiol Ethnomed 2:34

Demo A, Petrakis C, Kefalas P, Boskou D (1998) Nutrient antioxidants in some herbs and Mediterranean plant leaves. Food Res Int 31(5):351–354

Di Tizio A, Łuczaj ŁJ, Quave CL, Redžić S, Pieroni A (2012) Traditional food and herbal uses of wild plants in the ancient South-Slavic diaspora of Mundimitar/Montemitro (Southern Italy). J Ethnobiol Ethnomed 8

Dias MG, Camoes MFGFC, Oliveira L (2009) Carotenoids in traditional Portuguese fruits and vegetables. Food Chem 113(3):808–815

Dias MI, Barros L, Alves RC, Oliveira M, Santos-Buelga C, Ferreira IC (2014) Nutritional composition, antioxidant activity and phenolic compounds of wild *Taraxacum* sect. *Ruderalia*. Food Res Int 56:266–271

Díaz-Fernández PM, Ramos Miras JJ, San José A, Pascual Gallego V, López-Almansa JC, Muñoz-Gallego CL (2008) Puesta en valor de recursos genéticos vegetales y su aplicación a estrategias de desarrollo rural. Estudio etnobotánico de *Montia fontana* L. en la provincia de Ávila. CONAMA 9. Congreso Nacional de Medio Ambiente

Dogan Y (2012) Traditionally used wild edible greens in the Aegean Region of Turkey. Acta Soc Bot Pol 81(4):329–342

Dogan Y, Baslar S, Ay G, Mert HH (2004) The use of wild edible plants in Western and Central Anatolia (Turkey). Econ Bot 58(4):684–690

Doležal M, Velíšek J, Famfulíková P (2001) Chemical composition of less known wild fruits. In: Pfannhauser W, Fenwick GR, Khokhar S (ed.) Biologically-active phytochemicals in food. Analysis, Metabolims, Bioavailability and Function. Royal Society of Chemistry, London, pp 241–244

Dommers Vehling J (1936) Apicius, cookery and dining in imperial Rome. A bibliography, critical review and translation of the ancient book known as *Apicius de re Coquinaria*. Walter M. Hill, Chicago

Egea I, Sánchez-Bel P, Romojaro F, Pretel MT (2010) Six edible wild fruits as potential antioxidant additives or nutritional supplements. Plant Foods Hum Nutr 65(2):121–129

Egea-Gilabert C, Niñirola D, Conesa E, Candela ME, Fernández JA (2013) Agronomical use as baby leaf salad of *Silene vulgaris* based on morphological, biochemical and molecular traits. Sci Hortic 152:35–43

El-Mallah MH, El-Shami SM, Hassanein MM (2003) Detailed studies on some lipids of *Silybum marianum* (L.) seed oil. Grasas y Aceites 54(4):397–402

El-Mokasabi FM (2014) Floristic composition and traditional uses of plant species at Wadi Alkuf, Al-Jabal Al-Akhder, Libya. American-Eurasian J Agric Environ Sci 14(8):685–697

Ertuğ F (2000) An ethnobotanical study in Central Anatolia (Turkey). Econ Bot 54(2):155–182

Ertuğ F (2004) Wild edible plants of the Bodrum Area (Mugla, Turkey). Turk J Bot 28:161–174

Ertuğ F (2014) Etnobotanik. In: Güner A, Ekim T (eds) Resimli Türkiye Florasi (Ilustrated flora of Turkey), vol 1. Ali Nihat Gökyigit Vakfi, Flora Arastirmalari Dernegi and Türkiye Is Bankasi Kültür Yayinlari yayini, Istanbul, pp 319–420

Erturk Y, Ercisli S, Tosun M (2009) Physico-chemical characteristics of wild plum fruits (*Prunus spinosa* L.). Int J Plant Prod 3(3)

Esiyok D, Ötles S, Akcicek E (2004) Herbs as a food source in Turkey. Asian Pac J Cancer Prev 5(3):334–339

Espírito-Santo D, Galego L, Goncalves T, Quintas C (2012) Yeast diversity in the Mediterranean strawberry tree (*Arbutus unedo* L.) fruits' fermentations. Food Res Int 47(1):45–50

Etkin NL (1996) Medicinal cuisines: diet and Ethnopharmacology. Int J Pharmacogn 34(5):313–326

European Medicines Agency (2011) Assessment report on *Capsella bursa-pastoris* (L.) Medikus, herba. European Medicines Agency, European Union, London

European Parliament and Council (2006) Regulation (EC) No. 1924/2006 of the European Parliament and of the Council on nutrition and health claims made on foods. Off J Eur Union L12/3:18. (18/01/2007)

European Parliament and Council (2011) Regulation (EU) No 1169/2011 of the European Parliament and of the Council of 25 October 2011, on the provision of food information to consumers. Off J Eur Union L304/18:63. (22/11/2011)
Fadda A, Mulas M (2010) Chemical changes during myrtle (*Myrtus communis* L.) fruit development and ripening. Sci Hortic 125(3):477–485
Fajardo J (2008) Estudio etnobiológico de los alimentos locales de la Serranía de Cuenca. PhD Thesis. Universidad de Castilla La Mancha
Fajardo J, Verde A, Rivera D, Obón C (2000) Las plantas en la cultura popular de la provincia de Albacete. Instituto de Estudios Albacetenses, Albacete
Fajardo J, Verde A, Rivera D, Obón C (2007) Etnobotánica en La Serranía de Cuenca. Las plantas y el hombre. Diputación de Cuenca, Cuenca
FAO/WHO (2001) Human vitamin and mineral requirements. Report of a joint FAO/WHO expert consultation, Bangkok, Thailand. Food and Nutrition Division. FAO, Rome
FAO/WHO (1998) Carbohydrates in human nutrition. Report of a joint FAO/WHO expert consultation, Rome, Italy, 14–18 April 1997. FAO, Rome
Fernández Ocaña AM (2000) Estudio etnobotánico en el Parque Natural de las Sierras de Cazorla, Segura y Las Villas. Investigación química de un grupo de especies interesantes. PhD Thesis. Universidad de Jaén
Ferrández JV, Sanz JM (1993) Las plantas en la medicina popular de la comarca de Monzón. Instituto de estudios Altoaragoneses (Diputación de Huesca), Huesca
Font Quer P (1962) Plantas medicinales. El Dioscórides renovado. Ed. Labor, Barcelona
Franke W (1982) Vitamin-C in Sea Fennel (*Crithmum maritimum*), an edible wild plant. Econ Bot 36(2):163–165
Franke W, Kensbock A (1981) Vitamin-C content of native wild growing vegetables and greens. Ernahrungs-Umschau 28(6):187–191
Froehlicher T, Hennebelle T, Martin-Nizard F, Cleenewerck P, Hilbert JL, Trotin F, Grec S (2009) Phenolic profiles and antioxidative effects of hawthorn cell suspensions, fresh fruits, and medicinal dried parts. Food Chem 115(3):897–903
Galán R (1993) Patrimonio etnobotánico de la provincia de Córdoba: Pedroches, Sierra Norte y Vega del Guadalquivir. PhD Thesis. Universidad de Córdoba
Gálvez M, Martín-Cordero C, Houghton PJ, Ayuso MJ (2005) Antioxidant activity of methanol extracts obtained from *Plantago* species. J Agric Food Chem 53(6):1927–1933
Ganhão R, Estévez M, Kylli P, Heinonen M, Morcuende D (2010) Characterization of selected wild Mediterranean fruits and comparative efficacy as inhibitors of oxidative reactions in emulsified raw pork burger patties. J Agric Food Chem 58:8854–8861
García Jiménez R (2007) Etnobotánica Leonesa. Municipio de Palacios del Sil. PhD Thesis. Universidad Complutense de Madrid
García-González JA (2008) Los lotófagos y el loto homérico. Baetica 30:171–188
García-Herrera P (2014) Plantas silvestres de consumo tradicional. Caracterización de su valor nutricional y estimación de su actividad antifúngica. PhD Thesis. Universidad Complutense de Madrid
García-Herrera P, Sánchez-Mata MC, Cámara M, Tardío J, Olmedilla-Alonso B (2013) Carotenoid content of wild edible young shoots traditionally consumed in Spain (*Asparagus acutifolius* L., *Humulus lupulus* L., *Bryonia dioica* Jacq. and *Tamus communis* L.). J Sci Food Agric 94:1914–1916. (J Sci Food Agric 93:1692–1698. Erratum published)
García-Herrera P, Morales P, Fernández-Ruiz V, Sánchez-Mata MC, Cámara M, Carvalho AM, Ferreira ICFR, Pardo-de-Santayana M, Molina M, Tardío J (2014a) Nutrients, phytochemicals and antioxidant activity in wild populations of *Allium ampeloprasum*, a valuable underutilized vegetable. Food Res Int 62:272–279
García-Herrera P, Sánchez-Mata MC, Cámara M, Fernández-Ruiz V, Díez-Marqués C, Molina M, Tardío J (2014b) Nutrient composition of six wild edible Mediterranean Asteraceae plants of dietary interest. J Food Compos Anal 34:163–170
Garidel PJ (1715) Histoire des plantes qui naissent aux environs d'Aix, et dans plusieurs autres endroits de la Provence. Joseph David, Aix-en-Provence

Gatto MA, Ippolito A, Linsalata V, Cascarano NA, Nigro F, Vanadia S, Di Venere D (2011) Activity of extracts from wild edible herbs against postharvest fungal diseases of fruit and vegetables. Postharvest Biol Technol 61(1):72–82

Ghirardini M, Carli M, Del Vecchio N, Rovati A, Cova O, Valigi F, Agnetti G, Macconi M, Adamo D, Traina M, Laudini F, Marcheselli I, Caruso N, Gedda T, Donati F, Marzadro A, Russi P, Spaggiari C, Bianco M, Binda R, Barattieri E, Tognacci A, Girardo M, Vaschetti L, Caprino P, Sesti E, Andreozzi G, Coletto E, Belzer G, Pieroni A (2007) The importance of a taste. A comparative study on wild food plant consumption in twenty-one local communities in Italy. J Ethnobiol Ethnomed 3(1):22

Gil Pinilla M (1995) Estudio etnobotánico de la Flora aromática y medicinal del término municipal de Cantalojas (Guadalajara). PhD Thesis. Universidad Complutense de Madrid

Gjorgieva D, Kadifkova-Panovska T, Baceva K, Stafilov T (2010) Some toxic essential metals in medicinal plants growing in R. Macedonia. American-Eurasian J Toxicol Sci 2(1):57–71

Gjorgieva D, Kadifkova-Panovska T, Baeva K, Stafilov T (2011) Metalic trace elements in medicinal plants from Macedonia. Middle-East J Sci Res 7(1):109–114

Gómez-Ortega C (1784). Continuación de La Flora española ó Historia de las plantas de España, que escribía don Joseph Quer, vol 5–6. Joachin Ibarra, Madrid

González JA, García-Barriuso M, Amich F (2010) Ethnobotanical study of medicinal plants traditionally used in the Arribes del Duero, Western Spain. J Ethnopharmacol 131(2):343–355

González JA, García-Barriuso M, Amich F (2011a) The consumption of wild and semi-domesticated edible plants in the Arribes del Duero (Salamanca-Zamora, Spain): an analysis of traditional knowledge. Genet Resour Crop Evol 58(7):991–1006

González JA, García-Barriuso M, Amich F (2011b) Ethnoveterinary medicine in the Arribes del Duero, Western Spain. Vet Res Commun 35(5):283–310

González Turmo I (1997) Comida de rico, comida de pobre. Evolución de los hábitos alimentarios en el occidente andaluz (siglo XX). Universidad de Sevilla, Sevilla

González-Hernández MP, Romero R, Rodríguez-Guitián M, Rigueiro A (2004) Medicinal use of some plants in Galicia (NW Spain). Acta Hort (ISHS) 629:63–75

González-Tejero MR (1989) Investigaciones etnobotánicas en la provincia de Granada. PhD Thesis. Universidad de Granada

González-Tejero MR, Casares-Porcel M, Sánchez-Rojas CP, Ramiro-Gutiérrez JM, Molero Mesa J, Pieroni A, Giusti ME, Censorii E, De Pasquale C, Della A, Paraskeva-Hadjichambi D, Hadjichambis AC, Houmanie Z, El-Demerdash M, El-Zayat M, Hmamouchi M, ElJohrig S (2008) Medicinal plants in the Mediterranean area: synthesis of the results of the project Rubia. J Ethnopharmacol 116:341–357

Greenfield H, Southgate DAT (2003) Food composition data. Production, management and use. FAO, Rome

Guarrera PM (2006) Usi e tradizioni della flora italiana. Aracne, Roma

Guenaoui C, Mang S, Figliuolo G, Neffati M (2013) Diversity in *Allium ampeloprasum*: from small and wild to large and cultivated. Genet Resour Crop Evol 60(1):97–114

Guil-Guerrero JL (2001) Nutritional composition of *Plantago* species (*P. major* L., *P. lanceolata* L., and *P. media* L.). Ecol Food Nutr 40(5):481–495

Guil-Guerrero JL, Rodríguez-García I (1999) Lipids classes, fatty acids and carotenes of the leaves of six edible wild plants. Eur Food Res Technol 209(5):313–316

Guil-Guerrero JL, Torija-Isasa ME (1997) Nutritional composition of leaves of *Chenopodium* species (*C. album* L., *C. murale* L. and *C. opulifolium* Shraeder). Int J Food Sci Nutr 48(5):321–327

Guil-Guerrero JL, Torija-Isasa ME (2002) Edible wild plants. In: Majundar DK, Govil JN, Singh VK (eds) Recent progress in medicinal plants, vol VIII. SciTech Publising LLC, Texas, pp 431–466

Guil-Guerrero JL, Torija-Isasa ME, Giménez-Martínez JJ (1996a) Nutritional composition of rock samphire (*Crithmum maritimum* L.). Alimentaria 272:65–72

Guil-Guerrero JL, Torija-Isasa ME, Giménez-Martínez JJ, Rodríguez-García I (1996b) Identification of fatty acids in edible wild plants by gas chromatography. J Chromatogr A 719:229–235

Guil-Guerrero JL, Rodríguez-García I, Torija-Isasa ME (1997a) Nutritional and toxic factors in selected wild edible plants. Plant Foods Hum Nutr 51(2):99–107

Guil-Guerrero JL, Torija-Isasa E, Giménez Martínez JJ (1997b) Composición centesimal de plantas silvestres comestibles. Alimentaria 280:59–64

Guil-Guerrero JL, Giménez-Giménez A, Rodríguez-García I, Torija-Isasa ME (1998a) Nutritional composition of *Sonchus* species (*S. asper* L, *S. oleraceus* L and *S. tenerrimus* L). J Sci Food Agric 76(4):628–632

Guil-Guerrero JL, Giménez-Martínez JJ, Torija-Isasa ME (1998b) Mineral nutrient composition of edible wild plants. J Food Compos Anal 11:322–328

Guil-Guerrero JL, Campra-Madrid P, Torija-Isasa ME (1999a) Mineral elements determination in wild edible plants. Ecol Food Nutr 38(3):209–222

Guil-Guerrero JL, Giménez-Martínez JJ, Torija-Isasa ME (1999b) Nutritional composition of wild edible crucifer species. J Food Biochem 23(3):283–294

Guil-Guerrero JL, Rebolloso-Fuentes MM, Torija-Isasa ME (2003) Fatty acids and carotenoids from stinging nettle (*Urtica dioica* L.). J Food Compos Anal 16(2):111–119

Gurses O, Artik N (1984) A research on the composition of *Beta vulgaris* var. *cicla, Malva neglecta* Wallr., *Portulaca oleracea* L. and *Spinacia oleracea* L. Gida Dergisi 9(2):83–87

Guzmán MA (1997) Aproximación a la etnobotánica de la provincia de Jaén. PhD Thesis. Universidad de Granada

Haciseferoğullari H, Özcan MM, Arslan D, Unver A (2012) Biochemical compositional and technological characterizations of black and white myrtle (*Myrtus communis* L.) fruits. J Food Sci Tech Mys 49(1):82–88

Hadjichambis AC, Paraskeva-Hadjichambi D, Della A, Giusti ME, De Pasquale C, Lenzarini C, Censorii E, González-Tejero MR, Sánchez-Rojas CP, Ramiro-Gutiérrez JM, Skoula M, Johnson C, Sarpaki A, Hmamouchi M, Jorhi S, El-Demerdash M, El-Zayat M, Pieroni A (2008) Wild and semi-domesticated food plant consumption in seven circum-Mediterranean areas. Int J Food Sci Nutr 59(5):383–414

Hanelt P, Institute of Plant Genetics and Crop Plant Research (2001) Mansfeld's encyclopedia of agricultural and horticultural crops. Springer, Berlin

Heiss AG (2014) Ceremonial foodstuffs from prehistoric burnt-offering places in the Alpine region. In: Chevalier A, Marinova E, Peña-Chocarro L (eds) Plants and people: choices and diversity through time. Oxbow Books, Oxford, pp 343–353

Henneberg W, Stohmann F (1860) Beiträge zur Begründung einerrationellen Fütterung der Wiederkäuer I & II. Schwetschke, Braunschweig

Hiçsönmez U, Erees FS, Özdemir C, Özdemir A, Çam S (2009) Determination of major and minor elements in the *Malva sylvestris* L. from Turkey using ICP-OES techniques. Biol Trace Elem Res 128(3):248–257

Holland B, Unwin ID, Buss DH (1991) Vegetables, herbs and spices. Fifth supplement to McCance and Widdowson's the composition of foods. Royal Society of Chemistry, Cambridge

Hussain J, Muhammad Z, Ullah R, Khan F, Khan IU, Khan N, Ali J, Jan S (2010) Evaluation of the chemical composition of *Sonchus eruca* and *Sonchus asper*. J Am Sci 6(9):231–235

Isler O (1971) Carotenoids. Springer, Basel

IUCN (2005) A guide to medicinal plants in North Africa. IUCN Centre for Mediterranean Cooperation, Málaga

Jabłońska-Ryś E, Zalewska-Korona M, Kalbarczyk J (2009) Antioxidant capacity, ascorbic acid and phenolics content in wild edible fruits. J Fruit Ornam Plant Res 17(2):115–120

Jage H (1979) Portulacaceae. In: Conert HJ, Schultlze-Motel W, Wagenitz G (eds) Gustav Hegi Illustrierte Flora von Mitteleuropa. Paul Parey, Berlin, pp 1183–1221

Kiliç CS, Coşkun M (2007) *Capsella bursa-pastoris* (L.) Medik. (Cruciferae) Askorbik asit içeriği üzerinde karşilaştirmali bir çalişma. J Fac Pharm Ankara 36(3):153–160

Kirleis W, Klooß S (2014) Ceremonial foodstuffs from prehistoric burnt-offering places in the Alpine region. In: Chevalier A, Marinova E, Peña-Chocarro L (eds) Plants and people: choices and diversity through time. Oxbow Books, Oxford, pp 326–335

Konig J (1878) Chemie der Menschlichen Nahrungs-und Genussmittel. Springer, Berlin

Krstic-Pavlovic N, Dzamic R (1985) Astringent and mineral components in the leaves of nettle (*Urtica dioica*, L.) from many natural locations. Agrochemija 3:191–198
Kudritsata SE, Filman GM, Zagorodskaya LM, Chikovanii DM (1987) Carotenoids of *Urtica dioica*. Khim Prir Soedin 6(5):640–641
Kuhnlein HV (1990) Nutrient values of indigenous wild plant greens and roots used by the Nuxalk people of Bella Coole, British Columbia. J Food Compos Anal 3:38–46
Kültür S (2008) An ethnobotanical study of Kirklareli (Turkey). Phytologia Balcanica 14(2):279–289
Laghetti G (2009) Microevolution of *Scolymus Hispanicus* L. (compositae) in south Italy: from gathering of wild plants to some attempts of cultivation. In: Buerkert A, Gebauer J (eds) Agrobiodiversity and genetic erosion. Contributions in Honor of Prof. Dr. Karl Hammer. Kassel University Press, Kassel, pp 119–126
Laghetti G, Perrino P, Hammer K (1994) Utilization of *Silene vulgaris* (Moench) Garcke in Italy. Econ Bot 48(3):337–339
Laguna A (1555) Pedacio Dioscorides Anazarbeo, Acerca de la materia medicinal y de los venenos mortíferos. Edición facsímil 1991. Comunidad de Madrid, Madrid
Lastra JJ (2003) Etnobotánica en el Parque Nacional de Picos de Europa. Ministerio de Medio Ambiente, Parques Nacionales, Madrid
Latorre JA (2009) Estudio etnobotánico de la provincia de La Coruña. PhD Thesis. Universitat de Valencia
Le Floc'h E (1983) Contribution à une étude ethnobotanique de la flore tunisienne. Ministère de l'Enseignement Supérieur et de la Recherche scientifique, Tunis
Lentini F, Venza F (2007) Wild food plants of popular use in Sicily. J Ethnobiol Ethnomed 3:15
Leonti M (2012) The co-evolutionary perspective of the food-medicine continuum and wild gathered and cultivated vegetables. Genet Resour Crop Evol 59(7):1295–1302
Leonti M, Nebel S, Rivera D, Heinrich M (2006) Wild gathered food plants in the European Mediterranean: a comparison analysis. Econ Bot 60(2):130–142
Liu LX, Howe P, Zhou YF, Hocart C, Zhang R (2002) Fatty acid profiles of leaves of nine edible wild plants: an Australian study. J Food Lipids 9(1):65–71
Louhaichi M, Salkini AK, Estita HE, Belkhir S (2011) Initial assessment of medicinal plants across the Libyan Mediterranean Coast. Adv Environ Biol 5(2):359–370
Łuczaj Ł (2010) Changes in the utilization of wild green vegetables in Poland since the 19th century: a comparison of four ethnobotanical surveys. J Ethnopharmacol 128:395–404
Łuczaj Ł, Pieroni A, Tardío J, Pardo-de-Santayana M, Soukand R, Svanberg I, Kalle R (2012) Wild food plant use in 21st century Europe: the disappearance of old traditions and the search for new cuisines involving wild edibles. Acta Soc Bot Pol 81(4):359–370
Łuczaj Ł, Fressel N, Perkovic S (2013a) Wild food plants used in the villages of the Lake Vrana Nature Park (Northern Dalmatia, Croatia). Acta Soc Bot Pol 82(4):275–281
Łuczaj Ł, Zovko-Koncic M, Milicevic T, Dolina K, Pandza M (2013b) Wild vegetable mixes sold in the markets of Dalmatia (Southern Croatia). J Ethnobiol Ethnomed 9:2
Mahan LK, Escott-Stump S, Raymond JL (2012) Krause Dietoterapia, 13th edn. Elsevier, Madrid
Males Z, Zuntar I, Nigovic B, Plazibat M, Vundac VB (2003) Quantitative analysis of the polyphenols of the aerial parts of rock samphire—*Crithmum maritimum* L. Acta Pharm 53(2):139–144
Marakoğlu T, Arslan D, Özcan M, Haciseferoğullari H (2005) Proximate composition and technological properties of fresh blackthorn (*Prunus spinosa* L. subsp. *dasyphylla* (Schur.) Domin) fruits. J Food Eng 68(2):137–142
Marco C, Chauvet M, Mathez J, Ubaud J, Passama L, Garrone B, Molina J, Cornillon M, Martin P, Wotan JM, Walsh J (2003) Les salades sauvages. L'Ensalada champanèla. Les Ecologistes de L'Euzière, Sant Jean de Cuculles
Marouf ML (2005) Assessing the current use and significance of wild edible plants traditionally gathered in Lebanon: an ethnobotanical study. Master of Science in Environmental Sciences Thesis. American University of Beirut

Martínez-Lirola MJ, González-Tejero MR, Molero Mesa J (1997) Investigaciones etnobotánicas en el Parque Natural de Cabo de Gata-Níjar (Almería). Sociedad Almeriense de Historia Natural, Almería

Martínez-Para MC, Masoud TA, Torija-Isasa ME (1979) La ortiga en la alimentación. I) Composición centesimal y elementos minerales. An Bromatol 31(3–4):327–331

Martins D, Barros L, Carvalho AM, Ferreira ICFR (2011) Nutritional and in vitro antioxidant properties of edible wild greens in Iberian Peninsula traditional diet. Food Chem 125(2):488–494

McCollum EV (1957) A history of nutrition. Houghton Mifflin, Boston

Mendonça de Carvalho LM (2006) Estudos de Etnobotânica e Botânica Económica no Alentejo. PhD Thesis. Universidade de Coimbra

Menendez-Baceta G, Aceituno-Mata L, Tardío J, Reyes-García V, Pardo-de-Santayana M (2012) Wild edible plants traditionally gathered in Gorbeialdea (Biscay, Basque Country). Genet Resour Crop Evol 59(7):1329–1347

Menendez-Baceta G, Aceituno-Mata L, Molina M, Reyes-García V, Tardío J, Pardo-de-Santayana M (2014) Medicinal plants traditionally used in the Northwest of the Basque Country (Biscay and Alava), Iberian Peninsula. J Ethnopharmacol 152(1):113–134

Mercadante AZ, Rodríguez-Amaya DB (1990) Carotenoid composition and vitamin-A value of some native Brazilian green leafy vegetables. Int J Food Sci Technol 25(2):213–219

Mesa S (1996) Estudio Etnobotánico y Agroecológico de la comarca de la Sierra de Mágina (Jaén). PhD Thesis. Universidad Complutense de Madrid

Mhamdi B, Aidi Wannes W, Hosni K, Bellila A, Chahed T, Kchouk ME, Marzouk B (2007) *Borago officinalis* L. foliar fatty acids. Asian J Biochem 2:79–83

Molero Mesa J, Soguero Muñoz A, Jiménez JJ (2001) Estudio Etnobotánico del macizo central de Sierra Morena (Sierra de San Andrés) y extremo suroriental del histórico Campo de Calatrava de la provincia de Ciudad Real. Universidad de Granada, Granada

Molina N (2001) Estudio de la flora de interés etnobotánico en el municipio de Carcabuey (Córdoba). B. Sc. Thesis. Universidad de Córdoba

Molina M, Pardo-de-Santayana M, Aceituno L, Morales R, Tardío J (2011) Fruit production of strawberry-tree (*Arbutus unedo* L.) in two Spanish forests. Forestry 84:419–429

Molina M, Pardo-de-Santayana M, García E, Aceituno L, Morales R, Tardío J (2012) Exploring the potential of wild food resources in the Mediterranean region: natural yield and gathering pressure of the wild asparagus (*Asparagus acutifolius* L.). Span J Agric Res 10(4):1090–1100

Moll M (2005) Les plantes a Menorca: noms i usos. Institut Menorquí d' Estudis, Maó

Monardes N (1574) Primera y segunda y tercera parte de la Historia Medicinal de las cosas que se traen de nuestras Indias Occidentales que sirven en Medicina. Alonso Escribano, Sevilla

Montoro P, Tuberoso CI, Piacente S, Perrone A, De Feo V, Cabras P, Pizza C (2006) Stability and antioxidant activity of polyphenols in extracts of *Myrtus communis* L. berries used for the preparation of myrtle liqueur. J Pharmaceut Biomed Anal 41(5):1614–1619

Morales P (2011) Vegetales silvestres de uso alimentario: determinación de compuestos bioactivos y capacidad antioxidante. PhD Thesis. Universidad Complutense de Madrid

Morales P, Carvalho AM, Z-Mata MC, Cámara M, Molina M, Ferreira ICFR (2012a) Tocopherol composition and antioxidant activity of Spanish wild vegetables. Genet Resour Crop Evol 59:851–863

Morales P, Ferreira ICFR, Carvalho AM, Sánchez-Mata MC, Cámara M, Tardío J (2012b) Fatty acids profiles of some Spanish wild vegetables. Food Sci Technol Int 18(3):281–290

Morales P, Ferreira IC, Carvalho AM, Fernández-Ruiz V, Sánchez-Mata MC, Cámara M, Morales R, Tardío J (2013) Wild edible fruits as a potential source of phytochemicals with capacity to inhibit lipid peroxidation. Eur J Lipid Sci 115(2):176–185

Morales P, Ferreira ICFR, Carvalho AM, Sánchez-Mata MC, Cámara M, Fernández-Ruiz V, Pardo-de-Santayana M, Tardío J (2014) Mediterranean non-cultivated vegetables as dietary sources of compounds with antioxidant and biological activity. LWT-Food Sci Technol 55:389–396

Morales P, Fernández-Ruiz V, Sánchez-Mata MC, Cámara M, Tardío J (2015) Optimization and application of FL-HPLC for folates analysis in 20 species of Mediterranean wild vegetables. Food Anal Method 8:302–311

Morán I, Baldirà J, Marruecos L, Nogué S (2011) Toxicología clínica. Difusión Jurídica y Temas de Actualidad, Madrid
Mulet L (1991) Estudio etnobotánico de la provincia de Castellón. Diputación de Castellón, Castellón
Mulet L (1997) Flora tóxica de la Comunidad Valenciana. Diputación de Castellón, Castellón
Muntané J (1991) Aportació al coneixement de l'etnobotànica de Cerdanya. PhD Thesis. Universitat de Barcelona
Nassif F, Tanji A (2013) Gathered food plants in Morocco: the long forgotten species in ethnobotanical research. Life Sci Leafl 3(3):17–54
Nebel S, Pieroni A, Heinrich M (2006) *Ta chòrta*: wild edible greens used in the Graecanic area in Calabria, Southern Italy. Appetite 47(3):333–342
Nedelcheva A, Dogan Y, Guarrera P (2007) Plants traditionally used to make brooms in several European countries. J Ethnobiol Ethnomed 3(1):20
Neves JM, Matos C, Moutinho C, Queiroz G, Gomes GL (2009) Ethnopharmacological notes about ancient uses of medicinal plants in Trás-os-Montes (Northern of Portugal). J Ethnopharmacol 124(2):270–283
Nogué S (2010) Intoxicaciones agudas. Bases para el tratamiento en un servicio de urgencias. Área Científica Menarini, Badalona
Obón C (2006) La importancia de la conservación de las plantas comestibles locales en la Vega Baja (Alicante). In: Proyecto de Recuperación del Conocimiento Tradicional Agrícola de la Vega Baja del Segura "Raijos de la Vega". Asociación Natura Viva, Almoradí. pp 7–13
Oltra JE (1998) Fer herbes a Quatretonda. Col·lectiu Cultural Dorresment, Quatretonda
Osbaldeston TA (2000) The herbal of dioscorides the Greek. IBIDIS Press, Johannesburg
Özbucak TB, Akçin OE, Yalçin S (2007) Nutrition contents of some wild edible plants in Central Black Sea Region of Turkey. Int J Nat Eng Sci 1:11–13
Özcan MM, Haciseferoğullari H (2007) The strawberry (*Arbutus unedo* L.) fruits: chemical composition, physical properties and mineral contents. J Food Eng 78:1022–1028
Özcan M, Haciseferoğullari H, Marakoğlu T, Arslan D (2005) Hawthorn (*Crataegus* spp.) fruit: some physical and chemical properties. J Food Eng 69(4):409–413
Özcan MM, Unver A, Ucar T, Arslan D (2008) Mineral content of some herbs and herbal teas by infusion and decoction. Food Chem 106(3):1120–1127
Pallauf K, Rivas-Gonzalo JC, Del Castillo MD, Cano MP, Pascual-Teresa S (2008) Characterization of the antioxidant composition of strawberry tree (*Arbutus unedo* L.) fruits. J Food Compos Anal 21(4):273–281
Paludosi S, Pignone D (1997) Rocket: a Mediterranean crop for the world. Report of a workshop,13–14 December 1996, Legnaro (Padova), Italy. International Plant Genetic Resources Institute (IPGRI), Roma
Paoletti MG, Dreon AL, Lorenzoni GG (1995) Pistic, traditional food from Western Friuli, Ne Italy. Econ Bot 49(1):26–30
Parada M (2008) Estudi etnobotànic de L'Alt Empordà. PhD Thesis. Facultat de Farmàcia. Universitat de Barcelona
Parada M, Selga A, Bonet MA, Vallès J (2002) Etnobotànica de les terres gironines: natura i cultura popular a la plana interior de l'Alt Empordà i de les Guilleries. Diputació de Girona, Girona
Parada M, Carrió E, Vallès J (2011) Ethnobotany of food plants in the Alt Empordà region (Catalonia, Iberian Peninsula). J Appl Bot Food Qual 84:11–25
Paradores (2014) Jarandilla de la Vera y su Parador. http://www.parador.es/sites/default/files/parador/…/08/jarandilladelavera.pdf. Accessed 6 May 2014
Pardo-de-Santayana M (2003) Las plantas en la cultura tradicional de la antigua Merindad de Campoo. PhD Thesis. Universidad Autónoma de Madrid
Pardo-de-Santayana M (2004) Guía de las plantas medicinales de Cantabria. Salud y tradición popular. Stvdio, Santander
Pardo-de-Santayana M (2008) Estudios etnobotánicos en Campoo (Cantabria): conocimiento y uso tradicional de plantas. CSIC, Madrid

Pardo-de-Santayana M, Tardío J, Blanco E, Carvalho AM, Lastra JJ, San Miguel E, Morales R (2007) Traditional knowledge on wild edible plants in the Northwest of the Iberian Peninsula (Spain and Portugal): a comparative study. J Ethnobiol Ethnomed 3:27

Pehrsson P, Scheett A, Phillips K, Patterson K, Lukaski H (2011) Carotenoids and folate vitamers in indigenous plants consumed by Northern Plains American Indians. FASEB J 25(609):5

Pellicer J (2001) Customari botànic. Recerques etnobotàniques a les comarques centrals valencianes. Edicions del Bullent, Picanya

Pellicer J (2004a) Customari botànic (2). Recerques etnobotàniques a les comarques centrals valencianes. Edicions del Bullent, Picanya

Pellicer J (2004b) Customari botànic (3). Recerques etnobotàniques a les comarques centrals valencianes. Edicions del Bullent, Picanya

Pereira C, Barros L, Carvalho AM, Ferreira IC (2011) Nutritional composition and bioactive properties of commonly consumed wild greens: potential sources for new trends in modern diets. Food Res Int 44(9):2634–2640

Pereira C, Barros L, Carvalho AM, Ferreira IC (2013) Use of UFLC-PDA for the analysis of organic acids in thirty-five species of food and medicinal plants. Food Anal Method 6(5):1337–1344

Perera López J (2005) La toponimia de La Gomera. Un estudio sobre los nombres de lugar, las voces indígenas y los nombres de plantas, animales y hongos de La Gomera. AIDER, La Gomera

PFAF (2014) Plants for a Future: edible, medicinal and useful plants for a healthier world. http://www.pfaf.org. Accessed 2014

Picchi G, Pieroni A (2005) Atlante dei prodotti tipici: le erbe. Agra, RAI-Eri, Roma

Piera H (2006) Plantas silvestres y setas comestibles en la comarca Valle Ayora-Cofrentes. Mancomunidad Comarcal Valle de Ayora-Cofrentes, Ayora

Pieroni A (1999) Gathered wild food plants in the upper valley of the Serchio river (Garfagnana), Central Italy. Econ Bot 53(3):327–341

Pieroni A (2001) Evaluation of the cultural significance of wild food botanicals traditionally consumed in Northwestern Tuscany, Italy. J Ethnobiol 21(1):89–104

Pieroni A (2008) Local plant resources in the ethnobotany of Theth, a village in the Northern Albanian Alps. Genet Resour Crop Evol 55(8):1197–1214

Pieroni A, Price LL (2006) Eating and healing. Traditional food as medicine. Food Product Press, Haworth Press, New York

Pieroni A, Nebel S, Quave C, Münz H, Heinrich M (2002) Ethnopharmacology of liakra: traditional weedy vegetables of the Arbëreshë of the Vulture area in Southern Italy. J Ethnopharmacol 81:165–185

Pieroni A, Quave CL, Villanelli ML, Mangino P, Sabbatini G, Santini L, Boccetti T, Profili M, Ciccioli T, Rampa LG, Antonini G, Girolamini C, Cecchi M, Tomasi M (2004) Ethnopharmacognostic survey on the natural ingredients used in folk cosmetics, cosmeceuticals and remedies for healing skin diseases in the inland Marches, Central-Eastern Italy. J Ethnopharmacol 91(2–3):331–344

Pieroni A, Nebel S, Santoro RF, Heinrich M (2005) Food for two seasons: culinary uses of non-cultivated local vegetables and mushrooms in a South Italian village. Int J Food Sci Nutr 56(4):245–272

Polat R, Satil F (2012) An ethnobotanical survey of medicinal plants in Edremit Gulf (Balikesir—Turkey). J Ethnopharmacol 139(2):626–641

Polat R, Cakilcioglu U, Satil F (2013) Traditional uses of medicinal plants in Solhan (Bingöl-Turkey). J Ethnopharmacol 148(3):951–963

Poschenrieder C, Llugany M, Lombini A, Dinelli E, Bech J, Barcelo J (2012) *Smilax aspera* L. an evergreen Mediterranean climber for phytoremediation. J Geochem Explor 123:41–44

Quer J (1762) Flora española, vol 3. Joachin Ibarra, Madrid

Queralt I, Ovejero M, Carvalho ML, Marques AF, Llabres JM (2005) Quantitative determination of essential and trace element content of medicinal plants and their infusions by XRF and ICP techniques. X-Ray Spectrom 34(3):213–217

Rabal G (2000) "Cuando la chicoria echa la flor…" Etnobotánica en Torre Pacheco. Rev Murc Antropol 6:1–240

Raju M, Varakumar S, Lakshminarayana R, Krishnakantha TP, Baskaran V (2007) Carotenoid composition and vitamin A activity of medicinally important green leafy vegetables. Food Chem 101(4):1598–1605

Ranfa A, Maurizi A, Romano B, Bodesmo M (2014) The importance of traditional uses and nutraceutical aspects of some edible wild plants in human nutrition: the case of Umbria (Central Italy). Plant Biosyst 148(2):297–306

Redžić S (2006) Wild edible plants and their traditional use in the human nutrition in Bosnia-Herzegovina. Ecol Food Nutr 45(3):189–232

Rigat M, Bonet MA, García-Giménez S, Garnatje T, Vallès J (2009) Ethnobotany of food plants in the High River Ter Valley (Pyrenees, Catalonia, Iberian Peninsula): non-crop food vascular plants and crop food plants with medicinal properties. Ecol Food Nutr 48(4):303–326

Rigat M, Vallès J, Iglesias J, Garnatje T (2013) Traditional and alternative natural therapeutic products used in the treatment of respiratory tract infectious diseases in the Eastern Catalan Pyrenees (Iberian Peninsula). J Ethnopharmacol 148(2):411–422

Rivera D, Obón C (1991) La guía de INCAFO de las plantas útiles y venenosas de la Península Ibérica y Baleares (excluidas medicinales). INCAFO, Madrid

Rivera D, Obón C, Inocencio C, Heinrich M, Verde A, Fajardo J, Llorach R (2005) The ethnobotanical study of local mediterranean food plants as medicinal resources in Southern Spain. J Physiol Pharmacol 56(Suppl 1):97–114

Rivera D, Obón C, Heinrich M, Inocencio C, Verde A, Fajardo J (2006) Gathered Mediterranean food plants-ethnobotanical investigations and historical development. Forum Nutr 59:18–74

Rivera D, Matilla-Riquer G, Obón C, Alcaraz F (2012) Plants and humans in the Near East and the Caucasus: ancient and traditional uses of plants as food and medicine, a diachronic ethnobotanical review (Armenia, Azerbaijan, Georgia, Iran, Iraq, Lebanon, Syria, and Turkey). Editum, Murcia

Rodríguez-Berrocal J, Castro-Escobar MC, Tirado-Serrano J (1987) Aprovechamiento de recursos alimenticios naturales: II. Valoración nutritiva de algunas especies vegetales del clímax mediterráneo en biotopos de Sierra Morena. Arch Zootec 36(134):65–74

Romojaro A, Botella MA, Obon C, Pretel MT (2013) Nutritional and antioxidant properties of wild edible plants and their use as potential ingredients in the modern diet. Int J Food Sci Nutr 64(8):944–952

Rosati A (2001) Un possibile futuro per l'asparago selvatico. L'Informatore Agrario 7:89–92

Rosati A, Pepe R, Senatore A, Perrone D, Falavigna A (2005) Produttività dell'asparago selvatico. L'Informatore Agrario 8:75–77

Rottoli M (2014) Before the empire: prehistoric fruit gathering and cultivation in Northern Italy. In: Chevalier A, Marinova E, Peña-Chocarro L (eds) Plants and people: choices and diversity through time. Oxbow Books, Oxford, pp 163–169

Rsaissi N, Kamili EL, Bencharki B, Hillali L, Bouhache M (2013) Antimicrobial activity of fruits extracts of the wild jujube *Ziziphus lotus* (L.) Desf. Int J Sci Eng Res 4(9):1521–1528

Ruiz-Rodríguez BM (2014) Frutos silvestres de uso tradicional en la alimentación. Evaluación de su valor nutricional, compuestos bioactivos y capacidad antioxidante. PhD Thesis. Universidad Complutense de Madrid

Ruiz-Rodríguez BM, Morales P, Fernández-Ruiz V, Sánchez-Mata MC, Cámara M, Díez-Marqués C, Pardo-de-Santayana M, Molina M, Tardío J (2011) Valorization of wild strawberry-tree fruits (*Arbutus unedo* L.) through nutritional assessment and natural production data. Food Res Int 44(5):1244–1253

Ruiz-Rodríguez BM, De Ancos B, Sánchez-Moreno C, Fernández-Ruiz V, Sánchez-Mata MC, Cámara M, Tardío J (2014a) Wild blackthorn (*Prunus spinosa* L.) and hawthorn (*Crataegus monogyna* Jacq.) fruits as valuable sources of antioxidants. Fruits 69(1):61–73

Ruiz-Rodríguez BM, Sánchez-Moreno C, De Ancos B, Sánchez-Mata MC, Fernández-Ruiz V, Cámara M, Tardío J (2014b) Wild *Arbutus unedo* L. and *Rubus ulmifolius* Schott fruits are underutilized sources of valuable bioactive compounds with antioxidant capacity. Fruits 69(6):435–448

Saadoudi M, Hambaba L, Abdeddaim M (2012) Study of the glucidic fraction of *Celtis australis* L., *Crataegus azarolus* L., *Crataegus monogyna* Jacq., *Elaeagnus angustifolia* L. and *Zizyphus lotus* L. fruits. Int J Biol Vet Agric Food Eng 6:195–199

Saleh N, El-Hawary Z, El-Shobaki FA, Abbassy M, Morcos SR (1977) Vitamins content of fruits and vegetables in common use in Egypt. Z Ernahrungswiss 16(3):158–162

Salvatore S, Pellegrini N, Brenna OV, Del Río D, Frasca G, Brighenti F, Tumino R (2005) Antioxidant characterization of some Sicilian edible wild greens. J Agric Food Chem 53(24):9465–9471

San Miguel E (2004) Etnobotánica de Piloña (Asturias). Cultura y saber popular sobre las plantas en un concejo del Centro-Oriente Asturiano. PhD Thesis. Universidad Autónoma de Madrid

Sánchez López MD, García Sanz JA, Gómez Merino A, Zon Blanco S (1994) Plantas útiles de la comarca de la Manchuela. Colectivo de Escuelas Rurales de la Manchuela, Albacete

Sánchez-Mata MC, Cabrera-Loera RD, Morales P, Fernández-Ruiz V, Cámara M, Díez-Marqués C, Pardo-de-Santayana M, Tardío J (2012) Wild vegetables of the Mediterranean area as valuable sources of bioactive compounds. Genet Resour Crop Evol 59(3):431–443

Sánchez-Monge E (1991) Flora agrícola: taxonomía de las Magnoliofitas (Angiospermas) de interés agrícola, con excepción de las de aprovechamiento exclusivamente ornamental o forestal. Ministerio de Agricultura, Pesca y Alimentación, Madrid

Sánchez-Romero MJ (2003) Estudio de la flora de interés etnobotánico en el municipio de Rute (Córdoba). B. Sc. Thesis. Universidad de Córdoba

Scarborough J (1978) Theophrastus on herbals and herbal remedies. J Hist Biol 11(2):353–385

Serce S, Ercisli S, Sengul M, Gunduz K, Orhan E (2010) Antioxidant activities and fatty acid composition of wild grown myrtle (*Myrtus communis* L.) fruits. Pharmacogn Mag 6(21):9–12

Serio A, Chaves-Lopez C, Martuscelli M, Mazzarrino G, Mattia C D, Paparella A (2014) Application of central composite design to evaluate the antilisterial activity of hydro-alcohol berry extract of *Myrtus communis* L. LWT-Food Sci Technol 58(1):116–123

Signorini MA, Piredda M, Bruschi P (2009) Plants and traditional knowledge: an ethnobotanical investigation on Monte Ortobene (Nuoro, Sardinia). J Ethnobiol Ethnomed 5:6

Simopoulos AP (2004) Omega-3 fatty acids and antioxidants in edible wild plants. Biol Res 37:263–277

Somogyi JC (1974) National food composition tables. In: Southgate DAT (ed) Guidelines for the preparation of tables of food composition. Karger, Basel, pp 1–5

Souci SW, Fachmann W, Kraut H (2008) Food composition and nutrition tables. Medpharm Scientific Publishers, Stuttgart

Stef DS, Gergen I, Hărmănescu M, Stef L, Druga M, Heghedus-Mîndru G (2010) Screening of 33 medicinal plants for the microelements content. Anim Sci Biotechnol 43:127–132

Sumbul S, Ahmad MA, Asif M, Akhtar M (2011) *Myrtus communis* Linn.—A review. Indian J Nat Prod Resour 2(4):395–402

Tanji A, Nassif F (1995) Edible weeds in Morocco. Weed Technol 9:617–620

Tardío J (2010) Spring is coming: the gathering and consumption of wild vegetables in Spain. In: Pardo-de-Santayana M, Pieroni A, Puri R (eds) Ethnobotany in the New Europe: people, health and wild plant resources. Berghahn Books, Oxford, pp 211–238

Tardío J, Pascual H, Morales R (2002) Alimentos silvestres de Madrid. Guía de plantas y setas de uso alimentario tradicional en la Comunidad de Madrid. Ediciones La Librería, Madrid

Tardío J, Pascual H, Morales R (2005) Wild food plants traditionally used in the province of Madrid. Econ Bot 59(2):122–136

Tardío J, Pardo-de-Santayana M, Morales R (2006) Ethnobotanical review of wild edible plants in Spain. Bot J Linn Soc 152(1):27–72

Tardío J, Molina M, Aceituno-Mata L, Pardo-de-Santayana M, Morales R, Fernández-Ruiz V, Morales P, García P, Cámara M, Sánchez-Mata MC (2011) *Montia fontana* L. (Portulacaceae), an interesting wild vegetable traditionally consumed in the Iberian Peninsula. Genet Resour Crop Evol 58(7):1105–1118

Tejerina A (2010) Usos y saberes sobre las plantas de Monfragüe. Etnobotánica de la Comarca Natural. Itomonfragüe, Cáceres

Teofrasto (1988) [3rd. century BC]. Historia de las plantas. Introducción, traducción y notas por J. M. Díaz-Regañón. Editorial Gredos, Madrid

Tetik F, Civelek S, Cakilcioglu U (2013) Traditional uses of some medicinal plants in Malatya (Turkey). J Ethnopharmacol 146(1):331–346

The Local Food-Nutraceutical Consortium (2005) Understanding local Mediterranean diets: a multidisciplinary pharmacological and ethnobotanical approuch. Pharmacol Res 52:353–366

Triano EC, Ruiz Cabello E, Fernández Luque A, Gómez Miranda A, Jiménez Conejo A, Gutiérrez Campaña JA, Postigo JA, Castro Montes J, Sánchez Najarro JF, Marín Osuna JR, Martos M, Mérida Moral MD, Mérida Ramírez MJ, Moral R, Hinijosa R (1998) Recupera tus tradiciones. Etnobotánica del Subbético Cordobés. Ayuntamiento de Carcabuey, Carcabuey

Trichopoulou A, Vasilopoulou E, Hollman P, Chamalides Ch, Foufa E, Kaloudis T, Kromhout D, Miskaki Ph, Petrochilou I, Poulima E, Stafilakis K, Theophilou D (2000) Nutritional composition and flavonoid content of edible wild greens and green pies: a potential rich source of antioxidant nutrients in the Mediterranean diet. Food Chem 70:319–323

Tukan SK, Takruri HR, Al Eisawi DM (1998) The use of wild edible plants in the Jordanian diet. Int J Food Sci Nutr 49(3):225–235

Turan M, Kordali S, Zengin H, Dursun A, Sezen Y (2003) Macro and micro mineral content of some wild edible leaves consumed in Eastern Anatolia. Acta Agric Scand 53(3):129–137

Tutin TG, Heywood VH, Burges DM, Moore DH, Valentine SM, Walters SM, Webb DA (1964–1980) Flora Europaea, vols 1–5. University Press, Cambridge

Vallejo JR, Peral D, Carrasco MC (2009) Anotaciones al conocimiento etnobotánico y medicinal de los espárragos extremeños. Med Natur 3(1):41–46

Vanzani P, Rossetto M, De Marco V, Sacchetti LE, Paoletti MG, Rigo A (2011) Wild mediterranean plants as traditional food: a valuable source of antioxidants. J Food Sci 76:C46–C51

Vârban R, Vârban D (2012) Comparative study of the active ingredients content *Plantago lanceolata* L. ProEnvironment 5:248–250

Vardavas CI, Majchrzak D, Wagner KH, Elmadfa I, Katafos A (2006a) Lipid concentrations of wild edible greens in Crete. Food Chem 99(4):822–834

Vardavas CI, Majchrzak D, Wagner KH, Elmadfa I, Katafos A (2006b) The antioxidant and phylloquinone content of wildly grown greens in Crete. Food Chem 99(4):813–821

Vázquez FM, Suárez MA, Pérez A (1997) Medicinal plants used in the Barros Area, Badajoz Province (Spain). J Ethnopharmacol 55:81–85

Velasco C, Nieto JM, Conde R, Gil Y, Navas P, Silva T (1998) Una prospección etnobotánica en el Campo de Gibraltar. Almoraima 19:131–142

Velasco JM, Criado J, Blanco E (2010) Usos tradicionales de las plantas en la provincia de Salamanca. Diputación de Salamanca, Salamanca

Verde A, Rivera D, Obón C (1998) Etnobotánica en la sierras de Segura y Alcaraz: las plantas y el hombre. Instituto de Estudios Albacetenses, Albacete

Verde A, Fajardo J, Rivera D, Obón C (2001) Etnobotánica en el entorno del Parque Nacional de Cabañeros. Ministerio de Medio Ambiente, Parques Nacionales, Madrid

Verde A, Rivera D, Heinrich M, Fajardo J, Inocencio C, Llorach R, Obón C (2003) Plantas alimenticias recolectadas tradicionalmente en la provincia de Albacete y zonas próximas, su uso tradicional en la medicina popular y su potencial como nutracéuticos. Sabuco Rev Est Albacet 4:35–72

Vidrih R, Hribar J, Prgomet Z, Poklar Ulrih N (2013) The physicochemical properties of strawberry tree (*Arbutus unedo* L.) fruits. Croat J Food Sci Technol 5(1):29–33

Villar L, Palacín JM, Calvo C, Gómez García D, Montserrat G (1987) Plantas medicinales del Pirineo Aragonés y demás tierras oscenses. CSIC y Diputación de Huesca, Huesca

Villatoro-Pulido M, Moreno Rojas R, Munoz-Serrano A, Cardenosa V, Amaro López MA, Font R, Del Río-Celestino M (2012) Characterization and prediction by near-infrared reflectance of mineral composition of rocket (*Eruca vesicaria* subsp. *sativa* and *Eruca vesicaria* subsp. *vesicaria*). J Sci Food Agric 92(7):1331–1340

Wannes WA, Mhamdi B, Marzouk B (2009) Variations in essential oil and fatty acid composition during *Myrtus communis* var. *italica* fruit maturation. Food Chem 112(3):621–626

Wannes WA, Mhamdi B, Sriti J, Marzouk B (2010) Glycerolipid and fatty acid distribution in pericarp, seed and whole fruit oils of *Myrtus communis* var. *italica*. Ind Crop Prod 31(1):77–83
Wetherilt H (1992) Evaluation of *Urtica* species as potential sources of important nutrients. Food Sci Hum Nutr 29(15):25
Widdowson EM, McCance RA (1943) Food tables—their scope and limitations. Lancet 1:230–232
Wilman D, Riley JA (1993) Potential nutritive-value of a wide-range of grassland species. J Agric Sci 120:43–49
Wright CA (2012) Mediterranean vegetables: a cook's compendium of all the vegetables from the world's healthiest cuisine, with more than 200 recipes. Harvard Common Press, Boston
Yildrim E, Dursuna A, Turan M (2001) Determination of the nutrition contents of the wild plants used as vegetables in Upper Çoruh Valley. Turk J Bot 25:367–371
Zeghichi S, Kallithraka S, Simopoulos AP, Kipriotakis Z (2003) Nutritional composition of selected wild plants in the diet of Crete. In: Simopoulos AP, Gopalan C (eds) Plants in human health and nutrition policy. World Review of Nutrition and Dietetics, vol 91. Karger, Basel, pp 22–40
Zhou W (1998) Ethnobotany of *Capsella bursa-pastoris* (L.) Medic. J Plant Resour Environ 7(3):49–53

Index

M. de C. Sánchez-Mata, J. Tardío (eds.), *Mediterranean Wild Edible Plants,*
DOI 10.1007/978-1-4939-3329-7

MIX
Papier aus verantwortungsvollen Quellen
Paper from responsible sources
FSC® C105338
FSC
www.fsc.org

If you have any concerns about our products,
you can contact us on
ProductSafety@springernature.com

In case Publisher is established outside the EU,
the EU authorized representative is:
Springer Nature Customer Service Center GmbH
Europaplatz 3, 69115 Heidelberg, Germany

Printed by Libri Plureos GmbH
in Hamburg, Germany